2nd International Conference on Computational Fluid Dynamics in Research and Industry (CFDRI 2017)

IOP Conference Series: Materials Science and Engineering
Volume 243

Songkhla, Thailand
3 – 4 August 2017

ISBN: 978-1-5108-4894-8
ISSN: 1757-8981

Printed from e-media with permission by:

Curran Associates, Inc.
57 Morehouse Lane
Red Hook, NY 12571

Some format issues inherent in the e-media version may also appear in this print version.

Printed by Curran Associates, Inc. (2017)

For permission requests, please contact the Institute of Physics at the address below.

Institute of Physics
Dirac House, Temple Back
Bristol BS1 6BE UK

Phone: 44 1 17 929 7481
Fax: 44 1 17 920 0979

techtracking@iop.org

Additional copies of this publication are available from:

Curran Associates, Inc.
57 Morehouse Lane
Red Hook, NY 12571 USA
Phone: 845-758-0400
Fax: 845-758-2633
Email: curran@proceedings.com
Web: www.proceedings.com

Table of contents

Volume 243

2nd International Conference on Computational Fluid Dynamics in Research and Industry (CFDRI 2017)

3–4 August 2017, Songhkla, Thailand

Accepted papers received: 13 September 2017
Published online: 29 September 2017

Preface

Track 1

Track 2

Track 3

IOP Conf. Series: Materials Science and Engineering **243** (2017) 011001 doi:10.1088/1757-899X/243/1/011001

2nd International Conference on Computational Fluid Dynamics in Research and Industry 2017 (CFDRI 2017)

"Streamlining Technological Shift"

3rd – 4th August, Songkhla, Thailand

The 2nd International Conference on Computational Fluid Dynamics in Research and Industry 2017 (CFDRI 2017) was organised by Faculty of Mechanical and Manufacturing Engineering, Universiti Tun Hussein Onn Malaysia (UTHM), in collaboration with Prince of Songkla University (PSU). This event focused on the application of Computational Fluid Dynamics (CFD) in academic and industrial research. Since its inception, CFDRI conference always aims to bring together researchers from both academia and industries to exchange knowledge, experience and ideas in CFD study.

With the theme *"Streamlining Technological Shift"*, the second conference was successfully held on the 3rd and 4th August 2017 at BP Samila Beach Hotel & Resort, Songkhla, Thailand. Topics covered in this conference included:

Track 1
- CFD Methods
- Heat Transfer
- Fluid Structure Interaction
- Electronics
- Fluid Machines

Track 2
- Fluid Flow Characterization
- Turbulence Analysis
- Automotive Engineering
- Porous Media Flow
- Aerodynamics

Track 3
- Heat Exchangers
- Spray Analysis
- Industrial Environment
- Indoor Air Quality
- Combustion

More than 60 papers across the regions have been received and upon rigorous review, 52 papers have been accepted for publication in the *IOP Conference Series: Materials Science and Engineering*. In conjunction with the conference too, two Memorandum of Agreement (MOA) documents on research and academic programs had been signed between UTHM and PSU. These agreements will boost the collaboration between the two institutions at international level, apart from enhancing the exchange of expertise in the coming years.

We would like to thank all authors who had contributed to the conference and all reviewers for their efforts in reviewing the submissions, as well as the conference's sponsors. Immeasurable appreciation is extended to our inspiring keynote speakers and all members of the organising committee for their efforts in making CFDRI 2017 a success. We hope the connection that was built through this conference could lead us further to a better cooperation.

IOP Conf. Series: Materials Science and Engineering **243** (2017) 011001 doi:10.1088/1757-899X/243/1/011001

FOREWORD FROM CHAIRMAN OF CFDRI 2017

Dear respected colleagues, in the name of Allah, the Most Gracious and the Most Merciful, His blessing has enabled us to become a part of the 2nd International Conference on Computational Fluid Dynamic in Research and Industry (CFDRI 2017), organized by Universiti Tun Hussein Onn Malaysia and the Prince of Songkla University, Thailand.

First and foremost, I would like to warmly welcome all conference participants and delegates, who I presumed share similar interest: Computational Fluid Dynamics (CFD). We all love beautiful things, fluidity of movements, streamline flows, vibrant colors, all of which are part of our lives. Some we think we understood. Some we are trying to understand. Some we just awe and appreciate its splendour. The world is evolving with new technologies emerging every second. We, as part of the world scholars, scientist and researchers, are considered partly responsible for these evolution. We all hope that the CFD that we use to simulate from natural occurrence to universal hypothetical possibilities may help streamlining the world technological shift towards making our world a better place for us and for our future generations.

On behalf of CFDRI 2017 organising committee, I would like to express our appreciation to all the VIPs, our keynote speakers, presenters, editors, reviewers and all the authors for their contribution and participation in this conference. To the CFDRI 2017 organising committee from both UTHM and PSU, thank you for your outstanding work, soaring team working spirit and extraordinary commitment. None of these would be possible without you. I hope that your presence here for this conference will be gratifying and beneficial to your academic and research endeavours. Let's take this opportunity to get to know each other, to create international networking and to initiate research collaboration. Let's set this conference as our common ground, and this day as the seed of our blooming future.

Assoc. Prof. Dr. Mas Fawzi bin Mohd Ali
Chairman of CFDRI 2017

IOP Conf. Series: Materials Science and Engineering **243** (2017) 011001 doi:10.1088/1757-899X/243/1/011001

CFDRI 2017 ORGANISING COMMITTEE

Patron
Prof. Dr. Wahid bin Razzaly

Advisors
Prof. Dr. Ruzairi bin Abdul Rahim
Assoc. Prof. Dr. Shahruddin bin Mahzan

Chairman
Assoc. Prof. Dr. Mas Fawzi bin Mohd Ali

Co-chairman
Assoc. Prof. Dr. Perapong Tekasakul (PSU)

Vice Chairman I
Assoc. Prof. Dr. Ahmad Jais bin Alimin

Vice Chairman II
Assoc. Prof. Dr. Norasikin binti Mat Isa

Secretary I
Dr. Akmal Nizam bin Mohammed*

Secretary II
Dr. Shahrul Azmir bin Osman

Treasurer I
Dr. Aslam bin Abdullah

Treasurer II
Ms. Halimah Jan binti Fakir

Publication (Review, Editing, Publishing)
Dr. Normayati binti Nordin*
Dr. Mohd Fadhli bin Zulkafli*
Assoc. Prof. Dr. Amir bin Khalid*

Online System (Publication, Payment, Registration)
Assoc. Prof. Dr. Waluyo Adi Siswanto
Dr. Azwan bin Sapit

Publicity & Promotion
Dr. Ishkrizat bin Taib*
Dr. Nurhayati binti Rosly
Mr. Yaacub Zaki bin Ali
Mr. Noor Nasriq bin Selamat
Dr. Juntakan Taweekun (PSU)
Dr. Passakorn Vessakosol (PSU)

On-site Registration, Conference Kit & Lucky Draw Event
Dr. Nurul Fitriah binti Nasir
Mrs. Siti Mariam binti Basharie
Dr. Chayut Nuntadusit (PSU)
Dr. Kittinan Maliwan (PSU)
Mrs. Thanmunin Pawintanathon (PSU)

Event Schedule
Dr. Azmahani binti Sadikin
Dr. Wan Saiful Islam bin Wan Salim

Sponsorship
Dr. Mohammad Kamil bin Abdullah
Dr. Mohd Faizal bin Mohideen Batcha

Conference Venue, Seminar Rooms, F&B and Accommodation
Dr. Sofian bin Mohd*
Dr. Syariful Syafiq bin Shamsudin
Dr. Juntakan Taweekun (PSU)

Conference Kit & Lucky Draw Preparation
Dr. Shahrul Azmir bin Osman
Dr. Nurul Fitriah binti Nasir

Logistic, Technical & Multimedia
Dr. Mohd Azahari bin Razali
Mr. Mzahar bin Abd. Jalal
Mr. Mohamad Farid bin Sies
Dr. Wiriya Thongruang (PSU)
Dr. Makatar Wae-hayee (PSU)
Mr. Prayoon Duangsiri (PSU)

Special Tasks
Mr. Muhammad Sufyan Amir bin Paisal
Mr. Riyadhthusollehen bin Khairulfuaad
Ms. Nur Amani Hanis binti Roseman
Mr. Mohamad Syahmi bin Saringat

*Note: * Track Director*

IOP Conf. Series: Materials Science and Engineering **243** (2017) 011001 doi:10.1088/1757-899X/243/1/011001

KEYNOTE SPEAKERS

Assoc, Prof. Dr. Norzelawati binti Asmuin
Head, Industrial and Indoor Environment Research Group
Department of Energy & Thermofluids Engineering
Faculty of Mechanical and Manufacturing Engineering
Universiti Tun Hussein Onn Malaysia
86400 Parit Raja, Batu Pahat
Johor Darul Takzim, MALAYSIA
Email: norzela@uthm.edu.my

Norzelawati is the current Head of Industrial and Indoor Environment Research Group under the Centre for Energy and Industrial Environment Studies (CEIES). She is an Associate Professor in Mechanical and Manufacturing Engineering at the Universiti Tun Hussein Onn Malaysia (UTHM), Parit Raja, Batu Pahat, Johor. She earned her B.Eng. with Honours in Mechanical and Material Engineering from Universiti Kebangsaan Malaysia, Masters by Research in Aerospace Engineering from Universiti Putra Malaysia, and PhD in Gas Engineering from University of Salford, UK. She has published numerous journal and conference papers in CFD, Fine Spray Nozzle, Kitchen Hood Ventilation System, LEV and Sand-Blasting. Norzelawati has done consultation works with SCOMI Sdn. Bhd., ANSELL (Glove Manufacturing), Kolej Kemahiran Teknologi MARA and Biomedical Engineering UTM. She also works closely with Halton Asia Sdn. Bhd. and TDK Industries Sdn. Bhd. Her research interests include CFD, Fine spray, Multiphase Flow and Thermoacoustic. Her current research focuses on the tools of CFD on Bioengineering, especially Stent for Glaucoma. She was the previous chairman of CFDRI 2015. She is also a registered member of IEM and BEM.

Assoc. Prof. Dr. Perapong Tekasakul
Vice President, Research System and Graduate Studies
Department of Mechanical Engineering
Faculty of Engineering
Prince of Songkla University
Hat Yai, Songkhla 90112
Thailand
Email: perapong.t@psu.ac.th

Tekasakul is currently the Vice President for Research System and Graduate Studies at Prince of Songkla University (PSU), one of the nine National Research Universities in Thailand. His role includes research policy setting and driving, postgraduate study capability enhancement, technology transfer, research and innovation commercialization, and university-social engagement. He also chairs the Research and Innovation Committee of the Council of University Presidents in Thailand (CUPT) including 30 public universities. He and his CUPT committee members work closely with the Thai government and play a significant role in drafting the technology and innovation development plan in "Thailand 4.0" model and proposing a "University 4.0" concept for Thai Universities. Academically, he is an associate professor in Mechanical Engineering at Prince of Songkla University. His research interests include Solar Drying, Computational Fluid Dynamics, and Atmospheric Environment.

CONFERENCE PHOTOGRAPHS

Delegates attending the 2nd International Conference on Computational Fluid Dynamics in Research and Industry 2017 (CFDRI 2017), Songkhla, Thailand

MOA signing on research and academic programs between UTHM (represented by Vice Chancellor, Prof. Dr. Wahid bin Razzaly) and PSU (represented by Vice President, Assoc. Prof. Dr. Perapong Tekasakul)

CFDRI 2017 delegates from various institutions and nationalities

IOP Conf. Series: Materials Science and Engineering **243** (2017) 011001 doi:10.1088/1757-899X/243/1/011001

SPONSORS ACKNOWLEDGEMENTS

CFDRI 2017 committee wishes to express its gratitude to the following parties who, through their generosity and commitment, have helped to make this event possible:

Centre for Energy and Industrial Environment Studies (CEIES)
Faculty of Mechanical and Manufacturing Engineering
Universiti Tun Hussein Onn Malaysia
86400 Parit Raja, Batu Pahat
Johor Darul Takzim, Malaysia

Flow Analysis, Simulation, and Turbulence Research Group (FAST)
Faculty of Mechanical and Manufacturing Engineering
Universiti Tun Hussein Onn Malaysia
86400 Parit Raja, Batu Pahat
Johor Darul Takzim, Malaysia

Quantum Two Engineering
No.10-2-2 Medan Niaga Siswa Jaya
Taman Siswa Jaya
86400 Parit Raja, Batu Pahat
Johor Darul Takzim, Malaysia

IOP Conf. Series: Materials Science and Engineering **243** (2017) 011002 doi:10.1088/1757-899X/243/1/011002

Peer review statement

All papers published in this volume of *IOP Conference Series: Materials Science and Engineering* have been peer reviewed through processes administered by the proceedings Editors. Reviews were conducted by expert referees to the professional and scientific standards expected of a proceedings journal published by IOP Publishing.

IOP Conf. Series: Materials Science and Engineering **243** (2017) 012001 doi:10.1088/1757-899X/243/1/012001

The study of flow and heat transfer characteristics of impinging jet array mounting air-induced duct

K Yeranee[1], M Wae-hayee[1], I Piya[2], Y Rao[3] and C Nuntadusit[1]
[1] Department of Mechanical Engineering, Faculty of Engineering, Prince of Songkla University, Songkhla, Thailand
[2] Department of Mechanical Engineering, Princess of Naradhiwas University, Naradhiwas, Thailand
[3] School of Mechanical Engineering, Shanghai Jiao Tong University, Shanghai, China

E-mail: wmakatar@eng.psu.ac.th

Abstract. Impinging jet is widely employed in thermal industrial applications due to having high heat transfer coefficient in impingement region. One method to increase heat transfer on an impingement surface is to increase turbulence intensity in jet flow. The mounting of an air-induced duct at nozzle outlet is a passive method to increase entrainment air resulting on increasing turbulence intensity. The aim of this research is to study flow and heat transfer characteristics of array of impinging jets mounting air-induced ducts. The investigation model was jets discharging from pipe nozzle having an inner diameter of d=17.2 mm and a length of 200 mm. Nozzle arrangement were inline configuration having 5 rows x 5 columns. A jet-to-jet distance (S) was S=6d, 8d and a jet-to-plate distance (H) was H=6d. The inner diameter (D) and the length (L) of the air-induced ducts were D=4d and L=4d, respectively. The Reynolds number was fixed at Re=20,000. In addition, the impinging jets without mounting the air-induced ducts were also investigated for benchmarking with the case of mounting the air-induced ducts. In the study, a thin foil technique was used to measure heat transfer on the impingement surface, and a computational fluid dynamic (CFD) using ANSYS, Fluent (V.15.0) was also adopted. The results showed that the effect of mounting air-induced duct can enhance entrainment air into the jet flow resulting on increasing of heat transfer of impinging jets on target surface, and the effects of mounting air-induced duct on increasing heat transfer in case of larger jet-to-jet distance (S/d=8) was more effective than that of smaller jet-to-jet distance (S/d=6).

1. Introduction

This century, efforts have been made to produce more efficient thermal equipment by employing various methods of heat transfer enhancement. Heat transfer enhancement techniques can be separated into two categories: passive and active methods. Passive method requires no external power while active method requires external power to activate fluid flow which causes to enhance heat transfer.

Heat transfer on a surface is principle transferring of thermal energy between fluid and solid surface. More commonly, an impinging jet is used in thermal engineering application or thermal industrial processes due to having high heat transfer coefficient in impingement region. Previous works have been extensively studied on flow and heat transfer characteristics of impinging jet. Heat transfer rate on an impinged surface is governed by momentum of jet impinges on a target surface and a turbulence intensity of jet just before impingement. Generally, after jet discharging from nozzle, the spreading jet results in the reducing of axial velocity and the increasing of turbulence intensity in a jet flow. The optimal matching between axial velocity and turbulence intensity can be found at the end of potential core occurring in the rage of 5 to 8 time of nozzle diameter from jet outlet, depending on nozzle shape and jet Reynolds number. Subsequently, the maximum heat transfer at stagnation region is gained [1-5].

Many researchers have devoted to enhance heat transfer on an impingement surface by increasing turbulence intensity of jet flow such as attaching mesh screens [6] or triangular tape [7] at the jet outlet and inserting twisted tape [8-11] or guide vane into nozzles [12, 13]. So, an important factor to enhance turbulence intensity of jet flow is to increase entrainment of ambient fluid.

An expanding of jet outlet is a simple method to increase the entrainment of ambient fluid into the jet flow [14-19]. Generally, this method is adopted to increase the mixing and spreading of jet in combustion of industrial applications [20, 21]. However, it is rare to adopt for enhancing heat transfer on impingement region. Keawchoothong et al. [22, 23] studied a single impinging jet associated with short pipe called "air-induced duct" to increase heat transfer on impingement surface. They illustrated an ambient air was increasingly sucked through air-induced duct for enhancing turbulence intensity into a jet flow. These results can be confirmed that the heat transfer on an impingement surface increased due to using of air-induced duct. However, the applying of air-induced duct in an array of impinging jets is yet to study, and it is very interesting to adopt for heat transferring in large surface.

The aim of this research is to experimentally and numerically investigate flow and heat transfer characteristics of an impinging jet array mounting air-induced ducts. The results will be compared to the case of conventional impinging jet under the same of mass flow rate.

2. Experimental setup and method

2.1. Experimental model and parameters

The experimental model of impinging jet from nozzle with air-induced ducts is shown in Figure 1. The jets discharged from nozzle pipes and perpendicularly impinged on a target surface. The air-induced ducts were assembled at the end of pipe nozzle. Both centerlines of pipe nozzles and air-induced ducts were concentric. Nozzle arrangements were inline configuration having 5 rows x 5 columns. An origin of the Cartesian coordinates was located at the center of the jet exit. The Z-axis was on the axial of jet; X-axis and Y-axis were normal to the axial of jet in horizontal and vertical direction, respectively. The details of experimental parameters are summarized in Table 1.

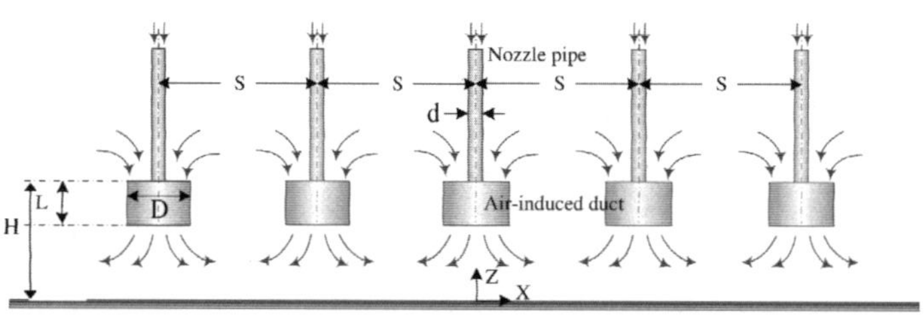

(a) Pipe nozzle with air-induced duct

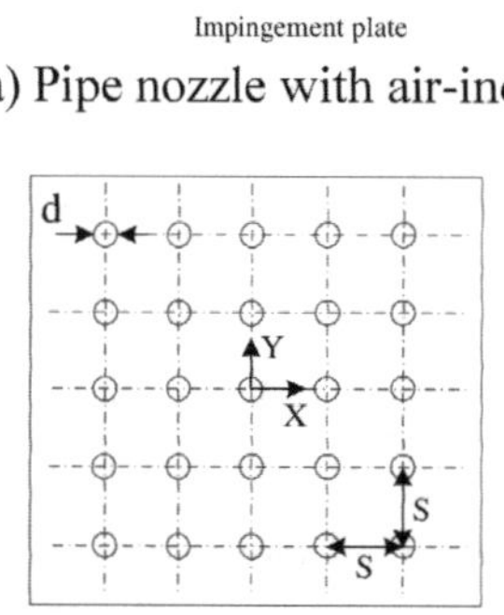

(b) Nozzle pipe arrangement

Figure 1. Experimental models of impinging jet array

Table 1. The details of experimental parameters

Descriptions	Variables
Inner diameter of pipe nozzle (d)	d=17.2 mm
The length of air-induced duct (L)	L=4d
The inner diameter of air-induced duct (D)	D=4d
The jet-to-plate distance (H)	H=6d
The jet-to-jet distance (S)	S=6d, 8d
Reynolds number (Re)	Re=20,000

2.2. Experimental setup

The diagram of experimental setup is shown in Figure 2. The blower accelerated the air which flow through the orifice flow meter and temperature controlled chamber equipped with 12-kW heater. The temperature of air jet was controlled by a temperature controller and a power controller at 27 °C. The flow rate of air jet was controlled by adjusting rotating speed of blower with an inverter. The turbulent jet discharging from round pipe with inner diameter of d=17.2 mm and length = 200 mm impinges on the wall. The pipe length was long enough to ensure the flow being fully developed flow at pipe exit. Air-induced ducts were assembled at pipe nozzle outlet. Three layers of mesh plate were mounted in jet chamber to uniform the jet temperature at pipe exit. The pipe nozzle and impingement wall were movable to adjust required positions.

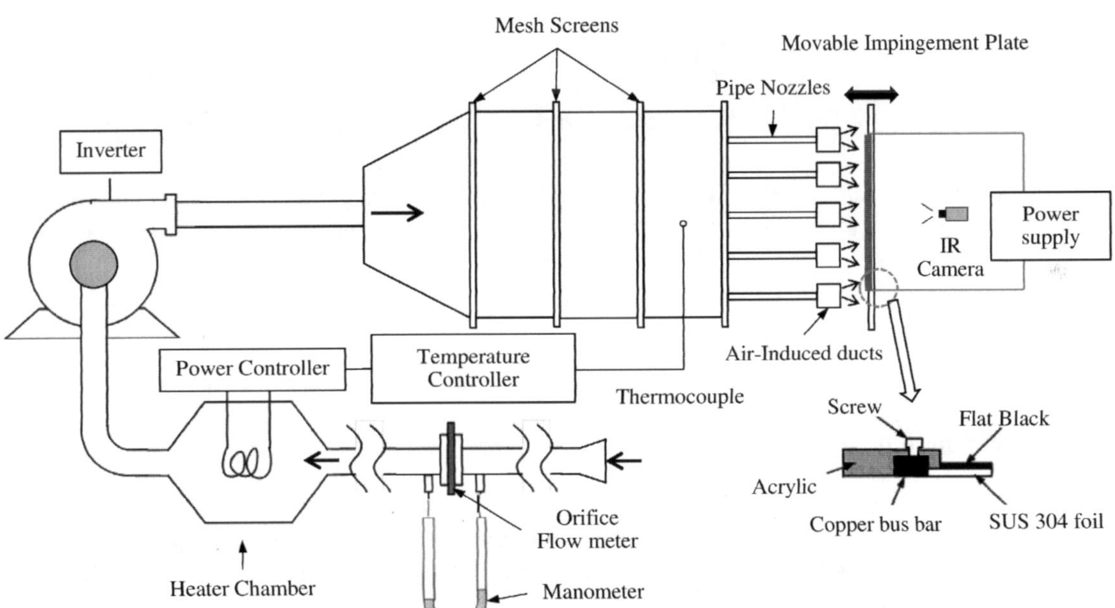

Figure 2. The diagram of experimental setup

2.3. Data reduction

The impingement plate was made of plastic plate (830 mm × 830 mm and 15 mm in thickness) and was opened with rectangular hollow on the center. A stainless steel foil with 300 mm × 300 mm and 0.03 mm in thickness was used to be impingement surface. It was tightly stretched between two copper bus bars over the rectangular hollow.

The temperature distributions on the impingement surface were measured from the rear side of impingement surface using an infrared camera. The temperatures on both impingement side and measurement side can be considered being the same since the stainless foil was sufficiently thin.

IOP Conf. Series: Materials Science and Engineering **243** (2017) 012001 doi:10.1088/1757-899X/243/1/012001

The air jet discharging from the pipe nozzle impinged on the heated surface with constant heat flux for cooling. The wall temperature on the impingement surface was measured by using the infrared camera, and the local heat transfer coefficient can be subsequently evaluated from

$$h = \frac{\dot{q}_{input} - \dot{q}_{loss,r} - \dot{q}_{loss,c}}{T_w - T_{aw}} \tag{1}$$

where $\dot{q}_{loss,r} = \sigma\varepsilon(T_w^{\ 4} - T_s^4)$ and $\dot{q}_{loss,c} = h_{loss,c}(T_w - T_s)$ were the heat losses to the environment by radiation and natural convection from the rear side of jet impingement surface, respectively; T_w was the local wall temperature; T_{aw} was the local adiabatic wall temperature (without heat flux); σ was the Stefan-Boltzman constant; ε was the emissive coefficient of captured side; T_s was the surrounding temperature; and $h_{loss,c}$ was the heat transfer coefficient calculated from natural convective heat transfer from the vertical surface to the surrounding.

The input heat flux was calculated from

$$\dot{q}_{input} = IV / A \tag{2}$$

where I was the electrical current from supplier unit, V and A were the voltage across the bus bars and the area of the stainless steel foil, respectively.

The local Nusselt number was calculated from

$$Nu = hd / k \tag{3}$$

where d was the inner diameter of the pipe nozzle and k was a thermal conductivity of the air jet.

The average Nusselt number on impingement surface can be calculated from

$$\overline{Nu} = \bar{h}d / k \tag{4}$$

where $\bar{h}$ was the average heat transfer coefficient calculated from wall average temperature.

3. Numerical Simulation

3.1. Computational model and boundary conditions

ANSYS FLUENT (version 15.0) was adopted in the numerical simulation. The computational model divided in three sections and boundary types are shown in Figure 3. The first section was included the main pipe nozzles diameter (d) =17.2 mm, and length is 200 mm. The air-induced ducts were assembled at the end of main pipe nozzle in the second section. The length (L), inner diameter (D) of air-induced duct and the jet-to-plate distance (H) were fixed at L=4d, D=4d and H=6d, respectively. An impingement surface of impinging jet was included in the third section. The jet-to-jet distance (S) was varied at S=6d and 8d. The details of boundary conditions were summarized in Table 2.

Table 2. The details of boundary conditions

Boundary condition	Define
Constant velocity inlet	27.21 ms^{-1} (Re=20,000)
Front and side surface	Symmetry
Top and bottom surfaces	Wall

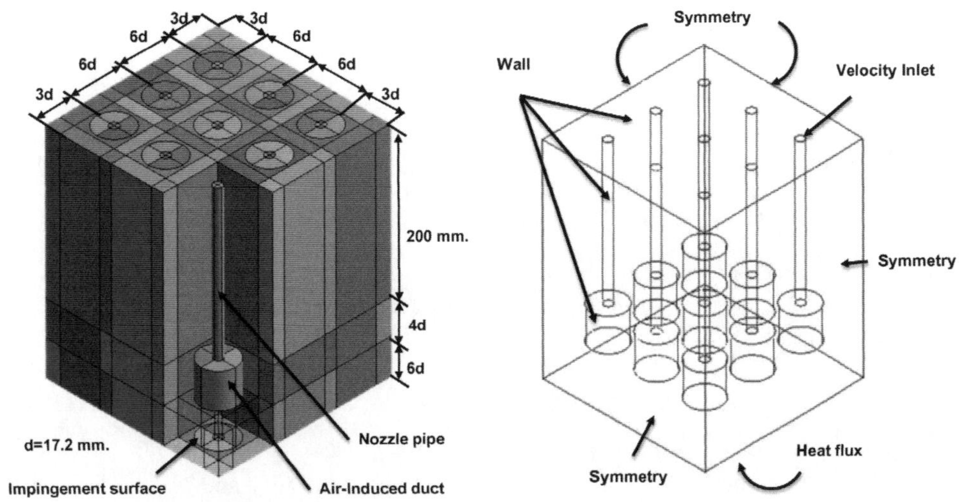

(a) Computational model (b) Boundary types

Figure 3. Computational model and boundary types

3.2. Grid generation and grid dependency

The rectangular grid was mainly applied in this numerical model. The grids in region having high velocity gradient as near impingement surface were finely controlled. The generating grid on Z-X plane at center of nozzle is shown in Figure 4. The number of generated grid was varied to achieve an accurate solution by considering the dimensionless wall distances (y^+) of the first node less than 1, calculated from

$$y^+ = \frac{y_1 u_\tau}{\nu} \tag{5}$$

where y^+ was the distance of the first node to the wall, u_τ was shear velocity, y_1 is the distance to the nearest wall and ν is the kinematic viscosity of air.

The effects of grid dependency on y^+ distribution on the target surface passing jet centerline are shown in Figure 5. In this model, the number of grids at 5,495,941 elements was selected to numerical simulation.

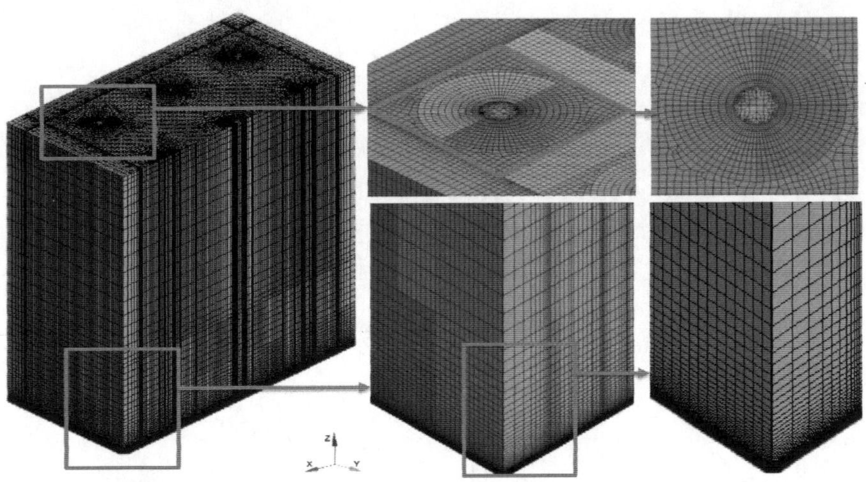

Figure 4. Generated rectangular grid for the numerical model

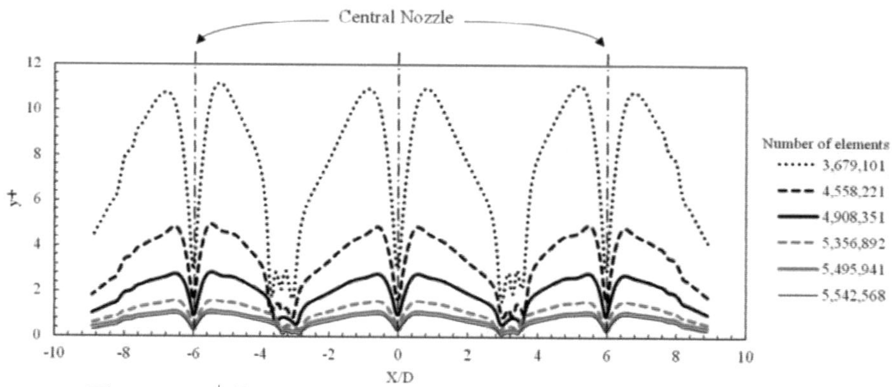

Figure 5. y^+ distributions on an impingement surface at center jet

3.3. Calculation method and algorithm

Computations were conducted by solving Reynolds averaged continuity and Navier-Stokes equations under existing boundary conditions. The SST k-ω turbulence model was selected for applying in this simulation. Since, this turbulence model has been adopted in solving many numerical simulations of jet impingement problems [24]. It excellently predicted the solutions of jet impingement problems with moderate computation cost. The SIMPLE algorithm was used with second order upwind scheme for all spatial discretization. The convergence of iterative solution was insured when the residual of all the variables was less than the specified values. The specified value was 1×10^{-5} for continuity and momentum equations.

4. Results and discussion

4.1. Flow characteristics

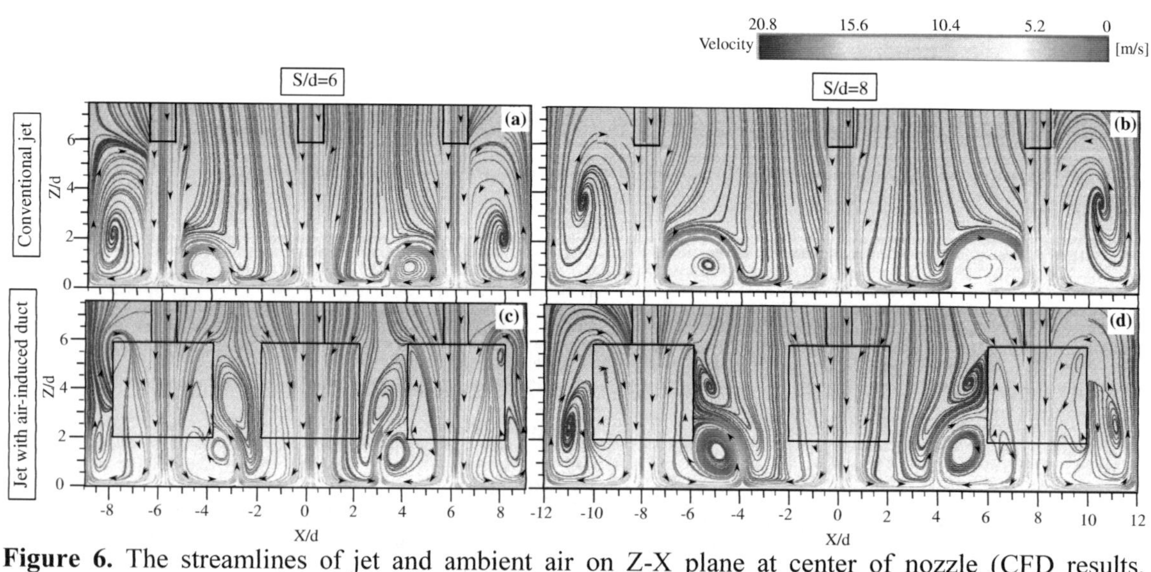

Figure 6. The streamlines of jet and ambient air on Z-X plane at center of nozzle (CFD results, Re=20,000)

The streamlines on Z-X plane passing the center of nozzle are shown in Figure 6. For the case of conventional jet at S/d=6 as shown in Figure 6(a), the ambient air was entrained and mixed in the jet flow, and the wall jet turned around becoming circulation flows above impingement surface. The size of these circulation flow became larger when jet-to-jet distance was larger (S/d=8) as shown in Figure 6(b). For the case of jet with air-induced ducts at S/d=6 as shown in Figure 6(c), the ambient air turned into the air-induced ducts. This entrainment of ambient air into the jet flow was more than the case of conventional jets. In addition, for the case of S/d=6 as shown in Figure 6(c), circulation flows occurred between air-induced ducts. These circulation flows seem that they blocked the entrainment of ambient air into the wall jet flow. When jet-to-jet distance become larger (S/d=8) as shown in Figure 6(d), the size of these circulation flows became smaller and the ambient air was entrained into wall jet flow that is similar with that case of conventional jet (Figure 6(b)).

From this flow study, it can be illustrated that the effect of mounting air-induced duct can enhance the entrainment of ambient air into the jet flow more that the case of jet without mounting air-induced duct (Conventional jets). For jet mounting air-induced duct at lager jet-to-jet distance (S/d=8), the ambient air entrained into the wall jet flow more than the case of smaller jet-to-jet distance (S/d=6). It would case to increase the heat transfer on the surface that would be discussed.

4.2. Heat transfer characteristics

The contours of Nusselt number distributions on the impingement surface are shown in Figure 7(a)-7(d). Small circle marks were identified the locations of nozzles, and the larger ones were identified the locations of air-induced duct. For all cases, Nusselt numbers were high in stagnation regions and were low in areas of interval jets. Generally, Nusselt number distributions for the case of S/d=6 were higher than the case of S/d=8. This is from the effect of interference of adjacent jets for the case of smaller jet-to-jet distance resulting on increasing turbulence intensity of jet before impingement as mentioned in previous work [25]. For the case of jet with air-induced duct for both S/d=6 and 8, areas of high Nusselt number around impingement regions were larger than those for conventional jets. This can be attributed that the mounting of air-induced duct can increase the entrainment of ambient air into the jet flow resulting on increasing of turbulence and heat transfer.

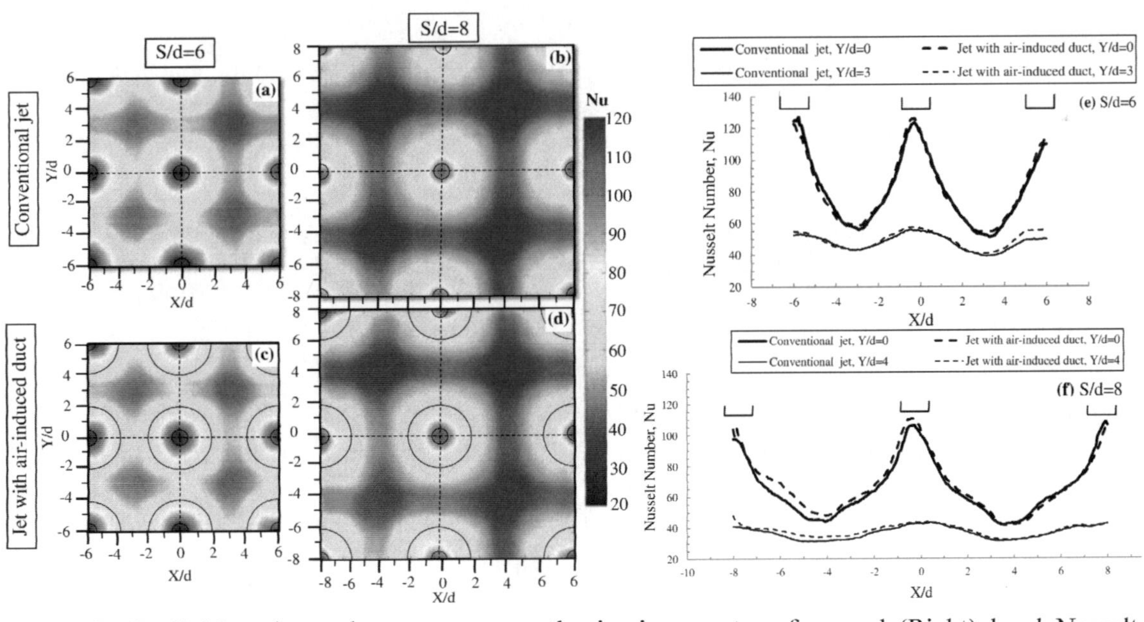

Figure 7. (Left) Nusselt number contours on the impingement surface and (Right) local Nusselt number distributions along X/d axis at jet center and interval jets (Experimental results, Re=20,000)

Nusselt number distributions along X/d are shown in Figure 7(e) and 7(f). The peak of Nusselts number for every nozzle took place at the stagnation points and gradually decreased along radial direction. For S/d=6 as shown in Figure 7(e), the Nusselt number distributions of jet with air-induced duct were almost similar when compare with the case of conventional jet whereas for the case S/d=8, the Nusselt number distributions of jet with air-induced duct were higher than those of conventional jets.

Average Nusselt numbers calculated using equation (4) by averaging temperature in area of -6≤Y/d≤6 and -6≤X/d≤6 for S/d=6 and in area of -8≤Y/d≤8 and -8≤X/d≤8 for S/d=8 are shown in Figure 8. Generally, the average Nusselt numbers for all cases of S/d=6 were higher than those of S/d=8. The average values for the cases of jet with air-induced duct were higher than those of the conventional jets: 1.12% for S/d=6 and 5.41% for S/d=8. It can be noticed that the effect of mounting air-induced duct on increasing of heat transfer in case of larger jet-to-jet distance (S/d=8) was more effective than that of smaller jet-to-jet distance (S/d=6).

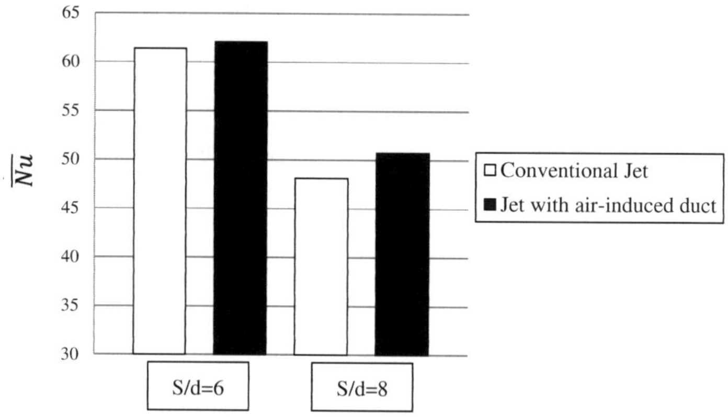

Figure 8. Average Nusselt number on the impingement surface (Experimental results, Re=20,000)

5. Conclusions

In this paper, flow and heat transfer characteristics of impinging jet array mounting air-induced duct were studied experimentally and numerically. The results can be concluded as follow:

- The effect of mounting air-induced duct can enhance entrainment air into the jet flow resulting on increasing of heat transfer of impinging jets on target surface, especially in area of around stagnation regions.
- The effects of mounting air-induced duct on increasing heat transfer in case of larger jet-to-jet distance (S/d=8) was more effective than that of smaller jet-to-jet distance (S/d=6).

Acknowledgements

This study is support by the Thailand Research Fund (TRF, No. MGR5980082)

References

[1] Viskanta R 1993 Heat transfer to impinging isothermal gas and flame jets *Exp. Therm Fluid Sci.* **68** 111-134
[2] Ashforth-Frost S and Jambunathan K 1996 Effect of nozzle geometry and semi-confinement on the potential core of a turbulent axisymmetric free jet *Int. C. Heat Mass Transfer* **23** 155-162
[3] Ashforth-Frost S, Jambunathan K and Whitney C F 1997 Velocity and turbulence characteristics of a semiconfined orthogonally impinging slot jet *Exp. Therm Fluid Sci.* **14** 60-67
[4] Martin H 1977 Heat and mass transfer between impinging gas jets and solid surfaces *Adv. Heat Transfer* **13** 1-60
[5] Jambunathan K, Lai E, Moss M A and Button B L 1992 A review of heat transfer data for single circular jet impingement *Int. J. Heat Fluid Flow* **13** 106-115

[6] Zhou D W and Lee S J 2004 Heat transfer enhancement of impinging jets using mesh screens *Int. J. Heat Mass Transfer* **47** 2097-2108
[7] Gao N, Sun H and Ewing D 2003 Heat transfer enhancement of Impinging jets using mesh screens *Int. J. Heat Mass Transfer* **46** 2557-2569
[8] Huang L and El-Genk M S 1998 Heat transfer and flow visualization experiments of swirling multi-channel and conventional impinging jets *Int. J. Heat Mass Transfer* **41** 583-600
[9] Wen M Y and Jang K J 2003 An impingement cooling on a flat surface by using circular jet with longitudinal swirling strips *Int. J. Heat Mass Transfer* **46** 4657-4667
[10] Nuntadusit C, Wae-hayee M, Bunyajitradulya A and Eiamsa-ard S 2012 Heat transfer enhancement by multiple swirling impinging jets with twisted-tape swirl generators *Int. C. Heat Mass Transfer* **39** 102-107
[11] Nuntadusit C, Wae-hayee M, Bunyajitradulya A and Eiamsa-ard S 2012 Visualization of flow and heat transfer characteristics for swirling impinging jet *Int. C. Heat Mass Transfer* **39** 640-648
[12] Alekseenko S V, Bilsky A V, Dulin V M and Markovich D M 2007 Experimental study of an impinging jet with different swirl rates *Int. J. Heat Mass Transfer* **28** 1340-1359
[13] Bakirci K and Bilen K 2007 Visualization of heat transfer for impinging swirl flow *Exp. Therm Fluid Sci.* **32** 182-191
[14] Selerowicz W C, Szumowski A P and Meier G E A 1991 Self-excited compressible flow in a pipe-collar nozzle *J. Fluid Mech.* **228** 465-458
[15] Hussain A K M F and Hasan M A Z 1983 The whistler-nozzle phenomenon *J. Fluid Mech.* **134** 431-458
[16] Zeng Y, New T H and Chng T L 2001 Flow behavior of turbulent nozzle jets issuing from beveled collars *Exp. Therm Fluid Sci.* **35** 1555-1564
[17] Husain H S and Hussain F 1999 The elliptic whistler jet *J. Fluid Mech.* **397** 23-44
[18] Hasan M A Z and Hussain A K M F 1982 The self-excited axisymmetric jet *J. Fluid Mech.* **115** 59-89
[19] Zeng Y, New T H and Tsai H M 2009 On the use of notched collars on an axisymmetric jet *Exp. Therm Fluid Sci.* **33** 1029-1034
[20] Nathan G J, Mi J, Alwahabi Z T, Newbold G J R and Nobes D S 2006 Impacts of a jet's exit flow pattern on mixing and combustion performance *Prog. Energy Combust Sci.* **32** 496-538
[21] Nathan G J, Hill S J and Luxton R E 1998 An axisymmetric 'fluidic' nozzle to generate jet precession *J. Fluid Mech.* **370** 347-380
[22] Kaewchoothong N, Wae-hayee M, Vessakosol P, Niyomwas B and Nuntadusit C 2014 Flow and heat transfer characteristics of impinging jet from expansion pipe nozzle with air entrainment holes *Adv. Mater. Res.* **931** 1213-1217
[23] Kaewchoothong N, Wae-Hayee M, Vessakosol P and Nuntadusit C 2015 Effect of modified expansion pipe nozzle on heat transfer enhancement with impinging jet *Rangsit J. Arts Sci.* **5** 49-62
[24] Zuckerman N and Lior N 2006 Jet impingement heat transfer: physics, correlation, and numerical modeling *Adv. Heat Transfer* **39** 565-631
[25] Hollworth B R and Berry R D 1978 Heat transfer from arrays of impinging jets with large jet-to jet spacing *J. Heat Transfer* **100** 352-357

The effect of conical dimple spacing on flow structure and heat transfer characteristics of internal flow using CFD

O Yemin[1], M Wae-Hayee[1*], P Narato[1], K Yerane[1], K Abdullah[2] and C Nuntadusit[1]

[1] Energy Research Centre and Department of Mechanical Engineering, Faculty of Engineering, Prince of Songkla University, Hat Yai, Songkhla 90112, Thailand

[2] Department of Energy and Thermofluid Engineering, Faculty of Mechanical and Manufacturing Engineering, UniversitiTun Hussein Onn Malaysia, Johor, Malaysia

E-mail: wmakatar@eng.psu.ac.th.

Abstract. In the present study, heat transfer and flow characteristics simulations over the surface of conical dimple were investigated. Single dimple row with inline arrangement was formed on the internal surface of the 3-D rectangular wind tunnel model. The air flow was perpendicular to the centre line of every dimple and the printed diameter of dimples on the surface was D=26.4mm. The depth of dimple on the surface of wind tunnel was H/D=2. The space between dimple-to-dimple was varied for S/D=1.125, 1.25,1.5, and 2. The Reynold number based on the hydraulic diameter of internal air flow was 20,000 depending on the wind tunnel hydraulic diameter. The numerical computation was applied with a Shear Stress Transport (SST) k- ω turbulence model. The average Nusselt number for the S/D=1.125 case is the highest. When the spacing becomes increase, the value of average Nusselt number tends to decrease.

1. Introduction

Dimple is one of the most effective structures for heat transfer enhancement in the industrial applications, such as the cooling of gas turbine blade and high pressure disk, the cooling of microelectronic components, and automotive, heating and refrigerating. Although these approaches are the effective method to improve the heat transfer performance; however, the increasing of fluid stream pressure drop should be concerned. The dimple surface is one of the effective methods to improve the heat transfer rates without significant pressure drop [1, 2]. Normally, the dimpled surface generates the vortex flow within its hole and the argumentation of heat transfer is obtained.

In past decades, lots of efforts have been made to improve efficiency and performance of thermal equipment accompanying a reduction in their size, weight and cost. The heat transfer can be enhanced by using either active and/or passive techniques. There are multifarious passive techniques such as swirl flow devices [3], surface tension devices, rough surfaces [3, 4], pin fins [3], ribbed tabulators [3] and surfaces with dimple [5, 6]. These are used for enhancing heat transfer in different utilizations. Dimple is regarded as one of the most effective heat transfer enhancement because of its profits for low pressure losses comparing with other devices and enhancing for heat transfer [7].

A wind tunnel investigation on an array of hemispherical dimple in staggered arrangement has been conducted by Manmood et al. [8]. In this work, they informed the formation of large vortexes pair on the dimple surface. Won et al. [9], Ligrani et al. [10], Won and Ligrani [11] and Shin Et al. [12] have reported about flow structure of dimple.

IOP Conf. Series: Materials Science and Engineering **243** (2017) 012002 doi:10.1088/1757-899X/243/1/012002

Previous studies, a CFD technique has been carried out popularly to investigate flow and heat transfer of dimple array in channel flow because of more accurate prediction of results by adopting of excellent turbulence model. Rao et al. [5] studied the flow and heat transfer of hemispherical and tear drop dimples by using CFD technique. The dimensions of hemispherical dimples were similar to the case of Mahmood et al. [8]. The results have been reported the heat transfer characteristics on the dimple surface in detail and proved CFD and experimental results for measuring overall heat transfer. Moreover, flow and heat transfer of dimple array in channel flow was investigated by Kim and Shin [13], Yoon et al. [14], Elyyan et al. [15] Xie et al. [7] using CFD technique. To reduce calculation load for computational modelling, the evaluated area of heat transfer was applied on some parts of dimple array.

The flow and heat transfer characteristics of teardrop dimple and protrusion was investigated and compared with hemispherical dimple and protrusion by Xie et al. [7]. They showed the thermal performance for teardrop dimple and protrusion is higher than hemispherical dimple and protrusion for lower Reynolds number and thermal performance increase gradually as the centre moves downwards for the teardrop dimple/protrusion.

Therefore, many researchers have been studied about different shapes shuch as spherical, teardrop and square shape [16] but conical shape have not been studied yet. Conical shape has many advantag for forming or creating easier than spherical or tear-drop shape. In this work, heat transfer effects and flow characteristics on a surface of conical dimple were investigated. Single row of conical dimples were designed over the internal surface of wind tunnel. The wind tunnel was arranged to get enough fully developed flow. Computational Fluid Dynamic (CFD) was used to visualize flow characteristics of dimples and compare with some experimental results.

2. Method

2.1. Model of dimple

The model of conical dimple row was formed over the inner bottom surface of the rectangular wind tunnel as shown in figure 1. The air with fully developed flow passed through the test section. The origin of Cartesian coordinate system was existed at the centre of middle dimple. X-axis is the along the flow direction of wind tunnel, Y-axis is the rectangular wind tunnel height direction and Z-axis is normal direction to the flow.

Figure 2 shows the details of conical dimple with projected diameter was D=26.4 mm. The depth of dimple, from the surface to the dimple bottom tip, was 0.2D [5,13 and 17]. The spacing of dimple-to-dimple was adjusted at S=1.125D, 1.25D, 1.5D and 2D. The Reynolds number of air inside the wind tunnel is Re_H=20,000 that was based on the hydraulic diameter of wind tunnel [5,8].

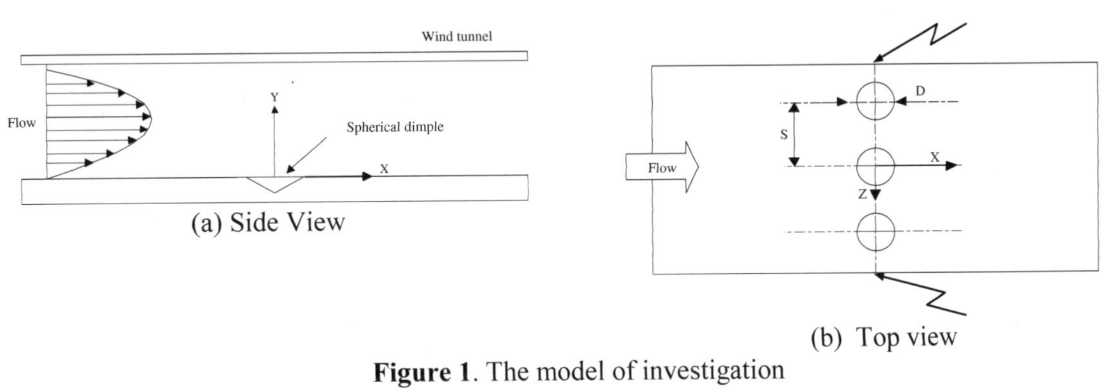

Figure 1. The model of investigation

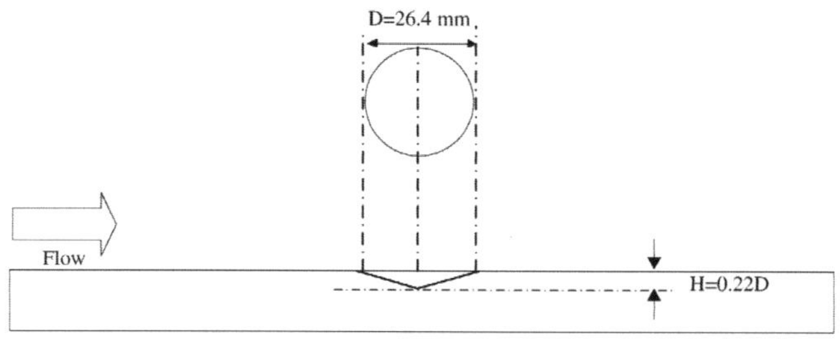

Figure 2. The details of dimple

2.2. Wind tunnel

The experimental setting for wind tunnel was utilized to prove and compare with CFD results. By using TLC technique, the heat transfer and temperature on the surface of smooth wall test section was measured. The wind tunnel measurement details were explained in previous works of M. Wae-Hayee et al. [18, 19]. The schematic diagram of wind tunnel simulation model was shown in figure 3. The wind tunnel geometries and dimensions for simulation were identical with the experimental design. The rectangular wind tunnel has three parts: the upstream of test section (1700 mm) that was sufficient distance to get fully developed flow, test section (280mm) where was formed conical dimples and downstream of test section where the air vent from the wind tunnel. The wind tunnel height was 26.4 mm (1D). The width of wind tunnel was varied with the spacing of dimple-to-dimple to reduce moderate computational loads of simulations.

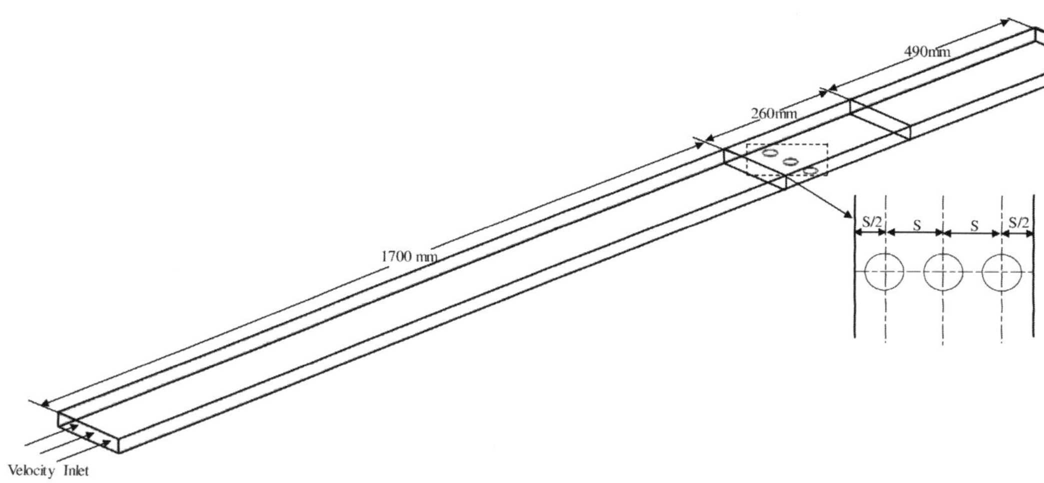

Figure 3. The wind tunnel with conical dimple

The generated grid details used in this work were illustrated in figure 4. Cutting plane along centreline of dimple was shown to expose the internal grid system. The majority of mesh elements were even rectangular. Uneven square elements were composed of dimple area and their surrounding regions. The elements were concentrated near the wall. Grid dependency in space between upper and lower wall is very important, especially, the thickness of the first layer above the wall to resolve the flow field within the viscous sublayer. To consider grid dependence in this work, y+ distributions on the heat transfer surface were lesser that 1. The selected element was 7,294,941 elements.

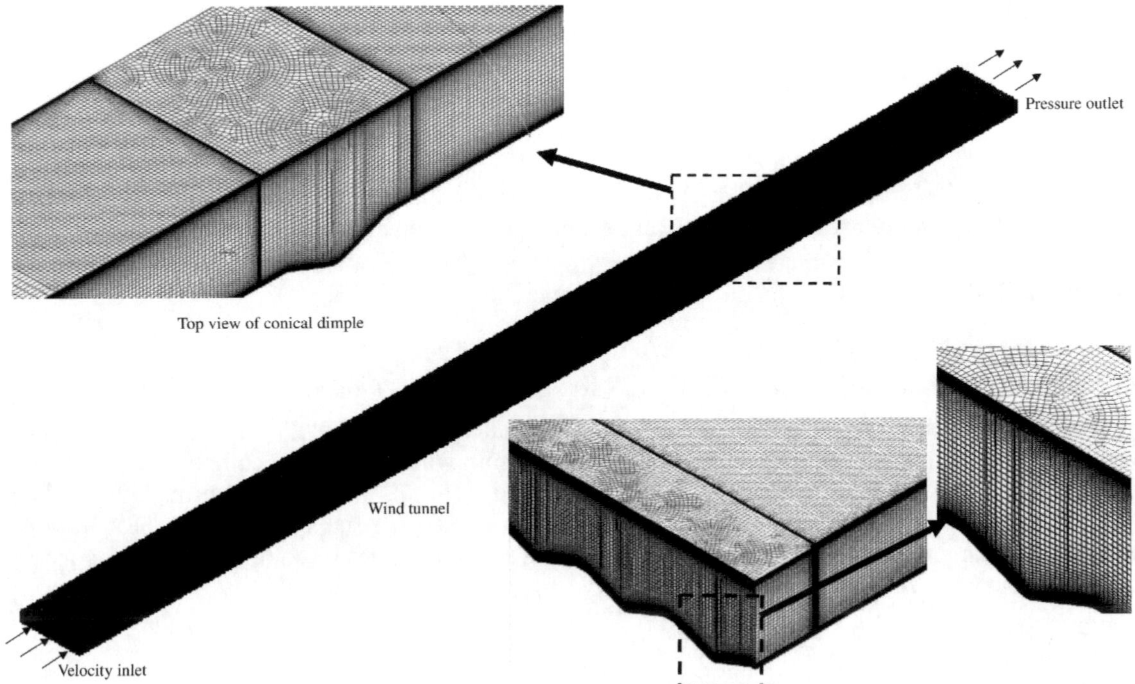

Figure 4. Generated grids

2.3. Assumptions and boundary conditions

In numerical model, the upper and the lower wall were set to have no slip condition. All walls, except bottom wall of test section, were insulated for performing as adiabatic condition. The fluid was considered as an incompressible flow with constant thermal properties and there is no gravitational effect in this simulation. At the air inlet, the uniform velocity was entered by getting Reynolds number at Re_H=20,000. The pressure outlet was set at 1 atm. Both of lateral walls were set as symmetry with the temperature of air was at 25°C. At the heat transfer surface formed with dimples was set at constant heat flux (150 W/m^2).

2.4. Numerical calculation method

The numerical 3-D model which based on the finite volume method was adopted to solve governing equations with boundary conditions. The details of equation can be found in [20]. The fluid flow and heat transfer were solved using a Shear Stress Transport (SST) k-ω turbulence model due to accurate prediction with moderate computation cost [20]. This turbulence model had been used in relative studies [5, 7 and 18].

The solution method was based on Semi-Implicit Method for Pressure-Linked Equations (SIMPLE) algorithm with second order upwind for all spatial discretization. To consider solutions to be convergent, the root mean square (RMS) residual of continuity and energy equation were lesser than 10^{-8} and that of momentum equation was lesser than 10^{-5} [5, 18].

2.5. Nusselt number calculation

The heat transfer coefficient, h, could be calculated from:

$$h = \frac{\dot{q}}{T_{wall} - T_{air}} \tag{1}$$

Where, $\dot{q}$ was heat flux, T_{wall} was wall temperature and T_{air} was air temperature.

The Nusselt number, Nu, was calculated from:

$$Nu = \frac{hD_H}{k} \tag{2}$$

where D_H was the hydraulic diameter of the tunnel and k was a thermal conductivity of the air.

3. Result and discussions

3.1. Verification of simulation

The well-known Dittus-Boelter correlation, $Nu_o = 0.23\,\mathrm{Re}^{0.8}\,\mathrm{Pr}^{0.4}$ where Re and Pr are Reynolds number and Prandtl number, were used for comparing internal heat transfer in the smooth channel. The average Nusselt numbers versus Reynold numbers for smooth wind tunnel was compared between the correlation, the experimental results, and the CFD method from this work is shown in figure 5. The current data agrees well with the correlation and the experimental results and overall heat transfer increase according to increasing of Reynolds number. The discrepancies of comparison were found in the same range with the work of Rao et al. [5].

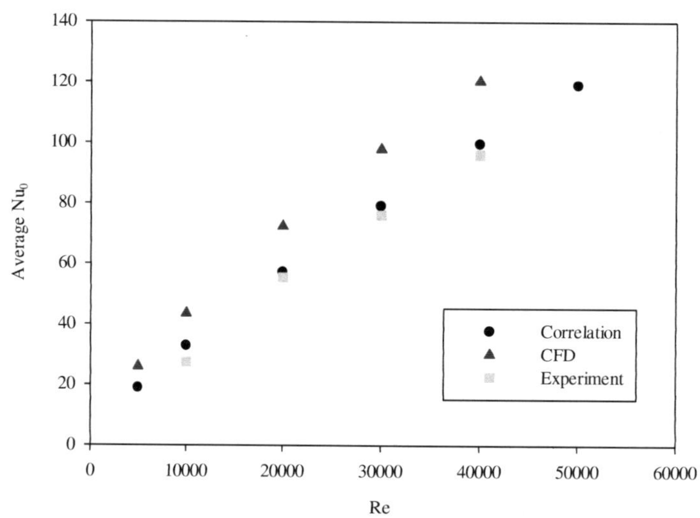

Figure 5. Average Nusselt numbers versus Reynold numbers of internal heat transfer correlation and the CFD method in this work

3.2. Flow characteristics

Streamlines flow over dimple surface were shown in figure 7. The occurrence of circulation flow in the dimple cavity was described obviously at the upstream part of the dimple cavity. The streamline inside the dimple cavity was shown in figure 6. This figure could also be proved that the circulation flows were happened at upstream half of the dimple cavity. The static circulation flow acted as thermal insulator resulting in low heat transfer on upstream portion of dimple surface.

For the case of 1.125D, the streamline which pass through the centre of middle dimple cavity (Z/D=0) where the flow condition inside the dimple was shown in figure 6. The stationary circulation flow appeared in upstream portion of dimple and attachment flow appeared in downstream portion of

IOP Conf. Series: Materials Science and Engineering **243** (2017) 012002 doi:10.1088/1757-899X/243/1/012002

dimple. This effect enhances heat transfer augmentation which had been described in literature reviews.

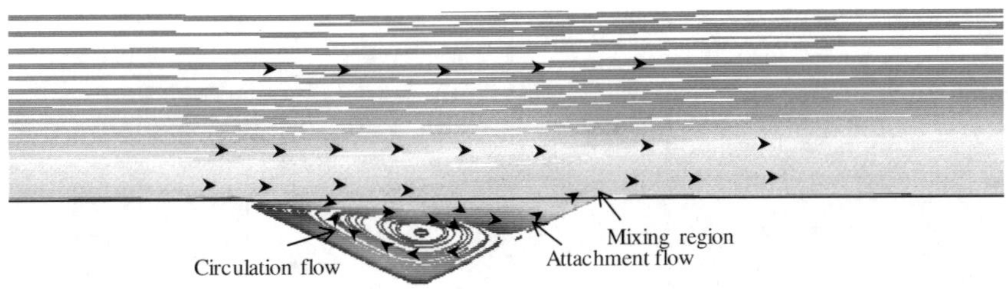

Figure 6. Streamline at the centre of plan of middle dimple (S/D=1.125D, Re=20,000).

In figure 7, the characteristics of upstream streamline (before passing dimple) were straight whereas those of downstream streamline (immediately after passing dimple, X/D≈0.5) were separated from its centre- line and tend to lateral side. This is from the effect of the longitudinal vortex pair occurrence after flow passing dimple which had also been reported in the literatures [5,7 and 23]. Here, it can be noted that for the case of S/D=1.125D as shown in figure 7 (a), the flow tendency to lateral sides of this vortex pair seems to be slighter than those other cases due to confinement from near neighbouring vortex pair in condition of short dimple-to-dimple spacing.

The conditions of the contour of Nusselt number on the surface of test section were illustrated in figure 8. High distribution of Nusselt number can be seen at the downstream portion of dimple cavity due to the attachment flow especially near the rear rim of dimples that was previously described in figure 6. But low Nusselt number distribution can be found at the upstream portion of dimple cavity because of stationary circulation flow. Increasing or decreasing Nusselt number on the dimple surface can be caused because of both attachment and circulation flow effects that was also studied in literature [5,7]. For all case, the Nusselt number distribution on the surface of smooth wall along downstream of every dimple have one peak for X/D=0.5 case and other X/D cases have two peak regions that was caused by the longitudinal vortex pair effect resulting in stronger circulating over the heat transfer surface of test section of wind tunnel. Simultaneously, these double peak regions area can be found along the downstream of smooth surface and this effect can remove more heat.

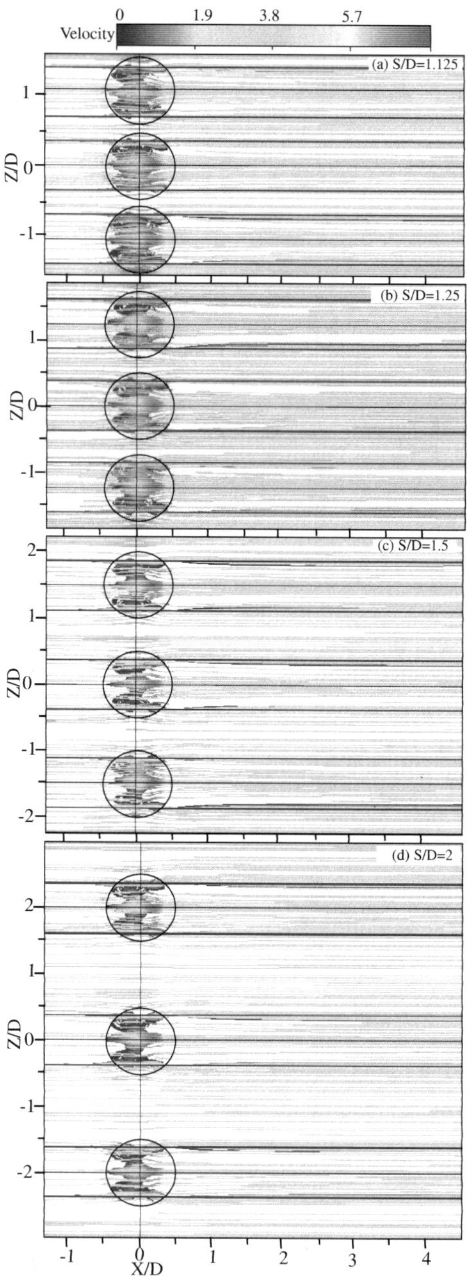

Figure 7. Streamline above the surface of dimples.

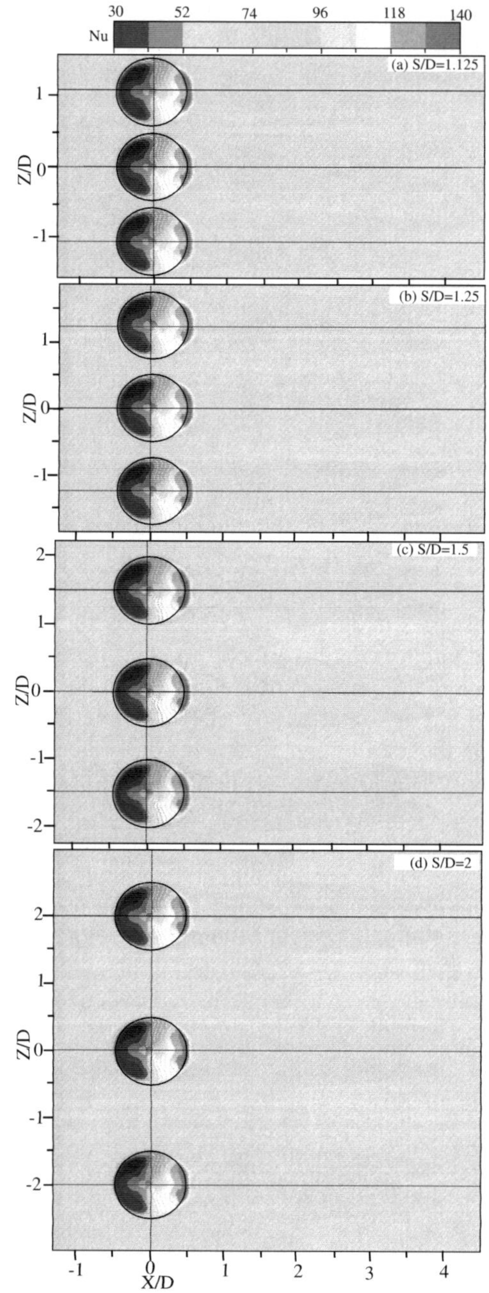

Figure 8. Nusselt number distributions on the surface (Re_H=20,000).

Span wise direction of Nusselt number distributions was shown in figure 9. Commonly for all S/D cases, high Nusselt number can be observed at just downstream of dimple cavity and became lower and lower along the direction of downstream. For every S/D cases, the single peak regions at every X/D=0.5 were extremely high and area of these peak regions were large because of the effect of attachment flow and double peak regions that was caused by the effect of neighbouring vortex pair between spacing

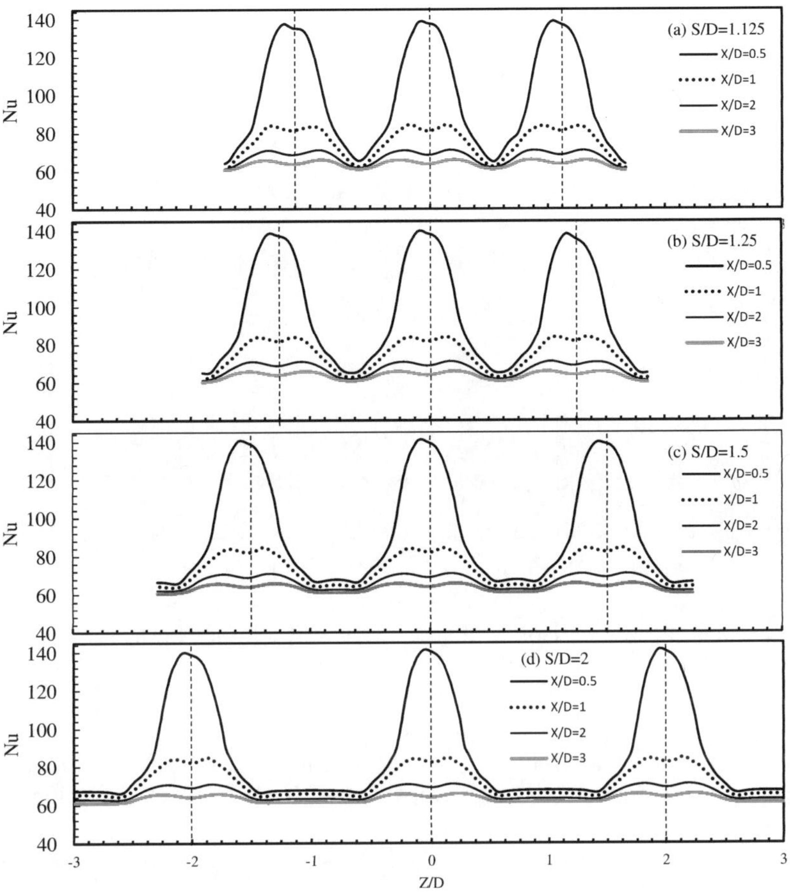

Figure 9. Nusselt number distributions in span wise direction (Dash straight line is the centre of dimple)

The average Nusselt number comparison for smooth and dimple surface was shown in figure 10. Every dimple surface's average value was higher than smooth surface. The trend of average Nusselt number became lower and lower when the spacing became larger and larger but the values were the same for S/D=1.25 and 1.5 cases.

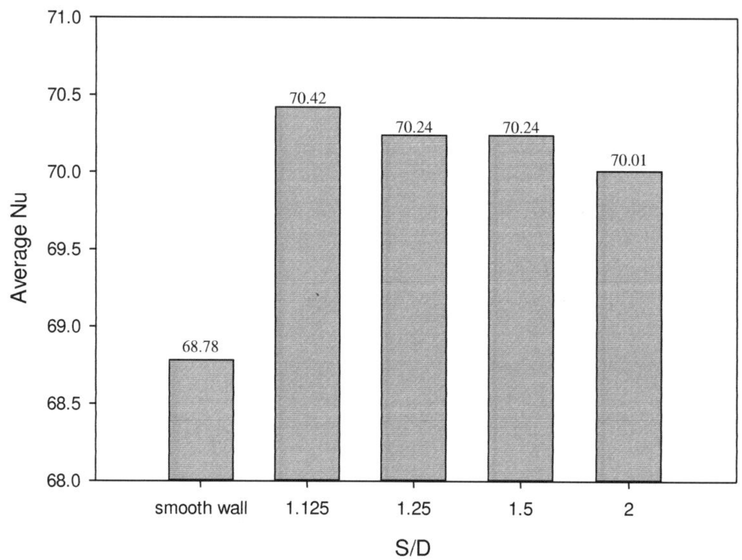

Figure 10. Comparison of Average Nusselt number for spherical and conical dimple

4. Conclusions

In this work, investigation of flow structure and heat transfer over the test section of wind tunnel for conical case that was adjusted by different spacings (1.125≤S/D≤2). The conclusions of this work are described as follow.

1. High Nusselt number regions were observed at the attachment flow region. At this place, the Nusselt number value are over 140.
2. Along the downstream of dimple, the pattern of Nusselt number was found into two peak regions (X/D>0.5) for all S/D case.
3. The flow for S/D=1.125 case, the circulation flow was seemed to be stronger than other cases. This effect can enhance heat transfer.
4. For the Nusselt number average, the dimple surfaces were higher than smooth surface. Moreover, the trend of average Nusselt number was lower when the spacing was increased.

Acknowledgement

The research grant was supported by the Research and Development Office (RDO), Prince of Songkla University, grant No. ENG590725S.

References

[1] Heo S C, Seo Y H, Ku T W and Kang B S 2011 Formability evaluation of dimple forming process based on numerical and experimental approach *J. Mech. Sci. Technol.* **25** 429–439

[2] Lan J, Xie Y and Zhang D 2011 Effect of leading edge boundary layer thickness on dimple flow structure and separation control *J. Mech. Sci. Technol.* **25** 3243–51

[3] Ligrani P M, Oliveira M M and Blaskovich T 2003 Comparison of heat transfer augmentation techniques *AIAA J.* **41** 337–362

[4] Kurniawan R, Kiswanto G and Ko T J 2017 Surface roughness of two-frequency elliptical vibration texturing (TFEVT) method for micro-dimple pattern process *Int. J. Mach. Tools Manuf.* **116** 77–95

[5] Rao Y, Li B and Feng Y 2015 Heat transfer of turbulent flow over surfaces with spherical dimples and teardrop dimples *Exp. Therm. Fluid Sci.* **61** 201–209

[6] Vorayos N, Katkhaw N, Kiatsiriroat T and Nuntaphan A 2016 Heat transfer behavior of flat plate having spherical dimpled surfaces *Case Stud. Therm. Eng.* **8** 370–377

[7] Xie Y, Qu H and Zhang D 2015 Numerical investigation of flow and heat transfer in rectangular channel with teardrop dimple/protrusion *Int. J. Heat Mass Transf.* **84** 486–496

[8] Mahmood G I, Hill M L, Nelson D L, Ligrani P M, Moon H K and Glezer B 2000 Local Heat Transfer and Flow Structure on and Above a Dimpled Surface in a Channel *J. Turbomach.* **123** 115–123

[9] Won S Y, Zhang Q and Ligrani P M 2005 Comparisons of flow structure above dimpled surfaces with different dimple depths in a channel *Phys. Fluids* **17** 0451051–9

[10] Ligrani P M, Harrison J L, Mahmmod G I and Hill M L 2001 Flow structure due to dimple depressions on a channel surface *Phys. Fluids* **13** 3442–51

[11] Won S Y and Ligrani P M 2007 Flow characteristics along and above dimpled surfaces with three different dimple depths within a channel *J. Mech. Sci. Technol.* **21** 1901–09

[12] Shin S, Lee K S, Park S D and Kwak J S 2009 Measurement of the heat transfer coefficient in the dimpled channel: effects of dimple arrangement and channel height *J. Mech. Sci. Technol.* **23** 624–630

[13] Kim K Y and Shin D Y 2008 Optimization of a staggered dimpled surface in a cooling channel using Kriging Model *Int. J. Therm. Sci.* **47** 1464–72

[14] Yoon H S, Park S H, Choi C and Ha M Y 2015 Numerical study on characteristics of flow and heat transfer in a cooling passage with a tear-drop dimple surface *Int. J. Therm. Sci.* **89** 121–135

[15] Elyyan M A and Tafi D K 2008 Large eddy simulation investigation of flow and heat transfer in a channel with dimples and protrusions *J. Turbomach.* **130** 0410161–9

[16] Acharya S and Zhou F 2012 Experimental and computational study of heat/mass transfer and flow structure for four dimple shapes in square internal passage *J. Turbomach.* **134** 0610281–13

[17] Isaev S A, Schelchkov A V, Leontiev A I, Baranov P A and Gulcova M E 2016 Numerical simulation of the turbulent air flow in the narrow channel with a heated wall and a spherical dimple placed on it for vortex heat transfer enhancement depending on the dimple depth *Int. J. Heat Mass Transf.* **94** 426–448

[18] Wae-Hayee M, Tekasakul P, Eiamsaard S and Nuntadusit C 2015 Flow and heat transfer characteristics of in-line impinging jets with cross-flow at short jet-to-plate distance *Experimental Heat Transf*er. **28** 511–530

[19] Wae-Hayee M, Tekasakul P, Eiamsaard S and Nuntadusit C 2014 Effect of cross-flow velocity on flow and heat transfer characteristics of impinging jet with low jet-to-plate distance *J. Mech. Sci. Technol.* **28** 2909–17

[20] Versteeg H K and Malalasekera W 2007 *An Introduction to Computational Fluid Dynamics*, Second Edition, Pearson Prentice Hall

[21] Mahmood G I and Ligrani P M 2002 Heat transfer in a dimpled channel: combined influences of aspect ratio, temperature ratio, Reynolds number, and flow structure *Int. J. Heat Mass Transf.* **45** 2011–20

IOP Conf. Series: Materials Science and Engineering **243** (2017) 012003 doi:10.1088/1757-899X/243/1/012003

Three-dimensional finite volume modelling of blood flow in simulated angular neck abdominal aortic aneurysm

Y A Algabri[1], S Rookkapan[2] and S Chatpun[1]
[1]Institute of Biomedical Engineering, Faculty of Medicine, Prince of Songkla University, Hat Yai, Songkhla, Thailand
[2]Department of Radiology, Faculty of Medicine, Prince of Songkla University, Hat Yai, Songkhla, Thailand

E-mail: algabriyousif80@gmail.com

Abstract. An abdominal aortic aneurysm (AAA) is considered a deadly cardiovascular disease that defined as a focal dilation of blood artery. The healthy aorta size is between 15 and 24 mm based on gender, bodyweight, and age. When the diameter increased to 30 mm or more, the rupture can occur if it is kept growing or untreated. Moreover, the proximal angular neck of aneurysm is categorized as a significant morphological feature with prime harmful effects on endovascular aneurysm repair (EVAR). Flow pattern in pathological vessel can influence the vascular intervention. The aim of this study is to investigate the blood flow behaviours in angular neck abdominal aortic aneurysm with simulated geometry based on patient's information using computational fluid dynamics (CFD). The 3D angular neck AAA models have been designed by using SolidWorks Software. Consequently, CFD tools are used for simulating these 3D models of angular neck AAA in ANSYS FLUENT Software. Eventually, based on the results, we summarized that the CFD techniques have shown high performance in explaining and investigating the flow patterns for angular neck abdominal aortic aneurysm.

1. Introduction

Aneurysm is an enlargement or dilation of vessels, in another word, a bulging or ballooning the wall of blood vessel [1]. The aortic aneurysm is commonly occurring at the region of thoracic, thoracoabdominal, and abdominal [2]. Abdominal aortic aneurysm (AAA) forms and grows in the abdominal aorta which usually located between the renal arteries and the iliac bifurcation in the infrarenal abdominal aorta [1]. The healthy abdominal aorta diameter is between 15 and 24 mm in humans based on gender, body weight, and age. The aneurysm results in aorta dilation, typically 50 % greater than the normal diameter of healthy abdominal aorta [3]. There are several symptoms that accompany the aneurysm rupture such as sudden abdomen or back pain, sever dizziness, clammy and wet skin, and rapidity of heart rates [3]. The aneurysm rupture is a serious illness with 65% to 85% rate of mortality in many cases where almost half of death cases happened before admitting to the surgery. About 1.3% of deaths amongst men between 60 - 85 years old caused by AAA in developed countries while in USA is considered as 13th of deaths leading cause [3]. Moreover, in Netherlands's hospitals for period between 1972 and 1992, there were obvious increases of AAA patients due to the detection improvement in ultrasound [4]. Therefore, in many developed countries health care reported this aneurysm is a major health problem. Therefore, it is significant for aneurysm to be treated when it is found before gradually increase widening. Furthermore, endovascular aneurysm repair (EVAR) is considered a less invasive surgical procedure and has become a preferable treatment for many eligible patients due to minor preoperative risks, lesser mortality, and long-term healing just like open surgery [5,6]. However, some of those eligible patients are not accepted for the EVAR procedure treatment. Due to complicated morphological aneurysm for example short aortic angular neck or highly proximal aortic angulated neck [7]. Shortly, there are few points to be considered when we study on neck angulations of AAA and their measurements such as knowing the exact location where angle occurs, the angle of neck formed in the area beneath the renal artery, proximal neck of an aneurysm and the bifurcation aorta. In contrast, the ideal neck angle of aorta is to be less than 60° if the angle is greater than that it leads to some implantation difficulties, endoleak and as well as the possibility for distal device migration. Computational fluid dynamics (CFD) is increasingly used in cardiovascular medicine that can play an important role in investigating and solving various complicated physiological fluid flows and demonstrating their potentials [8]. The aim of this study is to use CFD technique to investigate the flow pattern in the simulated angular neck abdominal aortic aneurysm.

2. Materials and methods

2.1. Geometry

In this study, we used a geometry which was based on idealized fusiform angular neck abdominal aortic aneurysm (AAA) of patient. The 3-dimensional simulated geometry was constructed by using computer-aided software SolidWorks (Solid Works Corp, Concord, MA) as shown in figure 1. The aneurysm size of 55 mm represented an average risk of rupture [9, 10]. Therefore, the four geometries were designed with the same concepts but the only difference was the neck angle (**θ**) of aneurysm which was technically set with an increment of 10 degrees as 60°, 70°, 80° and 90° [2].

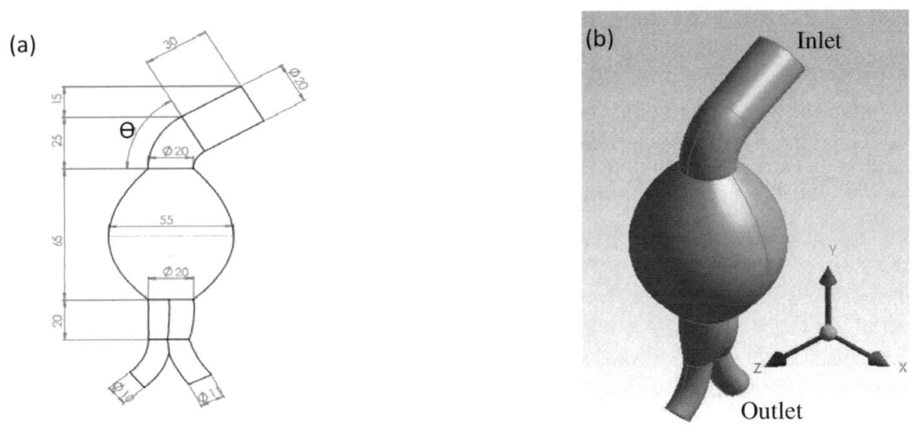

Figure 1. (a) Angular neck AAA simulated geometry and (b) 3-dimensional model

2.2. Meshing

The four-idealized angular neck AAA geometries were computationally meshed by using ANSYS FLUENT v16.2 (ANSYS Inc., Canonsburg, PA, USA). Tetrahedral meshing and a minimum of 1 mm body element size were applied. Cyclic quality checking and smoothing were done to reach satisfying mesh quality. Five different element numbers were used to determine the mesh quality were 55,304, 71,379, 391,306, 972,832 and 1,029,702. The blood properties and steady-state boundary conditions were applied for all mesh independency simulations. As shown in figure 2, the maximum velocity tended to be plateau when increasing the numbers of elements which indicated for velocity convergence and mesh independence.

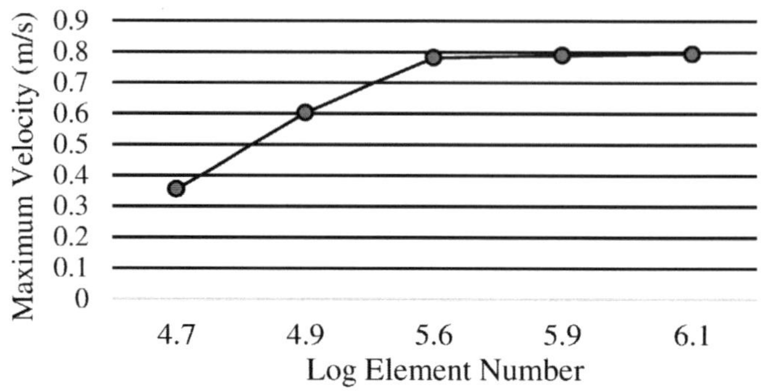

Figure 2. Mesh independence test based on maximum velocity

2.3. Governing equations

The fluid model solved numerically based on 3-dimensional incompressible Navier-Stokes equations along with the equations of momentum conservation (1) and mass conservation (2).

$$\rho\left(\frac{\partial u}{\partial t}\right) + \rho(u\cdot\nabla)u = -\nabla P + \mu\nabla^2 u, \tag{1}$$

$$\nabla\cdot(u) = 0, \tag{2}$$

where u, P, ρ, μ, and ∇ represent velocity, pressure, density, dynamic viscosity, and gradient symbol, respectively.

2.4 Simulation setup

The blood flow through the aorta was assumed to be incompressible and Newtonian fluid [8, 11]. The density and dynamic viscosity were considered constant with value of 1,060 kg/m^3 and 0.0035 Pa·S, respectively [12]. This viscosity was at approximately 45% of hematocrit value [13]. For all models, the steady-state simulation was performed with an inlet velocity of 0.28 m/s and outlet pressure of zero as well as the aortic wall was assumed to be no-slip and rigid [14]. The Reynolds number at the aorta inlet was calculated as of 1,696 based on inlet velocity and inlet aorta diameter, so flow was treated as laminar model. The CFD simulations were solved by using a commercial finite volume solver, ANSYS FLUENT v16.2 (ANSYS Inc., Canonsburg, PA, USA). All simulations were conducted on workplace computer with 3.40 GHz quad-core processor with 16.0 GB RAM.

3. Results and discussion

3.1. Flow behaviors at proximal aneurysm neck angulation

The flow patterns in angular neck and aneurysm sac were presented as velocity contours with more details in three different areas. The velocity contours on horizontal cross-sectional ZX-plane for the location below proximal aortic aneurysm neck is presented as in figure 3(a). The volume of blood flow was forming a half circular shape of velocity contour with low velocity in angular neck 60 and 70 degrees. Furthermore, it is observed that when the degree of angular neck increases, the velocity gradually increases in the centre of aorta. The chart in figure 3(b) shows the increment of average velocity in the four aortic neck angulations.

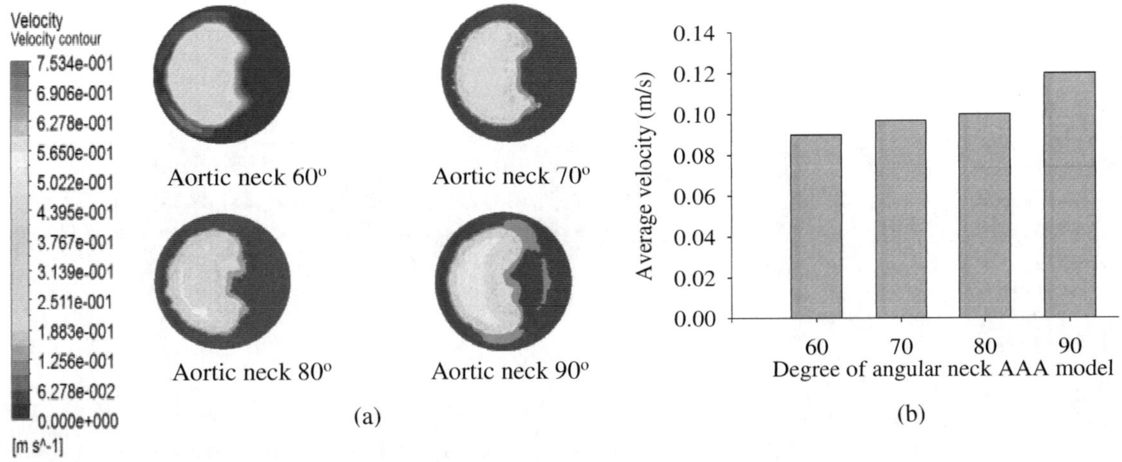

Figure 3. (a) Velocity contours at proximal neck of aneurysm sac, (b) average velocity at each neck at ZX-plane

However, to predict the exact change of velocity flow fields at the neck, the reference points were given at proximal aneurysm sac as revealed in figure 4(a) for this purpose. The average velocity values obtained by using the points starting from the near inner wall of the neck and reach to the outer wall of aneurysm are plotted in figure 4(b). It is clearly demonstrated that the velocity rose until it reached maximum value about 0.4 m/s and kept in the same range in the central region of aortic neck with little increase in 80° and 90° angles before it dropped to initial value zero at the outer wall of the distal neck.

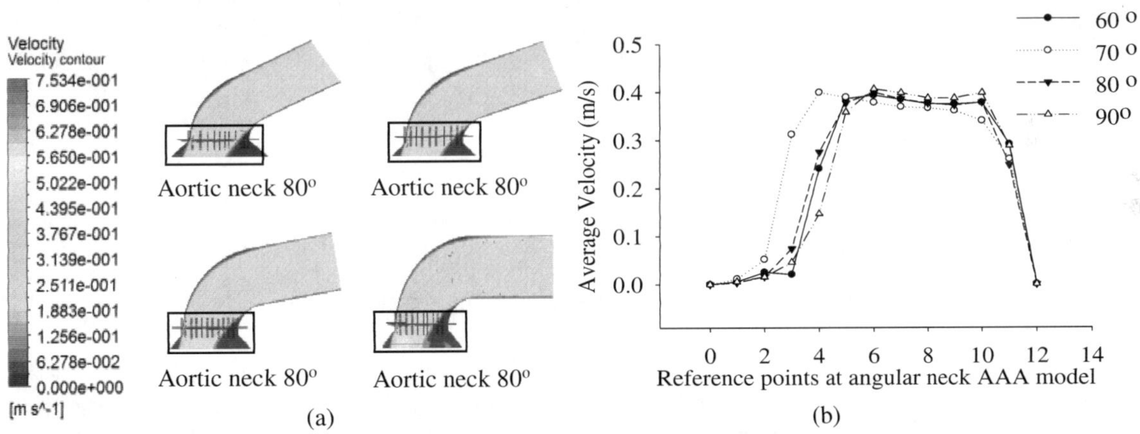

Figure 4. (a) 12-Point regions indicated by boxes in 4 geometries, (b) velocity values at 10-Points in angular neck AAA. On the X-axis, point 0 represents inner wall and point 12 represents outer wall.

3.2. Flow pattern through angular neck AAA model

The flow as illustrated in figure 5 on the longitudinal cross-section YZ-plane presents a fluffy bar of velocity from the proximal aneurysm neck cross the aneurysm sac towards the distal region of abdominal aortic aneurysm. At the same time, the velocity contours showed that the thin strip of high velocity is increasing gradually in the four angular neck aortas based on the change of degree of angular neck AAA.

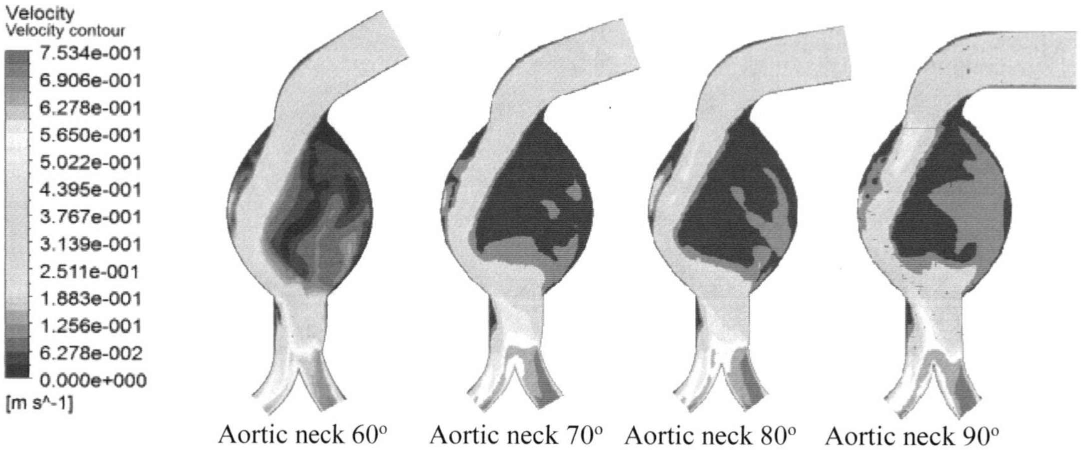

Figure 5. Velocity contours at the longitudinal YZ-plane of the angular neck AAA

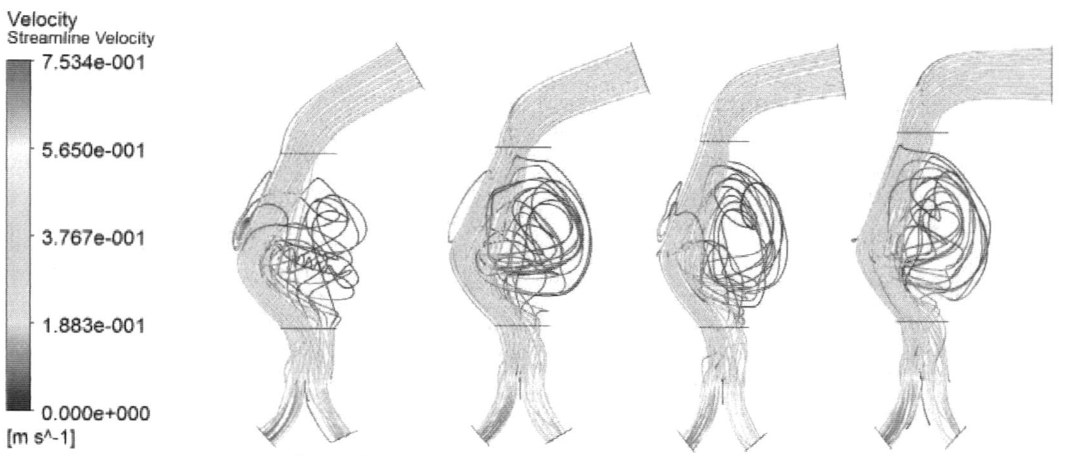

Figure 6. Velocity streamlines of angular neck AAA

The velocity streamlines are presented in Figure 6. The different recirculation vortexes were detected at the aneurysm in all aortic angular neck. However, when the angular neck of aorta rises, it is observed that the pattern of vortex inside aneurysm sac is debilitated and the velocity field appears to be more intensive near the middle of aorta.

Even though, this simulation was performed with idealized fusiform models but the simulation was based on realistic patient's information [15]. As findings in our study, the numerical simulation of the four geometries of highly angular neck AAA using computational fluid dynamics (CFD) has confirmed that the significant influence of aortic neck angles on the velocity flow field because of the degree of angular neck change impacts on the flow downstream [2]. Moreover, we calculated the flow rate ($Q = v_{ave} \cdot A$) at the left and right iliac outlets as shown in figure 7(a) based on the average velocity (v_{ave}) at the outlets and the cross-sectional area (A) of each iliac aorta. Since the inlets and outlets of the simulated angular neck AAA geometries were the same with the value of each outlet (left iliac = 11 mm and right iliac = 10 mm), the cross-sectional area of outlet was $A_{left\ iliac} \cong 9.5x10^{-5} m^2$ and $A_{right\ iliac} \cong 7.85x10^{-5} m^2$. The average velocity values for all angular neck AAA models at both outlets were presented in table 1. It was found that the average velocity at the left iliac outlet was higher than the average velocity at the right iliac outlet. The plotted chart in figure 7(b) illustrated the difference of flow rates at the outlet of each iliac aorta for each degree of angular neck AAA model. Although, the difference was as shown very minimum but we still can observe the influence of angular neck degrees on the flow rates at the outlet regions for each model. Eventually, it is proper to mention that the findings of this study have some limitations because it was performed under steady-state fluid flow condition, while the blood naturally is transient flow.

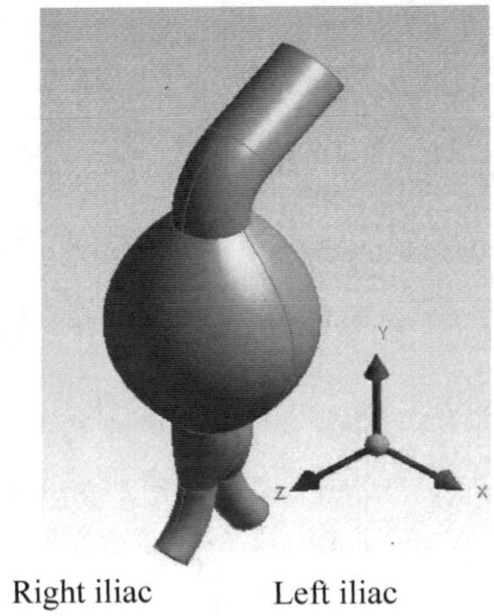

(a)

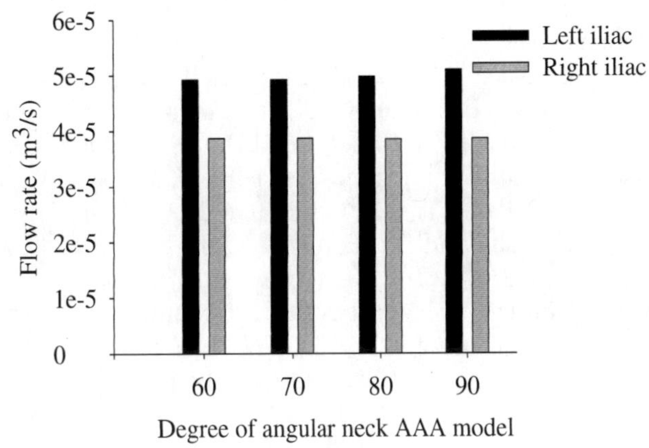

(b)

Figure 7. (a)Three dimensional model and (b) Flow rate at each iliac aorta outlet

Table 1. Average velocity values at both iliac aorta outlets.

	Average velocity (m/s)	
Neck	Left iliac outlet	Right iliac outlet
60°	0.5171	0.4921
70°	0.5176	0.4923
80°	0.5239	0.4904
90°	0.5271	0.4913

4. Conclusion

Four angular neck abdominal aortic aneurysm (AAA) models were simulated by using CFD technique to explain blood flow behaviours. As the results, by applying the steady state and fluid conditions on the three-dimensional angular neck AAA models, we can foresee the behaviours of the blood flow through the angular neck, aneurysm sac and as well the flow rate at the iliac aortas. Moreover, the study found that there was a correlation between the degrees of angular neck AAA and the pattern of the blood flow through the aorta. Therefore, the three-dimensional finite volume modelling and CFD technique can provide better visual details for blood flow in cardiovascular system which is useful for the cardiovascular intervention.

Acknowledgment
The authors would like to acknowledge the Thailand's Education Hub for Southern Region of ASEAN Countries (TEH-AC) scholarship given to Mr.Yousif Ali Yousif Algabri from the graduate school, Prince of Songkla University. We would like to thank also Mr. Kirttayoth Yeranee, Master student in the Mechanical Engineering department, Faculty of Engineering, Prince of Songkla University for his kind help in designing the 3D geometries.

References
[1] Kruchten T Van 2015 (Delft: Delft University of Technology)
[2] Yeow S L and Leo H L 2016 *Comput. Math. Methods. Med* **2016** 3830123
[3] Sakalihasan N, Limet R and Defawe O 2005 *Lancet.* **365** 1577-89
[4] Reitsma JB, Pleumeekers H J C M, Hoes A W, Kleijnen J, De Groot R M, Jacobs M J, Grobbee D E and Tijssen J G P 1996 *Eur.J. Vasc. Endovasc.Surg.* **12** 446-51
[5] Greenhalgh R M, Brown L C, Kwong G P, Powell J T and Thompson S G 2004 *Lancet.* **364** 843–848
[6] Prinssen M, Verhoeven E L, Buth J, Cuypers P W, van Sambeek M R, Balm R and Blankensteijn J D 2004 *N. Engl. J. Med.* **351** 1607-1618
[7] Morrison T M, Meyer C A, Fillinger M F, Fairman R M, Glickman M H, Cambria R P, Farber M A, Naslund T C, Fail P S, Elmore J R, White R A and Dall'Olmo C A and Williams D M 2013 *J. Vasc. Surg.* **57** 24S
[8] Morris P D, Narracott A, von Tengg-Kobligk H, Soto D A S, Hsiao S, Lungu A, Evans P, Bressloff N W, Lawford P V, Hose D R and Gunn J P 2016 *Heart.* **102** 18-28
[9] Meecham L, Evans R, Buxton P, Allingham K, Hughes M, Rajagopalan S, Fairhead J, Asquith J R and Pherwani A D 2015 *Eur.J. Vasc. Endovas*c. **49** 28-32
[10] Duncan J L, Harrild K A, Iversen L, Lee A J and Godden D J 2012 *BMJ* **2958** 1-9
[11] Shek T L, Leonard W T, Nabovati A and Amon C H 2012 *J. Biomech. Eng.* **134** 121002
[12] Frauenfelder T, Lotfey M, Boehm T and Wildermuth S 2006 *Cardiovasc. Intervent. Radiol.* **29** 613–23
[13] Carty G, Chatpun S and Espino D M 2016 *J. Med. Bio. Eng.* **36** 396–409
[14] Taylor T W and Yamaguchi T 1994 *J. Biomech. Eng.* **116** 87-97
[15] Li Z and Kleinstreuer C 2007 *Eur. J. Mech. B-Fluid.* **26** 615-631

Numerical simulations on flow and heat transfer in ribbed two-pass square channels under rotational effects

N. Kaewchoothong, K. Maliwan and C. Nuntadusit*
Department of Mechanical Engineering, Faculty of Engineering, Prince of Songkla Univesity, HatYai, Songkhla 90110, Thailand

E-mail: chayut@me.psu.ac.th

Abstract. The main objective of this research is to study the flow and heat transfer characteristics in a rotating two-pass square channel with ribbed walls. In this study, the channel length-to-hydraulic diameter ratio of the rotating two-pass square channel (L/D_h), the rib height-to-hydraulic diameter ratio (e/D_h), rib angle of attack (α) and the rib pitch-to-height (p/e) ratio are fixed at 11.33, 0.13, 60° and 10, respectively. The test fluid is air having the flow rate in terms of constant Reynolds number (Re) of 10,000. The rotation numbers (R_o) are varied from 0.1 to 0.4. The details of the local heat transfer distribution and the flow field of the rotating two-pass square channel are numerically studied by using commercial software ANSYS Fluent (ver.15.0). The results show that the ribbed walls enhance the heat transfer rate significantly. Under rotation, the average Nu in the first pass with radial outward flow is increased while that in the second pass is decreased, and also found that maximum heat transfer rate is observed for rotation number of 0.4 which is higher about 10-20% when compared with the other rotation number cases.

1. Introduction

Gas turbine engine is one kind of widely used in the powerful machineries which were applied in the aircraft propulsion, land-based power generation system and industrial applications. In order to achieve the higher thermal efficiency and power output. One of ways is by increasing the turbine inlet temperature (TIT) which current gas turbine engines can operate at TIT up to about 1,700°C. Thus, it need effective cooling technologies to augment heat transfer rate with less amount of cooling air, such as rib-roughened channel cooling, impingement cooling, film cooling, pin-fin cooling and dimple or protrusion cooling as shown in Fig. 1. Rib turbulated cooling in internal channel is the most commonly adapted approach in gas turbine blade. Han and his co-workers [1-7] have performed a large amount of research in this field including the discussions of aspect ratio, channel shape, Reynolds number, rib configurations, etc.

Walker and Zausner [8] used commercial software ANSYS CFX to predict local heat transfer distribution in a ribbed two-pass square channel and indicated that SST k-ω was an optimal turbulence model for engineering applications. Bonhoff et al. [9] numerically studied a two-pass channel with 45° ribs using the commercial solver ANSYS FLUENT with Reynolds stress model (RSM) and standard k-ε model. After that, Ekkad and Han [10] studied the detailed heat transfer distributions in two-pass square channels with the angles of 60°, 90° and 60° discrete rib turbulators. They reported that the heat transfer rate for rib-roughened walls was larger than smooth walls due to the secondary flows induced the effect of the bend and rib turbulators. In addition, Shen et al. [11]

numerically investigated smooth and ribbed two-pass square channel under rotation number between 0.4 and 0.6 at constant Reynolds number fixed at 12,500. They explained that enhanced heat transfer areas were moved to radial inward channel from the sidewalls of radial outward channel as divider-to-tip wall distance was increased. Rotational effect in turbine blade affects the flow and heat transfer characteristics considerably. In general, rotation effect induces Coriolis and centrifugal forces that produce cross-stream secondary flow in the rotating channel. The effect of rotation on the first-pass and second-pass channels is different due to the different outward and inward flow.

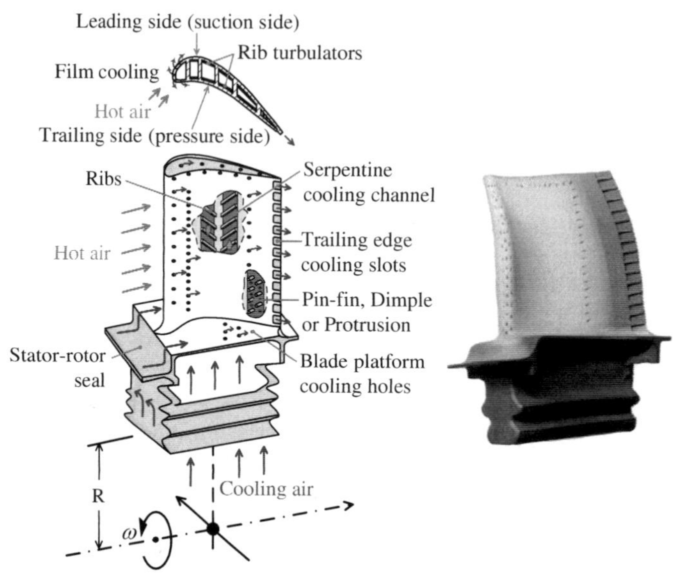

Figure 1. Advance gas turbine rotor blade cooling passage network and defined coordinate system

From the review of obtainable literature on internal flow and heat transfer rate under rotation conditions, it can be noted that the heat transfer characteristics should be investigated more detailed in order to design more effective gas turbine blade cooling. Additionally, few researchers numerically studied both flow characteristics and heat transfer rate overall surface in cooling channel. Thus, the present study focused on the detailed examination of the effect of rotation on the flow behavior and local heat transfer distributions inside the two-pass square channel where the ribs are attached on the two opposite walls. The angle of 60° inclined ribs in two-pass square channel is numerically performed using commercial software ANSYS Fluent (ver.15.0) which all cases are carried out under high rotational number of up to 0.4.

2. Numerical simulation and methods

2.1. Physical model

The geometry of a two-pass square channel with the angle of 60° inclined ribs on the opposite walls for this study is shown in Fig. 2. The coordinate system of x, y, z represents the spanwise, normal to heat transfer wall and streamwise directions, respectively, where the rotating direction of the two-pass square channel is clockwise of x-axis direction. The cross section ($W \times H$) of a channel was square of 15 mm × 15 mm with aspect ratio AR of 1. In this study, the ribs with square cross section were used to locate on two opposite walls in rotating durection. The rib height-to-hydraulic diameter ratio (e/D_h), the length-to-diameter ratio (L/D_h) of each of the first and second passages, and the rib pitch-to-height (p/e) ratio were fixed at 0.133, 11.33 and 10, respectively. The air inlet of channel was at R/D_h=22.0.

For boundary conditions in these numerical simulation, the flow rate was determined at the entrance in the radial outward channel, which the value of flow rate is defined by Reynolds number (*Re*) of 10,000. Inlet temperature and turbulence intensity of a channel entrance was imposed at 298.15 K and 5.0%, respectively. The flow exit of the channel was determined to be maintained at a constant pressure of 1.0 atm. All wall surfaces of channel were no-slip conditions with zero velocity component. The constant heat flux of about 300 W/m^2 was defined on trailing and leading surfaces. The rotation numbers of the two-pass square channel were varied between 0.1 to 0.4 into consideration similar with the aircraft blade condition [12].

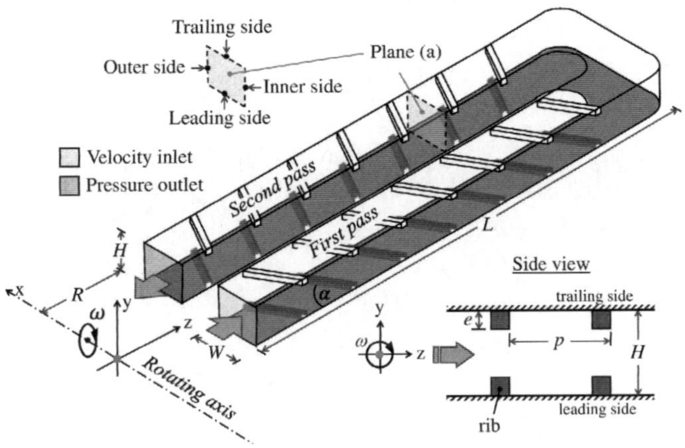

Figure 2. Geometry of a rotating two-pass square channel.

2.2. Computational details

In this numerical study, the 3D flow field and heat transfer on surface of the two-pass square channel are studied by CFD commercial software ANSYS Fluent (ver.15.0). The computational model was operated by solving Reynolds averaged continuity and Navier-Stokes equations. The Reynolds stress model (RSM) was applied in solving turbulent flow [13-15]. For numerical calculations, the SIMPLE algorithm was used for the pressure-velocity coupling in order to successful computations of governing equations. Second order upwind scheme was designated for special discretization of all equations. The solutions were considered to be convergent when the residual values for continuity, velocities and energy were less than 1x10^{-6}, 1x10^{-4} and 1x10^{-7}, respectively [16-17].

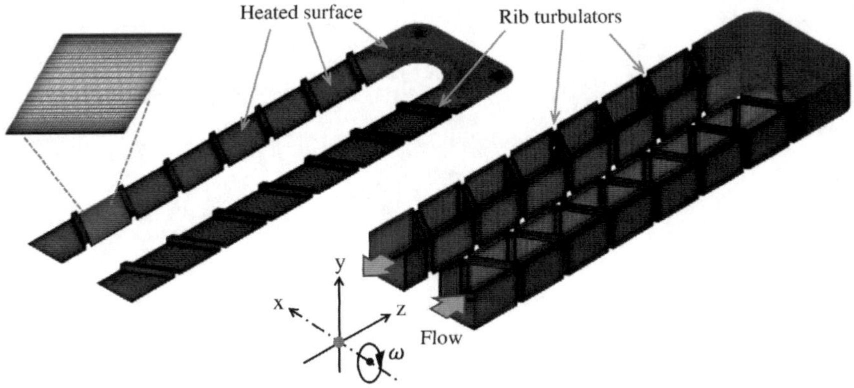

Figure 3. Grid used for computational domain.

Computational domain was generated by structured hexahedral element as shown in Fig. 3. The more intensive grids were used near the walls, bend and ribbed surfaces of rotating channel. In all domains, the number of grid more than 3.0 million could be considered of the two-pass square channel, which the grid spacing of the first layer normal to the wall corresponding to y^+ values was less than 1.0.

2.3. Data reduction

There are three parameters of interest included Reynolds number (*Re*), Nusselt number (*Nu*) and Rotation number (R_o). The Re is determined as

$$\mathrm{Re} = \frac{V_i D_h}{\nu} \tag{1}$$

where V_i is the inlet velocity, D_h is the hydraulic diameter of channel, which it can be calculated from D_h=2*WH*/(*W*+*H*) when *W* and *H* were width and height of channel (D_h=*H*), and ν is the kinematic viscosity of air. The local Nu is calculated from Eq. (1).

$$Nu = \frac{hD_h}{k} \tag{2}$$

where *h* is the convective heat transfer coefficient of air and k is the thermal conductivity.

The Nusselt number ratio (Nu/Nu_o) is calculated using the correlation of Dittus and Boelter [18] for fully developed flow, a smooth wall with rotation to normalized Nusselt number.

$$Nu/Nu_o = \frac{Nu}{0.023\mathrm{Re}^{0.8}\,\mathrm{Pr}^{0.4}} \tag{3}$$

where *Pr* is the Prandtl number of air (*Pr*=0.71).

The average Nusselt number ratio ($\overline{Nu}/Nu_o$) is considered by averaging local Nusselt number values (from Eq. (1)) over the heat transfer surface.

$$\overline{Nu}/Nu_o = \frac{\left(\sum_{i=1}^{n} Nu_i\right)/n}{Nu_o} \tag{4}$$

where Nu_i is the local Nusselt number at a measured point, n is the total number of measured point over the heat transfer surface.

The R_o is determined from Eq. (5).

$$R_o = \frac{\omega D_h}{V_i} \tag{5}$$

where ω is the rotating speed.

3. Results and discussion

3.1. Validation of CFD model

The computational model for a flow and heat transfer enhancement of the two-pass square channel is predicted using each turbulence model to validate with experimental results under stationary channel of A.K. Sleiti [17] as shown in Fig. 4. Comparison of calculation heat transfer on the first-pass leading surface between the 5th and 6th ribs is considered at Re=10,000 under stationary case. The results show that the predictions using RSM was the most accurate of about 5%. RTD k- ω and SST k- ω predicted the trend exactly but over predicted Nu/Nu_o near the rib top by 15% and 25%, respectively. Whereas, the worst are predictions both the standard (STD) k- ε and RNG k- ε of scalable wall function option in this numerical study. Thus, the RSM model is the best choice for this flow model.

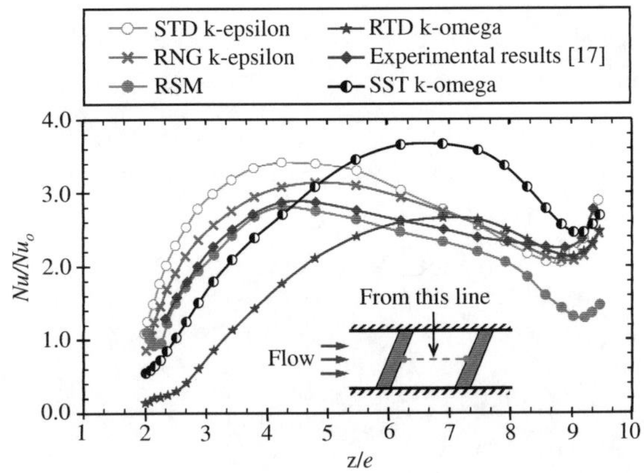

Figure 4. Numerical validation of heat transfer along the centerline of the floor in stationary channel.

3.2. Flow structures

Figure 5 shows the comparisons of 3D flow-field velocity contours and the velocity vector on each plane of channel under rotation number (R_o) of 0.0 (Fig. 5(a)) and 0.2 (Fig. 5(b)). Plane (i), (ii), (iii) and (iv) are on the middle between ribs in the first and second channels, and plane (v) is selected in the center of the bend section. As shown in Fig. 5(a) for the stationary case (R_o=0.0), the results show that angle of 60° ribbed channel of each plane in the first-pass channel created simultaneously the large vortex near the inner and the outer sides, which it is induced by the effect of inclined ribs. On the bend region, similar flow behavior occurs large vortex near the inner wall and the outer wall due to the combined effect of the U-bend and ribs in the first pass channel can induce the small counter rotating flow near inner side. After the bend, the flow field in second-pass channel is similar with the flow before the bend in first-pass channel because the rib-induced counter rotating vortex pairs.

For rotating case, the complex secondary flow occurred near the trailing surface in the first-pass channel (see Fig. 5(b)) is created by the angle of inclined ribs and the effect of rotation which is affected to increase the heat transfer enhancements on the walls. On the bend section, it is observed that a counter rotating vortex pair creates near the top side of the bend due to the combined effect of the centrifugal-induced vortex and the Coriolis-induced complex secondary flow in the upstream. In addition, the secondary flow pattern in the second-pass channel is rather complicated due to the combined effect of the bend and the rib-roughened walls, which the flow structure in the bend region is influenced by the angle of 60° inclined ribs, curvature of the bend and the shape of the ribs upstream and downstream of the bend.

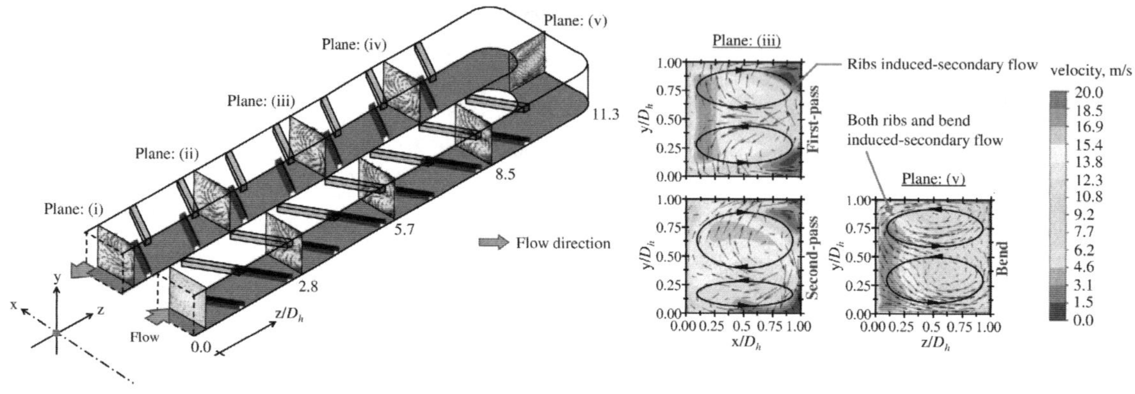

(a) Stationary case (R_o=0.0)

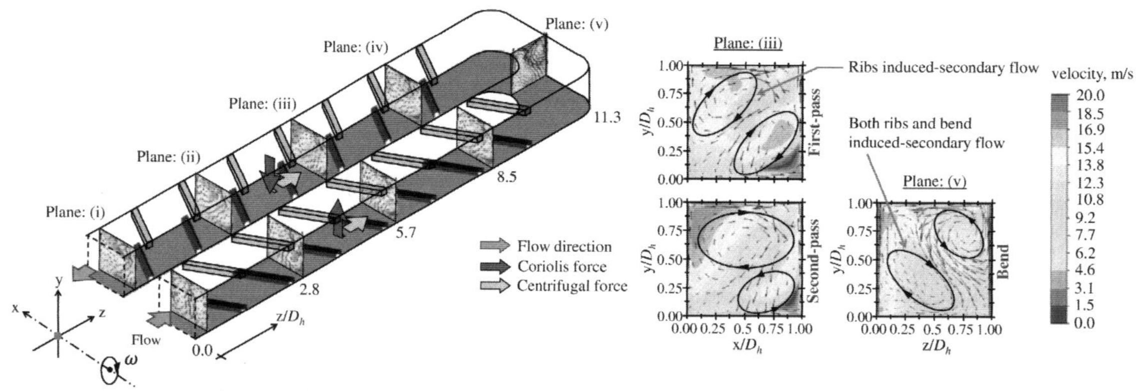

(b) Rotating case (R_o=0.2)

Figure 5. 3D flow-field velocity contours and vectors on each plane of the two-pass square channel.

3.3. Heat transfer distributions

The Nusselt number distribution pattern for the ribbed two-pass square channel is shown in Fig. 6. The results show that trends of Nu/Nu_o distributions are significantly high near the inlet due to the flow entrance effect and decreases along the main flow direction until the bend of channel. In second pass, similar heat transfer distribution can be noted at the different sections cause by the deflection of main fluids by the bend and the rib-induced circulation, which it declines gradually.

For rotational case, it is observed that the heat transfer rate is significantly enhanced in the first-pass trailing surface and distribution patterns between the ribs are different from the stationary case due to the combined effects of the Coriolis and centrifugal forces which can create the main fluids shift and attach on the trailing surface. Similar heat transfer ratio distribution pattern in second-pass leading surface is gradually reduced. Additionally, the computational results are also concluded that the combined effects between the Coriolis and centrifugal forces which happen due to the rotation of flow channel can be enhanced heat transfer on the ribbed surface in two-pass square channel larger than the smooth channel.

The average Nusselt number ratios in each region between the ribs on leading and trailing surfaces of the ribbed walls within the two-pass square channel are shown in Fig. 7. On the leading side, the overall $\overline{Nu}/Nu_o$ tends to increase in the first pass of the channel due to the effect of angle of inclined ribs. After that, heat transfer rate decreases slightly in the second pass channel after the bend due to the effect of the circulation which can induces a pair of counter rotating flows in the cross section of the channel. These circulation flows cause the flow to attach on the first-pass leading surface stronger than the second-pass leading surface. On the trailing side, the results show that the average heat transfer ratios are significantly increased in the first pass and gradually decrease in second pass after the bend of two-pass square channel. The results also show that increasing of the rotation number gives the highest heat transfer rate on the first-pass and second-pass surfaces when compared with the stationary case. This is because of the stronger rotation-induced secondary flow vortices and the combined effects between the Coriolis and centrifugal forces (see Fig.5).

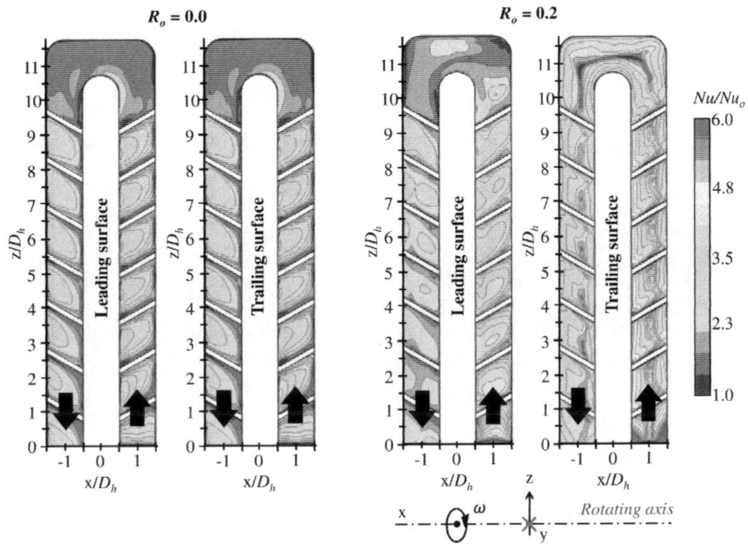

Figure 6. Detailed heat transfer distributions on leading and trailing surfaces of the two-pass square channel.

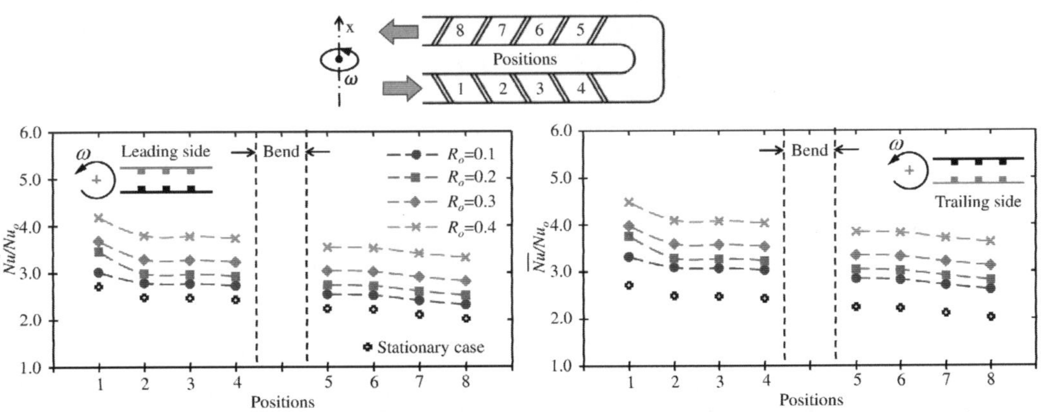

Figure 7. Regional average Nusselt number ratio between ribs for various rotation number. (Left: on leading side, Right: on trailing side)

4. Conclusions

In the present study, 3D flow field and heat transfer distributions were numerically performed in order to investigate the effect of the rotation number on flow and heat transfer characteristics in the two-pass square channel. The main results of this research can be summarized as follows.

1. Effect of the rotation creates heat transfer enhancement differences between the leading and trailing walls for ribbed channels.

2. The maximum value of average Nusselt number ratio is remarked for rotation number of 0.4 which is higher than about 10-20% when compared with the other rotation number cases.

3. Under the effect of rotation, the Coriolis force will increase the heat transfer rate in the first-pass channel for radial outward flow while decrease the heat transfer rate in the second pass as radial inward flow.

Acknowledgment
This research was financially supported by Faculty of Engineering, Prince of Songkla University, Thailand.

References

[1] Han J C, Glicksman L R and Rohsenow W M 1978 An investigation of heat transfer and friction for rib-roughened surfaces *International Journal of Heat and Mass Transfer* **21**(8) 1143-1156
[2] Han J C 1984 Heat transfer and friction in channels with two opposite rib roughened walls *Journal of Heat Transfer* **106**(4) 774-781
[3] Han J C 1988 Heat transfer and friction characteristics in rectangular channels with rib turbulators *Journal of Heat Transfer* **110**(2) 321-328
[4] Han J C and Park J S 1988 Developing heat transfer in rectangular channels with rib turbulators *International Journal of Heat and Mass Transfer* **31**(1) 183-195
[5] Park J S, Han J C, Huang Y, Ou S and Boyle R J 1992 Heat transfer performance comparisons of five different rectangular channels with parallel angled ribs *International Journal of Heat and Mass Transfer* **35**(11) 2891-2903
[6] Azad G S, Uddin M J, Han J C, Moon H K and Glezer B 2002 Heat transfer in a two-pass rectangular rotating channel with 45-deg angle rib turbulators *Journal of Turbomachinery* **124**(2) 251-259
[7] Huh M, Liu Y H, and Han J C 2009 Effect of rib height on heat transfer in a two pass rectangular channel (*AR* = 1:4) with a sharp entrance at high rotation numbers *International Journal of Heat and Mass Transfer* **52**(19-20) 4635-4649
[8] Walker D and Zausner J 2007 RANS evaluations of internal cooling passage geometries: ribbed passage and a 180 degree bend *ASME Paper GT2007-278*
[9] Bonhoff B, Tomm U and Johnson B V 1996 Heat transfer predictions for U-shaped coolant channels with skewed ribs and with smooth walls *ASME Paper 96-TA-7*
[10] Srinath V Ekkad and Han J C 1996 Detailed heat transfer distributions in two-pass square channels with rib turbulators *International Journal of Heat and Mass Transfer* **40**(11) 2525-2537
[11] Shen Z, Xie Y, Zhang D and Xie G 2014 Numerical Calculations on Flow and Heat Transfer in Smooth and Ribbed Two-Pass Square Channels under Rotational Effects *Mathematical Problems in Engineering* ID981376-1-7
[12] Shen Z, Xie Y and Zhang D 2015 Numerical predictions on fluid flow and heat transfer in U-shaped channel with the combination of ribs, dimples and protrusions under rotational effects, *International Journal of Heat and Mass Transfer* **80** 494-512
[13] Gibson M M and Launder B E 1978 Ground effects on pressure fluctuations in the atmospheric boundary layer *Journal of Fluid Mechanics* **86** 491-511
[14] Launder B E 1989 Second-moment closure: present and future *International Journal of Heat and Fluid Flow* **10** 282-300
[15] Launder B E, Reece G J and Rodi W 1975 Progress in the development of a Reynolds-stress turbulence closure *Journal of Fluid Mechanics* **68** 537-566
[16] Sewall E A and Tafti D K 2007 Large eddy simulation of flow and heat transfer in the developing flow region of a rotating gas turbine blade internal cooling duct with Coriolis and buoyancy forces *Journal of Turbomachinery* **130**(1) 011005-011005-7
[17] Sleiti K and Kapat J S 2005 Fluid flow and heat transfer in rotating curved duct at high rotation and density ratios *Journal of Turbomachinery* **127**(4) (2005) 659-667

[18] Dittus P W and Boelter L M K 1985 Heat transfer in automobile radiators of the tubular type *International Communications in Heat and Mass Transfer* **12** 3-22

Nomenclature

AR	: Channel aspect ratio (-)
D_h	: Hydraulic diameter of channel (m)
e	: Rib height (m)
H	: Channel height (m)
h	: Heat Transfer coefficient (W/m^2 K)
k	: Thermal conductivity of air (W/m K)
L	: Length of the channel (m)
n	: Total number of pixel over the heated surface. (-)
Nu	: Nusselt number (-)
$\overline{Nu}$	: Average Nusselt number (-)
Nu_i	: local Nusselt number at a pixel (-)
Nu_o	: Nusselt number of the smooth wall (-)
p	: Pitch of ribs (m)
Pr	: Prandtl number (-)
R	: Radius of rotation (m)
Re	: Reynolds number (-)
R_o	: Rotation number (-)
T	: Temperature of air (°C)
V_i	: Inlet velocity (m/s)
W	: Channel width (m)

Greek symbols

ν	: Kinematic viscosity of air (m^2/s)
ω	: Rotational speed (rad/s)
ρ	: Density of air (kg/m^3)

Effect of inclined angle of pin arrays on flow and heat transfer characteristics in flow channel

P Narato[1], M Wae-hayee[1], P Vessakosol[1], and C Nuntadusit[1]

[1] Department of Mechanical Engineering, Faculty of Engineering, Prince of Songkla University, Songkhla, Thailand

E-mail:chayut@me.psu.ac.th

Abstract. The aim of this research is to study effect of inclined angle of pin arrangement on flow and heat transfer characteristics in a rectangular flow channel with array of pins. In this study, the cylindrical of pins which having diameter of D=10 mm was mounted in the flow channel with staggered arrangement. The pin-to-pin distance was fixed at S_y=2D in spanwise direction and at S_x=2.5 in streamwise direction. The effect of pin inclined angle was investigated at θ = 60°, 90°, 120°, and 135° for constant Reynolds number at Re=32,860. The flow and heat transfer were investigated using ANSYS ver.15.0 (Fluent) to simulate 3-D steady turbulence flow. The realizable k - ε turbulence model was applied in this study. The results show that the pin inclined angle with θ = 120° and 135° can enhance the heat transfer on the upstream and downstream of the pins when compared to the case of pin angle θ = 90°. While, the pin inclined angle with θ = 60° gives the lowest heat transfer rate overall surface. The mechanism of heat transfer enhancement for inclined pin can be explained with increasing velocity and reduced wake flow region behind the pin.

1. Introduction

The pin array is normally used for increasing heat transfer on surface in engineering applications, such the cooling of gas turbine blades and electronic devices and enhancing heat transfer in compact heat exchangers.The heat transfer enhancement for pin array can be divided in two factors. The first factor is increasing of heat transfer surface of pin. The second factor is increasing of heat transfer coefficient on surface due to turbulence promotion near endwall of pins [2]. The heat transfer enhancementfor pin arrays with long height is dominated by the enhancement of pin surface area while the endwall effect is not significant [3]. However, the mechanism of heat transfer enhancement of short pins due to thermal boundary disturbance on endwall is larger than addition of heat transfer surface on pin surface. Pins are commonly cylindrical shape protruding from the heat transfer surface. The heat transfers fluid is forced to flow pass the array of pins for cooling or heating. Figure 1 shows the flow pattern near the endwall of the pin. The horseshoe vortex that promotes the turbulent was generated on endwall in front of the pin. A Pair of wake flow and separate flow is observed behind the pin. The separate flow will disturb/destroy the boundary layer and will promote the turbulent flow behind the pin. This will affect to the heat transfer enhancement in region in front of the pin and behind the pin.

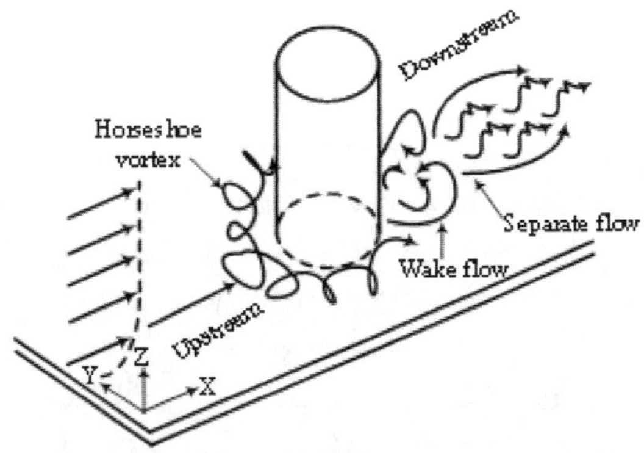

Figure 1. Schematic of flow features around cylinder pin

Chyu [4] studied the effect of pin shape on heat transfer enhancement focusing only for case of mounting pin normal to the heat transfer surface. Inline and staggered arrays with and without base fillet are studied. The heat transfer performance of straight cylinder (without fillet cylinder) with staggered arrays was higher than that of a fillet-cylinder. Pandit et al. [5] demonstrated three-types of pin fin arrays with staggered arrangement of circular, triangular, hexagonal and diamond shapes on increasing heat transfer in flow channel. The result shows that heat transfer from the diamond pin fin is higher than that all shape. Rao et al. [6] studied on heat transfer characteristics of pin arrays attached on surface with dimples in a channel flow. The pin arrays with dimple surface can enhance heat transfer performance by up to 19% when compared to pin arrays on surface with no dimples. Lawson et al. [7] studied the effects of spanwise and streamwise spacing on heat transfer with mounting pin arrays for a range of Reynolds number between 5,000 and 30,000. As a result, the reduction of spanwise and streamwise pin spacing increases heat transfer enhancement due to flow acceleration. The heat transfer enhancement with low streamwise pin spacing was highly significant than low spanwise pin spacing. Choi et al. [8] studied the influence of inclined short single cylinder on heat transfer surface and flow field around a cylinder. The heat transfer on the upper endwall is enhanced due to the interaction between the horseshoe vortex and the jet-like flow. However, the heat transfer on the lower endwall is reduced due to the large wake region behind the inclined cylinder. However, the effect of angle of inclined for short cylinders on flow and heat transfer characteristic is still not clearly understood for inclined pin arrays.

The objective of this study is investigating the flow and heat transfer characteristics of inclined pin arrays on heat transfer surface with simulation. The effects of pin inclined angle to heat transfer surface were investigated for case of $\theta = 60^{\circ}$, 90°, 120°, and 135° at constant Reynolds number of airflow at Re=32,860 for comparison.

2. Numerical Method

2.1. Geometric Model

Figure 2 shows computation domains and mesh arrangement for prediction of inclined pin arrays in flow channel. The computational domains with a width of 40 mm and a height of 32mm were divided into three parts; upstream of test section, test section, and downstream of test section. First part of domain (upstream of test section) with a length of 1,480 mm (25.6D_h) had a sufficient length to provide a fully developed flow through the test section. Second part of domain (test section) with a length of 370 mm (6.4D_h) was mounted inclined pin arrays and third part of domain (downstream of test section) with a length of 400 mm (6.9D_h) was connected to flow outlet. The hydraulic diameter of $D_h = 57.8$ mm is determined based on the rectangular of the wind tunnel.. The cylinder pins which having diameter of D = 10 mm were mounted with staggered arrangement in flow channle. The pin-to-pin distance was fixed at S_y=2D in spanwise direction and at S_x=2.5 in streamwise direction as shown in Fig. 3(a). The size of heat transfer surface was with length of 32D and width of 4D. In the simulation, the effects of inclination angle between pin centreline to mounted surface were investigated in the range of $\theta = 60^o$, 90°, 120°, and 135°. The Reynolds number of internal flow was fixed at Re = 32,860. For all computations, the y^+ values on heat transfer surfaceare approximately 1.0 for all case investigated.

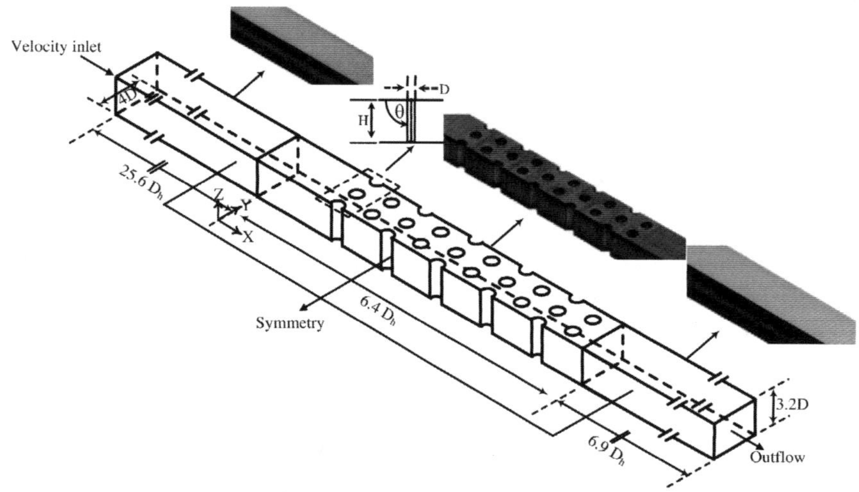

Figure 2. Computational domain and grids

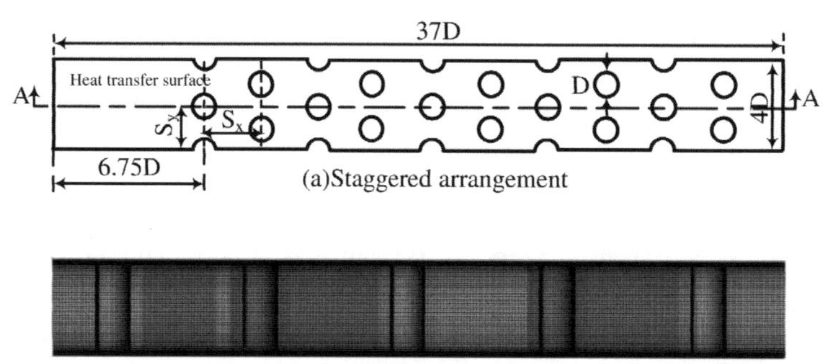

(a)Staggered arrangement

(b) Section A-A of mesh

Figure 3. Configuration of inclined Pins arrangement and cross-section of grids

2.2. Solution methodology

For the boundary conditions, the inlet was identified as uniform velocity profile and outlet was specified as outflow boundary condition (zero gradient in flow direction). The two side of computational domain in spanwise direction were identified as symmetry boundary condition. The top and bottom of computation domain were set wall boundary condition. A non-slip boundary condition was applied over the wall surface. The turbulence intensity was specified as 5% at the inlet section, and a uniform heat flux was fixed on the top surface.

The Numerical simulation was used to solve Reynolds Averaged continuity equation, Navier-Stokes equations and energy equation for obtaining three-dimensional flow field and heat transfer on the surface. The model of fluid flow and heat transfer were analyzed using the realizable k - ε turbulence model. The solution method was based on Semi-Implicit Method for Pressure-Linked Equations (SIMPLE) algorithm with second order upwind for all spatial discretization.The root mean square (RMS) residual of continuity and energy equation is less than 10^{-5} and that of momentum equation is less than 10^{-4}, the computed domain is considered convergent.

2.3. Data reduction

The Reynolds number was calculated by maximum velocity (V_C) at centre of channel and the equivalent hydraulic diameter (D_h) at the entrance, defined as:

$$Re = \frac{V_C D_h}{v} \tag{1}$$

The local heat transfer coefficient of the test channel is defined by:

$$h = \frac{\dot{Q}_{net}}{(T_w - T_m)} \tag{2}$$

This $\dot{Q}_{net}$ is the net heat flux on heated surface. T_w is local temperature on heat transfer surface. T_m is mean temperature of the airflow.to obtain the mean temperature of the airflow, T_m, the following equation is used

$$T_m = \frac{(T_i + T_o)}{2} \tag{3}$$

This T_i is the fluid temperature at inlet. T_o is the fluid temperature at outlet.

The local Nusselt number on heat transfer surface is defined by:

$$Nu = \frac{hD_h}{k} \tag{4}$$

The average Nusselt number of the smooth channel with fully-developed flow was calculated by Dittus-Boelter correlation as follow:

$$Nu_0 = 0.023\, Re^{0.8}\, Pr^{0.4} \tag{5}$$

Where, Pr is Prandtl number of air.

2.4. Turbulence models

To examine the effect of pin inclined angle on flow and heat transfer characteristics of pin arrays in flow channel, commercial software ANSYS FLUENT 15.0 has been utilized. For a turbulence closure, the realizable k - ε turbulence model combined with a standard wall function from Launder and Spalding [9] was used as a turbulence closure. The realizable k - ε turbulence model was used to predict complex turbulent flows with flow swirling and separation, and was used to simulate the flow and heat transfer in the channel with pin fins [10, 11]. Figure 4 shows a comparison of the numerical data with the experimental data for the inclined pin at $\theta = 90^o$. The overall average Nusselt number, $\overline{\overline{Nu}}$, was defined as follows:

$$\overline{\overline{Nu}} = \sum_{i=1}^{n} \overline{Nu}_y / n \tag{6}$$

Here $\overline{Nu}_y$ is spanwise average Nusselt number. It can be seen that the numerical data of the average Nusselt number of the inclined pin arrays in flow channel is in good agreement with experimental data. At Re=20,400, the experimental $\overline{\overline{Nu}}$ value of inclined pin at $\theta = 90^o$ is about 2.82% higher than the case of numerical. The numerical results showed the same trend as the experiments obtained.

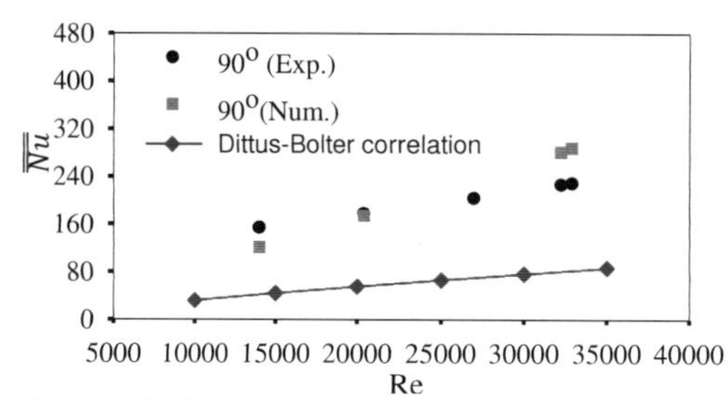

Figure 4. Comparison of the computational data with the experimental data, $\overline{\overline{Nu}}$ varying with Re

3. Results and discussion

3.1. Heat transfer characteristics

Nusselt number contour on heat transfer surface obtained from simulation are shown in Fig. 5. The results show for four different pin inclined angle at constant Re=32,860. In the Fig.5, the air flows from left hand to right hand in X-axis direction. There are 10 rows of inclined pins in the test section for simulation. As can be seen from the contours, the local Nusselt number for case of $\theta = 90^o$ is high for small area in upstream region in front of pin. This is due to the effect of horse shoe vortex around the pin while, the Nusselt number in downstream region is more distributed with large area. This agrees with the results obtained from other studies [7, 12 and 13]. The local Nusselt number in downstream region tends to increase from third rows to seventh rows but it tends to decrease next to eighth rows. For pin inclination to upstream, for case of $\theta = 60^o$, the local Nusselt number in upstream of pin tends to decrease when reduce the pin inclined angle from $\theta = 90^o$ to 60°. The local Nusselt number in downstream of pin increased in small area at first row, second row and third row when reduce the pin angle from θ = 90° to 60°. However, the overall Nusselt number in downstream of pin deceases and smaller than the case of θ = 90° when the pin inclined angle at θ = 60°. For pin inclination to downstream, for case of θ = 120° and 135°, the local Nusselt number in upstream of pin tends to increase at the third row onwards and the local Nusselt number in downstream of pin tends to significantly increase with large area, when increase the pin inclined angle from θ = 90° to 120° and 135°. The inclined pin for case of θ = 135° give the largest heat transfer enhancement in upstream and downstream of pin row.

Figure 6 shows local Nusselt number along the X-axis at Y/D = 0 obtained from Fig. 5. The local Nusselt number was not shown for region of heat transfer surface under the pins. For case of pin angle θ = 90°, the local Nusselt number in upstream region tends to increase as going close to the pin in each of the row of pin and it tends to decrease when increasing X/D. In downstream region, the local Nusselt number tends to increase from pin in each of the row of pin and it tends to decrease as increasing X/D. For pin inclined angle to upstream at θ = 60°, the local Nusselt number in upstream of pin tends to decrease in each of the row of pin when compared to case of θ = 90°. For case of pin inclined to downstream, for case of θ = 120° and 135°, the local Nusselt number in upstream and downstream of pin tends to increase in each of the row of pin when compared to case of θ = 90°.

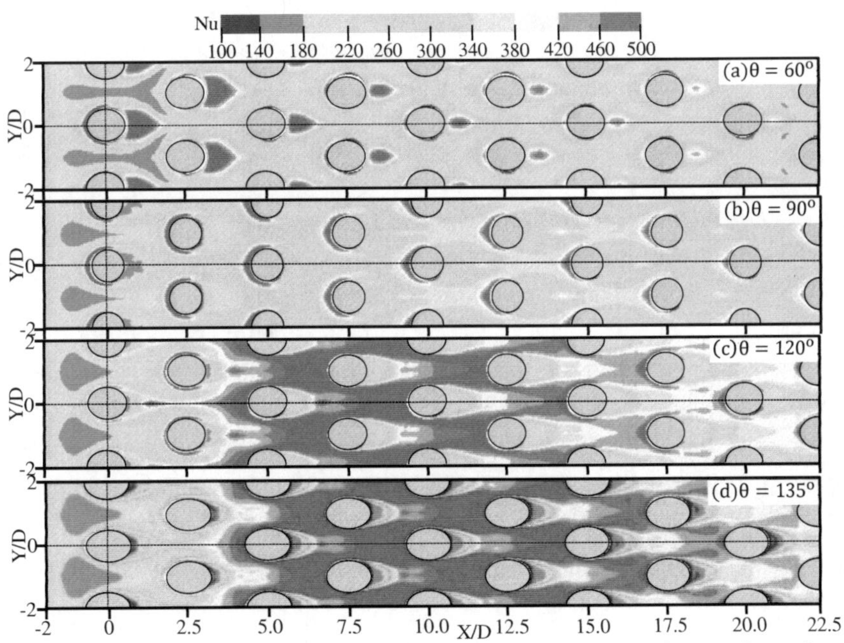

Figure 5. Nusselt number distributions on heat transfer surface for different pin inclined angles

The overall average Nusselt number for all of pin inclined angle are plotted in Fig. 7. The overall average Nusselt number is calculated by averaging the local Nusselt number obtained from Fig. 5. The pins array with pin inclined angle θ = 120° and 135° can enhance the overall average Nusselt number about 28.24% and 34.33% when compared to pin with angle θ = 90°, respectively.

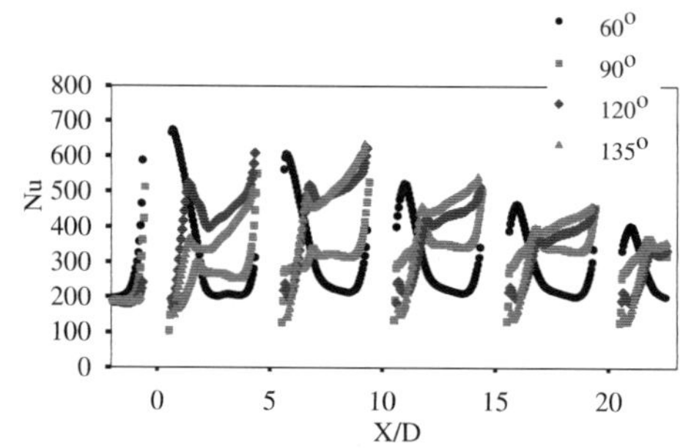

Figure 6. Local Nusselt number along the X-axis at Y/D=0

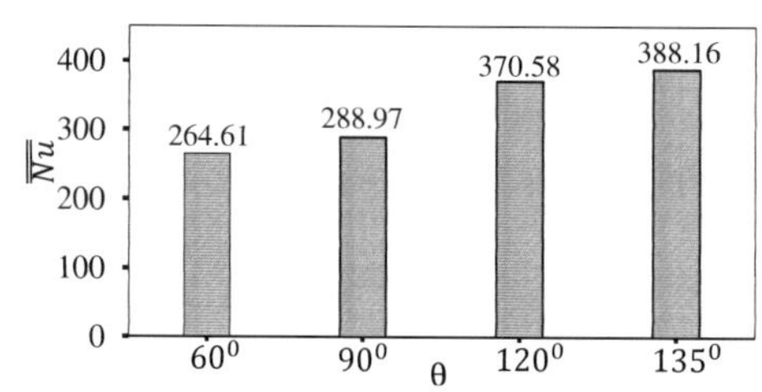

Figure 7. Overall average Nusselt number for different pin inclined angles

3.2. Flow characteristics

Figure 8 compares the streamline on X-Y plane above the heat transfer surface about 1 mm obtained from CFD results for different pin inclined angles. The air flows from left hand to right hand passing pin arrays. For case of pin inclined angle $\theta = 60^{o}$, there are also footprints two counter rotating vortices between the downstream of pins. For case of pin angle $\theta = 90^{o}$, there are footprint of two wake flows behind each pin. The velocity near the heat transfer surface becomes small for large area behind the pins. For case of pin inclined angle $\theta = 120^{o}$ and 135°, the footprint of wake flows can be also found behind the pin. However, the area of low velocity region is smaller when compare to case of pin angle $\theta = 90^{o}$. The area of wake flow behind the pin was reduced with increasing pin inclined angle from $\theta = 90^{o}$ to 120° and 135°.The enhanced velocity in promote the heat transfer on surface with increasing pin inclined angle from $\theta = 90^{o}$ to 120° and 135°.

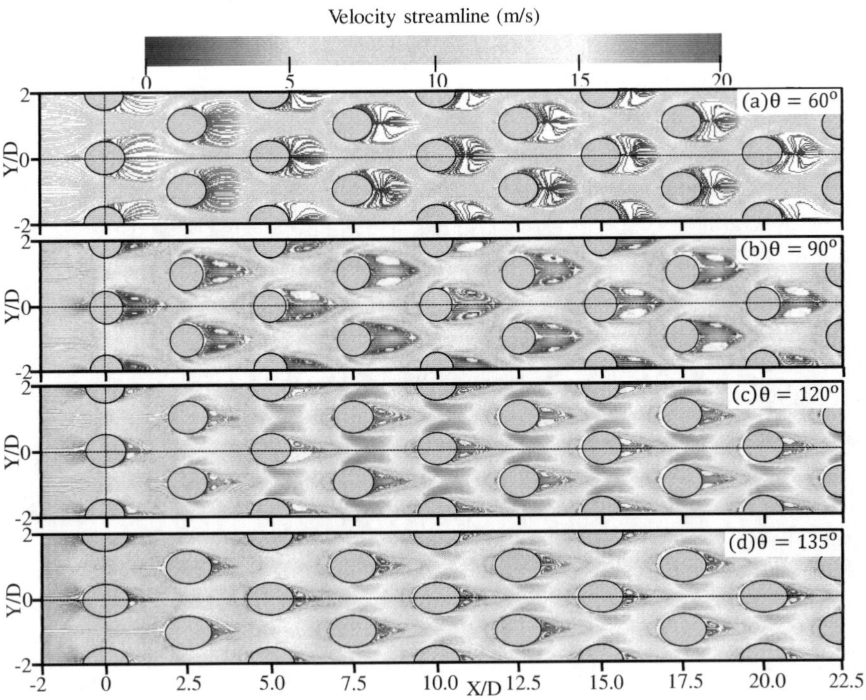

Figure 8. Streamlines near the heat transfer surface (1.0-mm under heat transfer surface) of inclined pins on the X-Y plane

Figure 9 shows the velocity contour on Z-X plane passing the pin centre obtained from CFD results compared with different pin inclined angle. The region with low velocity behind the pin can be obviously detected for case of pin angle $\theta = 90^{o}$. This region corresponds to wake flow region behind the pin. However, the region with low velocity cannot be seen for case of $\theta = 135^{o}$. This flow characteristic will promote the heat transfer behind the pins.

Figure 10 shows the velocity streamline on Y-Z plane at different location X/D=10, 11.25 and 12.5. For case of pin inclined angle $\theta = 60^{o}$, there are a pair of counter-rotating vortices behind each pin on Y-Z plane at X/D=10, 11.25 and 12.5. The counter-rotating vortices in downstream of this inclined pin may destroy the wake flow behind the pins. For case of pin angle $\theta = 90^{o}$, the counter-rotating vortices cannot be found. For case of pin inclined angle $\theta = 120^{o}$ and 135°, the counter-rotating vortices can be found behind the pin on Y-Z plane at X/D=12.5. The vortices locate near the bottom surface, which is far from the heat transfer surface.

IOP Conf. Series: Materials Science and Engineering **243** (2017) 012005 doi:10.1088/1757-899X/243/1/012005

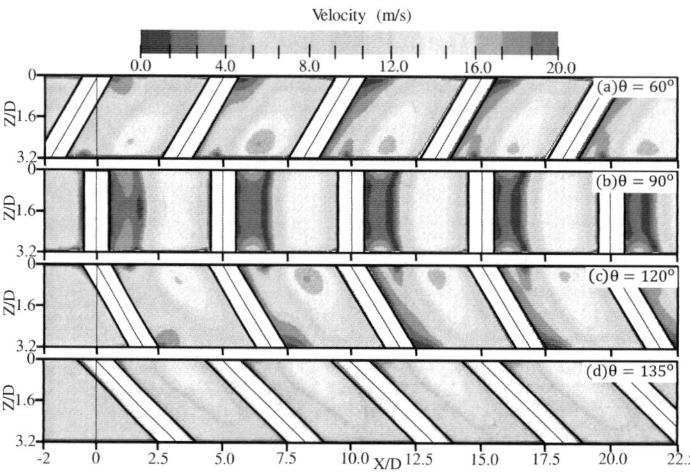

Figure 9. Comparisons of the streamwise velocity on Z-X plane passing the centre of wind tunnel for various pin inclined angles

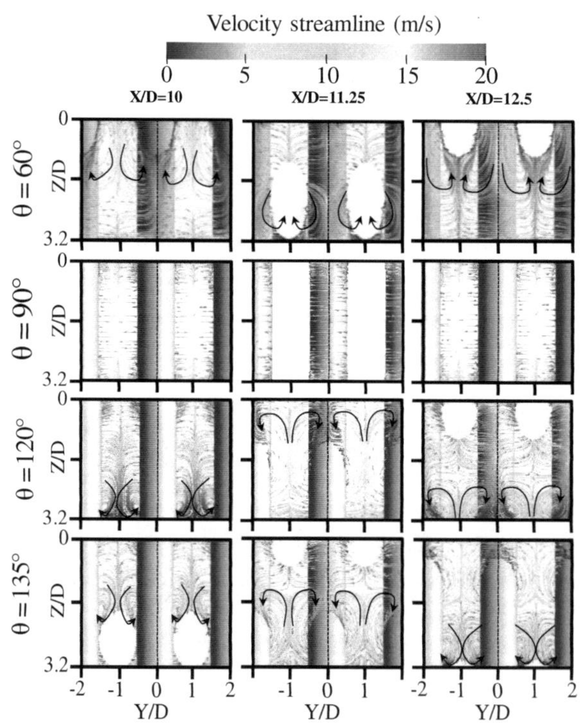

Figure 10. Comparisons of the streamlines on Y-Z plane with at some X/D for different pin inclined angles

4. Conclusions

In this study, the flow and heat transfer for inclined pin arrays attached in wind tunnel was studied with numerical simulation. The effects of pin inclined angle were investigated in the range of $\theta = 60^o$, 90°, 120°, and 135° at constant Reynolds number Re=32,860 for comparison. The main results obtained from this study can be concluded as follows:

- The pin inclined angle affects strongly on the heat transfer characteristics on surface with pin arrays. The pin inclined angle at $\theta = 120^o$ and 135° can improve the heat transfer in region in front of pin and can enhance the heat transfer behind pin. Particularly for the case of $\theta = 135^o$, the overall average Nusselt number is higher than the case of $\theta = 90^o$ about 34.33%.
- For case of pin inclined angle $\theta = 60^o$, 120° and 135°, a pair of counter-rotating vortices was generated behind each pin. The counter-rotating vortices may reduce the wake flow behind the pins.
- For case of pin angle $\theta = 90^o$, the wake with low velocity region can be detected behind each pin. This results in low heat transfer region on the surface behind the pin. When increase the pin angle from $\theta = 90^o$ to 120° and 135°, the region of wake flow become smaller and start to increase strongly velocity behind the pins at near the heat transfer surface. Therefore, the heat transfer rate is higher than the case of $\theta = 90^o$.

References

[1] Lau S C, Kim Y S and Han J C 1985 Effects of Fin Configuration and Entrance Length on Local Endwall Heat/Mass Transfer in a Pin Fin Channel ASME Paper 85-WA/HT-62
[2] Chyu M K and Natarajan V 1996 Heat Transfer on the Base Surface of Three-Dimensional Protruding Elements *Int. J. Heat Mass Transfer* **39** 2925-2935
[3] Zukauskas A A 1972 Heat Transfer from Tube in Cross Flow *Adv. Heat Transfer* **8** 116-133
[4] Chyu M K 1990 Heat Transfer and Pressure Drop for Short Pin-Fin Arrays with Pin-Endwall Fillet *ASME J. Heat Transfer* **112** 926-932
[5] Pandit J, Thompson M, Ekkad S V and Huxtable S T 2014 Effect of pin fin to channel height ratio and pin fin geometry on heat transfer performance for flow in rectangular channels *Int. J. Heat Mass Transfer* **77** 359-368
[6] Rao Y, Wan C, and Xu Y 2012 An experimental study of Pressure loss and heat transfer in the pin fin-dample channels with various dimple depths *Int. J. Heat Mass Transfer* **55** 6723-6733
[7] Lawson S A, Thrift A A, Thole K A and Kohli A 2011 Heat transfer from multiple row arrays of low aspect ratio pin fin *Int. J. Heat Mass Transfer* **54** 4099-4109
[8] Choi I K, Kim T, Song S J and Lu T J 2007 Endwall heat transfer and fluid flow around an inclined short cylinder *Int. J. Heat Mass Transfer* **50** 919-930
[9] Launder B E and Spalding DB 1974 the numerical computation of turbulent flows *Comput.Meth.Appl.Mech.Eng.* **3** 269-389
[10] Xie G, Sunden B and Zhang W 2011 Comparisons of pins/dimples/protrusions cooling concepts for a turbine blade tip-wall at high Reynolds number *J.Heat Transfer* **133** 1-9
[11] Rao Y, Wan C, Xu Y and Zang S 2011 Spatially-resolved heat transfer characteristics in channels with pin fin and pin fin-dimple arrays *Int. J. Therm. Sci.* **50** 2277-2289
[12] Wang F, Zhang J and Wang S 2012 Investigation on flow and heat transfer characteristic in rectangular channel with drop-shaped pin fins *Propulsion and Power Research* **1** 64-70
[13] Chang S W, Yanng T L, Huang C C and Chiang K F 2008 Endwall heat transfer and Pressure drop in rectangular channels with attached and detached circular pin-fin array *Int. J. Heat Mass Transfer* **51** 5247-525

Effect of jet-mainstream velocity ratio on flow characteristics and heat transfer enhancement of jet on flat plate flow

N Puzu[1], S Prasertsan[1] and C Nuntadusit[1]
[1] Department of Mechanical Engineering, Faculty of Engineering, Prince of Songkla University, Songkhla, Thailand

E-mail: chayut@me.psu.ac.th

Abstract. The aim of this research was to study the effect of jet-mainstream velocity ratio on flow and heat transfer characteristics of jet on flat plate flow. The jet from pipe nozzle with inner diameter of D=14 mm was injected perpendicularly to mainstream on flat plate. The flat plate was blown by mainstream with uniform velocity profile at 10 m/s. The velocity ratio (jet to mainstream velociy) was varied at VR=0.25 and 3.5 by adjusting velocity of jet flow. For heat transfer measurement, a thin foil technique was used to evaluate the heat transfer coefficient by measuring temperature distributions on heat transfer surface with constant heat flux by using infrared camera. Flow characteristics were simulated by using a computational fluid dynamics (CFD) with commercial software ANSYS Fluent (Ver.15.0). The results showed that the enhancement of heat transfer along downstream direction for the case of VR=0.25 was from the effect of jet stream whereas for the case of VR=3.5 was from the effect of mainstream.

1. Introduction

Heat transfer enhancement is one of key technology for energy saving. Especially, convective heat transfer is commonly found in the engineering application, e.g. engine cooling fin, electronics cooling system, heat exchanger etc. In the meanwhile, the method for heat transfer enhancement are various such as to install ribs, vanes, or pins on heat transfer surface to produce mainstream vortex on the surface. Many researchers presented the effect of vortex relate with the heat transfer of the surface.

M. Fiebig [1] studied the heat transfer enhancement by using wing-type vortex generators experimentally and numerically. The results showed that heat transfer rate was higher than the case of plane surface. However, the increasing of flow friction due to attachment wing-type vortex generators causes additional pressure loss and the need of more pumping power. For minimizing pressure loss penalty, vortex generating jet (VGJ) are attractive method by injecting jet into the boundary layer flow. Generally, the injecting jet in mainstream is used for film cooling application e.g. gas turbine engine blade cooling [2-5]. Ahn et al. [6] numerically investigated the interaction between jet and mainstream effect on film cooling effectiveness in gas turbine cooling application. The result showed that film cooling effectiveness depended on velocity ratio (jet to mainstream velociy) and the configuration of jet holes. Maio and Wu [7] numerically investigated the blowing and hole's shape are influencing film cooling effectiveness. In case of heat transfer enhancement, Qayoum et al. [8] used piezo-actuator for displacing the diaphragm to generate vortex and reported the heat transfer coefficient of downstream surface become increase. Jabbal and Zhong [9] shown that the heat transfer rate of underneath vortex surface increased significantly. Akdag et al. [10] experimentally investigated the cooling of constant heat flux surface by using thermocouple to measure surface temperature. The temperatures are varying with distance from the jet hole and the nearest position temperature is lowest. Whereas understanding of flow characteristics of vortex generating jet is important, the provide results from previous studies are limited.

The objective of this study was to investigate the flow and heat transfer characteristics of a vortex generating jet on a flat plate flow. The jet discharged from pipe nozzle with normal to the heat transfer surface. The effects of jet-to-mainstream velocity ratio on flow and heat transfer characteristic was focussed.

2. Experimental setup and method

The experimental setup is shown in Fig. 1. An open wind tunnel having bell shape nozzle was used to generate mainstream with uniform velocity profile over a flat plate. A VGJ nozzle was installed on the flat plate. The dimensions of bell shape nozzle exit of wind tunnel were 400 mm x 400 mm. The measured mainstream turbulence at the nozzle exit was nearly isotropic with turbulence intensity of 3.5% and a decay rate, $\partial Vm/\partial(X/D)$, of 0.175%, approximately. The difference of turbulence intensity and mean velocity along the nozzle exit was lesser than 2.0% respecting their mean value. The jet pipe was made of aluminum tube having the inner diameter of D=14 mm and the length of 10D for achieving fully developed flow of jet at the exit. The inlet of nozzle pipe was connected to blower supplier with flexible tube for reducing noise vibration.

The details of test section are shown in Fig. 1. The test section was made of acrylic plate with sharp edge (480 mm × 800 mm and 20 mm in thickness) and was opened with rectangular hollow. A stainless-steel foil with 300 mm × 300 mm and 0.03 mm in thickness was used for measuring heat transfer coefficient. It was tightly stretched between two copper bus bars over the rectangular hollow.

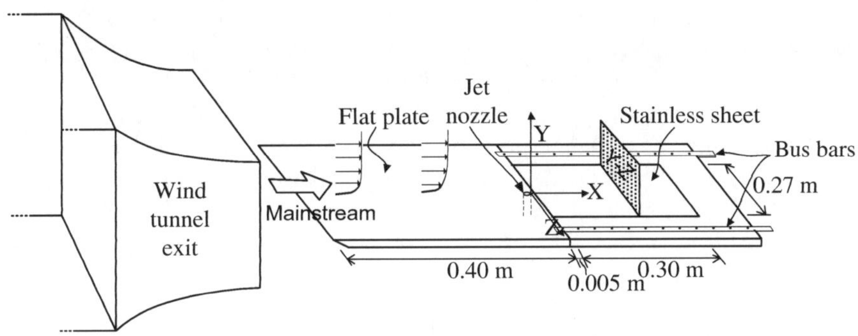

Figure 1. Detail of test section

The temperature distributions on the heat transfer surface were measured from the rear side of stainless foil using an infrared camera. The temperatures on both flow side and measurement side can be considered being the same since the stainless foil was sufficiently thin. The heat transfer surface was heated to obtain constant heat flux by supplying electrical current from power supply through the copper bus bars. The bottom side of heat transfer surface was sprayed with black paint which has emissivity at 0.95.

The temperature of mainstream and jet were controlled at $26.0 \pm 0.2^{\circ}$C. The mainstream velocity at wind tunnel exit was fixed at 10 m/s. The jet velocity from pipe nozzle was varied corresponding to jet-to-mainstream velocity ratio was varied at VR=0.25 and 3.5.

For data reduction, heat generated in heat transfer surface was defined as

$$\dot{Q}_{input} = I^2 R \tag{1}$$

IOP Conf. Series: Materials Science and Engineering **243** (2017) 012006 doi:10.1088/1757-899X/243/1/012006

Where, I and R were electrical current and electrical resistant of stainless foil. The heat losses from the rear side of heat transfer surface were considered from,

$$\dot{Q}_{losses} = \dot{Q}_{conv} + \dot{Q}_{rad} \tag{2}$$

where $\dot{Q}_{conv}$ and $\dot{Q}_{rad}$ were the heat losses from the rear side of stainless sheet to the environmental air by free convection and radiation, respectively. The local convective heat transfer coefficient can be obtained from following equations:

$$h = \frac{\dot{Q}_{input} - \dot{Q}_{losses}}{(T_w - T_{aw})A} \tag{3}$$

where, A was the area of heat transfer surface; T_w and T_{aw} were the local wall temperature (with wall heat flux) and local adiabatic wall temperature (with no heat flux) on heat transfer surface.

The convective heat transfer coefficient ratio, h/h_0 is used to represent the ability of heat transfer enhancement that compare between the case with jet and the case without jet. Here, h_0 was the convective heat transfer coefficient for the case with no jet.

3. Numerical Simulation

3.1. Computational model and calculation detail

In order to understand the effect of velocity ratio on flow characteristics of vortex generating jet, three-dimensional flow fields were simulated by using ANSYS FLUENT (version 15.0). The solutions for turbulence flow were achieved by solving steady incompressible Reynolds Averaged Navier-Stokes (RANS) equation through the finite volume method. The SIMPLEC (Semi-Implicit Method for Pressure Linked Equations-Consistent) algorithm was used for reveal velocity and pressure field in order to reduce the calculating time. In the present study, SST k-ϖ turbulence model was used.

The computational domain was divided into three regions as shown in Fig. 2. The first region is from nozzle exit to flat plate. The second region is on flat plate before the jet nozzle. The third region is on flat plate with constant heat flux. The grid was concentrated near the flat plate surface. The y+ on flat surface was less than 2 for all cases.

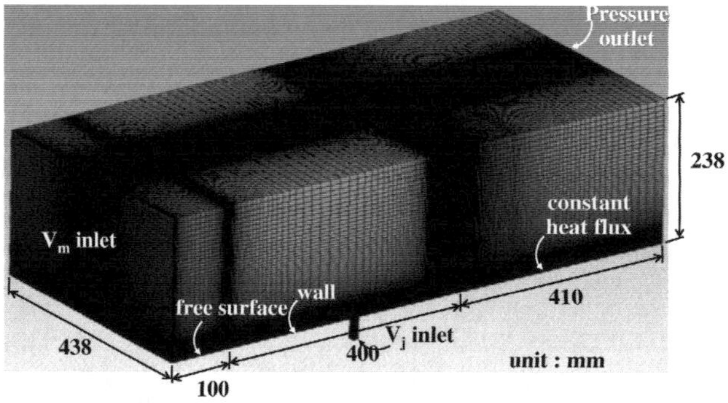

Figure 2. Computational model and grid generation

3.2. Computational verification
In order to verify computational results, the near wall velocity profiles that were measured with hot wire anemometry were compared with the simulation results. Figure 3 shows the comparison of velocity profiles above flat surface from experiment and simulation with no jet. The results show good agreement of both experimental and numerical results.

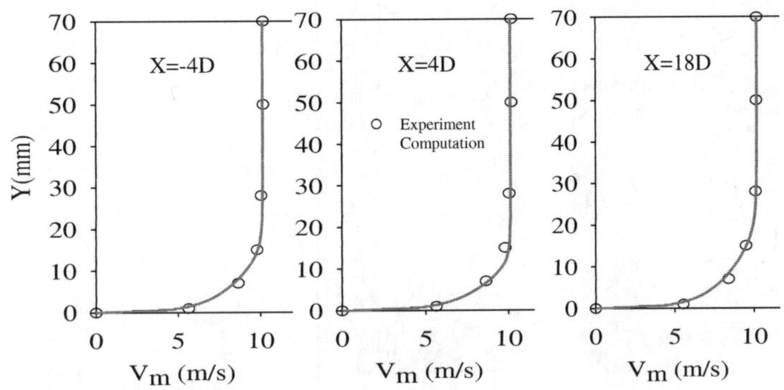

Figure 3. Comparisons of experimental velocity profile over flat plate of test section with computational result (no jet flow)

4. Results and discussions

4.1. Heat transfer
The h/h_0 from experimental results are shown in Fig. 4. For the case of VR=0.25, it is found that the area of high heat transfer peaks appeared into two region along downstream of jet. The characteristics of these two peaks were elongated parallel, along downstream direction. The highest heat transfer area was occurred at X/D=1-3 with h/h_0 up to 1.35 and gradually decrease along downstream direction. For closed area behind jet exit between X/D=0.36~1, h/h_0 became slightly increase when compare to the case of without jet. The jet effect on h/h_0 increasing in the surrounding area Z/D=-4~-2, 2~4 was insignificant. With the increasing of VR to 3.5, the heat transfer enhancement region expanded more in spanwise and streamwise directions. The highest heat transfer area (h/h_0=1.6) was initial immediately after jet exit to X/D=3, and its contour geometry became candle flame shape. Moreover, the effect of jet stream can increase h/h_0 on the surrounding area.

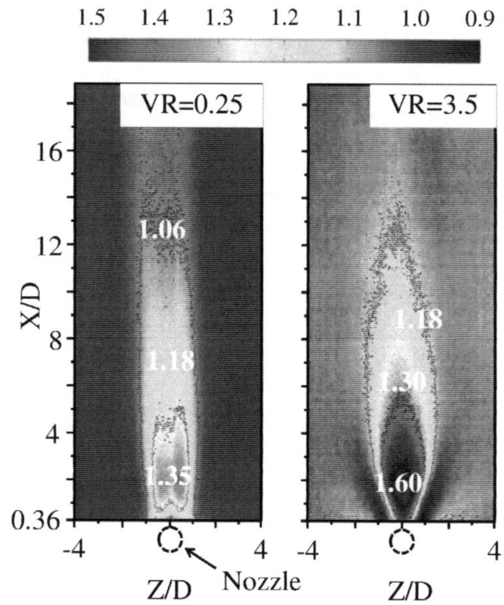

Figure 4. Effect of VR on contours of heat transfer coefficient ratio for vortex generating jet

4.2. Flow characteristics

The results of jet-mainstream mixing flow from simulation result are shown in figure 5. The two cases of VR=0.25 and VR=3.5 cases, jet stream path was considered fluid flow through jet exit and mainstream path was restricted pass normal plane with 2 mm height locate at upstream of jet exit position. After jet exit, jet stream lift away from surface for VR=3.5 case and move close with heat transfer surface for VR=0.25 case. For flow path of mainstream, it seem that mainstream flow covered over jet stream on VR=0.25 case and entrained upward mixing with jet stream for VR=3.5 case. From this effect, heat transfer surface was dominated from mainstream.

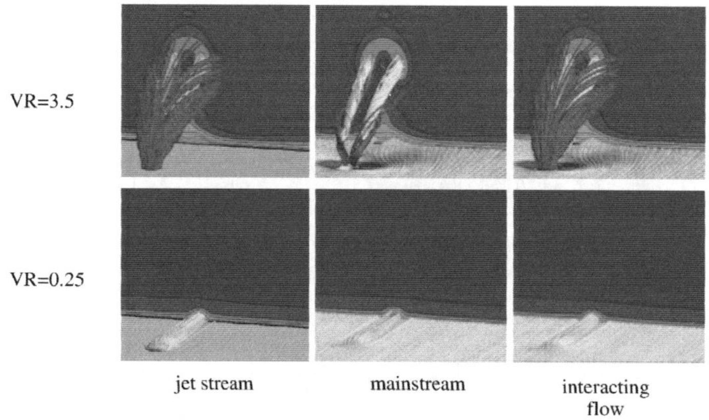

Figure 5. Flow path for jet, mainstream and interaction between jet and mainstream at different VR

4.3. Swirling strength

The swirling strength and vector of flow on a plane after jet exit are shown in Fig. 6. The counter rotating vortex pairs (CRVP) rotate close with surface and lift away for case VR=0.25 and VR=3.5, respectively. However, CRCP on the case VR=3.5 generate small rotation flow adjacent with surface and stronger than VR=0.25 case. From this result, the heat transfer rate becomes higher.

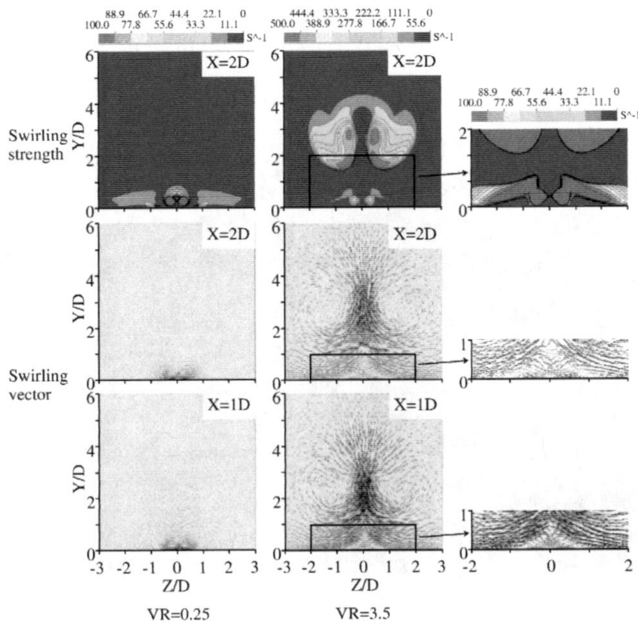

Figure 6. Swirling strength and vector of flow after jet exit

The flow mechanism after jet exit can be clearly explained from streamline on close parallel plane in Fig. 7. Figure 7 shows 3D streamline on ZX plane above heat transfer surface 1.0 mm at VR=0.25 and 3.5. For case of VR=0.25, it is clearly see the interaction between the jet flow and mainstream produce two of counter rotating flow. This two of rotating flow entrained the mainstream to attach on the heat transfer surface. This will result in two peaks of heat transfer behind the jet hole for this case. For case of VR=3.5, CRVP moved away from surface and introduced small vortices under upwash side. This phenomenon led heat transfer peak behind the jet exit.

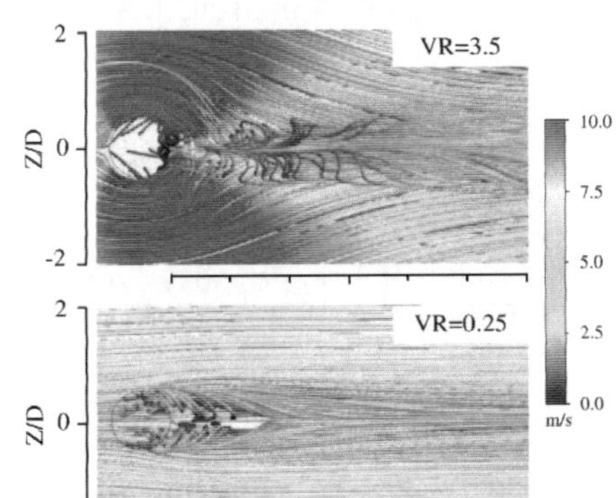

Figure 7. Plane streamline above heat transfer surface at 1 mm

4.4. Heat transfer mechanism on downstream surface

The temperature distribution closing to surface on normal planes after jet exit are shown in Fig. 8. The effect of CRVP to distribution of temperature at the centre of jet exit along downstream direction was different with beside. The different of thermal boundary layer thickness at close jet exit position was higher than away position. On VR=0.25 case, the thickest and lowest point accorded with area between and two high heat transfer, respectively. For VR=3.5 case, the thermal boundary layer thickness was the thinnest along jet centre. This result occurred from mainstream that flow around jet stream and generate swirl flow to impinge heat surface.

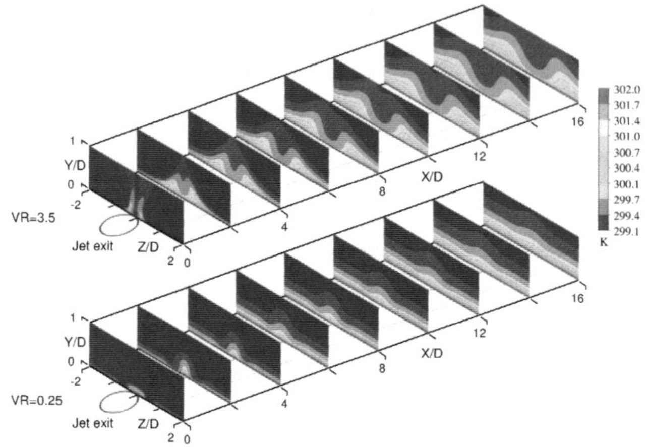

Figure 8. The temperature distribution closing to surface on perpendicular planes

5. Conclusion

In this paper, heat transfer enhancement on a flat plate in case of using a vortex generating jet from pipe nozzle was investigated for case VR=0.25 and 3.5. The flow characteristics were explored with numerical simulation. The main findings are shown as follow:

1. High heat transfer area on VR=0.25 case occur from CRVP of jet stream to move close heat surface along downstream direction.
2. High heat transfer area on VR=3.5 case occur from CRVP of mainstream that flow around

References

[1] Fiebig M 1998 *Trans IChemE* **76** 108-123
[2] Aris M S, McGlen R, Owenc I and Sutcliffe C J 2011 *Appl. Therm. Eng.* **31** 2230-2240
[3] Smulsky Y I, Terekhov V I and Yarygina N I 2012 *Int. J. Heat Mass Transfer* **55** 119-130
[4] Cho H H, Wu S J and Kwon H J 2000 ribs *J Turbomach* **122** 579–586
[5] Compton A D and Johnston P J 1992 *AIAA J* **30(3)** 640–647
[6] Ahn J, Jung I S and Lee J S 2003 *Int. J. Heat Fluid Flow* **24** 91–99
[7] Miao J-M and Wu C-Y 2006 *Int. J. Heat Mass Transfer* **49** 919–938
[8] Qayoum A, Gupta V, Panigrahi P K and Muralidhar K 2010 *Int. J. Heat Mass Transfer* **53** 5035–5057
[9] Jabbal M and Zhong S 2008 *Int. J. Heat Mass Transfer* **29** 119–130
[10] Akdag U, Cetin O, Domiral D and Ozkul I 2013 *Int. C. Heat Mass Transfer* **49** 96-103

IOP Conf. Series: Materials Science and Engineering **243** (2017) 012007 doi:10.1088/1757-899X/243/1/012007

Comparative study of Roe, RHLL and Rusanov fluxes for shock-capturing schemes

Malina Darwis, Kamil Abdullah* **and Akmal Nizam Mohammed**
Flow Analysis, Simulation and Turbulence Research Group,
Universiti Tun Hussien Onn Malaysia
86400 Parit Raja Johor, MALAYSIA

*E-mail: mkamil@uthm.edu.my

Abstract. Computational Fluid Dynamics provides approximations through iterative calculation to simulate and predict the solution and has been extensively used in the investigation of shockwave. The paper aims to compare the capability of Roe, RHLL and Rusanov flux function in capturing shock phenomena. Two cases have been presented in the paper; one-dimensional case and two-dimensional cases. In general, all the considered flux function capable to capture general pattern of the shockwave phenomena. However, the results shows that it is important to pay great attention to the diffusive component of the fluxes. Strong diffusive component of the flux will helps to bring stability in the solution but at the same time prevent the fluxes to capture small changes of the fluid properties.

1. Introduction

Shockwave is one of propagating disturbance in fluid medium. The disturbance occurs when a wave or the fluid medium travels at higher than a local speed of sound. Similar to the ordinary wave, shockwave carries energy while propagating through the medium. However, the shockwave is uniquely characterized by an abrupt change in the fluid medium properties (pressure, temperature and density) which also known as variable discontinuity. The unique characteristics of the shockwave which trigger discontinuity in flow variables attracts a lot of attention from the world of academics and research. The increasing attention demand for better understanding of shock behavior which now has becomes specialize focus areas in flow investigation. The shockwave investigation can be done through empirical and analytical approach. Although shockwave can be examined empirically, there are very limited cases can be covered by this approach due to sophisticated experimental setup requirement particularly in producing high speed flow. Among the established experimental setup in shockwave investigation is the shock tube. The experimental limitation has set-off extensive analytical work on shock wave investigation [1]. Computational Fluid Dynamics (CFD) in which fluid flow problems are resolved numerically using model equations that representing the governing laws of conservation has now become common in the shockwave investigation. The CFD provides approximations through iterative calculation to simulate and predict the solution as closely to real-world results as possible.

In general, these numerical methods can be classified into the major categories of finite difference, finite volume, and finite element methods [2]. Among these three, finite volume methods has been widely used in the shockwave investigation. In finite volume method, the control volume of the

IOP Conf. Series: Materials Science and Engineering **243** (2017) 012007 doi:10.1088/1757-899X/243/1/012007

interest area is define and also known as flow domain. The control volume is divided into a discrete number of cells that form a grid (structured, unstructured or hybrid). Information on fluid properties passes through the domain based on the designated boundaries and initial conditions. Changes that occur across a cell will influence the properties of other cells in the flow direction. All the changes is governed by the governing equation that will ensure any changes is coherent in terms of its physics.

The information on fluid properties that travels from cell to cell along the domain are track in terms of its changes from the domain inlet to its outlet. The transferred data between the cells is recovered in the same manner as if solving a series of Riemann problems. Thus, the set of solution equations that perform such calculations are often known as Riemann solvers [3]. The equations themselves are called flux functions. A lot of flux function has been introduce for the purpose of simulating shockwave. The present paper aims to compare between Roe [4], RHLL [5] and Rusanov [6] flux function for the one-dimensional and two-dimensional cases.

2. Methodology

The present papers considered two cases involving one-dimensional and two-dimensional. All three considered flux functions will be apply for both cases to allow comparison between results. The one-dimensional case make use of shock tube case while the two-dimensional make use of shock diffraction case.

2.1. One Dimensional

Figure 1 shows the overview of one dimensional flow domain of the Sod's shocktube. Total of 80 cells have been generated within $-5.0 \leq X \leq 5$ and $0 \leq Y \leq 1.0$ flow domain. The domain made of ideal gas with $\gamma = 1.4$. The domain is divided into two regions separated by a thin diaphragm. The diaphragm represent the disturbance in the flow (shockwave) separating the pre-shock region and post-shock region. The initial conditions of these regions are defined in Table 1. The results are taken at t = 1.7 with CFL = 0.8.

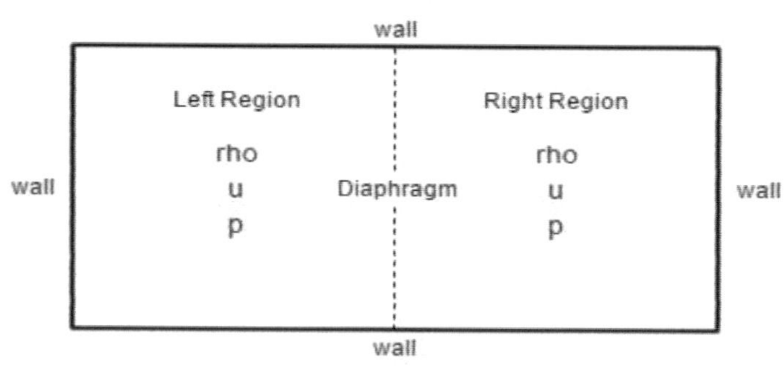

Figure 1. Computational domain of Sod's Shockwave tube considered for one dimensional case.

Table 1. Initial condition for Sod's shockwave.

Parameters	Left region	Right region
Velocity, u	0	0
Density, rho	1.0	0.12
Pressure, P	1.0	0.1

2.2. Two Dimensional

Figure 2 shows the overview of shock diffraction case for two-dimensional case of the present study. The case has been widely use to evaluate the stability of flux functions in simulating the shock. The flow domain is modelled as a square domain; -1.0 ≤ X ≤ 1.0 and 0.0 ≤ Y ≤ 1.0. The upper left boundary is designated as an inlet while the lower left and top boundary are set as wall. The other two sides; the right and bottom are set as an outlet as shown. The domain is builds by quadrilateral grid consists of 160,801 nodes. The inlet is set to have incoming normal flow to the boundary at Mach number of 5.09. Table 2 shows the initial conditions of the flow domain. The results for this case are taken at t = 0.18 and CFL = 0.95.

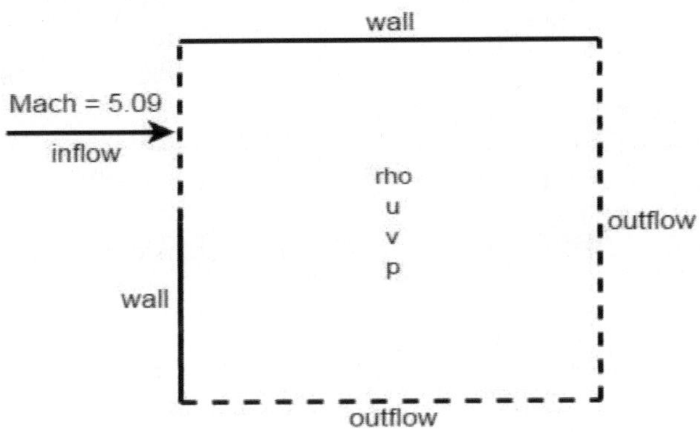

Figure 2. Computational domain for shock diffraction considered for one dimensional case.

Table 2. Initial conditions of shock diffraction problem

Parameters	Inlet Condition
Velocity, u	0
Velocity, v	0
Density, rho	1.0
Pressure, P	$1/\gamma$

3. Results and discussions

3.1. Sod's shocktube

Figure 3 shows the density variation of a Sod's shock tube case for Roe, RHLL and Rusanov flux functions. The exact solution of the case represented in solid line distinguish three phenomena; expansion wave (-2.5 ≤ X ≤ -0.5), contact discontinuity wave (X = 1.5) and shockwave (X = 3.0) as shown in the figure. The formation of these three non-linear waves represent the tendency of the flow to equalize the pressure inside the tube. The expansion wave propagates to the left region, X < 0 causes the expansion of the high pressure gas while the shockwave propagates to the right region, X > 0 causes the compression of the low pressure gas. The expansion and compression of gas are separated by contact discontinuity. Comparison between the solution of Roe, RHLL and Rusanov flux functions shows good solution stability with the absent of oscillation in the fluid properties profiles.

However, all solutions produce discrepancy in comparison with the exact solution at the aforementioned wave formation. Based on Figure 3, Roe solution produces the smallest discrepancy at the endpoint of expansion wave among all, while Rusanov solution produce the largest discrepancy. RHLL solution in the other hand produces nearly identical profile in comparison with the Roe solution. The larger discrepancy produces by Rusanov solution indicated by rounded corner at both endpoint of the expansion wave. Similar pattern can also be observed in the velocity profile presented in Figure 4. From the comparison, Roe scheme gives the best performance in solving the shock tube problem.

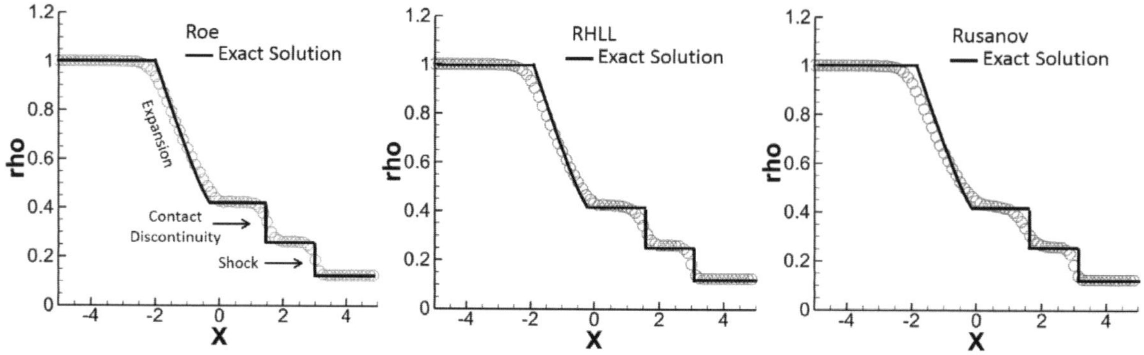

Figure 3. Density distribution for Roe, RHLL and Rusanov for one-dimensional case

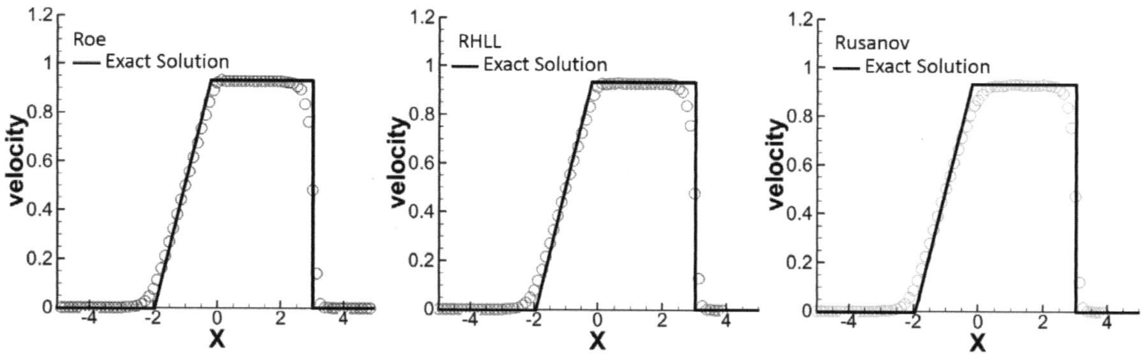

Figure 4. Velocity distribution for Roe, RHLL and Rusanov for one-dimensional case

3.2. Shock diffraction

Figure 5 shows the density contour plot for all considered flux functions for the case of shock diffraction. Several flow phenomena can be distinguish from the results. At the top side of the domain a clear contour separating the region mark as ES is region of reflected expansion wave due to the existence of the upper wall. SS is the secondary shock which occur after the slipstream (SL). The slipstream created from the flow separation from the lower left side. The separation also created a main vortex marked as V in the figure.

In comparison between Roe, RHLL and Rusanov flux, all flux functions are capable to capture the general pattern of the density distribution throughout the domain. However, Roe flux as shown in Figure 5 (marked as A) indicates to have concern in terms of stability. Such occurrence cannot be observed in the case of RHLL and Rusanov. The instability observed in the case of Roe appears more likely when the shock is aligned with the mesh. This is the reason for the instability to occur only at the tip of the DS.

The stability discussed in the earlier paragraph is closed related to the dissipative nature of the flux. At the present of pressure difference along a shockwave, the fluid will be pushed to the low-pressure region where the flow speed will be accelerated. Consequently, the pressure in the low-pressure region gets even lower and causes instability in the solution. To prevent such condition, stronger dissipative component of the flux functions is required. RHLL and Rusanov have a much stronger dissipative component which dampen the instability. Although strong dissipative component provides the required stability, too strong dissipative component also prevent the flux function to capture smaller change of the fluid properties.

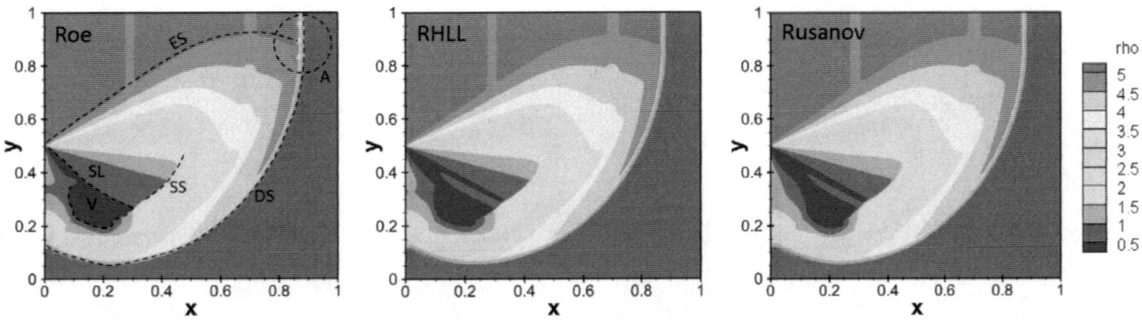

Figure 5. Density distribution for Roe, RHLL and Rusanov for two-dimensional case

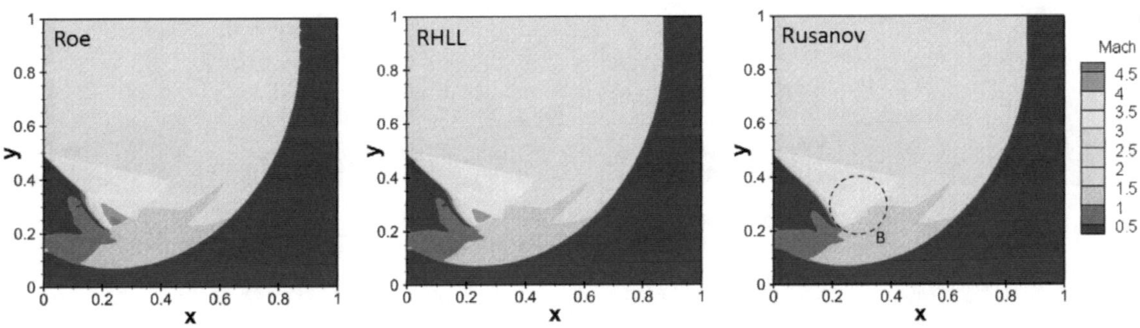

Figure 6. Mach number distribution for Roe, RHLL and Rusanov for two-dimensional case

Figure 6 shows the Mach number distribution for Roe, RHLL and Rusanov flux function. All flux function generally produce similar pattern of Mach number distribution. However, the figure obviously indicates the strength of the dissipative component of Rusanov flux function. The existence of high Mach number at the secondary shock region (marked as B) in the case of Roe become insignificant in the case of RHLL and non-exist for the case of Rusanov. Although, dissipative nature of the flux is important, too strong dissipative component will make the solution incapable to capture details of the flow phenomenon.

4. Conclusion

A comparative study of shock capturing schemes is carried out by using Roe, RHLL and Rusanov flux functions. Based on the results obtain, the effect of each scheme properties can be conclude as follow:

a) Roe scheme has the property of good capturing contact discontinuity. However, this property causes the Roe scheme to suffer shock instability in multidimensional problem.
b) Rusanov scheme is a dissipative scheme. This property causes the Rusanov scheme to produce very diffusive solution, thus unable to capture contact discontinuity.
c) RHLL scheme has both properties of Roe and Rusanov scheme. These properties make RHLL scheme capable in suppress the shock instability and at the same time avoid diffusive solution.

Acknowledgment
The author would like to thank both Universiti Tun Hussien Onn Malaysia (UTHM) and Ministry of Higher Education of Malaysia for the financial support through Fundamentals Research Grant Scheme (Vot 1544).

References
[1] Anderson J D 2004 *Modern Compressible Flow with Historical Perspective 3rd edn.* (McGraw-Hill)
[2] Le Veque R J 2002 *Finite Volume Methods for Hyperbolic Problems. 1st edn.* (Cambridge Text in Applied Mathematics)
[3] Toro E F Riemann 1999 *Solvers and Numerical Methods for Fluid Dynamics: A Practical Introduction. 2nd edn.* (Springer-Verlag)
[4] Roe P L 1981 Approximate Riemann Solvers, Parameter Vectors and Difference Schemes (*Journal of Computational Physics* Vol. 43) pp. 357-372
[5] Nishikawa H and Kitamura K 2008 Very Simple, Carbuncle-Free, Boundary-Layer Resolving, Rotated-Hybrid Riemann Solvers (*Journal of Computational Physics* Vol. 227) pp. 2560-2581
[6] Rusanov VV 1970 Article II. On Difference Schemes Of Third Order Accuracy For Nonlinear Hyperbolic (*Systems Journal of Computational Physics* Vol. 5) pp. 507-516

IOP Conf. Series: Materials Science and Engineering **243** (2017) 012008 doi:10.1088/1757-899X/243/1/012008

Numerical investigation of flow parameters for solid rigid spheroidal particle in a pulsatile pipe flow

Joffin Varghese[1, a] **and Jayakumar J S**[1, b]
[1]Department of Mechanical Engineering, Amrita School of Engineering, Amritapuri, Amrita Vishwa Vidyapeetham, Amrita University, India

E-mail:[a] joffin93@gmail.com, [b]jsjayan@gmail.com

Abstract. Quantifying, forecasting and analysing the displacement rates of suspended particles are essential while discussing about blood flow analysis. Because blood is one of the major organs in the body, which enables transport phenomena, comprising of numerous blood cells. In order to model the blood flow, a flow domain was created and numerically simulated. Flow field velocity in the stream is solved utilizing Finite Volume Method utilizing FVM unstructured solver. In pulsatile flow, the effect of parameters such as average Reynolds number, tube radius, particle size and Womersley number are taken into account. In this study spheroidal particle trajectory in axial direction is simulated at different values of pulsating frequency including 1.2 Hz, 3.33 Hz and 4.00 Hz and various densities including 1005 kg/ m^3 and 1025 kg/ m^3 for the flow domain. The analysis accomplishes the interaction study of blood constituents for different flow situations which have applications in diagnosis and treatment of cardio vascular related diseases.

1. Introduction

Bio medical stream issues are regularly extremely complex in view of multifaceted geometries, moving walls and high precision required to totally cover happening flow phenomena. The human body is a complex framework that requires transport of materials, for example, air, water, nutrients and minerals for survival and working. Nowadays, Computational fluid dynamics is emerging as a means of enhancing understanding of fluid flow in arteries and the associated forces. In the coming future, the study of arterial blood flow will lead to the prediction of individual haemodynamic flows in any patient, the development of diagnostic tools to quantify disease, and the design of devices that mimic or alter blood flow. Although extensive experimental, theoretical and numerical studies have been conducted in this area, to the best of our knowledge, there is yet no study that has revealed a direct relationship between blood fluid dynamics and the development of atherosclerosis. The purpose of this work is to perform a computer simulation method of blood flow using a spheroidal particle and modeled as a blood cell to investigate blood flow behaviour at the microscopic level. The motion of a single deformable red blood cell and the thrombogenesis can be studied as problems at the capillary level. When a particle, or a cloud of particles of low concentration, is introduced into an air stream the response of the particle depends on the relative velocity of the particle and the fluid. This relative velocity determines the drag, which is the sole force determining the motion of the particle if it is assumed that there is no particle interaction and further that the presence of the particles does not change the basic flow pattern. The cyclic nature of the heart pump creates pulsatile conditions in all arteries making the study of flow more complicated, which can quantified by Womersley or Witzig number. Abhinav et al. [1] suggested that for the Reynolds number ranging from 650 to 1200, flow in a straight tube is practically oscillatory and transverse velocities are small. Chandan Paul et al. [2] concluded that the velocity magnitude within the coil embolized aneurysm becomes negligible after coil insertion and the wall shear stress within that aneurysm decreases to a great extent for both Reynolds numbers 500 and 1500. Jae Sung Park et al. [3] concluded that the asymmetry of fluid velocity results in a lower pressure on the wall side than on the centreline side, leading the sphere to migrate away from the center line until the wall repulses it. Takayuki et al. [4] carried out the experimental and numerical investigation on an alumina particle in a vertical pipe flow. Sarath et al. [5] investigated the introduction of spherical particle in a pulsatile pipe flow and its trajectory was studied.

To resolve certain blood flow associated diseases, numerical investigations were done to find out inter constituent interactions of blood considering the wall shear stresses acting on these constituents. The methodology used for the numerical simulation was discussed and thereafter comes the details of geometry and meshing. The results and conclusions were discussed after the proper analysis of simulation results.

2. Methodology

2.1. Governing Equations and Mathematical Modelling

Flow modelling need the conservation principles of mass and momentum to be satisfied simultaneously. The solver was advanced based on Finite Volume Method in Computational fluid dynamics. The momentum equation for particle motion was added to the FVM solver to understand the transport mechanism of solid spheroidal particle in pulsatile flow. The simulation was carried out to find the trajectory of the particle in axial direction with variation of time. The following equations include Continuity equation and Navier Stokes equation which will govern the fluid flow inside the pipe.

$$\nabla . \vec{\boldsymbol{u}} = 0 \tag{1}$$

$$\rho \left(\vec{\boldsymbol{u}} . \nabla \vec{\boldsymbol{u}} + \frac{\partial \vec{u}}{\partial t} \right) = -\nabla P + \mu \nabla^2 \vec{\boldsymbol{u}} \tag{2}$$

It can be observed that the blood cells having the biconcave shapes are persistently experiencing disfigurements all through the stream, which can be approximated as a spheroidal molecule for the examination. The equivalent spheroidal diameter by equating the volume of a spheroid with that of a sphere of diameter, d has the form

$$d_v = \frac{2a}{AR^{0.5}} \tag{3}$$

The momentum equation for a particle can be solved to obtain the particle velocity and from that the migrated position of particle in axial directions. Shear forces acting from the walls of fluid conduit were taken into account, so that the force acting on the particle from the boundary layer can also be considered. Many forces are acting upon the particle having mass which is introduced in a pulsatile pipe flow and the motion of the particle can be predicted using the simulation using the three dimensional FVM unstructured code. The drag force acting on a spheroidal particle can be expressed by

$$F_D = 16 U_m e^3 \pi \mu a \left(e(\sqrt{1 - e^2}) - (1 - (2 sin^{-1} e)\, e^2 \right)^{-1} \tag{4}$$

The shear stress at the pipe wall is allied with slope of the velocity profile at the surface. The velocity profile stays unaltered in the hydro dynamically fully developed flow, the divider shear stress likewise stays consistent in that area. Wall shear stress, being highest at the pipe inlet having least boundary layer thickness decreases slowly to the fully developed value. The pressure drop is higher in the entrance regions of a pipe which is significant only for short pipes and is negligible for long ones. The wall shear stress for a laminar steady flow in a straight pipe is given by

$$\tau_{wall} = \frac{32\mu Q}{\pi D^3} \quad (5)$$

The trajectory of the particle is obtained by applying Newtons equations of motion after calculating the particle velocity. The instantaneous position of particle is updated during each time step.

$$s = s_0 + \int u_p dt \quad (6)$$

The frequency of pulsatile wave was varied by changing the Wormersley parameter. Also the flow domain density is altered by considering the density of particle introduced in the pulsatile flow.

2.2. Modeling of geometry and Meshing

The study was performed in a circular pipe domain of radius 0.02 m and length 0.1 m. Solidworks was used to model the pipe geometry. The pipe was created with appropriate dimensions and meshed properly for importing it into the code for validation purpose. The pipe geometry modelled using the Solidworks is shown in Figure 1. The left end is considered to be the entry or inlet portion to the simulation.

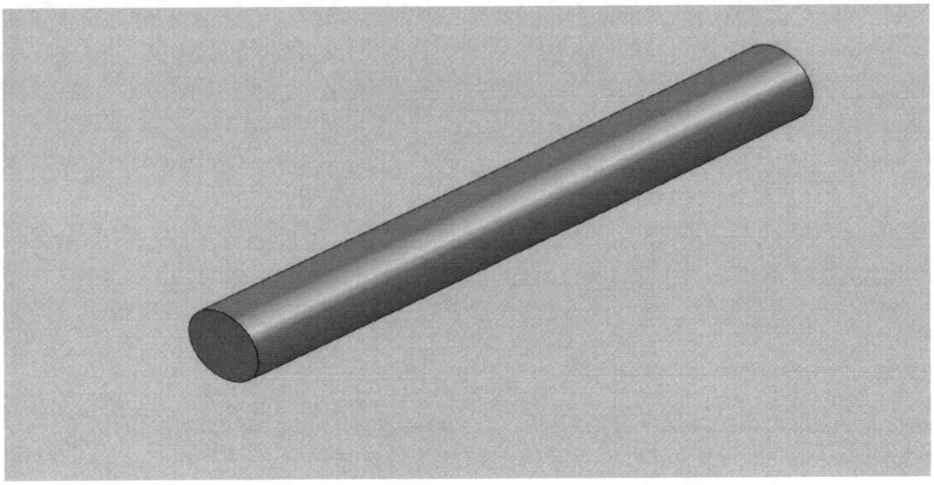

Figure 1. Pipe modeled in Solidworks model

After preparing the geometry, it was meshed using ANSYS ICEMCFD in order to make the computational domain into the cells or nodes, which are the discrete representation of the geometry on which the flow is solved. Boundary conditions are applied to these grouped boundary nodes for better convergence and accuracy of solutions. The mesh size of the flow domain is comprised of 18, 91,391 unstructured tetrahedron meshes in order to maintain the mesh quality of the 85 percentage elements above 0.85, the mesh generated is shown in Figure 2. The fluid flow is introduced with a solid rigid spheroidal particle with dimensions major axis and minor axis 0.005 m and 0.005 m respectively. When the major axis and minor axis of the spheroid become equal, the case become that of a sphere.

IOP Conf. Series: Materials Science and Engineering **243** (2017) 012008 doi:10.1088/1757-899X/243/1/012008

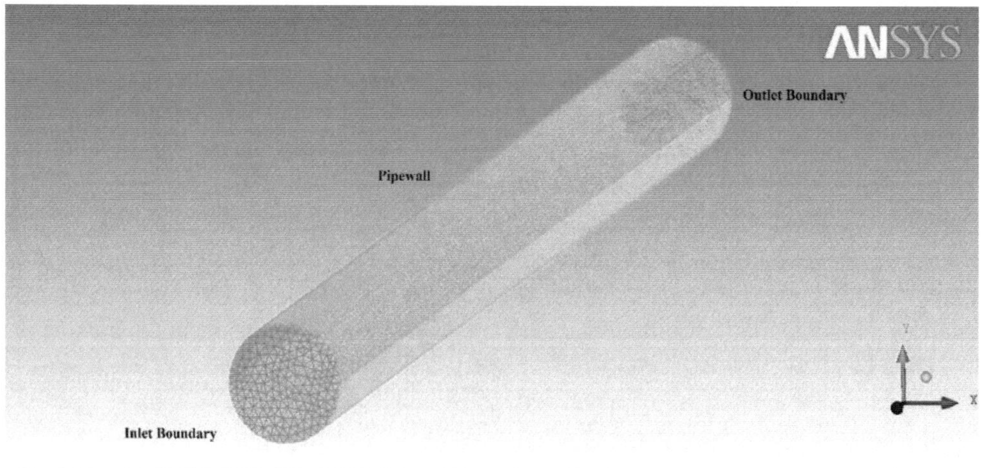

Figure 2. Pipe flow domain meshed in ANSYS ICEMCFD

2.3. Execution of the solver

The simulation was performed with Finite Volume Method (FVM) Unstructured solver and meshed geometry was imported into the solver in a readable format. The flow chart of execution of the code is shown in Figure 3.

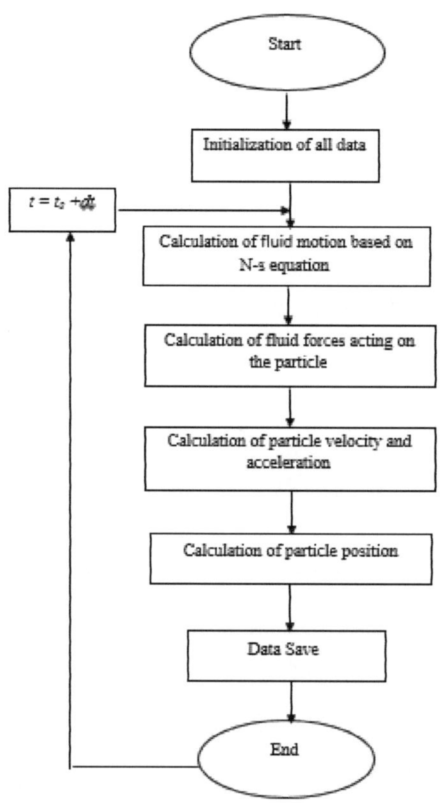

Figure 3. Flowchart of code execution

3. Results and Discussions

The base solver of FVM code was validated using certain benchmark solutions like lid driven cavity, flow past a cylinder and flow through pipe. The results of base solver were compared with some previous literature and correctness of the solver was proved.

The validation was carried out for axial migration of spheroidal particle in z direction. While performing the validation, the dynamic viscosity was taken as 0.001 Pa-s for the fluid considered for the analysis. For the simulation, geometry was modeled using Solidworks with flow set on to z direction. The average Reynolds number for the flow domain was taken as 1864 and at the initial case velocity was set to zero. The mass density of particle is about 3600 kg/m^3, which is close to the density of alumina particle. Simulations was carried out in steady flow conditions. Here, in this case the major axis and minor axis of spheroid turns to be equal (0.005 m).

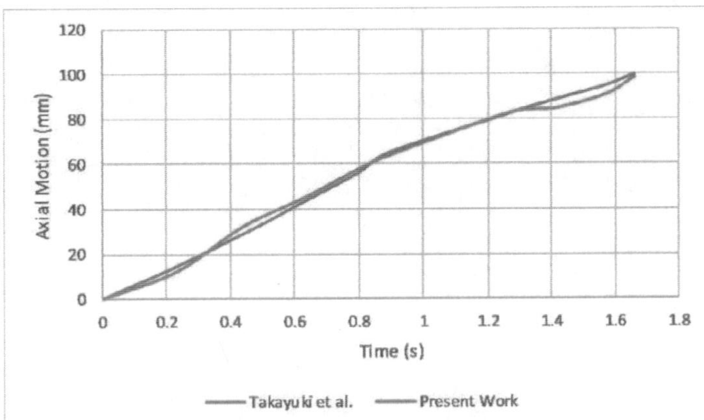

Figure 4. Comparison of axial motion for different pulsatile frequencies

The trajectories of spheroidal particle (spherical for validation) was compared with another spherical particle using experimental set up as studied by Takayuki et al. [4]. The equations for particle transport were incorporated with the base solver in order to modify the solver for obtaining desired outputs. The results for validation is shown in Figure 4. Comparing the results for both cases, the modified solver was validated since simulation results agrees with that of literature.

3.1. Variation of pulsating frequency

The geometry and meshing of the flow domain was done as per explained in the previous sections. The dimensional part of geometry was taken with respect to the geometric modeling section. The readable format for the mesh was imported into the FVM solver and simulations were carried out. The pulsating frequencies were provided for the flow as 1.20 Hz, 3.33 Hz and 4.00 Hz. The Reynolds number and viscosity were chosen as 1864 and 0.001 Pa-s. For each cases, other flow properties were kept constant and alumina particle was introduced into the flow.

The axial migration of spheroidal particle for different pulsating frequencies was plotted with respect to time was shown in Figure 5. Each cases showed some rises and downs in its trajectory for the spheroid, since the flow velocity reverses its direction over each cycle. From the plots, it is observed that the when the pulsating frequency of the flow increased, the migration rate also increases. The distance traveled by the particle which is introduced in the flow domain of higher pulsating frequency, which in turns increases the rate of migration of spheroid. For the present study, the alumina particle traverses more when the frequency was set to be 4.00 Hz.

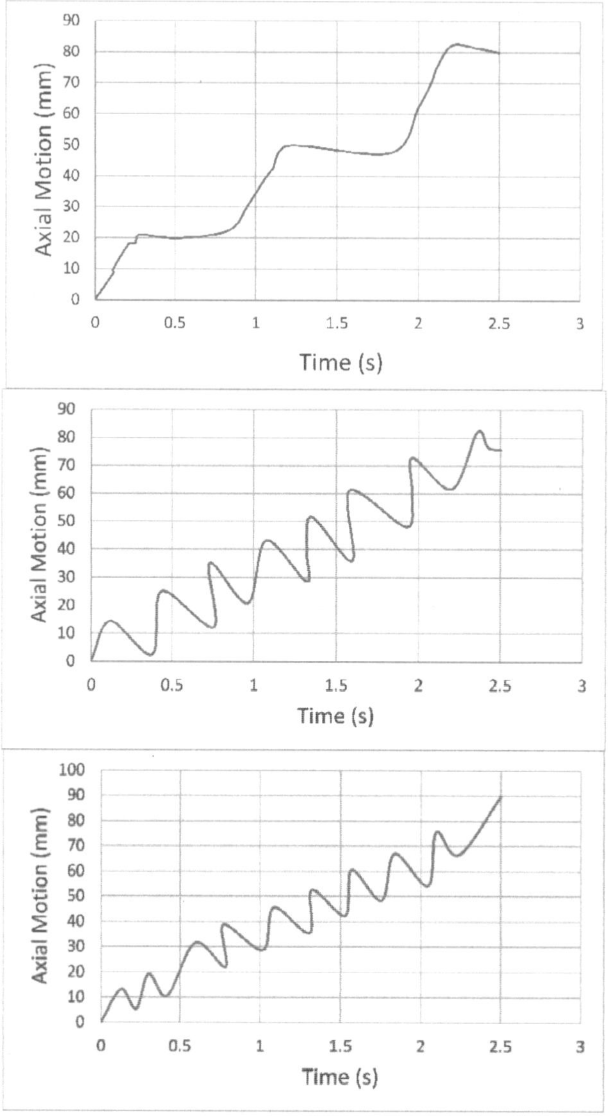

Figure 5. Comparison of axial motion for different pulsatile frequencies

3.2. Variation of fluid density

Similar to the studies in previous section, geometry and meshing of the flow domain was performed. The pulsating frequencies were kept constant along with viscosity of fluid used for simulation. For the flow as 1.20 Hz, 3.33 Hz and 4.00 Hz. The Reynolds number and viscosity were taken as 1864 and 0.001 Pa-s respectively. Simulation was carried out for fluid having density of 1005 kg/m^3 (nearer to density of water at normal temperature) and fluid of density 1025 kg/m^3 (average blood density).

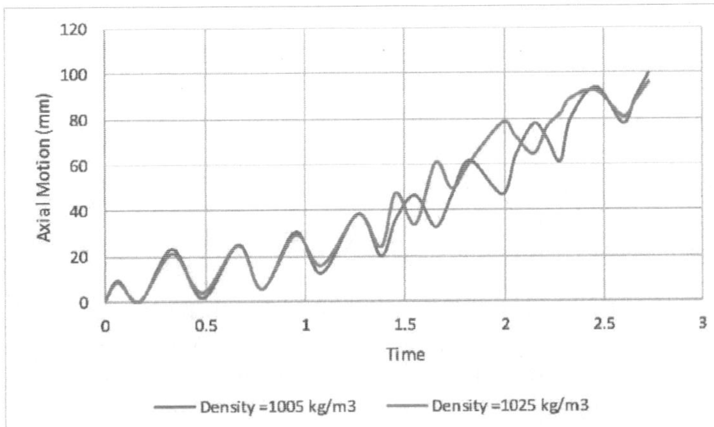

Figure 6. Comparison of axial motion for different fluid densities of flow

The trajectory of spheroidal particle for different fluid densities were plotted with respect to time was shown in Figure 6. Here also up and downs were observed in the plot for the motion of particle. It is evident that the when the density of the flow domain was increased, the migration rate decreased. The distance traversed by the spheroid which is introduced in the flow domain of higher density, shows lower migration rate. In this study, the alumina particle introduced in the fluid of density 1025 kg/m^3 undergoes less migration compared to the other case.

4. Conclusions

Simulations was done for axial motion of spheroidal particle by considering different pulsating frequencies. It can be concluded that when pulsating frequency increases, the distance travelled by the particle for a given particular time step will be more. The particle changes its direction over each cycles and high pulsatile flow holds high migration rate for particle, since the distance travelled by the body increases in every cycles.

The numerical simulations were carried out for spheroid considering fluid domains of different frequencies. The migration rate for particle introduced in high density fluid observed low migration rate. It happens since the particle experiences more resisting forces while introduced into the flow domain. The axial migration of spheroid is very complex when pulsatile nature is studied.

The present study can be extended to parameter study by including two phase algorithm to the base solver. Another wide study scope is enabled in the field of numerical simulation considering deformation of walls and particles.

Acknowledgement

We are grateful to Prof. K. Muralidhar, IITK, Mr. Sarath S and Mr. Nithin M for their valuable suggestions, advices, supports and encouragements regarding our work at all time steps, without which our work would have been a dream.

References

[1] Abhinav P, Rahul S, Panigrahi P K and Muralidhar K 2013 Chaotic flow in an aortic aneurysm *Journal of Applied Physics* **113** 1214909

[2] Chandan P, Malay K D and Muralidhar K 2015 Three Dimensional Simulation of Pulsatile Flow through a porous bulge *Springer Science Business Media Dordrecht Transp Porous Med* **107**(3) 843-870

[3] Jae S P, Suk-Heung S and Hyo-Il J 2009 Continuous focusing of micro particles using inertial lift force and vorticity via multi-orifice microfluidic channels *The Royal Society of Chemistry* **9**(7) 939-948

[4] Takayuki H, Hitoshi F, Masahiro K and Hirohito T 2009 Transport Phenomena of solid particles in pulsatile pipe flow *Advances in Mechanical Engineering* **2** 121326

[5] Sarath S, Jayakumar J S, Muralidhar K 2016 Numerical Simulation of motion of single rigid spherical particle in pulsatile flow *Fluid Machines and Fluid Power* 15-17

IOP Conf. Series: Materials Science and Engineering **243** (2017) 012009 doi:10.1088/1757-899X/243/1/012009

Thermal radiation heat transfer in participating media by finite volume discretization using collimated beam incidence

Harijishnu R[1, a] **and Jayakumar J S**[1, b]
[1]Department of Mechanical Engineering, Amrita school of Engineering, Amritapuri, Amrita Vishwa vidhyapeetham, Amrita University, India

E-mail: hariclt777@gmail.com [a], jsjayan@gmail.com [b]

Abstract. The main objective of this paper is to study the heat transfer rate of thermal radiation in participating media. For that, a generated collimated beam has been passed through a two dimensional slab model of flint glass with a refractive index 2. Both Polar and azimuthal angle have been varied to generate such a beam. The Temperature of the slab and Snells law has been validated by Radiation Transfer Equation (RTE) in OpenFOAM (Open Field Operation and Manipulation), a CFD software which is the major computational tool used in Industry and research applications where the source code is modified in which radiation heat transfer equation is added to the case and different radiation heat transfer models are utilized. This work concentrates on the numerical strategies involving both transparent and participating media. Since Radiation Transfer Equation (RTE) is difficult to solve, the purpose of this paper is to use existing solver buoyantSimlpeFoam to solve radiation model in the participating media by compiling the source code to obtain the heat transfer rate inside the slab by varying the Intensity of radiation. The Finite Volume Method (FVM) is applied to solve the Radiation Transfer Equation (RTE) governing the above said physical phenomena.

1. Introduction

The availability of high performance computing hardware and the introduction of user friendly interface have incited the change of CFD packages. Among them OpenFOAM which is the open source CFD software is used in major researches and Industrial applications. The OpenFOAM contains numerous solvers and utilities, having license free covering a wide range of problems and utilities created by its users with basic knowledge of the underlying Numerical methods and programming techniques. Angular discretization scheme of the Finite Volume Method (FVM) of radiative heat transfer is applied to emitting and scattering media by Talukdar et al. [1] with different optical thickness in a rectangular geometry and figured out that the FVM performs better than other Discrete Ordinate Method (DOM). Radiative flux, the average incident radiation, and diffuse incidence were found out by Coelho et al. [2]. Moreover the computation of collimated beam radiation have been proposed by Tuba Okutucu et al. [3] for short pulse laser irradiation in scattering and absorbing media. It is observed that as optical thickness increases, the backscattering signal is observed for a long time interval and reflectance signals get closer to each other. Chai et al. [4] observed that fluxes near to the hot surface will not change with respect to the angular grids and the fluxes at the centre of the enclosure are more than those at the ends. S.C Mishra et al. [5] figured out the variation of optical parameters by irradiating 3D geometry with a laser beam. The refractive index of the particles strongly influence the transmitted pulse shape of collimated beam. Ben Abdallah et al.

[6] have used discrete transfer radiation method for the analysis of radiation transfer in varying refractive index medium. Ananda Krishnan et al. [7] extended Discrete Transfer method to radiative transfer in a variable refractive Index semi-transparent medium and inferred that emissive power and radiant flux vary with the square of refractive index. Venkateshwar et al. [8] modeled two dimensional geometry of a slab and investigated the temperature distribution, transmissivity and critical angle. A collimated beam is passed through two media of different refractive index separated by a semi-transparent boundary which is validated in ANSYS Fluent. It is observed that the thin collimated beam entering through the opening get thick as it travels through the medium. In the present work, we use the Finite Volume Method (FVM) to discretize the Radiation Transfer Equation (RTE) in OpenFOAM to validate temperature distribution, transmitted angle and generate the collimated beam to find the rate of heat transfer across the slab by validating the work of Venkateshwar et al. [8].

2. Methodology

2.1. Mathematical Modelling

A radiant beam entering the participating medium get attenued by absorption and out scattering, which then regains energy due emission as well as by in-scattering from some direction into the direction of travel. The Radiation Transfer Equation (RTE) is governed by the integro-differential equation:

$$\frac{1}{c}\frac{\partial I}{\partial t}+\frac{1}{c}\frac{\partial I}{\partial s}=-kI(s,\hat{s},t)-\sigma I(s,\hat{s},t)+k(s,\hat{s},t)I_b+\frac{\sigma}{4\pi}\int I(s,\hat{s},t)\phi(s,\hat{s},t)d\psi \quad (1)$$

$$\frac{1}{c}\frac{\partial I}{\partial t}+\frac{1}{c}\frac{\partial I}{\partial s}=-\beta\, I(s,\hat{s},t)+k(s,\hat{s},t)I_b+\frac{\sigma}{4\pi}\int I(s,\hat{s},t)\phi(s,\hat{s},t)d\psi \quad (2)$$

Here C is velocity of light beam, I is Intensity of radiation in the direction s, k is Scattering coefficient, β is the extinction coefficient, σ is absorption coefficient, ϕ is Azimuthal angle and θ is Polar angle. The initial two terms on the right side of equation (1) represents the decrease in Intensity of radiation in a given direction as a result of absorption by the medium or it is scattered onto another direction. The third term with a positive sign implies an increase in the intensity of radiation. This term is due to thermal emission of radiant energy from the medium. I is the spectral radiation intensity at point that propagates along direction s, s is the coordinate along that direction.

The spectral radiance (I) of a body describes the amount of energy that it radiates with different wavelengths. It describes the variation of emission with temperature i.e. the emitted radiant power increases with temperature rise and the peak of the radiated spectrum is shifted towards shorter wavelengths with rising temperature.

$$T^4=\frac{\int Id\psi}{4\pi\sigma} \quad (3)$$

The transmitted beam is governed by Snell's law which is given as follows:

$$n_1 sin\theta_1 = n_2 sin\theta_2 \quad (4)$$

When a beam of light travels from a higher refractive index to lower refractive index, it deviates from its perpendicular axis at the boundary. At a particular angle of incidence, the beam reflects back to the same medium. This phenomenon is called total internal reflection and angle of incidence at which total internal reflection happens is called critical angle. The rate at which heat transfered is given by steffan-Boltzmann law which states that heat transferred in a particular direction s varies with the area perpendicular to the direction of heat flow and fourth power of radiative temperature. It is given as:

$$Q = \sigma A T^4 \tag{5}$$

The constant of proportionality is called steffan-Bolztmann's constant.

2.2. Initial and Boundary Condition
Initially, the source is kept at a position x=0 at the extreme left end of the slab. Initially the surface temperature is given as 300k. The Intensity of Radiation is set to zero such that when the radiation model is set to on in a different time step, the Intensity value become maximum initially and then gradually decreases. To give Intensity as initial boundary condition, there is Marshak Boundary condition where Intensity is given as a function of time, wavelength and position. The transmissivity is given as 0.8 and reflectivity as 0.152 as boundary condition initially. Another boundary condition is given in absorption-emission model where emissivity is designed as 0.2. The convergence criteria is set with gauss-seidel ieration of 10^{-4} tolerance where the solution converge by discretizing using Finite Volume Method at polar discretization =10 and azimuthal discretization=25.

2.3. Modelling and Computational Modelling
A two dimensional model of a slab model made of flint glass with refractive index 2 is designed as shown in Figure1. A light source is provided at a distance 80mm from the top surface. The length of the slab is taken as 100mm and height is 200mm such that light beam initiates from the source when the radiation model is on and get scattered along the whole length of the slab. At the semi-transparent layer at the extreme right end, both reflection and transmission occurs. The beam of light starts from the source and get transmitted through the boundary. The case was named as slabradiationFoam and the model was designed in blockMeshDict file of the solver. The solver used to model this two dimensional slab is buoyantSimpleFoam is a steady and incompressible solver. The solver buoyantSimpleFoam satisfies all the three basic mass, momentum and energy equations.

Figure 1. Slab model

Figure 2. Rectangular Mesh

The generated mesh is a rectangular mesh which is made uniform throughout the whole domain. It consist of 180000 elements and 192000 cells. The rectangular mesh is the best suited for modeling a slab and will obtain accurate results with greater clarity. A fine mesh in Figure 2 150x400x1 consisting of 180000 elements is created with aspect ratio 10. Concerning the FVM, the angular domain is discretized by polar and azimuthal control angles respectively. It is also considered to have azimuthal symmetry with reflectivity, transmissivity and scattering coefficient. The base solver is compiled by adding radiation model into it by combining the source code such that a new solver solving RTE for a slab model is designed. The code is then run for finite number of iterations where the diffusion of rays happens for a minute interval of time gap.

2.4. Grid Independency

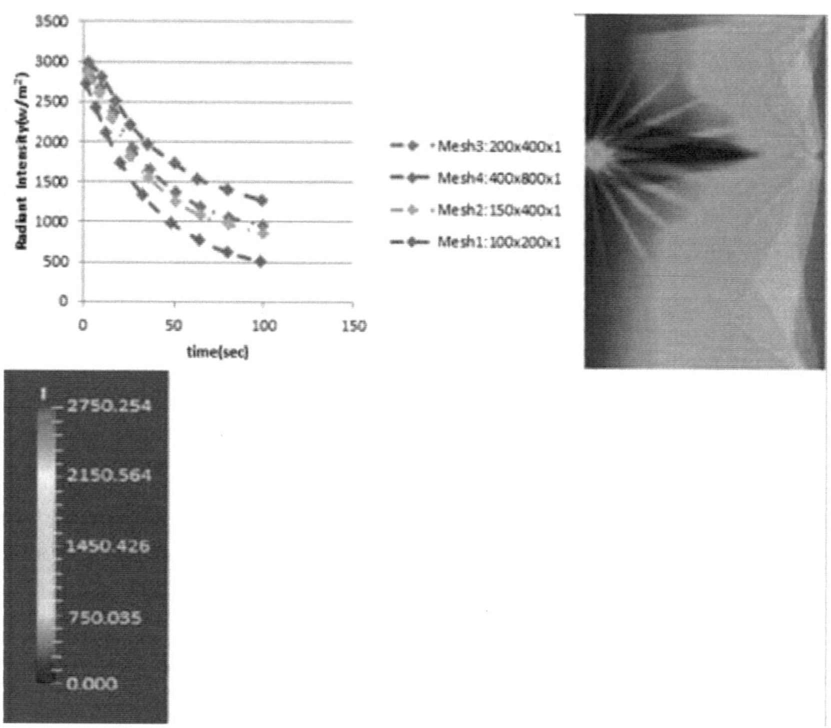

Figure 3. Variation of Intensity with time and contour plot

This paper considers the two dimensional steady state to reveal the effect of grid resolution on numerical results. Here we increase the grid resolution and then compare the results of two neighbouring results. If the results tend towards identical, the grid can be considered as grid independent. This will reduce the computational time and obtain reasonable result most efficiently. Mesh Independence is thus a way to find optimum grid size for a particular Intensity of radiation. As the number of grid points/mesh is increased, the error in the numerical solution would decrease and the agreement between the numerical and exact solution would be better. When the numerical solution obtained on different grid/mesh level agrees with a level of tolerance specifed, solution is grid converged. The computational domain was of 100 mm x 200 mm in dimension. A grid independency test is performed to ensure the accuracy of the results obtained by the modified solver. Various mesh sizes of the order 100x200x1, 150x400x1, 200x400x1, 400x800x1 were used to compute the variation of Intensity along the axial distance. The result as shown in Figure 3 tend to be almost similar for 150x200x1 grid system in comparison with 200x400x1 grid system. Hence mesh size of 150x400x1 was selected as the optimum number of mesh elements for analysis of Radiation heat transfer in participating medium.

IOP Conf. Series: Materials Science and Engineering **243** (2017) 012009 doi:10.1088/1757-899X/243/1/012009

2.5. Radiation Modelling

In order to run the Radiation model, the base solver is compiled by discretizing the Radiation Transport Equation (RTE) using Finite Volume Method. Thus new solver named "buoyantSimpleradiationFOAM" is developed which will run the code for finite number of iterations. This is being done by compiling using Wmake code thereby including RTE into the base solver. The model is made in the 0, constant and system directory and the output can be seen in the paraFoam window.

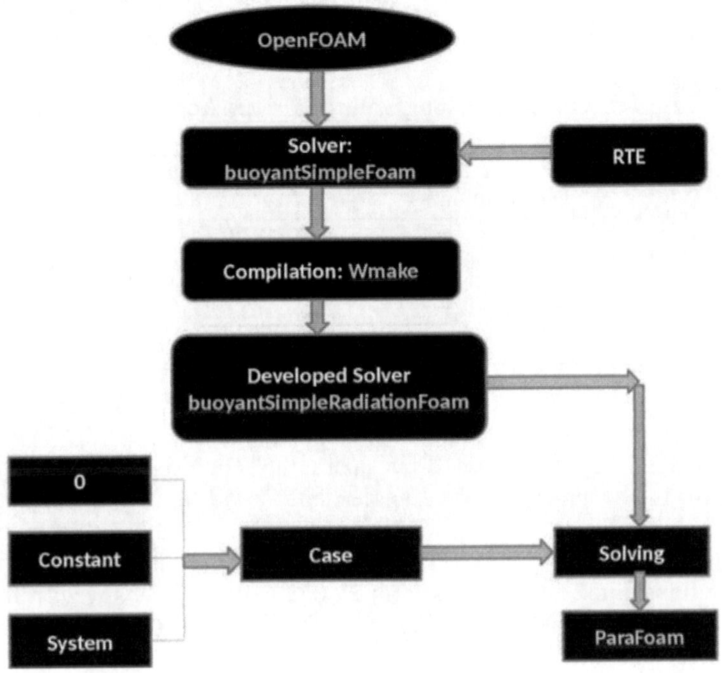

Figure 4. Flow chart of modelling radiation solver in OpenFOAM.

2.6. Material Specification

Table 1. Material Specification

Material Property	Numerical Value
Density	2.23g/cm^3
Refractive Index	2
Expansion coefficient	3x10-6
Thermal Conductivity	1.2 w/mk
Youngs Modulus	64Gpa
Viscosity of air	1.79x10-5kg/ms
Refractive Index of air	1

3. Result and Validation

3.1. Validation of Snells Law and Reflectivity

A collimated beam is passed through medium1 through a source provided at the inlet of the slab and finally strikes the boundary separating the two media. Some amount of energy of the beam crosses the boundary and enters another medium of different refractive index. The accuracy of angle at which the beam incident on the boundary depends on angular discretization for Finite volume method of radiative transfer equation. From Table 2, it is seen that when polar angle is kept fixed as 5 and azimuthal angle is allowed to vary, the transmitted angle decrease in a consistent manner.

Table 2. Effect of Transmitted Angle With Various Angular Discretization

Polar Discretization	Azimuthal Dicretization	Transmitted Angle	Error(%)
5	5	27.20	8.92
5	15	21.01	1.86
5	25	19.81	0.12
10	5	27.20	8.92
10	15	21.01	1.86
10	25	19.81	0.79

Also when polar angle is kept fixed as 10 and azimuthal angle is varied, the transmitted angle decrease upto 19.80. So the accuracy of strike angle depends on the azimuthal angle discretization rather than polar angle. According to snells law, transmitted angle for a beam emerging from medium 1 with refractive index 2 to medium 2 with refractive index 1 is 20.70 in Ansys Fluent. In the work of venkateswar et al., the transmitted angle varies from 27.022 to 19.70. Thus we can infer that the error percentage decreases gradually and is very less in OpenFOAM software as compared to the benchmarked result of venkateswar et al. [8].

Table 3. Effect of reflectivity and transmissivity With Various Angular Discretization

Polar Discretization	Azimuthal Dicretization	P	T	Error(%)
5	5	0.12493	0.87507	1.624
5	15	0.12390	0.87610	1.024
5	25	0.12386	0.87614	1.001
10	5	0.12378	0.87622	0.948
10	15	0.12310	0.78269	0.297
10	25	0.12289	0.77110	0.231

From Table 3, reflectivity(ρ) and transmissivity(τ) only vary with azimuthal angle discretization of FVM. Some amount of incident energy get transmitted and rest of them get reflected. The reflectivity and transmissivity at the boundary matches close to the calculated value of literature and error(%) is much less than 5 percentage.

3.2 Temperature distribution

The temperature variation is different for a collimated beam as compared to the diffused one. For a diffused beam of light, there is gradual decrease of temperature along the length of the slab. But for a collimated beam emerging from a source, there is an initial rise in temperature followed by a gradual

IOP Conf. Series: Materials Science and Engineering **243** (2017) 012009 doi:10.1088/1757-899X/243/1/012009

decrease. The temperature become maximum for I=2200(w/m^2) at the extreme end of the 2D slab model. The variation in temperature is due to the effect of back scattering followed by reflection of light at the boundary which overcome the loss of intensity and give rise to a sudden increase of temperature. It is also dependent on the angular discretization for the Finite volume method of radiation heat transfer equation.

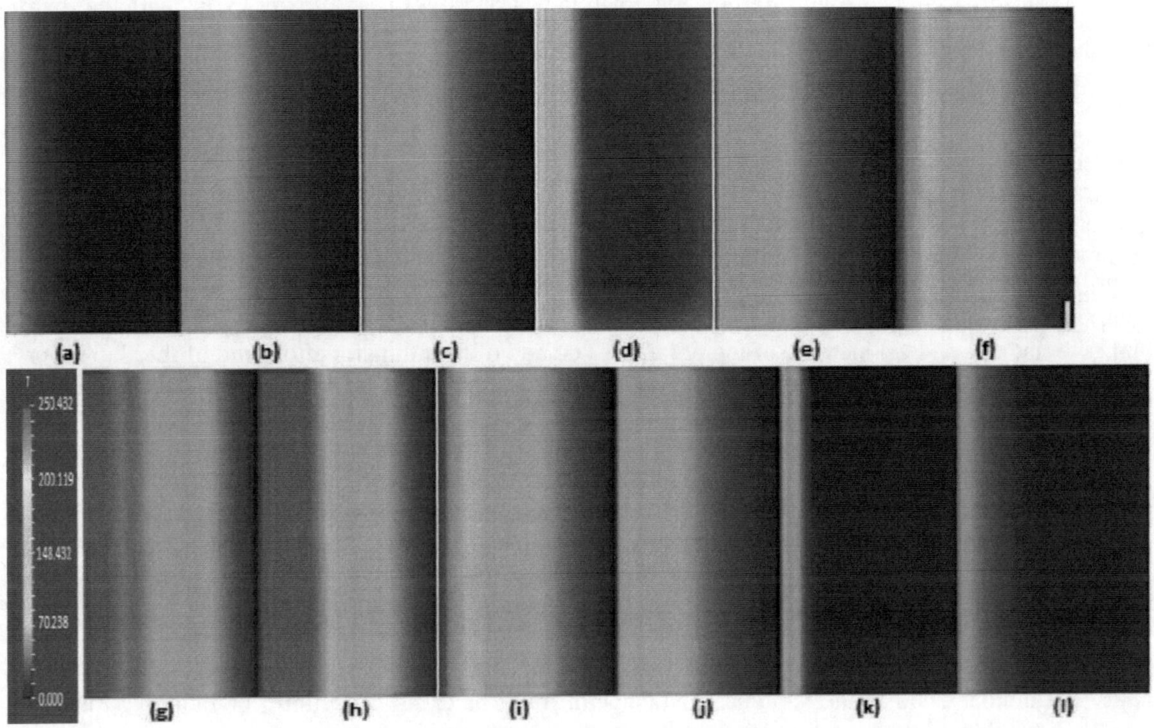

Figure 5. Variation of Temperature(T) along slab (a)T at t=9s (b)T at t=18s (c)T at t=25s (d)T at t=36s (e)T at t=48s (f)T at t=54s (g)T at t=66s (h)T at t=72s (i)T at t=81s (j)T at t=90 (k)T at t=98s (l)T at t=106s

3.3. Rate of Heat Transfer

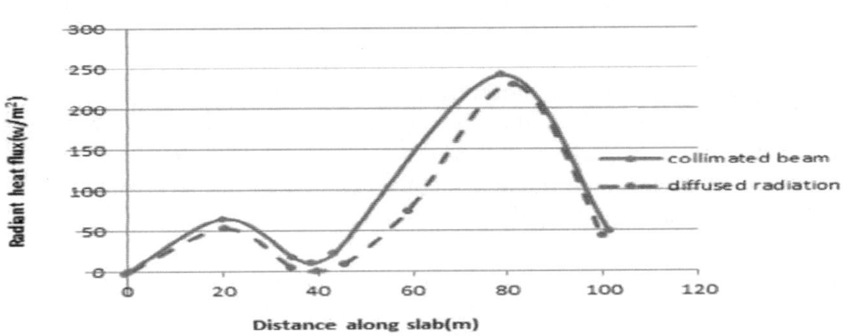

Figure 6. Variation of heat transfer rate.

From Figure 6 the heat transfer initially increases, reaches a maximum value and then decreases along the length of the slab for a diffused beam. But for a collimated beam, the decrease of heat energy happens very slowly and linearly until it reaches 50(w/m^2). The maximum rate of heat transfer occurs corresponding to the peak intensity of the collimated beam. At an intensity of 2750w/m^2, maximum value of heat transferred is 248w/m^2 for collimated incidence and is 232w/m^2 for diffused beam. This is in accordance with steffan- boltzman law. The rate of heat transfer varies with the fourth power of the temperature with constant area of slab. There is a sudden reduction of heat energy from 200w to 50w for a collimated beam because there is no backscattering along its direction as compared to the diffused beam when polar angle=5 and azimuthal angle=25.

4. Conclusion

The transmissivity and reflectivity matches the calculated value which does not influence the angular discretization. Thus modified solver is validated against the existing work and it is obtained that accuracy is much high for the modified solver. The temperature and heat flux first increases, reaches a maximum and then decreases across the length of the slab. This variation is due to the effect of backscattering followed by reflection of light at the boundary which overcomes the loss of intensity and give rise to a sudden increase of temperature. The rate of heat transfer at the end of the slab is very low. For a diffused beam, it varies linearly but for a collimated beam heat transfer decreases very slowly but linearly. The research can be done for different mediums with different refractive index thereby finding the heat transfer rate across each medium. The variation can be made by changing the position of the light source and incident angle by descretization of polar and azimuthal angle in the participating medium. The variation of emissivity with scattering albedo for a particular transmittance and reflectance are the further researches that can be done. Also one of the semi-transparent layer can be given additional heating to incorporate in the slab model which is used for the design of heat transfer mechanism in the medium.

References

[1] P.Talukdar 2006 Finite Volume Method with multi blocking procedure for radiative transport problems *Int. Journal of Heat and Mass transfer* **48** 4657-4666

[2] Pedro J C 2014 Discrete ordinate method for the solution of Radiative Transfer problems in participating media *Journal of Quantitative spectroscopy and Radiative transfer* **145** 121-146

[3] Tuba O and Yaman Y 2007 Participating media exposed to collimated short pulse irradiation *Int. Journal of Heat And Mass Transfer* **50** 4352-4359

[4] John C C 2004 Transient radiative transfer in 2D cylindrical medium with collimated pulse irradiation *Journal of Quantitative spectroscopy and Radiative Transfer* **83** 299-313

[5] Mishra S C and Rath P 2003 Transient radiative heat transfer in participating media of Numerical heat transfer *Int. Journal of Heat and Mass transfer* **88** 746-752

[6] Abdallah P B 2000 Thermal Emission of Semitransparent slab with variable spatial refractive index *Journal of Quantitative spectroscopy and Radiative transfer* **67** 185-198

[7] Krishna N A and Subhash C 2006 Discrete transfer method applied to radiative transfer in a variable refractive Index semi-transparent medium *Journal of Quantitative Spectroscopy and Radiative Transfer* **102**(3) 432-440

[8] Venkateshwar K, Chanakya G and Pradeep K 2016 Validation of snell's law and total internal reflection by radiative transfer equation in two dimensional geometry *Proceedings of the 6th International and 43rd National Conference on Fluid Mechanics and Fluid Power*

Effect of tip clearance on wall shear stress of an axial LVAD

Sarath S[1, a] **and Vikas R**[1, b]
[1]Dept of Mechanical Engineering, Amrita School of Engineering, Amritapuri, Amrita Vishwa Vidhyapeetham, Amrita University, India.

E-mail: sarathsasidharanofficial@gmail.com [a], vikasrajan@am.amrita.edu [b]

Abstract. Wall shear stress is a crucial parameter used for blood damage analysis, and typically a value of 400 Pa is set as a limit. Tip clearance is a major factor contributing to hemolysis and pump efficiency. In this study, different tip gap configurations are used to analyse the wall shear stress developed on the blade surface of a constant thickness blade design, and a varying thickness blade design using CFD analysis. It was found that, for a particular geometry, as the clearance gap reduces, flow rate over the high wall shear stress area decreases even though the high wall shear stress span is found to extend. For each design, the optimum clearance gap is iteratively attained, keeping the maximum WSS as a limiting factor. Thus a better pump designs is obtained, whose leakage flow patterns are lower than that of the initial design, hence also leading to higher pump efficiency.

1. Introduction

American Heart Association's Heart Disease and Stroke statistics 2003 update show that 552,000 new congestive heart failure cases are diagnosed each year in USA alone and death due to CHF is estimated to cross 52,827 each year [1]. The limited number of donor hearts for a heart transplantation points to the importance of mechanical devices for life support. A left ventricular assist device is a mechanical pump which pumps blood from the left ventricle of the heart and to aorta, which is the major artery arising from the heart, just as a healthy heart would do. The first VAD was implemented by Debakey in 1969 [1]. An axial flow LVAD consists mainly of 3 parts: Flow straightener, Diffuser, and Impeller blades. Flow straightener straightens blood flow to reduce shock losses. The varying pitch of the impeller blades leads to an increase in tangential speed along the length of the impeller. The flow exits the impeller into the stationary diffuser blades. These are curved in the opposite direction to the impeller blades. Their role is to dissipate the angular motion of the fluid so that it is directed linearly out of the pump, hence flow generates a large pressure head at the exit of the pump.

1.1. Design Considerations for an LVAD

Blood damage in LVADs occur mainly due to factors like, high or very low shear, recirculation, flow separation, rough pump geometry, and stagnation regions. Axial flow pumps work at a higher speed compared to centrifugal pumps causing RBCs to be exposed to higher degrees of shear stress, but they have the advantage of using lower power and having better implantability. Below are the two main criteria to be followed while designing an LVAD.

1.1.1. Hemolysis

Due to the high speed of rotational of the impeller, flow through the LVAD is turbulent in nature. The Reynolds number for the pump under study is nearly 4500. Blood cells are naturally evolved to flow in laminar regime. Hence, under turbulent flow conditions, they undergo structural changes. The high shear rates cause deformation of blood cells, red blood cells become spherical rather than disk-shaped and eventually lose their haemoglobin. This is called hemolysis.

According to Giersiepen - Heuser power law blood damage model [2], hemolysis and platelet activation are the function of both shear stress level (τ) and residence time or exposure time (t). Their expression for hemolysis index using power law model is shown in equation (1) below.

$$HI = 3.62 * 10^{-7} * \tau^{2.416} * t^{0.785} \tag{1}$$

1.1.2. Thrombosis

Low shear rate triggers the formation of clots or thrombus. Clots can grow in size, blocking the blood flow in the pump, or can even break off and be carried into the circulatory system. Recirculation regions, very low shear stress areas, flow separation, sharp edges etc. can lead to hemolysis, thrombosis as well as platelet activation.

Briefly, hemolysis occurs at very high shear stresses (above 150 Pa) and is a known complication of LVADs. Usually, a limit of 400 Pa is kept for the Wall Shear Stress (WSS). A region of very low shear stress below 1 Pa and subsequent flow stasis is critical for thrombus formation. The activation of platelets (volumes above shear stresses of 50 Pa) might lead to pump thrombosis or thromboembolic events, which are serious adverse events in LVAD patients. Von Willebrand factor is also referred to in the literature whose degradation occurs at low shear stress (3-10 Pa) [threshold of 9 Pa is used] and causes acquired Von Willebrand syndrome which affects bleeding complications. To guarantee successful operation of an LVAD, it has to be proven that these parameters caused by the LVAD are within acceptable limits. In this study, hemolysis is modelled as a function of WSS.

Several design modifications have been tried on LVAD design to optimize the design based on hemolysis and thrombosis constrains. Jingchun Wu et al. [5], used stator blade to reduce circumferential component of the flow, while LaRose et al. [6] used a multiple rotor design, with each rotor rotating opposite to each other. Huachun Wu et al. [7] compared a uniform thickness blade impeller with a variable thickeness blade impeller, and found lower WSS and better pressure distribution for the later. Schule et al. [4], compared the numerical and experimental results of flow inside the most widely used LVAD, the HM-II, and found that WSS was 585Pa, near the leading edge tip of impeller blade and these exceeded critical limits [400Pa]. Tip clearance causes blood flow losses that lead to reduction in the efficiency of the pump. Sinnott et al. [3], showed that including an impeller-housing clearance can affect the pressure rise across the pump, the shear losses and the flow rate. For high efficiency, low power consumption a smaller pump is required, but there is a problem of cavitation associated with it, hence the blade tip speed has to be limited to 10 m/s. It is evident from the literature that tip clearance contributes to high WSS which leads to hemolysis. The main objective of this work is to compare the WSS and leakage flow characteristics of the constant thickness blade design and varying thickness blade design and optimizing the designs respectively.

2. Methodology

Bladegen, a blade-design software program specifically for turbomachinery, was used to generate the pump geometry including the impeller blades. Turbogrid, the meshing tool used, was used to generate a computational mesh with acceptable element aspect ratios. CFX by ANSYS Incorporated was used to simulate the flow through the computational model of the LVAD. CFX solved the Reynolds-averaged Navier-Stokes equations in this three-dimensional computational domain. CFD-post was used to analyse the results obtained, using contours and isosurfaces.

2.1. Geometry
In this study we have analysed two impeller geometries, one with a constant/isochorous thickness of 0.5mm, denoted as "ISO scheme", and another with varying thickness along its length, denoted as "VAR scheme" [7]. The design consisted of 4 blades, with a chord length of 72mm for each. The inlet and exit vane angles were 12° and 75° respectively. The hub diameter and outer diameter were 14.8 and 18.5mm respectively. The blade tip clearance of the initial design for both schemes was 0.3mm. The impeller was designed based on the above specifications using Bladegen. The required parameters were entered into the BladeGen's initial meridional configuration dialog box. First geometry was generated with a constant thickness of 0.5mm as shown in Fig. 1 (a) (ISO scheme).

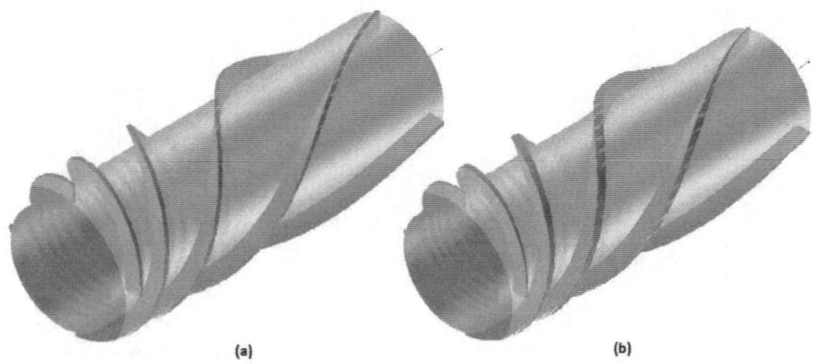

Figure 1. (a) Constant blade thickness geometry (ISO Scheme), (b)Varying blade thickness geometry (VAR Scheme)

The second geometry (VAR scheme) to be obtained had a varying thickness blade geometry. The blade thickness increased from 0.147mm at the leading edge to 0.998mm at 40% of the length, and then decreased to 0.092mm at trailing edge parabolically as in Fig.1 (b). This variation was incorporated using bladegen and the second geometry was also obtained. Separate geometries were generated for each of the above two schemes with different tip gap configurations for comparing WSS and leakage flow. Different normal tip clearance gap configurations, which have a constant tip clearance gap throughout the length, i.e 0.3 mm, 0.2 mm, 0.1 mm, 0.09 mm, 0.08 mm, 0.07 mm were generated. Varying tip clearance gap configurations, having linear tip gap variation from the leading edge to the trailing edge, i.e 0.1 mm at the LE to 0.09 mm at TE [0.1-0.09], 0.1mm at LE to 0.08mm at TE [0.1-0.08] etc. were also generated.

2.2. Meshing
The computational mesh was generated using Turbogrid. It uses unstructured hexahedral grids to obtain faster speed and higher meshing quality. Automated topology meshing method was used for easier meshing. A more refined o-grids were used in regions expected to have large gradients, that is, the tip clearance gap. Appropriate edge split was also done of high mesh error regions to minimize the error by changing the value of edge split factor. A major reduction in the tip clearance mesh error was rectified by increasing the number of elements on the tip gap. The grid independent results were obtained at 358460 and 391650 elements corresponding to ISO and VAR schemes for 0.3mm tip clearance configuration. The quality of mesh was maintained above 0.8.

2.3. Boundary Conditions and Parameters
Mass flow rate of 0.1055 kg/s was set as the inlet boundary condition, while the outlet boundary condition was set to constant pressure of 14600 Pa. The shroud wall was set as a no-slip counter rotating wall with 7000 rpm. The working fluid, blood, is considered as a Newtonian fluid with dynamic viscosity of 0.0035Pas and density of 1055kg/m3. The turbulence model used was k-ω.

IOP Conf. Series: Materials Science and Engineering **243** (2017) 012010 doi:10.1088/1757-899X/243/1/012010

3. Result and Discussion

The validation was done by finding the maximum WSS over the blade surface. Maximum WSS was found to be 373 Pa for ISO scheme and 343 Pa for VAR scheme which were according to the reference journal [7]. The maximum WSS occurred at the upstream third of the impeller blades as shown in Figure 2 as per the results by Schule et al. [4].

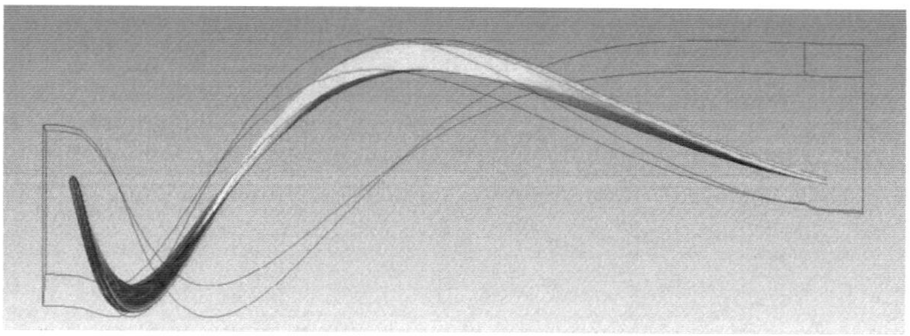

Figure 2. WSS along the impeller blade.

The WSS distribution over the impeller blade for each geometry was analysed. A polyline was created at blade tip for each of the geometries and WSS was plotted along the polyline as shown in Figure 3 and 4. Leakage flow was also analysed by creating isosurfaces of axial velocity, its length was also obtained for comparison of each case.

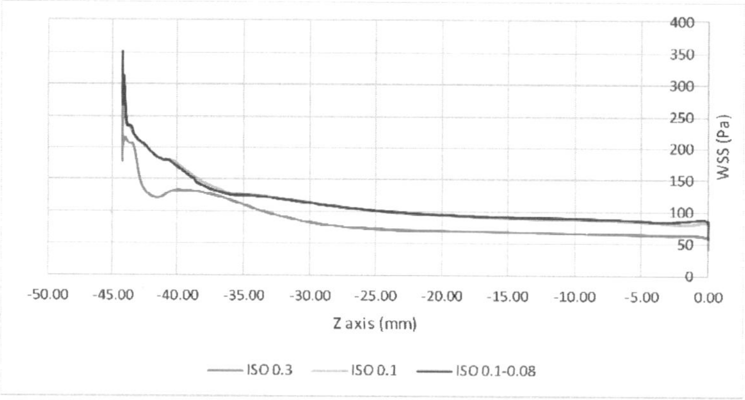

Figure 3. WSS along polyline in ISO scheme geometries

An optimum tip gap configuration, having minimum leakage flow, is found out for both schemes, corresponding to the tip gap configuration maintaining WSS below the critical limit for hemolysis, which is 400Pa. For ISO scheme the optimum tip gap configuration is found to be 0.1mm while for VAR scheme the tip gap configuration could be further reduced to 0.08mm. The length of the polyline for which the WSS values were greater than 150Pa, high WSS, was measured and tabulated. The optimum conditions are highlighted on Table 2. Further reduction of tip gap height increased the WSS above the critical limit and hence not tabulated. A decrease in the tip volume corresponding to high WSS is significant.

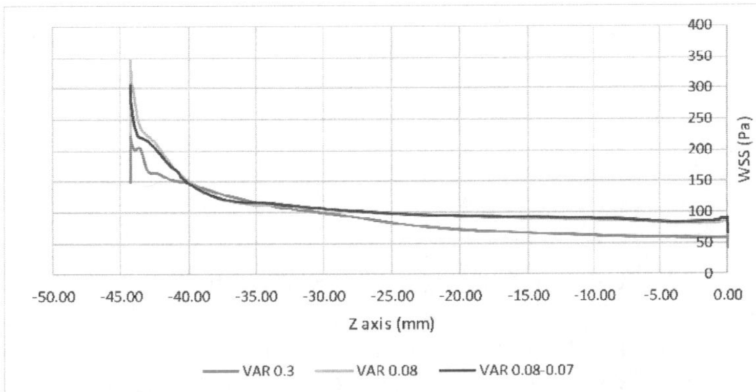

Figure 4. WSS along polyline in ISO scheme geometries

Table 1. Max WSS corresponding to different normal tip

Tip clearance (mm)	ISO Scheme		VAR Scheme	
	Max WSS (Pa)	High WSS span (mm)	Max WSS (Pa)	High WSS span (mm)
0.3	373	4.57	343	3.83
0.2	385	4.79	358	3.91
0.1	396	5.57	377	4.25
0.09	405	-	379	4.44
0.08	-	-	391	4.48
0.07	-	-	406	-

Table 2. Tip clearance volume for ISO scheme

Case	**Length**	**Area**	**V=A*t_{av}**
Iso 0.3	4.57	1.371	0.6855
Iso0.1	5.5754	0.55754	0.27877
Iso0.1-0.08	5.4595	0.518653	0.259326

Now, another modification on the blade tip was done to further optimize the design. A linear variation of the tip gap along the length of the blade is introduced on the previously optimized geometries. This was done by varying the trailing edge gap of the previously optimized design. For both schemes, different LE-TE configurations were tried, i.e 0.1-0.11, 0.1-0.9, 0.1-0.08. The results were analysed. Again optimization was done by maintaining maximum WSS below 400 Pa and the results are shown in Table 3 and 4. Using an average thickness, t_{av} , for ISO scheme as 0.5mm and of VAR scheme as 0.57mm, the volume available at tip gap for leakage to occur was calculated.

Table 3. Tip clearance volume for VAR scheme

Case	**Length**	**Area**	**V=A*t_{av}**
Var0.3	3.8379	1.15137	0.656281
Var0.08	4.4863	0.358904	0.204575
Var0.8-0.07	4.8424	0.377707	0.215293

It is found, out of the many LE-TE configurations, the linearly decreasing tip gap configurations lead to a decrease in the leakage flow, while linearly increasing tip gap increases leakage flow. The optimized design for ISO scheme (Iso0.1-0.08) has 62% lower tip clearance volume for leakage flow, while VAR scheme (Var0.8-0.07) has 67% lower tip clearance volume for leakage flow. The increase in pressure head, is 41.5% for ISO scheme while 66.7% for VAR scheme.

4. Conclusion

The WSS distribution along the blade of an axial LVAD was successfully simulated using the k-ω turbulence model. As the tip gap is reduced, for both the schemes, an increasing trend for WSS is observed, indicating the dependence of WSS on the clearance gap height. Higher WSS and leakage flow for ISO scheme compared to VAR scheme indicates strong dependence of both on the blade tip surface area also. Even though the WSS values increase as the tip gap reduces, the volume available for leakage flow decreases, which leads to a reduction in leakage flow, as leakage flow is a function of mass flow rate through clearance gap. The optimum clearance gap was found by keeping the maximum WSS as the limiting factor.

In any case, VAR scheme had a lower optimum clearance gap than that ISO scheme. This lead to lower leakage and hence higher efficiency in VAR scheme, also they had better pressure distribution along the pump. The study emphases the importance of using a variable thickness blade designs for an LVAD, which commercially available devices have not considered. Such tip gap modifications are very effective in reducing blood damage and also bringing down the overall size of the pump and hence better implantability on the human body. Further experimental analysis can be done to validate these results. Further modifications involving blade tip rounding and parabolic clearance configurations can also be studied.

References

[1] Esra S, Namik C, Fethi O A, Mehmet A A and Nilufer E A 2007 Flow simulation and optimization of a left ventricular assist device *ASME 2007 International Mechanical Engineering Congress and Exposition* vol 56 pp 98-45

[2] Jingchun Wu, Bradley E. Paden, Harvey S. Borovetz, and James F. Antak, 2010 Computational Fluid Dynamics Analysis of Blade Tip Clearances on Hemodynamic Performance and Blood Damage in a Centrifugal Ventricular Assist Device *Artif Organs* **34**(5) 402–411

[3] Sinnott M D and Cleary P W 2009 Effect of Rotor Blade Angle and Clearance On Blood Flow Through A Non-Pulsatile, Axial, Heart Pump *Seventh International Conference on CFD in the Minerals and Process Industries CSIRO, Melbourne, Australia*

[4] Chan Y S, Bente T, Bastian B, Michael L, Tamer K, Christian O P, Klaus A and Ulrich K 2016 Experimental and Numerical Investigation of an Axial Rotary Blood Pump *Artif Organs International Center for Artificial Organs and Transplantation and Wiley Periodicals, Inc.* **40**(11) 192-202

[5] Jingchun W, Michael R, David P and Shaun S 2014 Blood pump with separate mixed-flow and axial-flow impeller stages and multi-stage stators *US Patent Application Publication Pub. No: US 2014/0341726 A1*

[6] LaRose J, Charles R and Shambaugh J R 2006 Multiple rotor, wide blade, axial flow pump *US Patent Publication Pub. No: 2006/0245959 A1*
[7] Huachun W, Ziyan W and Xujun L 2011 Design and Simulation of Axial Flow Maglev Blood Pump *MECS, I.J. Information Engineering and Electronic Business* **2** 42-48
[8] Alexandrina U, Houston W and Paul A 2005 Computational design and experimental testing of a novel axial flow LVAD. ASAIO Journal vol 51

IOP Conf. Series: Materials Science and Engineering **243** (2017) 012011 doi:10.1088/1757-899X/243/1/012011

Study of conversion of ammonia from urea water solution droplets using CFD

Jyothis K[1, a] **and Vikas R**[1, b]
[1]Department of Mechanical Engineering, Amrita School of Engineering, Amritapuri, Amrita Vishwa Vidyapeetham, Amrita University, India
E-mail: jyothis8943@gmail.com [a] , vikasrajan@am.amrita.edu [b]

Abstract. This paper deals with the numerical analysis of the spray behaviour of urea water solution (UWS) droplets used in Selective catalytic reduction (SCR) process. The study uses the ANSYS- Fluent-14 as a basic framework for numerical simulations. The evaporation modelling is based on multi-component droplet evaporation approach along with the consideration of Stefan flow. The urea decomposition treated as direct thermolysis approach, where the modelling is based on the single kinetic rate approach by the proper fitting of pre-exponent. This developed model was used for spray simulation of UWS droplet evaporation to determine the urea to ammonia conversion efficiency. The obtained spray simulation results were compared with the available experimental data. The comparison shows the pre- exponent developed in the study is suitable for direct thermolysis approach to get promising results in determining the ammonia conversion efficiency.

1. Introduction

The Selective Catalytic Reduction (SCR) is the well-known technology in which nitrogen oxides in the diesel engine exhaust. The reduction of NOx is carried out with the aid of a reducing agent, mostly ammonia is used. The urea water solution (UWS) is mainly used to produce ammonia. The UWS is a solution that comprises urea and water. The ammonia (NH_3) is produced by thermal and chemical decomposition urea by spraying UWS droplets in exhaust gas [1]. In ideal case urea fully converted into ammonia but in practice, unconverted urea remains in the duct. The general steps involved in UWS droplet heating are given below.

- Water evaporation from UWS solution.
- Thermolysis of urea leading the formation of ammonia NH_3 and isocyanic acid (HNCO)
- Hydrolysis of isocyanic acid (HNCO) leading the formation of ammonia NH_3 and CO_2.

The Urea water solution injected into the hot exhaust, then it atomizes to fine droplets. The droplets undergo heating leads to temperature rise of droplet and this leads to evaporation of water. Due to higher vapour pressure of water compared to urea, rapid evaporation of water take place. This leads to further change in properties of UWS droplet such as mass fraction. From literatures studied we understood that for the multicomponent droplet modeling several approaches can be used for calculate the droplet dynamics. In Rapid mixing model (RM) approach assumes uniform temperature and species concentration. Hence droplet liquid properties same at a given instant of time for the droplets dynamics calculations. However, in Diffusion limit model (DL) assumes these properties are varied with time. The study of birchold [2] suggests that the rapid mixing model is sufficient for evaporation modeling, also it take less computational effect compared to the DL model. While the ramdan [3] report that the deviation between these two models is significant only at low temperatures. It also states that, at lowest temperature the DL model is better to capture droplet dynamics than the RM model. Vikas et al.[4] compared the evaporation rate of UWS droplet with that of water droplet, study shows a small decrease in the mass change but the trend follows the same pattern.

The evaporation of water from the droplet, resulting solid urea melts and starts to decompose thermally, which means thermolysis and hydrolysis of urea. There exist several suggestions regarding the exact temperature at which urea starts to thermally decompose. These studies indicate that different approaches can be used for the modeling of urea decomposition. Most of the studies proposed use Arrhenius type expression of urea decomposition modeling. The proposed model by strom et al. [5] states that after complete evaporation of water content present in the droplet, the remaining urea will not have any mass transfer with the surrounding gas phase until it is heated upto 425K. When it attains 425 K, the convective heat transfer take place from the gas phase to the droplet is used for vaporize urea without any sensible heat change within the droplet. However, the experiments results given by [6] pointed out that the notable decomposition of urea was observed to start at around 406K. The experiments on a single droplet by Wang et al.[7] conforming that the fast depletion of urea at temperatures above 423 K. Almost all of the above models is assumed that urea is evaporate from its solid phase and directly decompose into gaseous ammonia (NH_3) and isocyanic acid (HNCO). By the literature review, it is found that there is a scope for develop pre- exponent for Urea thermolysis to improve the prediction of variation in ammonia formation along the duct.

In the present study, develop an evaporation model for both UWS evaporation and urea decomposition. The developed model considers the urea decomposition is take place as direct decomposition process. The convective and diffusive species transport across the droplet is considered for heat and mass transfer modeling of UWS droplet evaporation. The urea decomposition process is modeled by using kinetic rate model. This study developed a Pre-exponent for Urea thermolysis modelling to improve the prediction of variation in ammonia formation along the duct. This study also made fine tuning of vapour pressure, diffusivity and specific heat values for urea and water from available experimental literatures. The numerical modeling is carried out by modify Inbuilt droplet evaporation and boiling laws with user defined functions (UDFs) developed in the present study. This developed model was used in UWS spray simulation modeling of UWS droplet evaporation to determine the urea to ammonia conversion effciency.

2. Methodology

In this study, the UWS droplet evaporation process considered to be quasi-steady. The droplet internal flow circulation and the radiative heat transfer were not considered. The evaporation model is based on the convection- diffusion approach by consider the effect of stefan flow and non-unity lewis number assumption. The heat and mass transfer processes in the gas phase near the droplet surfaces determine the amount of heat transferred to the droplet liquid phase. The internal circulation regarding the droplet characteristics is not considered in this study. The droplet physical properties are taken from the reference [8]. The vapour pressure values for urea and water taken from the reference [9].

2.1. Governing Equations

In the case of evaporation modelling, the calculation of heat and mass transfer at the surface of the droplet requires knowledge about mole fraction and component concentration at the surface. In this region, both liquid and gaseous phases are in thermal equilibrium. Hence, for the calculation of these quantities, Raoults law is used. The rate of mass transfer is determined by using the relation shown in Equation (1).

$$\frac{dm_p}{dt} = -\pi D_p \Gamma_{AB} \rho_{gmix} Sh_c \ln(1 + B_m) \tag{1}$$

Where, the splading mass transfer coefficient and Schmidit number is calculated as given by following equations (2) and (3) respectively.

$$B_m = \frac{Y_{i,s} - Y_{i,g}}{1 - Y_{i,s}} \tag{2}$$

$$Sh_c = 2 + 0.6\, Re^{\frac{1}{2}} Sc^{\frac{1}{3}} \tag{3}$$

The rate of droplet temperature change is determined by the energy balance as shown in equation (4).

$$\frac{dT_p}{dt} = -\frac{\dot{m}_{vap}}{m_p Cp_l}\left(\frac{Cp_{vap,l}(T_g - T_p)}{B_T} - \Delta H_{vap,l}\right) \tag{4}$$

Where, the splading heat transfer number is calculated as given in following expressions.

$$B_T = (1 + B_m)^{\phi} \tag{5}$$

$$\phi = \left(\frac{Sh_c}{Nu_c}\right)\left(\frac{Cp_{vap,l}}{Cp_{gmix,l}}\right)\left(\frac{1}{Le_{gmix,l}}\right) \tag{6}$$

The rate of droplet diameter change is computed from the change of the droplet mass as shown in Equation (7).

$$\frac{dD_p}{dt} = \left(\frac{6}{\pi}\frac{1}{\rho_p}\frac{dm_p}{dt}\right)^{\frac{1}{3}} \tag{7}$$

The urea decomposition process is modelled as volumetric reaction in the gaseous phase. For the description of the decomposition an Arrhenius expression is used and is shown in Equation (8).The activation energy constant for this reaction is taken from the study [10]. The pre-exponent is obtained by trial and error to match reasonably with the experimental data.

$$\frac{dm_u}{dt} = -\pi D_d A^{\left(\frac{-E_a}{RT_d}\right)} \tag{8}$$

$A=0.451\ kg/ms$ and $E_a=6.7 \times 10^4\ J/mol$.

2.2. Numerical Simulation Details

The numerical simulation study is carried by using the commercial CFD code ANSYS-Fluent. This study considers the single UWS droplet evaporation in a hot stream air flowing through a pipe. The Eulerian-Lagrangian approach available in ANSYS-Fluent is used as the basic framework for UWS spray simulation in the present study. The exhaust gas is considered as the continuous phase and the UWS droplets are considered as the discrete phase. In addition to solve transport equations for the continuous phase, we can simulate discrete second phase in a Lagrangian frame. The coupling between discrete phase and the continuous phase flow is made with two way coupling method. The droplet trajectories are computed in a Lagrangian frame using the Discrete Phase Model (DPM). The heat and mass exchange is modelled by using convection-diffusion approach. The urea decomposition is modelled by using single kinetic rate devolatilisation model. The urea decomposition is consider as volumetric reactions occurring in the gaseous phase, which is defined by using laminar finite-rate interactions between turbulence and chemistry the model constant is fitted to experimental value through repeated simulation. The developed heat and mass transfer model for UWS droplet evaporation including vapour pressure expressions can be implemented through user-defined C code.

Steady state numerical simulations are conducted for the validation spray analysis of urea water solution. The turbulence model used in the study is realizable k-epsilon model with wall treatment based on standard wall function method. The lower order implicit and higher order Runge-Kutta methods are adopted for capturing droplet trajectories. The pressure velocity coupling is solved by using coupled scheme. The second order upwind scheme is used for solving momentum, turbulence and species transport equations. The residual convergence criterion for continuity momentum turbulence equations 1e-3 and for species energy equations is 1e-6. Urea water solution spray simulation is conducted using the computational domain shown in Figure 2, which have similar dimensions as of kim et al [10] shown in figure 1. 124993 hexahedral cells are used for geometry meshing with minimum orthogonal quality of 0.8149. No significant changes in the species concentration are observed in the Eulerian domain with further mesh refinement.

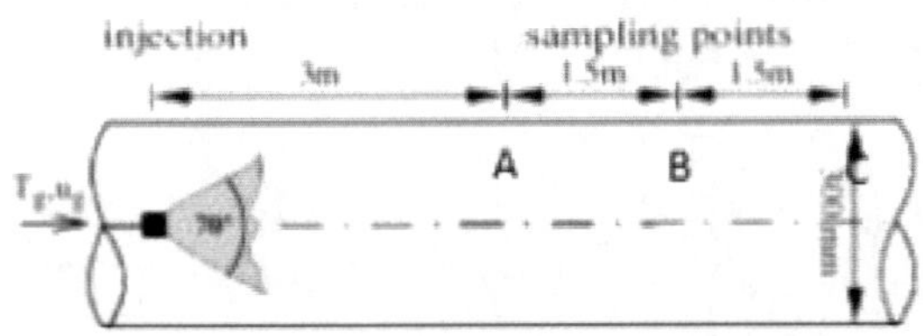

Figure 1. Sketch of kim's experimental setup.

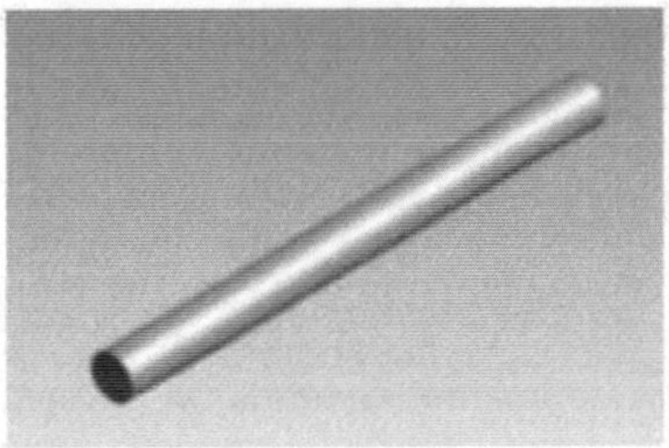

Figure 2. The flow domain developed for the present study.

The hot air is introduced into the duct at different velocities and temperatures as shown in Table (1). The cone type injection is used for the injection of UWS droplets. The injection parameters are given in table (2).

Table 1. Different cases of Ambient Air Temperatures and Air Velocities used in the study.

Air Temperature (K)	Air Velocity (m/s)
573	6.6
623	9.1
673	10.8

Table 2. Spray Parameters and Droplet Characteristics used in the DPM injection setup.

Parameters	Magnitude
Cone angle (digree)	70
Spread parameter	3.27
Mass flow rate (kg/s)	.00033
Velocity of droplets (m/s)	10.6
Minimum diameter (m)	5×10^{-6}
Mean diameter (m)	4.4×10^{-5}
Number of diameters	10
Number of particle streams	6

IOP Conf. Series: Materials Science and Engineering **243** (2017) 012011 doi:10.1088/1757-899X/243/1/012011

3. Results and Discussions

The ideal urea conversion is different with respect to the different flow rate and ambient conditions. The simulated results of the ammonia production are converted to the ammonia conversion efficiency. The ammonia conversion efficiency is defined as the ratio of the actual ammonia formed in the duct to the ideal possible ammonia conversion from the injected amount of urea. The conversion efficiency of ammonia depends upon the temperature of the exhaust gas and residence time. The residence time has the meaning that, how much time the urea conversion reaction occurs within the domain. The residence time is the function of flow velocity Vikas et al.[11].For the purpose of the validation, this study measure the values of ammonia content present at each sampling points A, B and C. The ammonia conversion efficiency obtained in this study is compared with experimental data [10]. The efficiency versus residence time plots for the ambient conditions of 573 K, 623 K and 673 K are shown in Figure 3, Figure 4, and Figure 5 respectively.

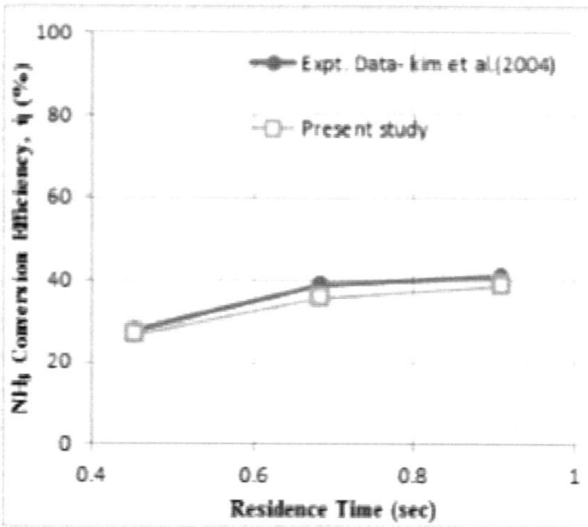

Figure 3. Ambient Temperature 573K

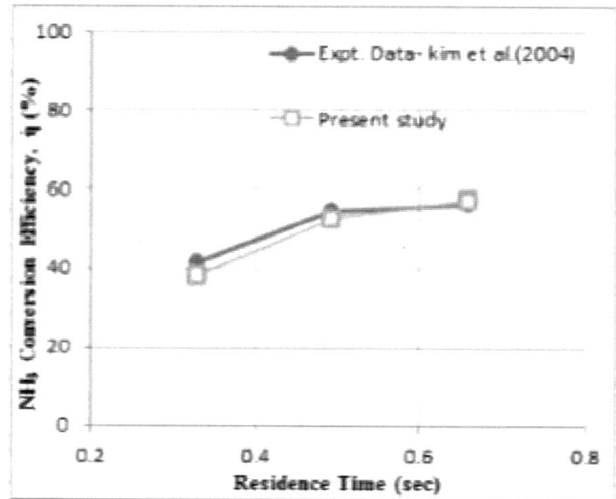

Figure 4. Ambient Temperature 623K

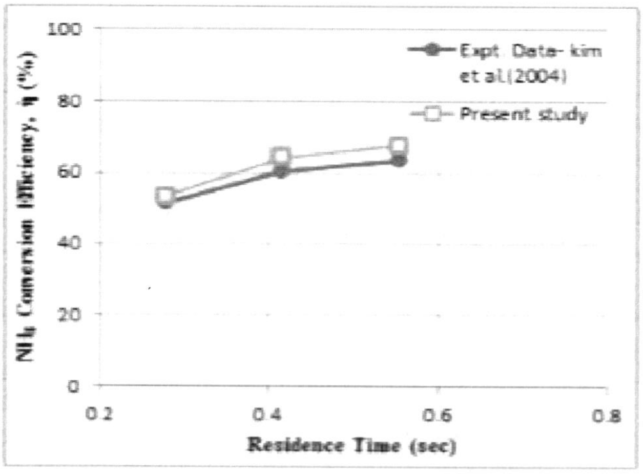

Figure 5. Ambient Temperature 673K

IOP Conf. Series: Materials Science and Engineering **243** (2017) 012011 doi:10.1088/1757-899X/243/1/012011

The results show generous agreements with the experimental data [10]. The obtained results show that the ammonia conversion efficiency increases initially (A to B) and then shows a reduction (B to C) along the flow direction. It can also see that the conversion efficiency is increases with increase in exhaust temperature. This is because higher evaporation rate of UWS droplets at higher temperatures, which leads to higher conversion of ammonia from UWS.

Figure 6 shows the temperature distribution in the duct for 673 K. In addition, the urea thermolysis reaction also causes temperature drop in the hot air, since during the direct thermolysis approach this heat is extracted from the surroundings at the droplet surface. The ammonia mass fraction contours in the duct at 623K obtained in the study is shown in Figure 7. This shows that the thermolysis reaction is active at a location downstream of the injection point. This is due to the time lag required for the initiation of urea evaporation.

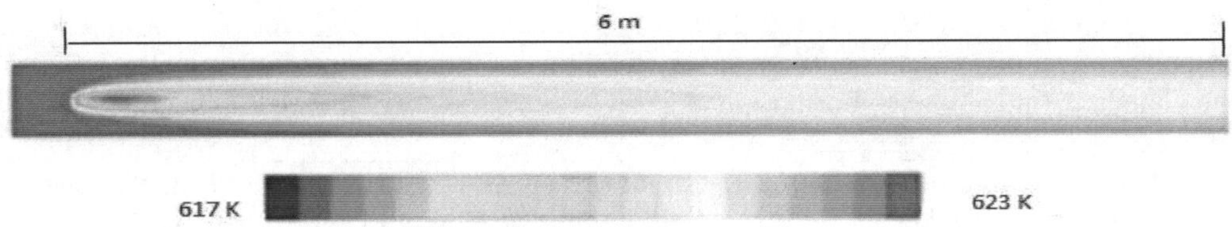

Figure 6. Temperature distribution contour

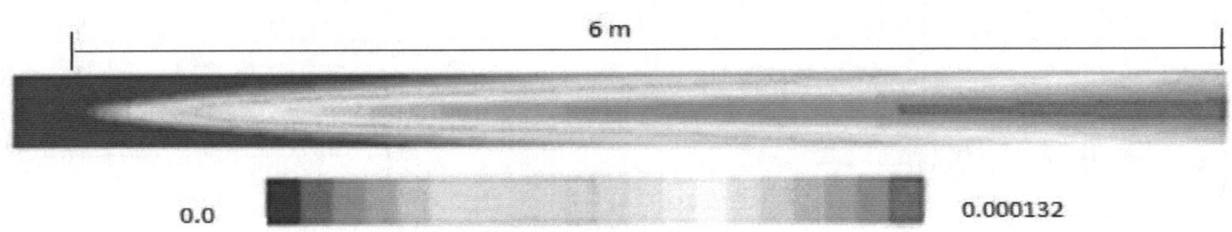

Figure 7. Mass fraction contour of ammonia.

4. Conclusion

This study develops a UWS evaporation model based on multi-component droplet evaporation approach along with the consideration of Stefan flow. The direct thermolysis approach is considered for the urea depletion process, where the modeling is based on single kinetic rate approach by proper fitting of pre- exponent. The overall comparison shows results show a reasonable match with the experimental data. This shows the predicted pre- exponent is more suitable to get promising results for the spray simulations of urea-water solution of the direct thermolysis approach to determine the ammonia conversion efficiency.

IOP Conf. Series: Materials Science and Engineering **243** (2017) 012011 doi:10.1088/1757-899X/243/1/012011

References

[1] Koebel M, Elsener M and Kleemann M 2000 Urea-SCR: A promising technique to reduce NOx emissions from automotive diesel engines. *J. Elsevier Catalysis Today* **59** 335-345

[2] Birkhold F, Meingast U, Wassermann P and Deutschmann O 2007 Modeling and simulation of the injection of urea-water-solution for automotive SCR DeNOx-systems *J.Applied Catalysis B: Environmental.* **70** 119-127

[3] Abu R E, Saha K and Li X 2011 Modeling the Depleting Mechanism of Urea- Water-Solution Droplet for Automotive Selective Catalytic Reduction Systems *AIChE Journal* **57** 3210-3225

[4] Vikas R, Mallikarjuna J M and Gnanesan V 2013 3-D Numerical Study of Effect of Injection Parameters Upon the Uniformity of Ammonia in Urea-SCR *SAE international Technical Paper*

[5] Strom H, Lundstro M A and Andersson B 2009 Choice of urea-spray models in CFD simulations of urea-SCR systems *J.Chem. Engng* **150** 69-82

[6] Emel Y V, Kabo G and Verevkin S 2006 Measurement and Prediction of Thermochemical Properties: Improved Increments for the Estimation of Enthalpies of Sublimation and Standard Enthalpies of Formation of Alkyl Derivatives of Urea *J. Chem. Eng. Data.* **51** 79-87

[7] Wang T J, Baek S W, Lee S Y, Kang D H and Yeo G K 2009 Experimental Investigation on Evaporation of Urea Water Solution Droplet for SCR Applications *AIChE Journal.* **55** 3267-3276

[8] Poling B E, Prausnitz J M and Connell O 2001 *Properties of gases and liquids* (New York: McGraw-Hill)

[9] Perman E P and Lovett T 1926 Vapour pressure and heat of dilution of aqueous solutions *Trans. Faraday Soc* **22** 119

[10] Kim J, Ryu S and Ha J 2004 Numerical Prediction on the Characteristics of Spray Induced Mixing and Thermal Decomposition of Urea Solution in SCR System *Proceedings of 2004 Fall Technical Conferences of the ASME International Combustion Engine Division, (Long Beach, California USA)*

[11] Vikas R, Mallikarjuna J M and Gnanesan V 2012 3-D Numerical study of effect of flow parameters upon the Uniformity of NH_3 in Urea-SCR. J. *Advanced Materials research* **505** 175-179

Parametric study and shrinkage modelling of natural rubber sheet drying using COMSOL multiphysics

C Ajani[1], A Kumar[1], S Curcio[2] and P Tekasakul[1]

[1]Energy Technology Research Center and Department of Mechanical, Engineering, Prince of Songkla University, Hat Yai, Songkhla 90112, Thailand
[2]Laboratory of Transport Phenomena and Biotechnology, Department of Computer Engineering, Modeling, Electronics and Systems, University of Calabria, 87036 Rende (CS), Italy

E-mail: clemajan@gmail.com

Abstract. Natural rubber is one of the major exporting cash crops in Thailand. Shrinkage in natural rubber sheet during drying creates uneven stresses in the rubber, hampers the quality and the water activity predisposes it to microbial activation. Hence, the effect of drying parameters on shrinkage has been considered in this work. The finite element concept coupled with the arbitrary Lagrangian-Eulerian (ALE) method was used to solve the two-dimensional coupled physics in the rubber sheet drying chamber and account for the shrinkage effect. The spatial domain (drying chamber), material domain (rubber sheet) was analysed. An isotropic linear elastic model was assumed for the rubber sheet for analysis. Three Case studies of different velocity, temperature, relative humidity and shrinkage coefficient were considered in the numerical study using COMSOL Multiphysics. It is concluded that increase in the operating parameters increases the shrinkage of the rubber. Therefore, rubber should dry at relatively lower operating parameters to improve its quality.

1. Introduction

Rubber sheets are important exporting product of Thailand. Thailand is the largest natural rubber (NR) producing and exporting country in the world at present. Total amount of NR products in year 2015, was around 4.5 million tons and the export was about 3.7 million tons, which accounted for 36% of global natural rubber production in same year. Commercially produced natural rubber is in the form of; ribbed smoked sheet (RSS), block rubber, rubber concentrated latex, and others [1]. Drying temperature, relative humidity, and air velocity, size of material, and initial moisture content are factors affect the quality of dried product, drying time and energy consumption. The optimization of rubber drying depends on four factors namely: drying temperature, relative humidity, and air velocity, and initial moisture content [2]. Optimization of the drying process and parameters is particularly required to enhance process efficiency in terms of energy usage and production time, without compromising the product quality (excessive shrinkage or Warping) [3]. Drying shrinkage can be defined as the volume reduction in porous materials such as rubber, concrete etc. suffers as a consequence of the moisture migration when exposed to a lower relative humidity environment than the initial one in its own pore system [4]. Simultaneous heat and moisture transfer accompanied by phase change, shrinkage, and/or absorption heat in some porous and composite materials (Rubber, concretes, wood) is a process, which occurs frequently in various engineering applications. Hassini et

IOP Conf. Series: Materials Science and Engineering **243** (2017) 012012 doi:10.1088/1757-899X/243/1/012012

al. reported that shrinkage effect cannot be neglected when moisture diffusivity in highly shrinking materials like natural rubber [5]. However, most of research on rubber drying solely considers the drying models without consideration of the rubber shrinkage during the drying process [6, 7, 8]. Shrinkage strongly depends on matrix mobility. It is proved that shrinkage is more significant during the constant and the falling rate periods, i.e. when matrix mobility is higher [9]. Katekawa and Silva, observed that shrinking in material volume actually corresponded to the volumetric amount of liquid water removed from the sample [10]. FEM is a versatile program applied in variety of engineering and science applications. Initially used in structural analysis, later gain relevance in multiphysics problems such as fluid structure interactions and other fluid flow problems such as complex industrial, mechanical, biomedical, aerospace applications etc. [11, 12, 13]. Perspective of the present work is to carry out a numerical study on the effect of operating parameters on the flow field and shrinkage of natural rubber sheet using COMSOL multiphysics.

2. Material and Method

2.1. Model Assumption

The basic assumptions in model formulation are: (a) The cubical symmetry, which restrict the analysis to a 2D geometry, (b) The natural rubber sheet sample is considered as a fictitious non-porous though it is a hygroscopic porous medium, (c) Transport in the sample is only a function of the water concentration, (d) Operating conditions in the chamber are assumed uniform, (e) Shrinkage characteristics are dependent on natural rubber moisture content.

2.2. Geometry and Material

A half 2D CAD geometry of the side view of the actual rubber drying chamber was designed in COMSOL Multiphysics v5.2 (Comsol AB, Stockholm, Sweden) as shown in figure 1. The same geometry is used for the simulation of the three sets of operating conditions. The side view as shown in figure 2 is considered for the shrinkage modelling. The labels 1, 2, 3 are the symmetry, 4 is the inlet, 5 is the outlet, 6 is the wall while 7, 8, 9 are the fluid structure boundary. The summary of the initial, boundary conditions and parameters used for the case studies are given in table 1, 2 and 3 respectively.

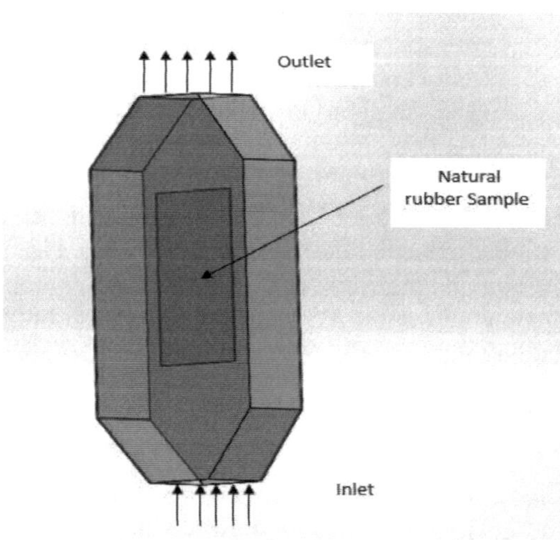

Figure 1. Schematic representation of the drying cell

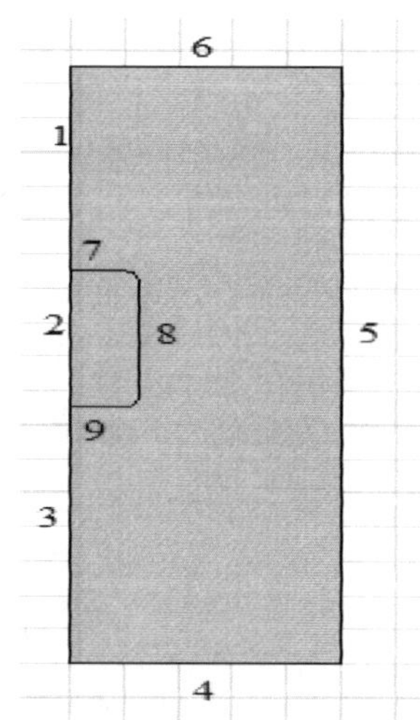

Figure 2. 2D geometry of drying cell

Table 1. Initial conditions used for the simulation

Medium	Variable	Value
Rubber	Moisture content on a dry basis (X_{b0})	4.06
Drying air	Temperature (T_0)	298 K
	Relative humidity (U_{x0})	50%
	Velocity in x and y directions (u_{x0}, u_{y0})	0 m/s
	Pressure in the drying chamber (P_0)	101, 325 Pa

Table 2. Boundary condition

Position	Transport	Expression
1, 2, 3	Symmetry	
4	Momentum	$u_x = 0, u_y = v_0$
	Heat	$T_2 = T_{air}$
	Mass	$C_2 = C_{air}$
5	Momentum	$\underline{u}/u_\tau = 1/kIn(\delta_w/l^*) + C^+$
	Heat	$T_2 = T_{air}$
	Mass	$C_2 = C_{air}$
6	Momentum	$p = p_{atm}$
	Heat	$\underline{n}.(-k_a \nabla T_2) = 0$
	Mass	$\underline{n}.(-D_a \nabla C_2) = 0$
7,8, 9	Momentum	$\underline{u}/u_\tau = 1/kIn(\delta_w/l^*) + C^+$
	Heat	$T = T_{air}$
		$T = T_{air}$
	Mass	$C_2 = C_w$

Table 3. Operating parameters exploited for the simulation

Variable	Case A	Case B	Case C
Inlet Velocity (m/s)	0.5	1.0	1.5
Inlet Temperature (°C)	45	50	55
Inlet Relative humidity (%)	40	50	60

2.3. Governing Equations

The analysis is limited to a 2D due to complexity of the governing equation. The drying air/fluid model solved numerically incompressible Navier-Stokes equations along with the equations of momentum conservation, mass conservation, Energy balance, Mass balance as in equations (1), (2), (3) and (4) respectively [6, 7, 14].

$$\rho_a \frac{\partial \underline{\mathbf{u}}}{\partial t} + \rho_a (\underline{\mathbf{u}}.\underline{\nabla})\underline{\mathbf{u}} = \underline{\nabla}\left[-p\underline{I} + \eta_a (\underline{\nabla \mathbf{u}} + (\underline{\nabla \mathbf{u}})^T) - 2/3\eta_a (\underline{\nabla.\mathbf{u}})\underline{I}\right] \tag{1}$$

$$\frac{\partial \rho_a}{\partial t} + \nabla.\left(\rho_a \underline{\mathbf{u}}\right) = 0 \tag{2}$$

$$\rho_a C_{pa} \partial T_2 / \partial t - \underline{\nabla}.(k_a \underline{\nabla T_2}) + \left(\rho_a C_{pa} \underline{\mathbf{u}}.\underline{\nabla T_2}\right) = 0 \tag{3}$$

$$\partial C_2 / \partial t + \underline{\nabla}.(-D_a \underline{\nabla C_2}) + \underline{\mathbf{u}}.\underline{\nabla C_2} = 0 \tag{4}$$

The transport in the rubber sheet is given by the mass balance equations in the rubber sample is as represented in equation (5) [13].

$$\partial C_W / \partial t + \underline{\nabla}.\left(-D_W \underline{\nabla} C_W\right) + I = 0 \tag{5}$$

where, p is the pressure within the drying chamber (Pa), $\underline{\mathbf{u}}$ is the averaged velocity field (m/s), C_2 is water concentration in air (mol/m^3), Cw is the water concentration in rubber (mol/m^3), D_W is capillary diffusion coefficient of water in rubber (m^2/s), I is volumetric rate of evaporation (mol/(m^3 s)), T is air temperature (K), C_{pa} is the air specific heat (J/(kg K)), k_a is the air thermal conductivity (W/(m K)), ρ_a is air density (kg/m3), η_a is the air viscosity (Pa s), t *is* time (s), D_a is the diffusion coefficient of water in air (m^2/s), and $\underline{I}$ is the fluctuating part of velocity field (m/s).

The rubber is assumed to be undergoing an isotropic shrinkage. Local total strain $\{d\varepsilon\}$ is a function of the mechanical strain $\{d\varepsilon_s\}$ the shrinkage strain $\{d\varepsilon_0\}$ due to the moisture content.

$$\{d\varepsilon\} = \{d\varepsilon_s\} + \{d\varepsilon_0\} \tag{6}$$

Total strains $\{d\varepsilon\}$ is actually a function of the total displacement $\{dU\}$

$$\{d\varepsilon\} = [A]\{dU\}$$

$$\{d\varepsilon_0\} = K_{ds} \cdot dC_w$$

dC_W is the variation in the moisture content and the hydrous compressibility factor is K_{ds} which is estimated from the previous work [14]. The values of 6.01 $\times 10^{-6}$, 7.01 $\times 10^{-6}$, 8.01 $\times 10^{-6}$ was exploited for the three cases.

2.4. Physics and Mesh

The temperature and concentration are coupled to the laminar flow for the spatial domain. Solid mechanics is used for the material domain while ALE method is used to account shrinkage. Laplace smoothing, boundary and initial condition are applied. The solid and fluid domains are computationally meshed using COMSOL Multiphysics v5.2 (Comsol AB, Stockholm, Sweden). The material and spatial domains are meshed using an extremely fine boundary layer mesh based on the COMSOL physics controlled mesh of minimum quality of 0.08.

2.5. Simulation Setup

The CFD simulations are solved by using commercial finite element solver, COMSOL Multiphysics 5.2 on a i5 dual-core computer. The solid mechanics and transport of diluted species module is used to solve the transport phenomena occurring in the rubber sheets to account shrinkage. Laminar flow module is exploited for solving the laminar flow as shown in equations (1) and (2). Equations (3) and (4) referred to the convection and conduction drying air was solved using heat transfer in fluids whereas the transport of diluted species module is used to solve equation (5). These equations are discretized using the second order for both the material and spatial frames. The relative and absolute tolerance are 0.01 and 0.001 respectively. Implicit time stepping scheme is used in the solution of the PDEs systems [14]. A combination of fully coupled and segregated solvers are applied in solving the time dependent study.

3. Result and Discussion

3.1. Flow behaviors at different operating parameters

The variation in the velocity flow patterns in the chamber for the three Cases are shown in figure 3(a), (b) and (c). Figure 3(c) shows the highest flow field and has more loading on the material domain. This is validated from previous research [14]. The coupled temperature and concentration in the drying chamber and the moisture content variation in the material domain for Case C are shown in figure 4(a), (b) and (c) respectively. Case C has the highest velocity, temperature and concentration variation in the drying chamber and thus could engender a faster evolution of moisture in the material domain of the three Case studies.

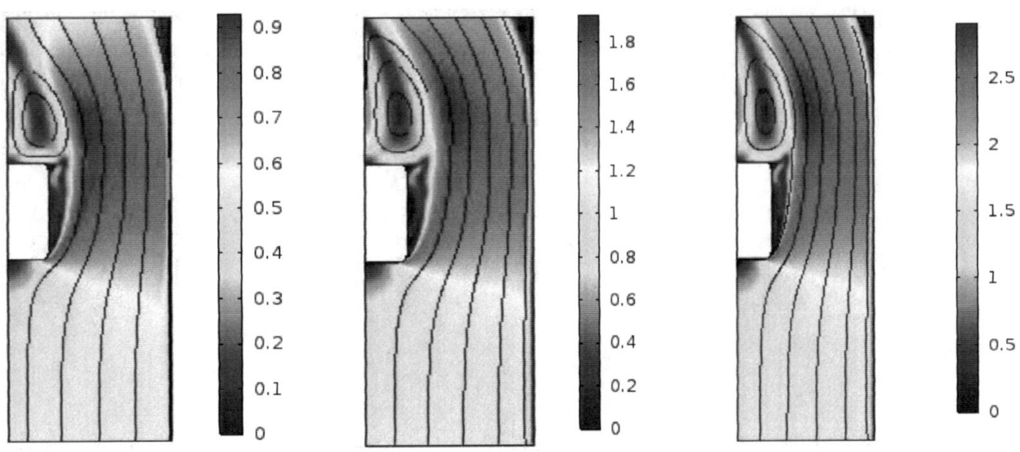

(a) (b) (c)

Figure 3. (a) Flow field for Case A, (b) Flow field for Case B (c) Flow field for Case C

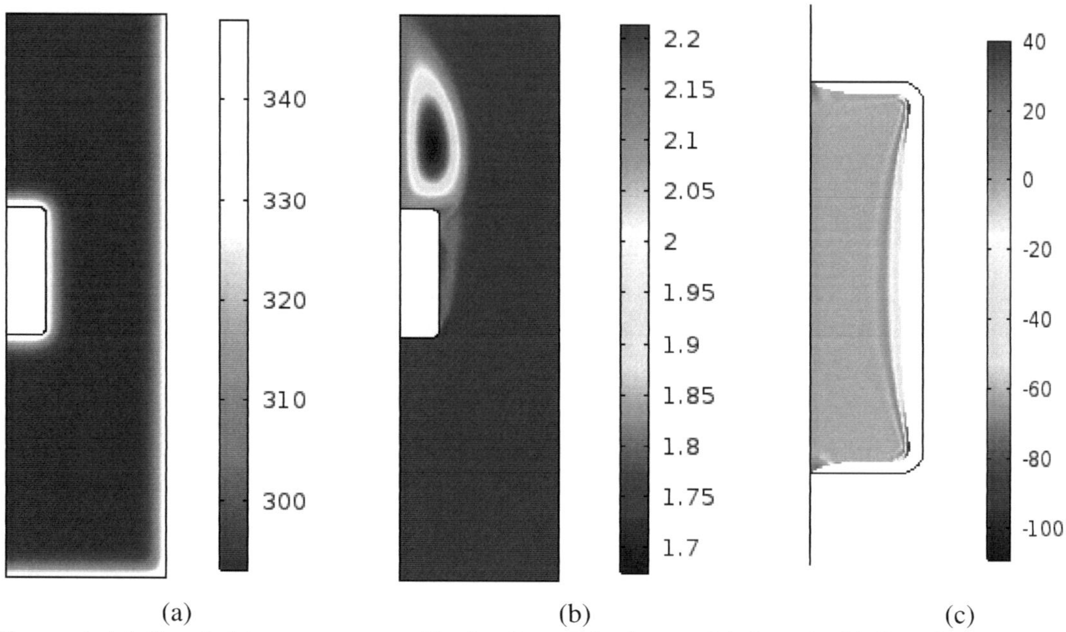

Figure 4. (a) Coupled temperature profile for Case C in the spatial domain (b) Coupled concentration profile for Case C in the spatial domain (c) Concentration profile for Case C in material domain

3.2. Shrinkage Behaviour

The von mises stress responsible for the shrinkage as shown in figure 5 (a), (b) and (c). Shrinkage followed a similar trend in all the Cases is also observed in same figure. This is due to an isotropic shrinkage model consideration. Figure 5 (c) has the highest shrinkage stress this mean, more strain on it. The horizontal shrinkage stress distributions from a 2D line plot at the height of 0.18m are shown in figure 6. The highest stresses are 1.96e8 N/m^2, 2.28e8 N/m^2 and 2.61e8 N/m^2 for all Cases A, B and C respectively. This implies that operating at higher operating parameter could engender higher shrinkage stress, thereby making the shrinkage more drastic and reducing the quality of the rubber. Although the rapid removal of the moisture content (at high operating condition) will reduce the predisposition of the rubber to microbial spoilage because of reduced water activity. This is similar with previous research on convective drying of porous agricultural products with shrinkage effect [14, 15], therefore model is validated.

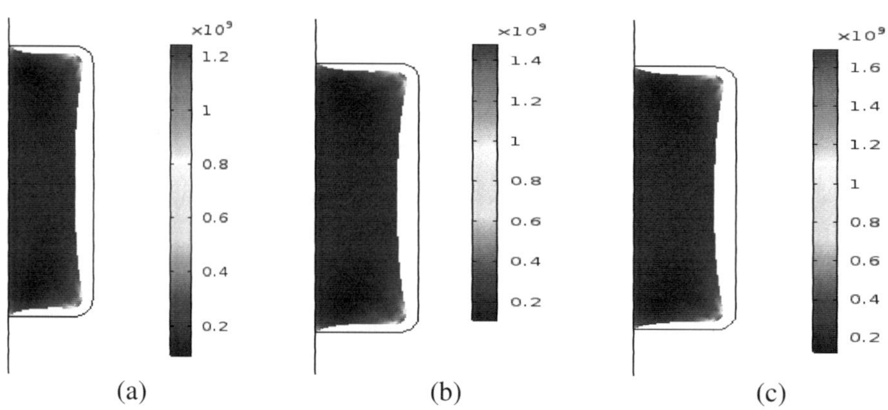

Figure 5. (a) Von mises shrinkage stress for Case A (b) Von mises shrinkage stress for Case B (c) Von mises shrinkage stress for Case C.

IOP Conf. Series: Materials Science and Engineering **243** (2017) 012012 doi:10.1088/1757-899X/243/1/012012

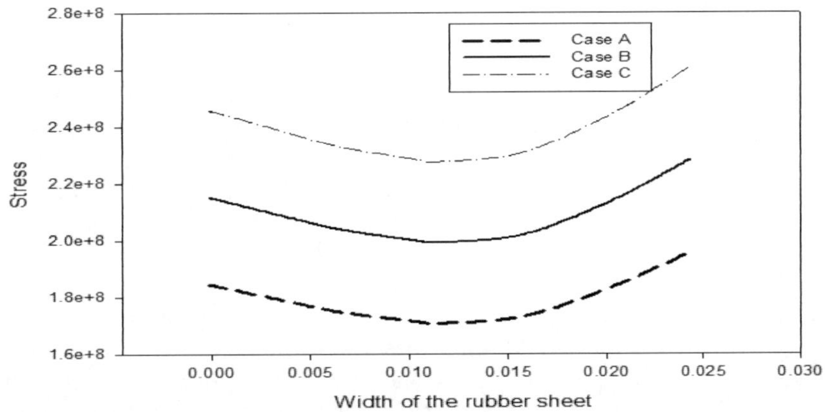

Figure 6. Shrinkage stress distribution on width of material at 0.18m height

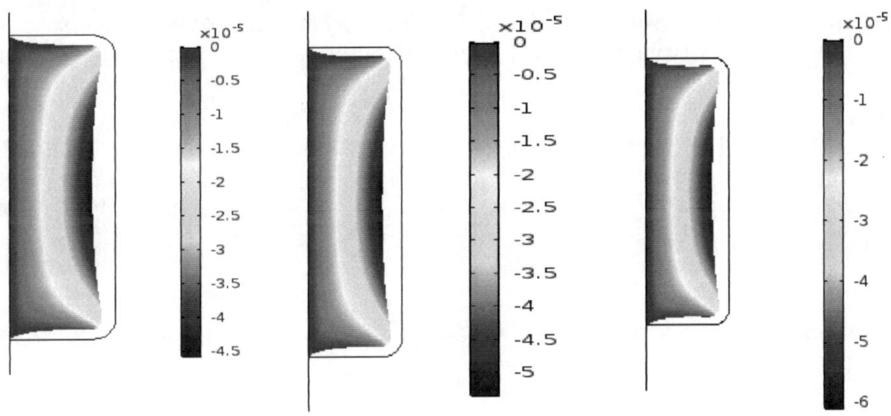

(a) (b) (c)

Figure 7. Shrinkage strain /displacement

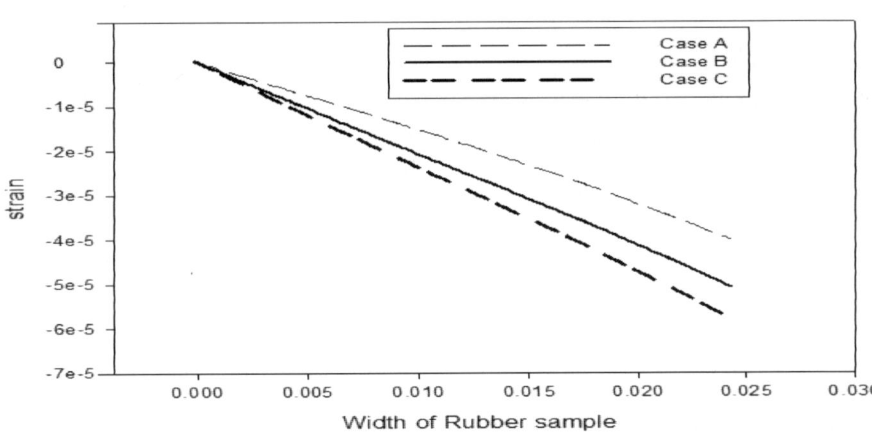

Figure 8. Shrinkage strain distribution on width of material at 0.18m height

Figure 7 (a), (b) and (c) show the resulting shrinkage strains /displacements due to the applied stresses in all Cases. It is observed that Case A has the least shrinkage while Case C has the highest shrinkage. A 2D line plot of the displacements /shrinkages of the rubber from their initial width at height 0.18m are shown in figure 8. The nature of the graphs for the three case studies further explains the shrinkage behaviours. The highest displacement/ shrinkage from the initial width at height 0.18m are -4.0264e-5m, -5.0702e-5m and -5.7943e-5m for Cases A, B and C respectively. The stresses developed in C is the highest as shown in figure 6, hence, the highest shrinkage/ displacement of -5.7943e-5m as shown in figure 8. The shrinkage in A is the milder compared to the others because of it reduced operating conditions.

4. Conclusion

A predictive model to study the effects of operating conditions and rubber mechanical properties on rubber shrinkage drying using COMSOL Multiphysics was formulated. The ALE method with the relevant domain integration was used to solve the equations. The obtained results confirmed that it is crucial to develop a model capable of predicting the actual variation of natural rubber shape and dimensions. This shows that deformation strongly influenced the transport phenomena and hence, cannot be neglected. The proposed approach is capable to calculate the effect of different variables, which could be used to ascertain either the natural rubber safety in terms of the microbial inactivation kinetics based on the water activity or the quality index based on the shrinkage stress distribution.

Acknowledgement

This research was supported by Higher Education Research Promotion and the Thailand's Education Hub for Southern Region of ASEAN Countries Project Office of Higher Education Commission. This research was also supported by Natural Rubber Innovation Research Institute, Prince of Songkla University (Grant No. ENG590305S).

References

[1] Agricultural Statistics of Thailand 2016 *Ministry of Agriculture & Co-Operatives, Bangkok,* Thailand
[2] Theppaya T and Prasertsan S 2004 *Dry. Technol.* 22 1637–60
[3] Kowalski S J and Pawlowski A 2011 *Chem. Eng. Sci.* 66 1893–905
[4] P. Mehta P M 2005 *Concrete: Microstructure, Properties, and Materials* 684
[5] Hassini L, Azzouz S, Peczalski R and Belghith A 2007 *J. Food Eng.* 79 47–56
[6] Promtong M and Tekasakul P 2007 *Appl. Therm. Eng* 27 2113–21
[7] Tekasakul P, Dejchanchaiwong R, Tirawanichakul Y and Tirawanichakul S 2015 *Dry. Technol.* 33 1124–37
[8] Dejchanchaiwong R, Tirawanichakul Y, Tirawanichakul S, Kumar A and Tekasakul P 2017 *Appl. Therm. Eng.* 112 761–70
[9] Karathanos V T and Belessiotis V G 1996 *J. Food Eng.* 29 167-83
[10] Katekawa M E and Silva M A 2007 *Dry. Technol.* 25 1659–66
[11] Bathe K-J 2006 *Finite element procedures* (Watertown, MA)
[12] Zienkiewicz O C and Taylor R L 2000 *The Finite Element Method Volume 1 : The Basis Methods* (Butter Heinemann) p708
[13] Bathe K J and Saunders H 1984 *J. Press. Vessel Technol.* 106 421
[14] Curcio S and Aversa M 2014 *J. Food Eng.* 123 36–49
[15] Curcio S and Aversa M 2009 *COMSOL Conf.* (Milán)

Effect of turning angle on performance of 2-D turning diffuser via Asymptotic Computational Fluid Dynamics

Y T Khong[1, a], N Nordin[1, b], S M Seri[1], A N Mohammed[1], A Sapit[1], I Taib[1], K Abdullah[1], A Sadikin[1] and M A Razali[1]

[1] Flow Analysis, Simulation and Turbulence Research Group, Center for Energy and Industrial Environment Studies, Faculty of Mechanical and Manufacturing Engineering, Universiti Tun Hussein Onn Malaysia, Johor, Malaysia

E-mail: yanteng.0714@gmail.com[a], mayati@uthm.edu.my[b]

Abstract. The present work aims to numerically investigate the effect of varying turning angle, ϕ = 30° - 90° on the performance of 2-D turning diffuser and to develop the performance correlations via integrating the turning angle using Asymptotic Computational Fluid Dynamics (ACFD) technique. Standard k-ε adopting enhanced wall treatment of $y^+ \approx 1.1$ appeared as the best validated model to represent the actual cases with deviation of ±4.7%. Results show that the pressure recovery, C_p and flow uniformity, σ_{out} are distorted of respectively 37% and 28% with the increment of turning angle from 30° to 90°. The flow separation starts to emerge within the inner wall, S=0.91L_{in}/W_1 when 45° turning diffuser is applied and its scale is enlarged by further increasing the turning angle. The performance correlations of 2-D turning diffuser are successfully developed with deviation to the full CFD solution approximately of ±7.1%.

1. Introduction

Turning diffuser is often introduced in applications such as HVAC [1 - 3], wind-tunnel [4 - 6], gas turbine cycle [7] and aircraft engine [8, 9] as an adapter to join the conduits of different cross-sectional areas or an ejector to decelerate the flow and raise the static pressure before discharging to the atmosphere. The performances of diffuser are measured primarily using outlet pressure recovery coefficient (C_p) and flow uniformity index (σ_{out}) that are so much affected due to the nature of its geometry by the existence of flow separation and dispersion of core and secondary flow.

Study of the geometrical effect on diffuser performance has been of fundamental interest to researchers in the area of fluid mechanics since decades and it continues to grow [1-12]. Turning angle (ϕ) is one of important geometrical parameters that is deemed to influence significantly the performance of turning diffuser. Sullerey et al. [13] found that the C_p of 55° turning diffuser was slighly lower than a straight diffuser. It may however improve comparable to the performance of straight diffuser by applying the turbulent intensity of minimum 3.4%. The ϕ = 90° [6, 14] and 180° [15] were found to affect notably the flow performance and often flow control devices such as guide vanes, screens, honeycomb and vortex generators were introduced to improve the flow.

Nordin [1] has recently established empirical correlations integrating the effects of both geometrical and operating parameters on the performance of 90° 2-D turning diffuser using Asymptotic Computational Fluid Dynamics (ACFD) technique. However, these correlations, i.e. Equations (1) and (2) have been developed not to integrate the effect of $\Delta\phi$:

IOP Conf. Series: Materials Science and Engineering **243** (2017) 012013 doi:10.1088/1757-899X/243/1/012013

$$C_{p\,ACFD} = 0.200 + \left(\left[\frac{Re_{in}}{Re_{in\,ref}}\right]^2 - 1\right)0.0059 - \left(\left[\frac{L_{in}/W_1}{L_{in}/W_{1ref}}\right]^1 - 1\right)0.6569 - \left(\left[\frac{W_2/W_1}{W_2/W_{1ref}}\right]^3 - 1\right)0.0288 \quad (1)$$

$$\sigma_{out\,ACFD} = 3.45 + \left(\left[\frac{Re_{in}}{Re_{in\,ref}}\right]^2 - 1\right)0.7591 - \left(\left[\frac{L_{in}/W_1}{L_{in}/W_{1ref}}\right]^1 - 1\right)0.2577 - \left(\left[\frac{W_2/W_1}{W_2/W_{1ref}}\right]^3 - 1\right)0.0357 \quad (2)$$

A CFD is a relatively new analytical technique established by Herwig et al. that applies Taylor Series expansion to produce correlations between all the relevant non-dimensional variables of the problem analysed by CFD [16-19]. This technique requires less number of solutions $n + 1$ than the usual-used linear regression and curve fitting techniques, 4^n number of solutions to develop correlations from the CFD simulations, where n represents non-dimensional variables.

The present work aims to numerically investigate the effects of varying turning angle ($\phi = 30°$, 45°, 55°, 75° and 90°) as shown in Figure 1 on the performance of 2-D turning diffuser. The ACFD technique is then applied to develop performance correlations via integrating the effect of turning angle, i.e. $C_p = f(L_{in}/W_1, W_2/W_1, \phi, Re_{in})$ and $\sigma_{out} = f(L_{in}/W_1, W_2/W_1, \phi, Re_{in})$.

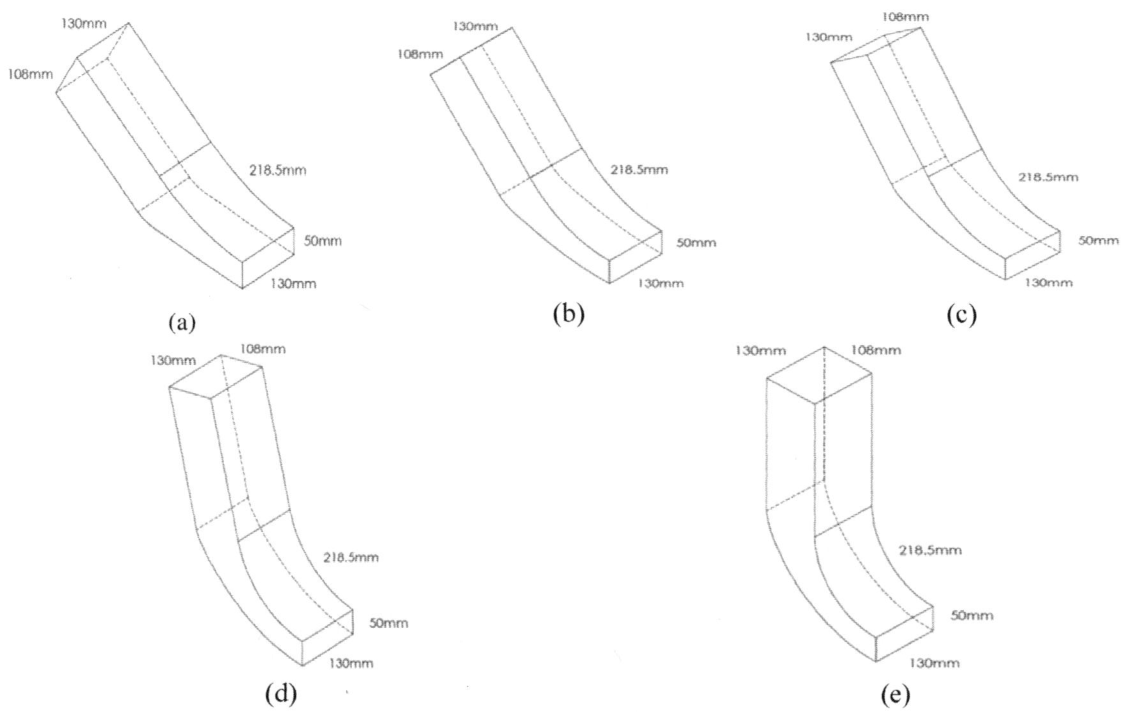

Figure 1. 2-D turning diffuser of (a) $\phi = 30°$, (b) $\phi = 45°$, (c) $\phi = 55°$, (d) $\phi = 75°$ and (e) $\phi = 90°$

2. Methodology

2.1. CFD Method

The geometrical domains as shown in Figure 1 were modelled using Solidworks. Three types of boundary conditions were imposed as depicted in Table 1. At the solid wall, the velocity was zero due to the no-slip condition. The inlet velocity, V_{in} respective to the Re_{in} = 6.382 x 10^4 was specified at 14.25 m/s. This corresponded to the turbulent intensity, I_{in} of 4.0%. At the outlet boundary, the pressure was set at the atmospheric pressure (0 gage pressure). The working fluid was air at 30°C with ρ = 1.164 kg/m^3 and μ = 1.872 x 10^{-5} kg/m.s.

Table 1. Boundary operating conditions

Inlet	Type of boundary	Velocity-inlet
	Velocity magnitude, V_{in} (m/s)	14.25 (Re_{in} = 6.382 x 10^4)
	Turbulent intensity, I_{in} (%)	4.0
	Hydraulic diameter, D_h (mm)	72.22
Outlet	Type of boundary	Pressure-outlet
	Pressure (Pa)	0 gauge pressure
Wall	Type of boundary	Smooth wall
	Shear condition	No-slip
Working fluid properties	Working fluid	Air
	Temperature (°C)	30
	Density, ρ (kg/m^3)	1.164
	Dynamic viscosity, μ (kg/m.s)	1.872 x 10^{-5}

The grid was generated using ANSYS ICEM CFD by adopting enhanced wall treatment with the size of wall-adjacent cell, $y^+ \approx 1.0$. The grid independency test was carried out as shown in Table 2 with Mesh 4 appeared as the optimum to produce accurate result with reasonable CPU time.

Table 2. Grid independency test

Mesh	Nodes	Pressure Recovery, C_p	Deviation, %
1	78213	0.1528	28.9
2	85938	0.1842	14.3
3	87272	0.1941	9.7
4	91520	0.2188	1.8
5	131660	0.2150	-

ANSYS Fluent was used as a platform for the simulation. The flow was assumed to be incompressible, three-dimensional (x, y and z direction), fully-developed, steady state and isothermal. The gravitational effect was negligible. The Reynolds Average Navier Stokes (RANS) equations as follows were solved.

Continuity equation:

$$\frac{\delta u}{\delta x}+\frac{\delta v}{\delta y}+\frac{\delta w}{\delta z}=0 \tag{3}$$

x- momentum equation:

$$u\frac{\delta u}{\delta x}+v\frac{\delta u}{\delta y}+w\frac{\delta u}{\delta z}=-\frac{\delta P}{\delta x}+\mu\frac{\delta^2 u}{\delta x^2}+\mu\frac{\delta^2 u}{\delta y^2}+\mu\frac{\delta^2 u}{\delta z^2}-\frac{\delta(\overline{u'^2})}{\delta x}-\frac{\delta(\overline{u'v'})}{\delta y}-\frac{\delta(\overline{u'w'})}{\delta z}+S_{Mx} \tag{4}$$

y- momentum equation:

$$u\frac{\delta v}{\delta x}+v\frac{\delta v}{\delta y}+w\frac{\delta v}{\delta z}=-\frac{\delta P}{\delta y}+\mu\frac{\delta^2 v}{\delta x^2}+\mu\frac{\delta^2 v}{\delta y^2}+\mu\frac{\delta^2 v}{\delta z^2}-\frac{\delta(\overline{u'v'})}{\delta x}-\frac{\delta(\overline{v'^2})}{\delta y}-\frac{\delta(\overline{v'w'})}{\delta z}+S_{My} \quad (5)$$

z- momentum equation:

$$u\frac{\delta w}{\delta x}+v\frac{\delta w}{\delta y}+w\frac{\delta w}{\delta z}=-\frac{\delta P}{\delta z}+\mu\frac{\delta^2 w}{\delta x^2}+\mu\frac{\delta^2 w}{\delta y^2}+\mu\frac{\delta^2 w}{\delta z^2}-\frac{\delta(\overline{u'w'})}{\delta x}-\frac{\delta(\overline{v'w'})}{\delta y}-\frac{\delta(\overline{w'^2})}{\delta z}+S_{Mz} \quad (6)$$

The applicability of standard k-ε (std k-ε), shear stress transport k-ω (SST k-ω) and Reynolds stress model (RSM) to close the RANS equations was verified. Pressure based solver with a robust pressure-velocity coupling scheme, SIMPLE was applied. The gradient was discretised by Green-Gauss Cell-based. As it involved high pressure gradients, pressure was discretised by PRESTO scheme. A 3rd order accuracy scheme, QUICK was used to discretise the convection terms, i.e. momentum, turbulent kinetic energy and turbulent dissipation rate owing to its proven capability to solve the flow in diffuser when hybrid mesh was applied. The convergence criterion was set to be 10^{-6}.

2.2. ACFD Method

The performance correlations of 2-D turning diffuser were developed using ACFD technique. This technique involved several steps as follows [1, 16 - 19]:

1) Identifying the dependent and independent variables
2) Linearizing the relationship between the dependent and independent variables
3) Applying the Taylor's series expansion
4) Determining the convergence point and gradients
5) Substituting all the constants to complete the correlations

Firstly, the dependent and independent variables were identified. The C_p and σ_{out} were dependent on Re_{in}, W_2/W_1, L_{in}/W_1 and ϕ. The preliminary relationships between the dependent and independent variables were written as follows:

$$C_{p\,ACFD} = f(\mathrm{Re}_{in}, W_2/W_1, L_{in}/W_1, \phi) \quad (7)$$

$$\sigma_{out\,ACFD} = f(\mathrm{Re}_{in}, W_2/W_1, L_{in}/W_1, \phi) \quad (8)$$

Five set of solutions were needed for each independent variable (Re_{in}, W_2/W_1, L_{in}/W_1, ϕ). The reference values were $Re_{in\,ref} = 6.382 \times 10^4$, $W_2/W_{1\,ref} = 2.16$, $L_{in}/W_{1\,ref} = 4.37$ and $\phi_{ref} = 90°$. The relationship of dependent and independent variable was linearized in which appropriate chosen reference values may ease the linearization procedure otherwise the modification to the term for instance by adding an exponent should be considered.

The performance correlations were developed by applying Taylor series based expansion:

$$\eta(\phi_1,\phi_2,\phi_3,\dots\phi_n) = \eta_{ref} + (\phi_1-\phi_{1\,ref})\frac{\partial\eta}{\partial\phi_1} + (\phi_2-\phi_{2\,ref})\frac{\partial\eta}{\partial\phi_2} + (\phi_3-\phi_{3\,ref})\frac{\partial\eta}{\partial\phi_3} + \dots + (\phi_n-\phi_{n\,ref})\frac{\partial\eta}{\partial\phi_n} \quad (9)$$

where,

η = dependent variables (C_p, σ_{out})

$\phi_1 = \left[\frac{Re_{in}}{Re_{in\,ref}}\right]^a$, $\phi_2 = \left[\frac{L_{in}/W_1}{L_{in}/W_{1\,ref}}\right]^b$

$$\phi_3 = \left[\frac{W_2/W_1}{W_2/W_{1\,ref}}\right]^c, \quad \phi_4 = \left[\frac{\phi}{\phi_{ref}}\right]^d$$

$\phi_1, \phi_2, \phi_3, \phi_4$ = Dimensionless independent groups.

a, b, c, d = Exponent values chosen to let all lines intersect and converge at one point in graph.

η_{ref} = Reference value of dependent variable (intersection/convergence point at y-axis)

$\phi_{1\,ref}, \phi_{2\,ref}, \phi_{3\,ref}, \phi_{4\,ref}$ = Reference values of dimensionless groups (intersection/ convergence point at x-axis)

$\frac{\partial\eta}{\partial\phi_1}, \frac{\partial\eta}{\partial\phi_2}, \frac{\partial\eta}{\partial\phi_3}, \frac{\partial\eta}{\partial\phi_4}$ = Gradients / slopes of the corresponding lines

Convergence point and gradients should be determined in order to solve Equation 9. The detail solution is presented in the following section.

3. Results and Discussion

3.1. CFD validation results

CFD validation was carried out by comparing the simulation results with the experimental results by Nordin [1]. Table 3 shows that the std k-ε appears as the most optimum model producing the least deviation to the experiment of 4.7%. Furthermore, as shown in Figure 2, comparable flow structure with almost similar onset flow separation between CFD and experiment is obtained by applying std k-ε adopted enhanced wall treatment.

Table 3. CFD validation

Solver Model	C_p	Deviation, %
Experiment [1]	0.2090	-
std k-ε	0.2188	4.7
SST k-ω	0.1573	24.7
RSM	0.1475	29.4

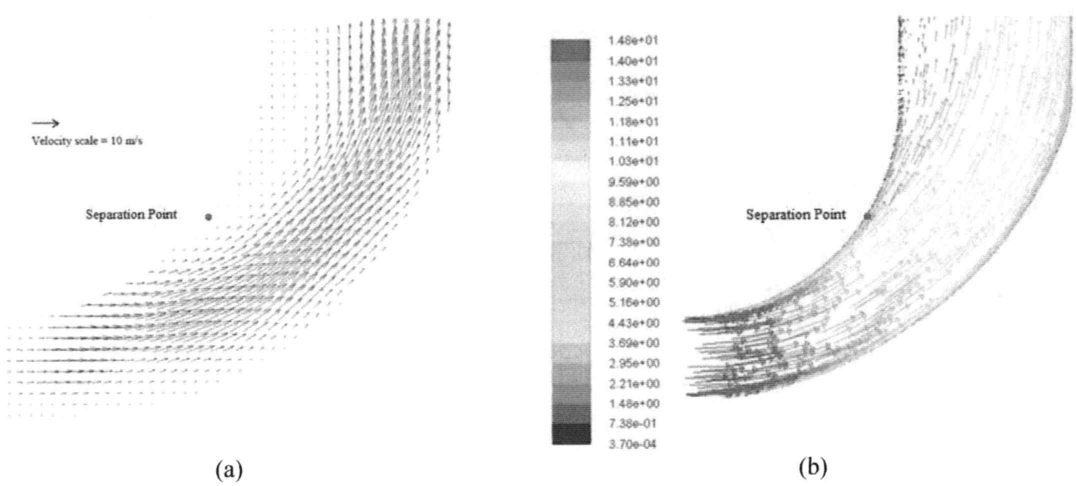

Figure 2. Flow structure (a) Experiment, PIV [1] (b) CFD, std k-ε + enhanced wall treatment

3.2. Effect of turning angle (ϕ)

Table 4 shows the effects of varying turning angle (ϕ) on the pressure recovery coefficient (C_p), flow uniformity index (σ_{out}) and onset flow separation (S). The C_p is affected by 37% with the increase of ϕ from 30° to 90° due to considerable flow separation occurs particularly within the 90° diffuser, S=0.61L_{in}/W_1 as shown in Figure 3(c). The flow separation is undesirable as it not only affects the C_p but also diminish the core flow area as shown in Figure 4. This then disturbs the flow uniformity of 28% with the increase of ϕ to 90°.

Table 4. Effect of turning angle on 2-D turning diffuser performances

Turning angle, ϕ (°)	Pressure recovery, C_p	Flow uniformity, σ_{out}	Separation Point, S
30	0.3469	2.9477	-
45	0.3394	3.0755	0.91L_{in}/W_1
55	0.3045	2.9698	0.85 L_{in}/W_1
75	0.2372	3.4278	0.70 L_{in}/W_1
90	0.2188	3.7043	0.61 L_{in}/W_1

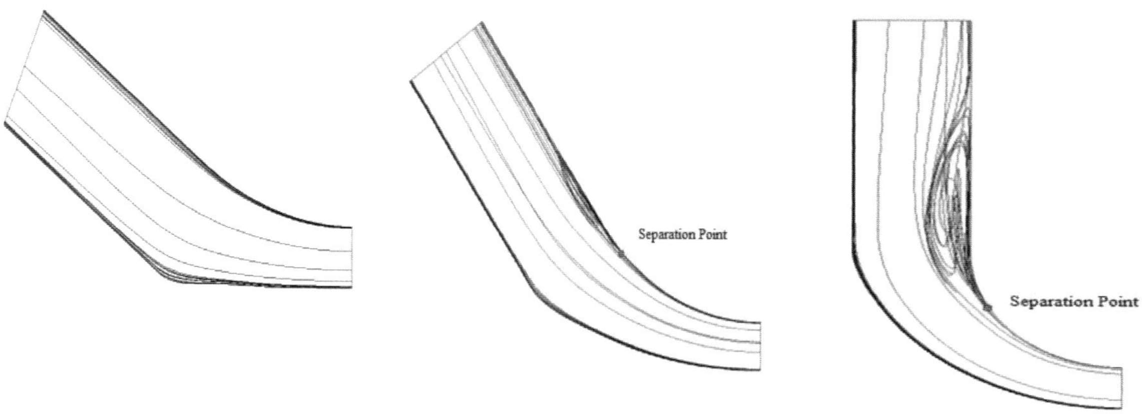

(a) (b) (c)

Figure 3. Flow separation (a) ϕ = 30°, (b) ϕ = 55°, (c) ϕ = 90°

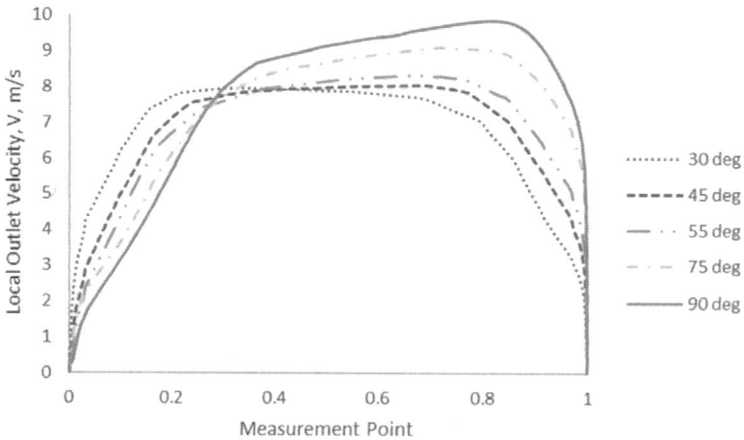

Figure 4. Outlet velocity profiles by varying ϕ = 30°-90°

IOP Conf. Series: Materials Science and Engineering **243** (2017) 012013 doi:10.1088/1757-899X/243/1/012013

3.3. Performance correlations via ACFD

Applying Taylor Series expansion to develop outlet pressure recovery (C_p) correlation:

$$C_{p\,acfd} = C_{p\,ref} + (\phi_1 - \phi_{1\,ref})\frac{\partial C_p}{\partial \phi_1} + (\phi_2 - \phi_{2\,ref})\frac{\partial C_p}{\partial \phi_2} + (\phi_3 - \phi_{3\,ref})\frac{\partial C_p}{\partial \phi_3} + (\phi_4 - \phi_{4\,ref})\frac{\partial C_p}{\partial \phi_4} \qquad (10)$$

ϕ_1, ϕ_2, ϕ_3 and ϕ_4 are dimensionless independent groups where the data for ϕ_1, ϕ_2, ϕ_3 are taken from previous work [1]. Table 5 shows the data for ϕ_4 where in order to fit all lines in a graph as shown in Figure 5 and converge at a point represented as $C_{p\,ref} = 0.200$, the exponent values $a = 1$, $b = 1.3$, $c = 3$ and $d = 4.5$ are chosen.

Table 5. $\phi_4 = [\phi/\phi_{ref}]^{4.5}$ and the corresponding $C_{p\,cfd}$ for various ϕ

ϕ	$\phi_4 = [\phi/\phi_{ref}]^{4.5}$	$C_{p\,cfd}$
30	0.0071	0.3469
45	0.0442	0.3394
55	0.1090	0.3045
75	0.4402	0.2372
90	1.0000	0.2188

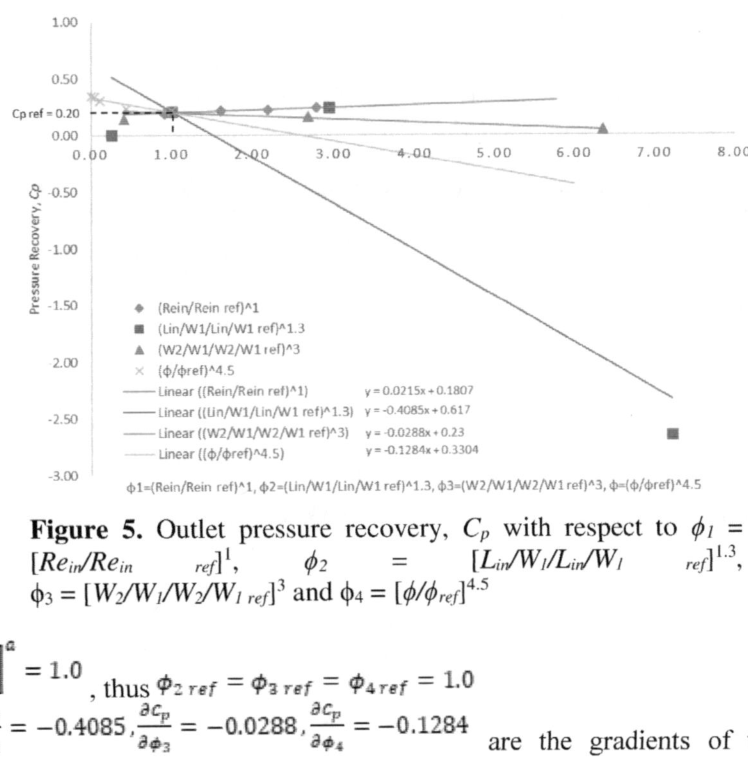

Figure 5. Outlet pressure recovery, C_p with respect to $\phi_1 = [Re_{in}/Re_{in\,ref}]^1$, $\phi_2 = [L_{in}/W_1/L_{in}/W_{1\,ref}]^{1.3}$, $\phi_3 = [W_2/W_1/W_2/W_{1\,ref}]^3$ and $\phi_4 = [\phi/\phi_{ref}]^{4.5}$

$\phi_{1\,ref} = \left[\frac{Re_{in\,ref}}{Re_{in\,ref}}\right]^a = 1.0$, thus $\phi_{2\,ref} = \phi_{3\,ref} = \phi_{4\,ref} = 1.0$

$\frac{\partial C_p}{\partial \phi_1} = 0.0215, \frac{\partial C_p}{\partial \phi_2} = -0.4085, \frac{\partial C_p}{\partial \phi_3} = -0.0288, \frac{\partial C_p}{\partial \phi_4} = -0.1284$ are the gradients of the corresponding lines.

Substituting all the constants in Equation (10) yielding,

$$C_{p\,acfd} = 0.200 + \left(\left[\frac{Re_{in}}{Re_{in\,ref}}\right]^1 - 1\right)0.0215 - \left(\left[\frac{L_{in}/W_1}{L_{in}/W_{1\,ref}}\right]^{1.3} - 1\right)0.4085 - \left(\left[\frac{W_2/W_1}{W_2/W_{1\,ref}}\right]^3 - 1\right)0.0288 - \left(\left[\frac{\phi}{\phi_{ref}}\right]^{4.5} - 1\right)0.1284 \qquad (11)$$

The efficacy of ACFD correlation to represent the full CFD solution is verified using a parity plot as shown in Figure 6. Satisfied agreement between the ACFD and CFD solutions is achieved within ± 5.9%. This basically applies for the range of 5.786 x $10^4 \leq Re_{in} \leq 1.775$ x 10^5, $4.37 \leq L_{in}/W_1 \leq 20$, $2.16 \leq W_2/W_1 \leq 4.0$, $30° \leq \phi \leq 90°$.

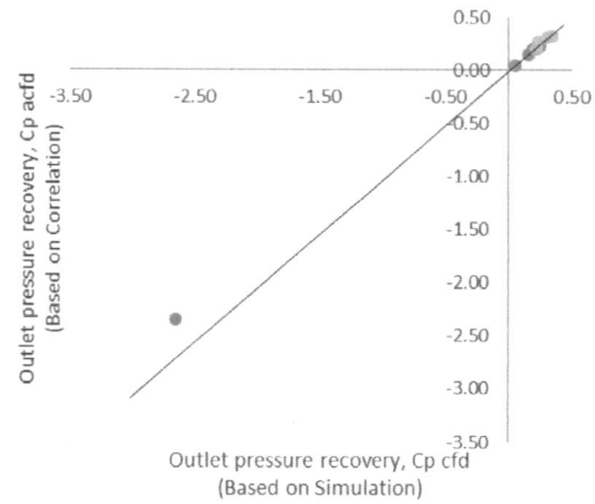

Figure 6. Parity plot showing agreement of CFD simulation ($C_{p\ cfd}$) with ACFD correlation ($C_{p\ acfd}$)

Applying Taylor series expansion to develop σ_{out} correlation:

$$\sigma_{out\ acfd} = \sigma_{out\ ref} + (\phi_1 - \phi_{1\,ref})\frac{\partial\sigma_{out}}{\partial\phi_1} + (\phi_2 - \phi_{2\,ref})\frac{\partial\sigma_{out}}{\partial\phi_2} + (\phi_3 - \phi_{3\,ref})\frac{\partial\sigma_{out}}{\partial\phi_3} + (\phi_4 - \phi_{4\,ref})\frac{\partial\sigma_{out}}{\partial\phi_4} \quad (12)$$

ϕ_1, ϕ_2, ϕ_3 and ϕ_4 are dimensionless independent groups where the data for ϕ_1, ϕ_2, ϕ_3 are taken from previous work [1]. Table 6 shows the data for ϕ_4 where in order to fit all lines in a graph as shown in Figure 7 and converge at a point represented as $\sigma_{out\ ref} = 3.45$, the exponent values $a = 2.9$, $b = 1.6$, $c = 1$ and $d = 0.1$ are chosen.

Table 6. $\phi_4 = [\phi/\phi_{ref}]^{0.1}$ and the corresponding $\sigma_{out\ cfd}$ for various ϕ

ϕ	$\phi_4 = [\phi/\phi_{ref}]^{0.1}$	$\sigma_{out\ cfd}$
30	0.8959	2.9477
45	0.9330	3.0755
55	0.9519	2.9698
75	0.9819	3.4278
90	1.0000	3.7043

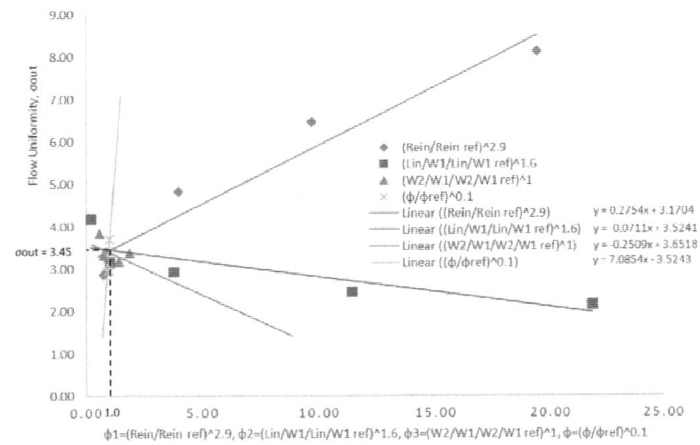

Figure 7. Flow Uniformity Index, σ_{out} with respect to $\phi_1 = [Re_{in}/Re_{in\ ref}]^{2.9}$, $\phi_2 = [L_{in}/W_1/L_{in}/W_{1\ ref}]^{1.6}$, $\phi_3 = [W_2/W_1/W_2/W_{1\ ref}]^1$ and $\phi_4 = [\phi/\phi_{ref}]^{0.1}$

$\phi_{1\,ref} = \left[\frac{Re_{in\,ref}}{Re_{in\,ref}}\right]^a = 1.0$, thus $\phi_{2\,ref} = \phi_{3\,ref} = \phi_{4\,ref} = 1.0$

$\frac{\partial\sigma_{out}}{\partial\phi_1} = 0.2754, \frac{\partial\sigma_{out}}{\partial\phi_2} = -0.0711, \frac{\partial\sigma_{out}}{\partial\phi_3} = -0.2509, \frac{\partial\sigma_{out}}{\partial\phi_4} = 7.0854$ are the gradients of the corresponding lines.

Substituting all the constants in Equation 12 yielding,

$$\sigma_{out\,acfd} = 3.45 + \left(\left[\frac{Re_{in}}{Re_{in\,ref}}\right]^{2.9} - 1\right)0.2754 - \left(\left[\frac{L_{in}/W_1}{L_{in}/W_{1\,ref}}\right]^{1.6} - 1\right)0.0711 - \left(\left[\frac{W_2/W_1}{W_2/W_{1\,ref}}\right]^{1} - 1\right)0.2509 + \left(\left[\frac{\phi}{\phi_{ref}}\right]^{0.1} - 1\right)7.0854 \quad (13)$$

The efficacy of ACFD correlation to represent the full CFD solution is verified using a parity plot as shown in Figure 8. Satisfied agreement between the ACFD and CFD solutions is achieved within ± 8.3%. This basically applies for the range of $5.786 \times 10^4 \leq Re_{in} \leq 1.775 \times 10^5$, $1.50 \leq L_{in}/W_1 \leq 30$, $1.20 \leq W_2/W_1 \leq 4.0$, $30° \leq \phi \leq 90°$.

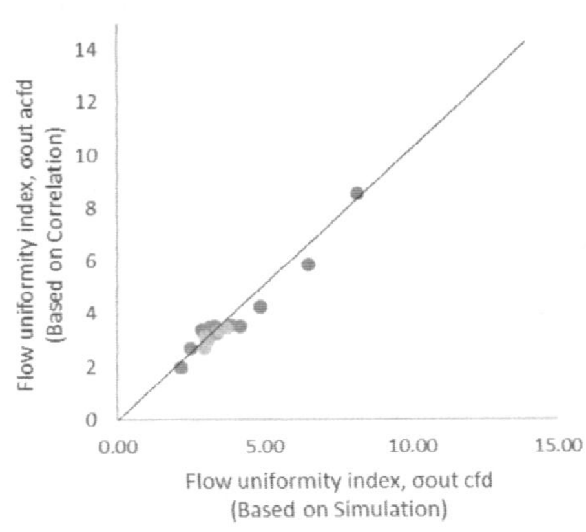

Figure 8. Parity plot showing agreement of CFD simulation ($\sigma_{out\ cfd}$) with ACFD correlation ($\sigma_{out\ acfd}$)

IOP Conf. Series: Materials Science and Engineering **243** (2017) 012013 doi:10.1088/1757-899X/243/1/012013

4. Conclusion

In conclusion, the research to investigate the effect of turning angle and develop performance correlations of 2-D turning diffuser via ACFD has been successfully carried out. The C_p and σ_{out} are distorted of respectively 37% and 28% with the increment of ϕ from 30° to 90°. The flow separation starts to emerge within the inner wall, S=0.91L_{in}/W_1 when 45° turning diffuser is applied and its scale is enlarged by further increasing the ϕ. The performance correlations by integrating turning angle are successfully developed to meet the full CFD solution within acceptable deviation of ±7.1%. These developed correlations are therefore reliable to be used by one in future to evaluate the performance of 2-D turning diffuser.

Acknowledgements

Immeasurable appreciation is extended to Flow Analysis Simulation and Turbulence Research Group (FAST) for all provided research supports and insight sharing. Also thanks to Mr. Rosman Tukiman (Assistant Engineer of CFD Laboratory, UTHM) for the technical-lab assist.

Nomenclature

2-D	Two dimensional
ϕ	Turning angle (°)
ϕ_{ref}	Reference turning angle (=90°)
$\sigma_{out\ ACFD}$	Flow uniformity index
$C_{p\ \mathrm{ACFD}}$	Outlet pressure recovery coefficient
L_{in}	Inner wall length (m)
L_{in}/W_{1ref}	Reference inner wall length to inlet throat width ratio (=4.37)
Re_{in}	Inflow Reynold number
$\mathrm{Re}_{in\ ref}$	Reference inflow Reynold number (=6.382 x 10^4)
W_1	Inlet width (m)
W_2	Outlet width (m)
W_2/W_{1ref}	Reference outlet inlet configuration (=2.16)

References

[1] Nordin N 2016 Performance investigation of turning diffusers at various geometrical and operating parameters *PhD Thesis Department of Mechanical Engineering* (Bandar Seri Iskandar: Universiti Teknologi PETRONAS)

[2] El-Askary W A and Nasr M 2009 Performance of a bend diffuser system: Experimental and numerical studies *Computer & Fluids* **38** 160-170

[3] Gan G and Riffat S B 1996 Measurement and computational fluid dynamics prediction of diffuser pressure-loss coefficient *Applied Energy* **54**(2) 181-195

[4] Lindgren B and V Johansson A V, 2002 Design and Evaluation of a Low-Speed Wind-Tunnel with Expanding Corners (Stockholm: Universitetsservice US AB)

[5] Nguyen C K, Ngo T D, Mendis P A and Cheung J C 2006 A flow analysis for a turning rapid diffuser using CFD *Comput. Wind Eng*

[6] Chong T P, Joseph P F and Davies P O A L 2008 A Parametric Study of Passive Flow Control for a Short, High Area Ratio 90 deg Curved Diffuser *J. Fluids Eng.* **130**(11)

[7] Wang Y C, Hsu J C and Lee Y C 2009 Loss characteristics and flow rectification property of diffuser valves for micropump applications *Int. J. of Heat and Mass Transfer* **52** 328-336

[8] Ibrahim I H, Ng E Y K, Wong K and Gunasekaran R 2008 Effects of centerline curvature and cross-sectional shape transitioning in the subsonic diffuser of the F-5 fighter jet *J. Mech. Sci. Technol.* **22**(10) 1993–1997

[9] Bourgeois J A, Martinuzzi R J, Savory E, Zhang C and Roberts D A 2011 Assessmentof Turbulence Model Predictions for an Aero-Engine Centrifugal Compressor *J. Turbomach.* **133**(1) 11025

[10] Nordin N, Karim Z A A, Othman S and Raghavan V R 2013 Design & development of low subsonic wind tunnel for turning diffuser application *Advanced Materials Research* **614-615** 586-591

[11] Nordin N, Karim Z A A, Othman S and Raghavan V R 2014 The performance of turning diffusers at various inlet conditions *Applied Mechanics and Materials* **465-466** 597-602

[12] Nordin N, Raghavan V R, Othman S and Karim Z A A 2012 Compatibility of 3-D turning diffusers by means of varying area ratios and outlet-inlet configurations *ARPN Journal of Engineering and Applied Sciences* **7**(6) 708-713

[13] Sullerey R K, Chandra B and Muralidhar V 1983 Performance comparison of straight andcurved diffusers *J. of Def. Sci.* **33** 195-203

[14] Eugene L Z P, Nordin N, Othman S and Raghavan V R 2011 A CFD preliminary study:Pressure losses and flow structure in turning diffuser by means of installing turning baffles *2nd International Conference of Mechanical Engineering in Putrajaya, Malaysia*

[15] Nguyen C K, Ngo T D, Mendis P A and Cheung J C K 2006 A flow analysis for a turning rapid diffuser using CFD *J. Wind Eng.* **108** 749-752

[16] Balaji C and Herwig H 2003 The use of ACFD approach in problems involving surface radiation and free convection *Int. Comm. Heat and Mass Transfer* **30** 251-259

[17] Balaji C, Holling M and Herwig H 2007 A general methodology for treating mixed convection problems using asymptotic computational fluid dynamics (ACFD) *Int. Comm. In Heat and Mass Transfer* **34** 682-691

[18] Balaji C, Holling M and Herwig H Determination of temperature wall functions for high Rayleigh Number flows using asymptotics: A systematic approach *Int. Journal of Heat and Mass Transfer* **50** 3820-3831

[19] Premachandaran B and Balaji C 2006 Conjugate mixed convection with surface radiation from a horizontal channel with protruding heat sources *Int. Journal of Heat and Mass Transfer* **49** 3568-3582

Computational fluid dynamic (CFD) analysis on ALUDRA SR-10 UAV with parachute recovery system

R Saim[1], S Mohd[1], S S Shamsudin[1], M F Zulkifli[2], Z Omar[1], Z Subari@Rahmat[2], M F Mohd Masrom, Y Zaki
[1] Aircraft System and Design Research Group (ASDR), Universiti Tun Hussein Onn Malaysia
[2] Aerodynamics And Propulsion Research (APR), Universiti Tun Hussein Onn Malaysia

E-mail: sofian@uthm.edu.my

Abstract. In an operation, belly landing is mostly applied as recovery method especially on research Unmanned Aerial Vehicle (UAV) such as Aludra SR-10. This type of landing method may encounter tough landing on hard soil and gravel which create high impact load on the aircraft. The impact may cause structural or system damage which costly to be repaired. Nowadays, Parachute Recovery System (PRS) recently used in numerous different tasks such as landing purpose to replace belly landing technique. Parachute use in this system to slow down flying or falling UAV to a safe landing by opening the canopy to increase aerodynamic drag. This paper was described the Computational Fluid Dynamic (CFD) analysis on ALUDRA SR-10 model with two different conditions i.e. the UAV equipped with and without parachute in order to identify the changes of aerodynamic characteristics. This simulation studies using solid models of aircraft and hemisphere parachute and was carried out by using ANSYS 16.0 Fluent under steady and turbulent flow and was modelled using the k-epsilon (k-ε) turbulence model. This simulation was limited to determine the drag force and drag coefficient. The obtained result showed that implementation of parachute increase 0.25 drag coefficient of the aircraft that is from 0.93 to 1.18. Subsequent to the reduction of descent rate caused by the parachute, the drag force of the aircraft increase by 0.76N. These increasing of drag force of the aircraft will produce lower terminal velocity which is expected to reduce the impact force on the aircraft during landing.

1. Introduction

Unmanned Aerial Vehicle (UAV) is a type of aircraft with no pilot on board and can be remotely controlled by a pilot at a ground control station or fly autonomously based on pre-programmed flight plans. UAV was use in various civil and military type of mission including reconnaissance surveillance, target tracking and high structure inspection [1]. Furthermore, UAV has also been widely used as experimental platform in various research groups in university to test and validate guidance, navigation, dynamic modelling and control approaches [2-5]. There is no standard to classified unmanned aircraft. However, UAV can be distinguished from one another by size, weight, endurance, range, altitude, mission and design approach. Classification of UAV according to the size can be divided into four subclasses that is micro air UAVs, small UAVs, medium UAVs and large UAVs. There are various type for landing method used in fixed wing UAV such as wheel landing, skid or belly landing, vertical-net system, cable assisted recovery, parachute recovery, parachute, vertical takeoff and landing (VTOL) and mid-air retrieval [6-8]. However, there are pros and cons for each method. ALUDRA SR-10 is an example of small research UAV used in civil and commercial sector for survey and mapping purpose. It is manufactured by Unmanned Systems Technology (UST) subsidiary of Composites Technology Research Malaysia (CTRM). SR-10 is launched by hand and designed to land intact by belly landing method. Belly landing is a recovery techniques where unmanned aircraft's fuselage make a direct contact to the ground without landing gear. Even though this type of landing method does not require an additional equipment, the hard landing by this landing could affect the aircraft structure and system as well as require a long runway for landing purpose. This type of landing method is not suitable for ALUDRA SR-10 which commonly used in constricted operations. Compared to other landing method, parachute recovery system is the most suitable type of landing system for ALUDRA SR-10 which can be used especially at constricted area and during emergency flight termination system.

The research on dynamics study of UAV and parachute is an interesting problems but it is challenging, complex and difficult due to limited study for both theoretical and experimental viewpoints [9]. Flying Wing the Golden Eagle, Cargo UAV and SBXC Glider is an example of small size UAV (below 10kg) which already developed and analyse using computational simulation. The purpose of this computational simulation is to evaluate the model's aerodynamic performance and also to enhance stability and performance of the aircraft [10-12].

The flow structures across the inflated parachute canopies help a better understanding of aerodynamics characteristics. Some of researcher prove that developing 3-D computational simulation is not only can determine the flow structure, also it can provide aerodynamic performance such as lift, drag and stability in order to meet performance requirement [13]. However, the current study of UAV and parachute is not consolidated instead these two study was examined separately. Thus, this simulation was conducted to determine an aerodynamic characteristics analysis on ALUDRA SR-10 with parachute system which carried out by using ANSYS 16.0 FLUENT. It is to determine the changes of aerodynamic characteristics on the aircraft as the parachute is installed to the aircraft for a safe landing or during flight failure.

2. Methodology

The geometric design of the main ALUDRA SR-10 can be divided into four main parts that is fuselage, rudder, horizontal stabilizer and wing. The general specification of this unmanned aircraft are shown in Table 1. In this study, two different condition of ALUDRA SR-10 was studied that is with and without parachute. In the first condition, aircraft was predicted to be free falling without additional equipment like parachute. On the other hand, an extra equipment that is parachute use for landing is taken as second condition study.

Table 1. ALUDRA SR-10 General Characteristics

Max Takeoff Weight (MTOW)	5kg
Structure Material	Composite reinforced
Length	1.237 m
Wingspan	2.2m
Endurance	1 hour
LOS Range	10 km
Max Cruise Speed	30 km/h
Launch	Hand Launch
Landing	Belly landing

As PRS was selected as a new landing method, preliminary approaches and calculations led to the design of hemisphere parachute with uncontrolled deployment method. In this project, 6m2 flat area of Opale hemisphere parachute with 3m/s descent rate was selected. This size of parachute is design to accommodate the drone with a total flight payload of 4 to 9kg and 1.6 lift coefficient [14]. Fully inflated parachute was designed based on Knacke's PRS Design Manual [15]. The 2.76m of nominal diameter, Do and 2.13m of inflated diameter shape, Dp of this parachute were determined by using the following equation

IOP Conf. Series: Materials Science and Engineering **243** (2017) 012014 doi:10.1088/1757-899X/243/1/012014

$$Nominal\ Diameter = \sqrt{\frac{4S}{\pi}} \quad (1)$$

Inflated diameter shape of hemisphere parachute, Dp can be calculated by;

$$\frac{D_p}{D_o} = 0.66 \quad (2)$$

Both model were designed using Solidwork v15 software that is the aircraft without parachute as shown in Figure 1(a) and fully inflated parachute setup mounted on the aircraft as in Figure 1(b) schematically. The design then were exported to ANSYS's Design Modeller. In this simulation, parachute was modelled as light weight nylon fabrics.

The stationary region of inlet and outlet are located far enough from the aircraft model to prevent recirculation of flow from affecting the results in this analysis. The inlet and outlet boundaries are located 3H from downstream and 10H from upstream of the origin of the aircraft (H is refer to the aircraft's length). For the external flow domain, the enclosure is set to be 2W where W is refer to the wingspan of the aircraft [16]. Proper sizing of the domain distance of upstream and downstream from origin is very important to prevent recirculation of the flow that will cause convergence problems.

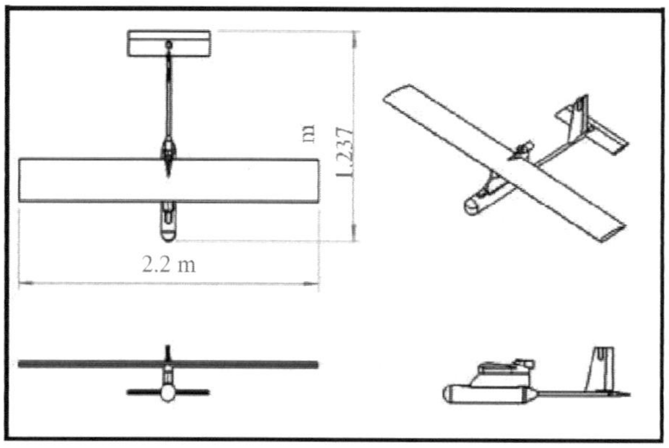

a. ALUDRA SR-10 UAV view without parachute

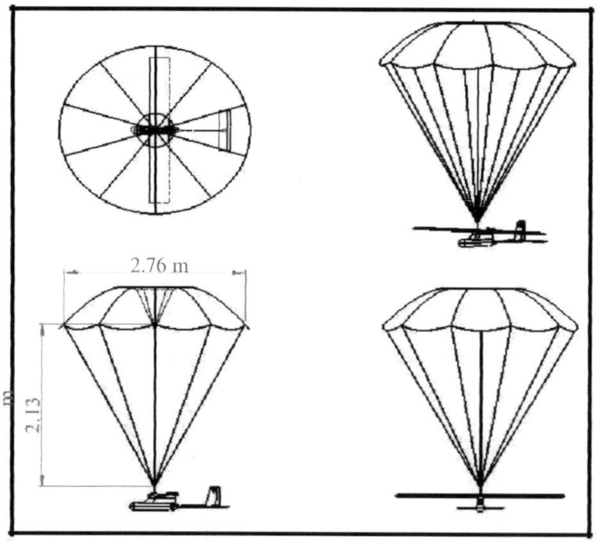

b. ALUDRA SR-10 UAV view without parachute

Figure 1. Geometry model of ALUDRA SR -10 UAV

2.1. Simulation Setup

The mesh was generated using Tetrahedrons element shape. A symmetry plane has been used in this study, in order to reduce the complexity of the solution. To ensure more accurate simulation of the flowfield, it is important to properly size the meshing especially near the wall of the geometry. Two prism of smooth transition inflation layer has been used for both geometry with 0.0471mm wall spacing for first condition and 0.1256mm wall spacing for second condition. The wall spacing obtained based on the desired dimensionless wall distance(y plus). Figure 2 illustrates the complete smooth transition meshing of ALUDRA SR-10 in two different condition i.e. with or without parachute.

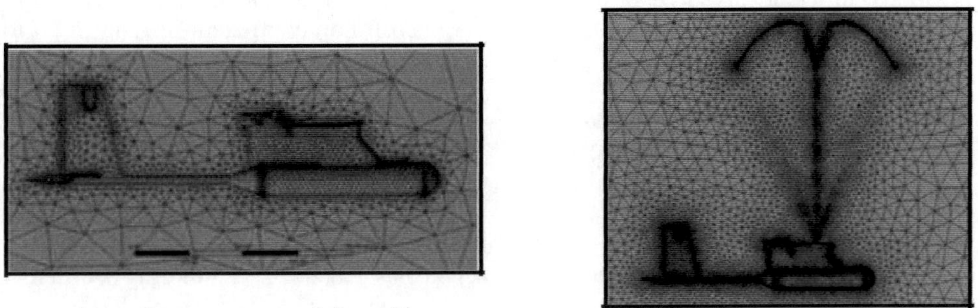

a. Aircraft geometry model meshing without parachute

b. Aircraft geometry model meshing with parachute

Figure 2. Geometry Mesh Size Control

2.2. Computational Fluid Dynamic (CFD) Simulation

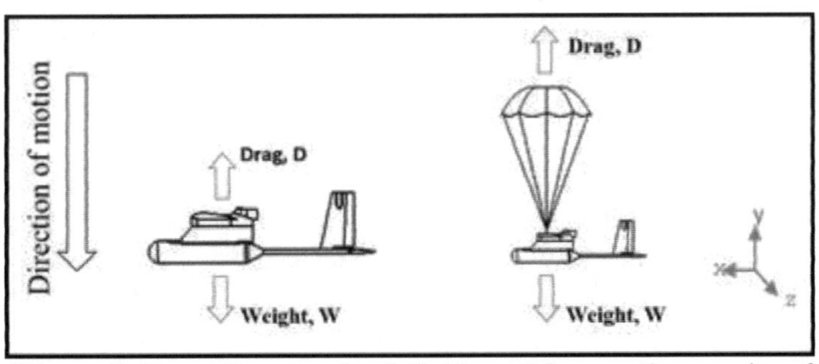

Figure 3. Force direction for definition of force acting on aircraft during descent

In this study, ALUDRA SR-10 aircraft was accelerate downward as the engine was turned off. Figure 3 illustrates the force acting on the aircraft as the aircraft descent toward ground. Parachute is considered as solid and the direction of air flow motion is directed from bottom to top in the positive y direction. In this simulation, CFD techniques were used to model a fixed shape of aircraft and parachute in three dimension view. The simulation were conducted under the free-flow conditions (at the atmospheric pressure and temperature) moved with the speed of 3.0 m/s which close to the experimental terminal descent rate [17]. The flow was considered as turbulent flow and was modelled using the standard two equation k-epsilon (k-ε) turbulent flow condition which transport two equation of turbulent kinetic energy and the turbulent dissipation rate.

3. Result and Discussion

Fluent analysis of an aerodynamic performance for full-scale Aludra SR-10 UAV was validated with the of computational methodologies study of several small size UAVs (below 10kg MTOW) in the positive x-direction of air flow. Lift and drag coefficient was measured based on aircraft at steady state level. Data of C_l /C_d ratio for studied aircrafts show the range of C_l /C_d ratio for most small UAVs is between seven to 16 [10-12]. Based on the computational aerodynamic performance simulation on ALUDRA SR-10, the value of C_l /C_d ratio for this aircraft is 13.67 with 0.41 lift coefficient and 0.03 of drag coefficient. This preliminary simulation allow for the further simulation as the aircraft model fulfil the average UAVs standard (below 10kg MTOW).

The preliminary simulation on parachute also has been conducted. However, only drag coefficient was studied and compare with previous research in parachute simulation. In this simulation, the value of steady drag coefficient measured is 0.79. By comparing with computational result of previous researcher, this value is acceptable since the average value of drag coefficient in range 0.78 to 0.87 [13]. This validation allow the further simulation on aircraft and parachute in positive y-direction.

3.1. Aerodynamic Characteristics of ALUDRA SR-10

The result of the main parameter study that is aerodynamic characterises which include drag coefficient and drag force have been obtained in this simulation. The parameter result has been tabulated and compared in Table 2. The changes of drag coefficient from 0.93 to 1.18 as the parachute is installed to the aircraft. The improvement of drag coefficient value causes an increases of 0.76N drag force of the aircraft. Based on the result, the parachute has generated 0.25 of drag coefficient.

Table 2. Simulation Aerodynamic Characteristics Result

	ALUDRA SR-10 Aircraft	
	Without parachute	*With parachute*
Drag Coefficient, C_D	*0.93*	*1.18*
Drag Force, F_D (N)	*2.73*	*3.49*

The simulation drag coefficient can be validate with the calculated drag coefficient by following equation [6, 7, 18]:

$$Drag\ Force, F_D = \frac{1}{2}\rho V^2 S_o C_D \quad (3)$$

where SO is the nominal area of the aircraft which is 0.5352 m2, is the density of air based on the altitude of the aircraft (1.236 kg/m3 at 100m altitude), V is the velocity at the inlet and FD is a drag force. Comparison result for both simulation and calculated drag force are tabulated in Table 3. Different of both simulation and calculated result whether with or without parachute give the small percentage different that is 1.49% and 1.16%.

Table 3. Drag Coefficient Comparison

Drag Force, F_D (N)	**ALUDRA SR-10 Aircraft**	
	Without parachute	*With parachute*
Simulation Result (N)	2.73	3.49
Calculation Result (N)	2.69	3.45
Percentage Different (%)	1.49	1.16

IOP Conf. Series: Materials Science and Engineering **243** (2017) 012014 doi:10.1088/1757-899X/243/1/012014

3.2. Pressure Distribution Result

Distribution of pressure on the aircraft and parachute are important to study especially to analyze the characteristics of parachute inflation. Changes in pressure distribution over the aircraft's body and parachute can be compared between the Figure 4 and Figure 5 where slight pressure reduction can be seen. The difference formation of pressure distribution on parachute either at the outer or inner surface region can be seen as in Figure 5. High static pressure formed at the inner surface region compared to the static pressure of the free airstream around the parachute due to the formation of turbulent airflow inside the parachute's inflated region.

These difference between inner and outer surface region of the parachute form a strong outwardly pressure that keeps the canopy inflated by Bernoulli's theory from high pressure region to the lower pressure region. The result of this differential pressure, provide an aerodynamic drag for parachute inflation [19]. As parachute moving through air, a vortex form and the air from the below part of the parachute races up along the curved fabrics towards the low pressure over the top to creating Coandă and Bernoulli effect lift. However, cause of periodic vortex shedding, this situation cannot be last forever and parachute will periodically descent [20].

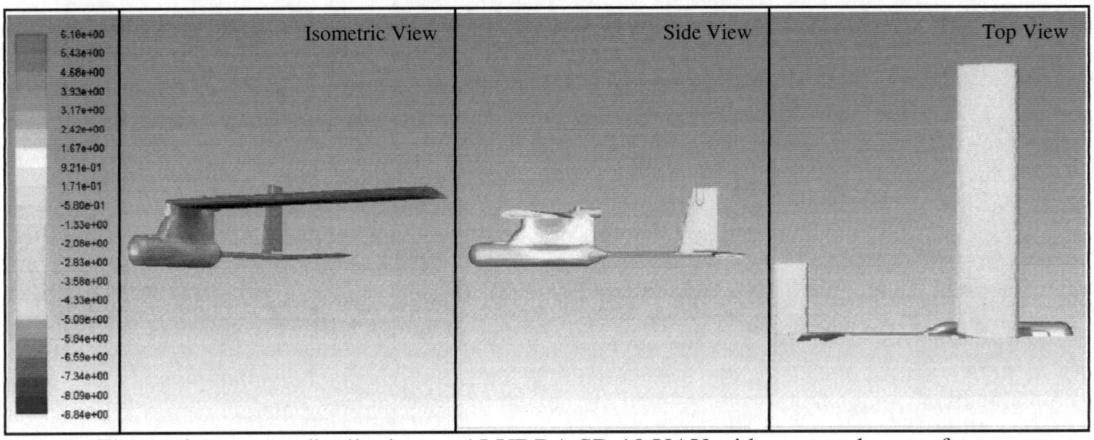

Figure 4. Pressure distribution on ALUDRA SR-10 UAV without parachute surface

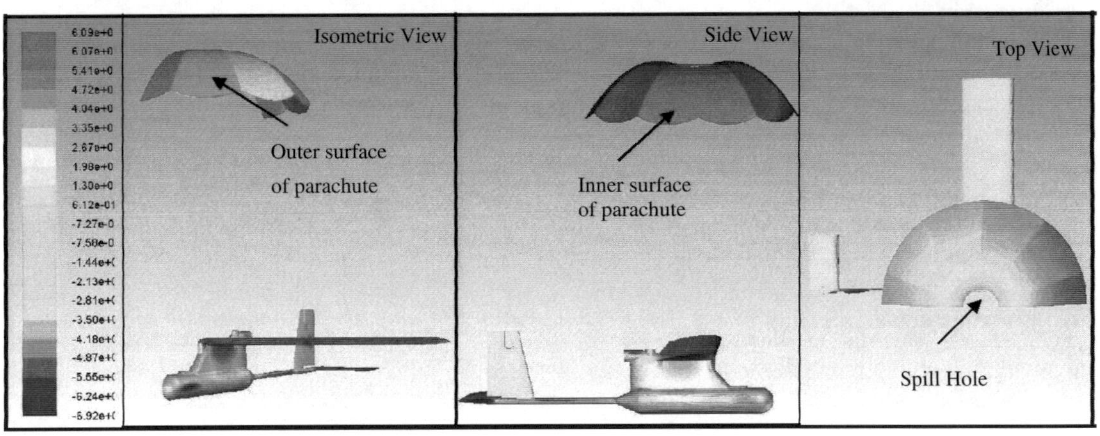

Figure 5. Pressure distribution on ALUDRA SR-10 UAV with

3.3. Velocity Distribution Result

Figure 6 provide a visualization of the airflow around the aircraft and parachute model in order to illustrate details of the wake structure. When air separate away from aircraft's body surface, its begin to swirl and form turbulent flow resulting in formation of a region of air around aircraft's top surface which has lower velocity than free stream velocity. This region called as Boundary layer region.

The airflow in front of the inflated parachute start to decelerate until zero as it approached the stagnation point. Again, a turbulent airflow occurs inside the inner region of parachute once it's passing through the stagnation point, I_o. These turbulent airflow are spilt into two pathway. Some of air directly passing through the spill hole and then moving downward along the parachute's surface to the edge of the parachute. Due to the compression of the streamline, the airflow around the edge of the parachute which cannot get through parachute will be accelerate. The airflow then will goes around and separate at the leading edge of parachute. At the leading edge of parachute, the airflow from inner and outer surface of parachute together forming a vortex flow called as Karman Vortex Trail[15].

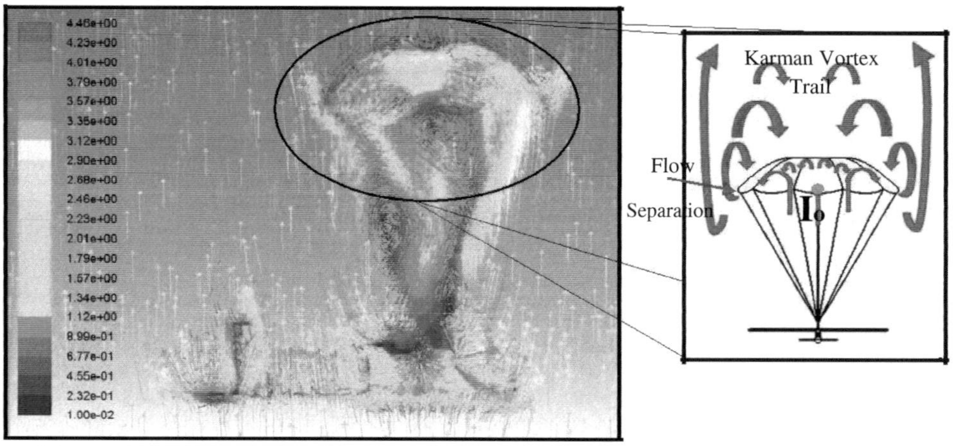

Figure 6. Airflow around ALUDRA SR-10 aircraft and inflated parachute

4. Conclusion

A computational simulation study has been performed to determine the aerodynamics characteristics of parachute for Aludra SR-10 UAV aircraft. Using periodic symmetry of the model configuration, parameter of aerodynamic characteristics including drag coefficient and drag force have been obtained and compared. Based on the result and discussion, parachute able to increase the vertical net force (drag force) up to 0.76N as the parachute decrease the descent rate of the aircraft. Also, implementation of PRS able to increase the drag coefficient approximately 0.25. These increasing of drag force tend to reduce the impact force on the aircraft due to low terminal velocity will be produce. This simulation will be validate with the flight test that will be conducted in the future research.

Acknowledgement

This project was supported by Exploratory Research Grant Scheme, ERGS under Grant Vot. No: E034 in order to conduct the research.

Reference

[1] Kendoul F 2012 Survey Of Advances In Guidance, Navigation, And Control Of Unmanned Rotorcraft System *Journal Of Field Robotics* **29** 315-378

[2] Cai G, Chem B M and Lee T H 2011 *Unmanned Rotorcraft System* (New York: Spriger Publishing Company)

[3] Beard R W and Mclain T W 2012 *Small Unmanned Aircraft; Theory And Practice* (New Jersey: Princeton University Press)

[4] Shamsudin S S 2013 *The Development Of Neural Network Based System Identification And Adaptive Flight Control For An Autonomous Helicopter System* (Christchurch: University Of Canterbury)

[5] Shamsudin S S, Chen X Q 2014 Recursive Gauss-Newton Based Training Algorithm For Neural Network Modelling Of An Unmanned Helicopter Dynamics *Ijista* **13** 56-80
[6] Austin R 2010 *Unmanned Aircraft Systems : Uavs Design, Development And Deployment*
[7] Fahlstrom and Paul G 2012 *Introduction To Uav Systems*
[8] Gundlach J 2013 *Designing Unmanned Aircraft Systems : A Comprehensive Approach*
[9] Dawoodian M and Dadvand A 2013 A Numerical And Experimental Study Of The Aerodynamics And Stability Of Horizontal Parachute *Irsn Aerospace Engineering*
[10] Aguilar A G 2016 *Analysis Of The Sbxc Glider Airframe* (California: Naval Postgraduate School Monterey)
[11] Kovanis A P, Skaperdas V and Ekaterinaris J A 2009 Design and Analysis Of A Light Cargo Uav *4th Ansa & Meta International Conference* vol 25 pp 228–237
[12] Navabalachandran J , Low J H *Reverse Engineering And Aerodynamic Analysis of A Flying Wing Uav* (Singapore: National University Of Singapore)
[13] Stein K, Benney R, Kalro V, Tezduyar T E, Leonard J and Accorsi M 2000 Parachute Fluid-Structure Interactions: 3-D Computation *Computer Methods In Applied Mechanics And Engineering* **190**(3–4) 373–386
[14] Opale Aero System Sarl from https://Www.Opale-Parachutes.Com/Gb/
[15] Knacke T W 1992 *Parachute Recovery Systems: Design Manual* (Santa Barbara: Para Publishing)
[16] Ansys 2015 Inc. Ansys Fluent User's Guide from Http://148.204.81.206/Ansys/150/Ansys%20fluent%20users%20guide.Pdf_
[17] John M 2005 *Properly Sizing Parachutes For Your Rockets* **1**(149) 1–7
[18] Sabu J 1997 *Computational Fluid Dynamics Modeling Of Parachute Clusters* (Adelphi:Army Research Laboratory)
[19] Cruz J R, Mineck R E, Keller D F and Bobskill M V 2003 Wind Tunnel Testing Of Various Diskgapband Parachutes *Aiaa 2003-2129, 17th Aiaa Aerodynamic Decelerator Systems Technology Conference And Seminar, 19–22 May 2003, Monterey, California*
[20] Xing G, Qing Z and Qian T 2013 Fluid-Structure Interaction Simulation Of Parachute In Low Speed Airdrop *Proceedings of the World Congress on Engineering* vol 3 pp 3-5

IOP Conf. Series: Materials Science and Engineering **243** (2017) 012015 doi:10.1088/1757-899X/243/1/012015

Purification of crude glycerol from transesterification reaction of palm oil using direct method and multistep method

N F Nasir[1], M F Mirus[1] and M Ismail [2]

[1]Centre for Energy and Industrial Environment Studies (CEIES), Universiti Tun Hussein Onn Malaysia, Batu Pahat, Johor, Malaysia
[2]Department of Chemical and Process Engineering, Faculty of Engineering and Built Environment, Universiti Kebangsaan Malaysia, Bangi, Selangor, Malaysia.

E-mail: fitriahnasir@gmail.com

Abstract. Crude glycerol which produced from transesterification reaction has limited usage if it does not undergo purification process. It also contains excess methanol, catalyst and soap. Conventionally, purification method of the crude glycerol involves high cost and complex processes. This study aimed to determine the effects of using different purification methods which are direct method (comprises of ion exchange and methanol removal steps) and multistep method (comprises of neutralization, filtration, ion exchange and methanol removal steps). Two crude glycerol samples were investigated; the self-produced sample through the transesterification process of palm oil and the sample obtained from biodiesel plant. Samples were analysed using Fourier Transform Infrared Spectroscopy, Gas Chromatography and High Performance Liquid Chromatography. The results of this study for both samples after purification have showed that the pure glycerol was successfully produced and fatty acid salts were eliminated. Also, the results indicated the absence of methanol in both samples after purification process. In short, the combination of 4 purification steps has contributed to a higher quality of glycerol. Multistep purification method gave a better result compared to the direct method as neutralization and filtration steps helped in removing most excess salt, fatty acid and catalyst.

1. Introduction

Glycerol is a chemical from the sugar alcohol group. It is a clear liquid oil and stable in normal condition. It consists of three hydroxyl hydrophilic groups for its solubility in water and its hygroscopic nature. Transesterification is a reaction of fat and oil with methanol with a catalyst which produced biodiesel and glycerol [1]. The stoichiometry of the reaction requires three moles of methanol and a mole of triglycerides to produce three moles of biodiesel and a mole of glycerol. Once the reaction ends, glycerol is separated by either precipitation or centrifugation and then pass through several neutralization processes. According to the study conducted by Haas et al. [2] in 2006, the glycerol price is US $ 0.33 per kg. The purified glycerol can be used in pharmaceutical, cosmetics and food industry. It is also used as food additives and labelled as E number which is E422 [3]. Recent developments have shown that glycerol is used in animal foods, carbon feedstock in fermentation, polymers, surfactant and lubricants [4]. Figure 1 shows the market for glycerol and it indicates that glycerol has played a crucial role in bio refineries [5].

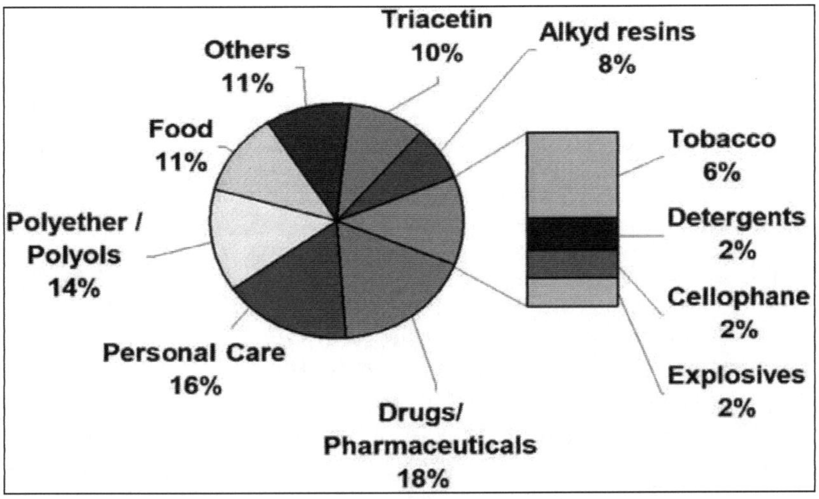

Figure 1. The market for glycerol in year 2002 [5].

Crude glycerol produced from the transesterification reaction has limited usage if it is not undergoing any purification process. It contains excess methanol, catalyst and soaps. Therefore, purification of glycerol is needed to remove all impurities and the glycerol purification process has been patterned in US Patent 4990695 [6]. However, conventional method for glycerol purification requires high cost and complex processes. An established method for glycerol purification by vacuum distillation technique has few disadvantages such as the high energy requirement and high maintenance. If purification use activated carbon, the technique is inefficient for elimination of other impurities. Another technique such as using chemical treatment has disadvantages of resulting in low glycerol yield when the repeated acidification process is carried out [7].

Busby and Grosvenor [8] carried out investigations on the purification of glycerol using the ion exchange technique. The result of their experiment showed that a high quality of glycerol (95-99%) can be produced using this technique. Carmona et al. [9] also used the ion exchange technique in the purification of glycerol. In their study, they determined the equilibrium and kinetic data for the purification reaction using ion exchange Amberlite-252. Another study using an ion exchanged method was conducted by Isahak et al. [10, 11] and Javani et al. [12].

In this study, the transesterification of palm oil with methanol have utilized sodium hydroxide as catalyst. The main product was biodiesel and side product was glycerol. The glycerol was considered as crude glycerol as it contains excess methanol, catalyst and soaps, which needs recovery and purification processes to obtain a better quality of the glycerol. Therefore, this study is vital in determining an easy method, hence practical, for purifying crude glycerol for further uses. A pure glycerol can be used in many applications such as moisturizer, plasticizer, emollient, lubricants, antifreeze, drugs, and foods. The objectives of this study are to produce biodiesel and crude glycerol from transesterification of refined, bleach deodorized palm oil and to purify the crude glycerol by two separation techniques; direct and multistep. Then, comparisons of these two techniques are performed for two types of samples; the sample self-produced through the transesterification process of palm oil (CGA) and the sample obtained from a biodiesel plant, Sime Darby Biodiesel Sendirian Berhad (CGB).

2. Materials and Methods

2.1 Transesterification reaction

RBD palm oil (Seri Murni Brand), methanol and sodium hydroxide were purchased from local groceries, Sigma Aldrich, and R&M Chemical, respectively. Amberlite IRN-78 (SUPELCO brand) and Amberlite 200C ion exchange resins were purchased from Sigma Aldrich. Palm oil and methanol are the main

IOP Conf. Series: Materials Science and Engineering **243** (2017) 012015 doi:10.1088/1757-899X/243/1/012015

reactant for transesterification reaction. Palm oil contains 1.128% free fatty acids. Both reactant were mixed at a molar ratio of oil to alcohol 15:1 with the homogeneous base catalyst (sodium hydroxide). The reaction time was set at 2 to 3 hours in a three-neck flask equipped with condenser, thermometer and magnetic stirrer. The mixing speed was set to 500-600 rpm. At the end of the process, the mixture was settled about half an hour. The settling process was required to separate biodiesel from glycerol.

2.2 Purification of glycerol with direct method

After separation of biodiesel and glycerol, the ion exchange process was carried out. The ion exchange process was carried out in a chromatography column. The column was loaded with ion exchange resins (cation and anion) Amberlite IRN-78 and Amberlite 200C type. The purpose of this process being conducted was to neutralize and remove free ions in the glycerol samples. Other than that, silica beads were also loaded on the chromatography column to remove excess moisture which may be existed in the glycerol samples. After the ion exchange process was performed, the methanol removal process was done using a rotary evaporator.

2.3 Purification of glycerol with multistep method

Multistep method was conducted by performing few procedures started with neutralization of crude glycerol, followed by filtration process, ion exchange process and methanol removal process. Neutralization of crude glycerol was conducted using phosphoric acid. The neutralization process has produced salt and therefore the salt need to be filtered using filter paper. After filtration, the samples were put in a chromatography column for ion exchange process and then put into the rotary evaporator for the methanol removal process. The methods in ion exchange and methanol removal process for multistep was identical to the direct method.

2.4 Analysis of purified glycerol samples

The samples were analysed by several types of analysis such as free fatty acids determination, pH value, Fourier Transform Infra-Red (FTIR), High Performance Liquid Chromatography (HPLC) and Gas Chromatography (GC) analysis. Table 1 provides the wavelengths of functional groups [13].

Table 1. Wavelength of functional groups [13].

Wavelength (cm^{-1})	Group	Classes
3200-3640	O-H	Water
3000-2800	C-H	Alkanes
1830-1800		
1780-1640	C=O	Esters
1650-1580	C=C	Alkenes
1475-1350	C-H	Alkanes
950-1300	C-O	Primer Secondary, Tertiary Alcohol
950-1300	O-H	
900-650	O-H	Aromatic compounds

Table 2 and Table 3 show the operational conditions for HPLC and GC, respectively.

IOP Conf. Series: Materials Science and Engineering **243** (2017) 012015 doi:10.1088/1757-899X/243/1/012015

Table 2. Operation condition of HPLC.

Operation conditions of HPLC	Value	
Mobile phase	Acetone/ Acetonitrile	
Permanent phase	C18 column	
Wavelength	254	
Mobile phase velocity	1 mL/min	
Gradient	Time (minutes)	Gradient Ratio (Acetone to Acetonitrile)
	0-1	60:40
	2-15	90:10
	16-26	95:5
	27-30	60:40

Table 3. Operation conditions of GC.

Operation conditions of GC	Value
Column type	DB-WAX
Length of capillary column	30 m
ID	0.25 mm
Film thickness of capillary column	0.25 µmg
GC Injection temperature	200 °C
Column detector temperature	300 °C

3. Results and Discussions

In this section, the results of pH values, free fatty acid content, FTIR, GC and HPLC analysis are presented and discussed.

3.1 pH values

The pH value analysis was conducted at room temperature and the results obtained can be compared in Table 4.

Table 4. Results of pH value analysis of samples before and after purification process.

Samples	pH value
Pure Glycerol	6.70
CGA-Unpurified Glycerol	9.20
CGA-Purified Glycerol with Direct Method	8.50
CGA-Purified Glycerol with Multistep Method	7.50
CGB-Unpurified Glycerol	6.0
CGB- Purified Glycerol with Direct Method	6.0

It is apparent from Table 4 that multistep method was able to purify the pH value of crude glycerol almost the value of the pure glycerol for CGA. It is noted that the purity of the samples obtained from biodiesel plant was 78%. Therefore, in this study, the purification of crude glycerol for CGB sample was conducted using direct method only. The pH value was found slightly acidic (pH 6) which was lower than pure glycerol (pH 6.7). After going through the purification process by the direct method, pH value for sample CGB was unchanged. This might be due to the addition of citric acid during the transesterification process in biodiesel plant. The addition of citric acid during the transesterification was to avoid soap formation, thus producing a high quality of biodiesel.

3.2 Free fatty acids content analysis

In order to access the free fatty acid (FFA) content in the samples, PORIM method was used [14]. Table 5 presents the FFA content of each sample.

Table 5. Results of FFA content analysis.

Samples	Free fatty acids content (%)
Pure Glycerol	0.052
CGA-Unpurified Glycerol	0.128
CGA-Purified Glycerol with Direct Method	0.077
CGA-Purified Glycerol with Multistep Method	0.055
CGB-Unpurified Glycerol	0.051
CGB- Purified Glycerol with Direct Method	0.025

As shown in Table 5, the results of acid value test indicate that the amount of acids was reduced after purifying of crude glycerol by both methods. The acid value in unpurified CGA samples was 0.128%, but after the purification process with direct and multistep methods, the acid values were reduced to 0.077% and 0.05%, respectively. For unpurified CGB samples, the acid value was found to be 0.051% and then it decreases to 0.025% after purification with direct method. The slight difference in the acid values, while preliminary, suggests that the change of the acid value happened when removing the mineral in ion exchange step. It also suggests that there was acid absorption process happened when glycerol passed through the column.

3.3 FTIR analysis

In order to determine the functional group in the glycerol samples, FTIR tests were used. Figure 2 shows the FTIR results for methanol and pure glycerol. The Figure is used as reference and comparison.

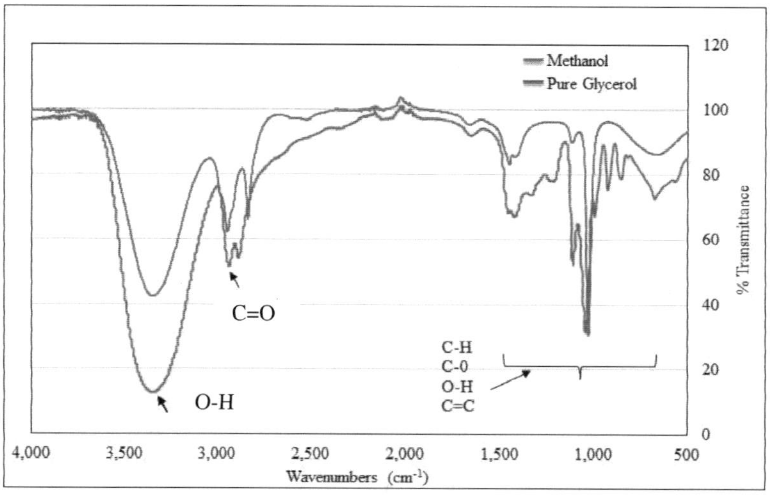

Figure 2. FTIR of methanol and pure glycerol.

The results of FTIR analysis for CGA samples using both methods is presented in Figure 3.

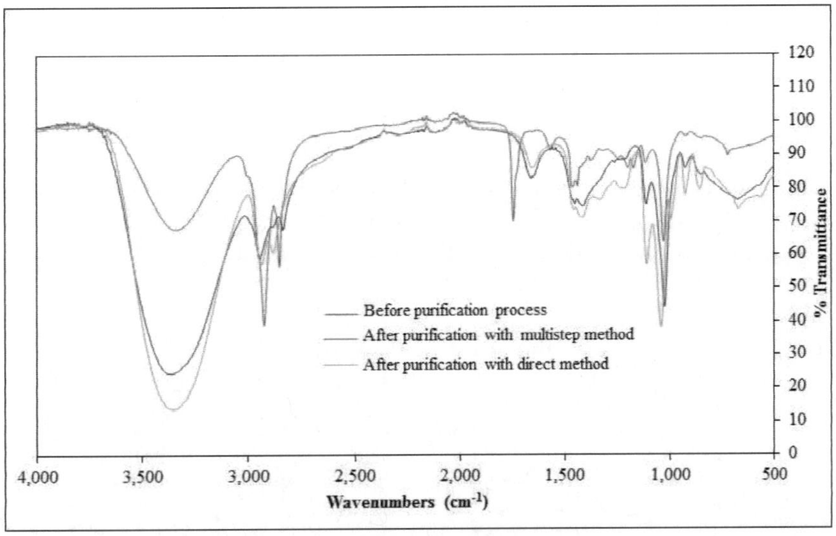

Figure 3. FTIR results of crude and purified glycerol for CGA samples.

For discussion of the results of FTIR analysis, Figs. 2, 3 and Table 1 are referred. From Figure 3, it is found that an ester peak (C=O) was detected in unpurified CGA samples at the wavelength of 1741 cm^{-1}. However, none of ester peak was detected in the purified CGA samples using both methods. This result confirms that fatty acid salt was existed in the crude glycerol samples, but then it successfully eliminated from the purification processes by both methods. All peaks for determining the existence of glycerol at each sample were clearly detected which confirms that glycerol was successfully separated from the biodiesel. The peaks are O-H at wavelength of 3200-3600 cm^{-1}, C-H, C-O and O-H at wavelength of 1500-650 cm^{-1}. Similar results were obtained for CGB samples as shown in Figure 4.

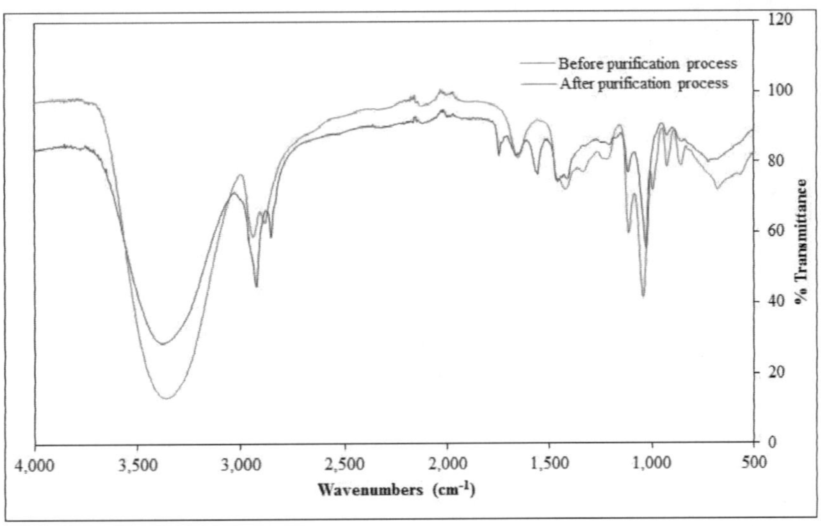

Figure 4. FTIR results of crude and purified glycerol for CGB samples.

Ester peak (C=O) was found at wavelength 1738 cm-1 for unpurified CGB samples. However, there was no peak detected for the ester group in CGB samples that purified using the direct method. This finding may be due to fatty acid salt was certainly presented in the unpurified samples and then removed by purification process. Besides, the main peaks for determining the glycerol samples are found such as O-H peaks at wavelength of 3200-3600 cm^{-1}, C-H, C-O and O-H at wavelength 1500-650 cm^{-1}.

3.4 GC analysis

Sample analysis using gas chromatography was conducted to determine the methanol existence as well as the purity of the glycerol. In this study, Figs. 5 and 6 show the results of GC analysis for both methods of purification. The standard retention time for glycerol and methanol are at minutes 32.75 and 2.57, respectively.

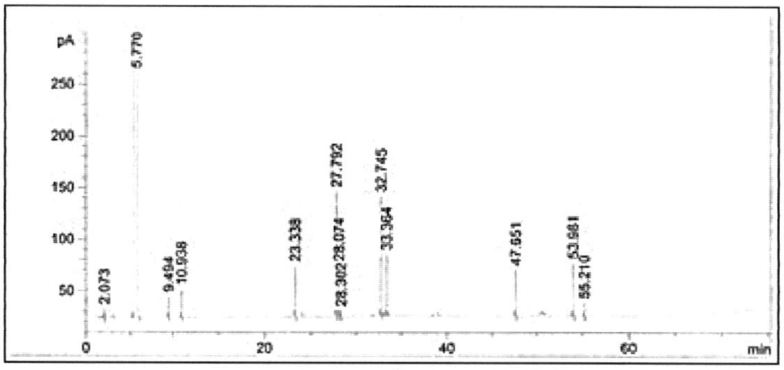

(a)

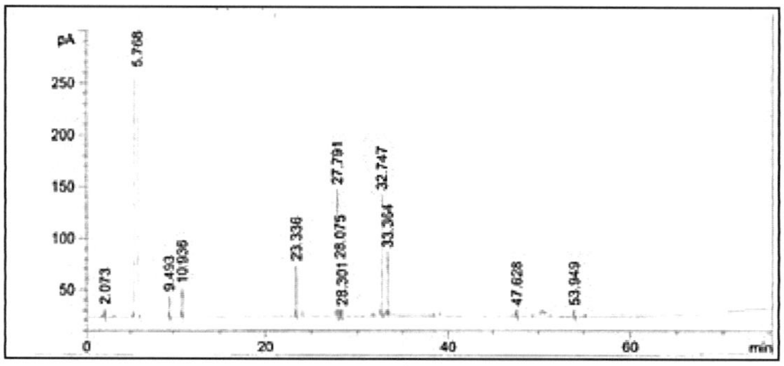

(b)

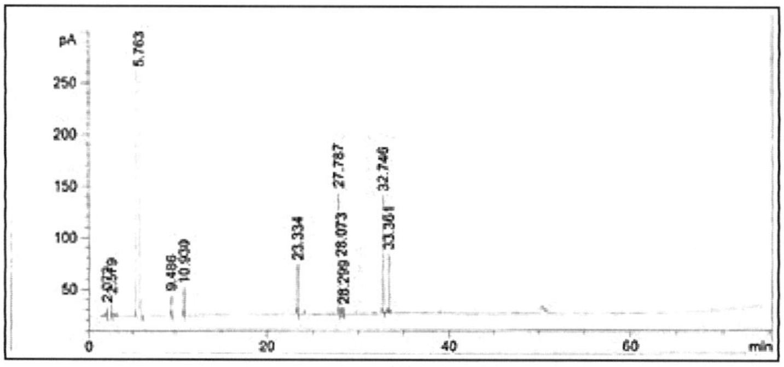

(c)

Figure 5. Result of GC analysis for CGA samples (a) before purification (b) after purification with direct method and (c) after purification with multistep method.

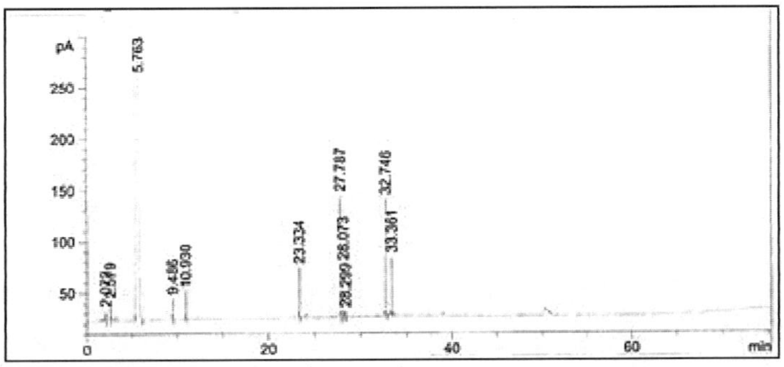

(a)

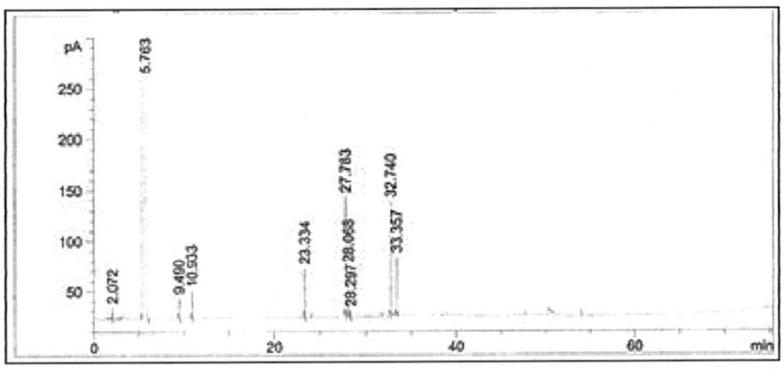

(b)

Figure 6. Result of GC analysis for CGB samples (a) before purification with direct method and (b) after purification with direct method.

When referred to Figs. 5 and 6, it is found that methanol peaks at minute 2.57 were not found in CGA and CGB samples after purification using both method. This result indicates that methanol was absent in the samples thus it confirms that the methanol has been removed in the purification processes by both direct and multistep methods.

3.5 HPLC analysis

Other than FTIR and GC analysis, HPLC analysis was also conducted in order to quantify the CGA and CGB samples after purified by both methods. Figs. 7 and 8 compare the results of HPLC analysis of CGA and CGB samples.

As Figure 7 shows, there is a significant retention time of glycerol at minute 2.79 with the 100% relative area. The results of CGA sample before purification (Figure 7 (b)) found that the retention time of glycerol at 2.80 min with relative area of 9.06 %. Glycerol peak was less tangible as the free fatty acids peak (at retention time of 5.79) has contributed up to 72.82% of relative area. When purified using the direct method, the relative area was reduced to 13.63%. There was significant difference when the relative area of glycerol increase to 100% for CGA samples when purified using the multistep method. Figure 8 shows the comparison of CGB samples when purified with direct and multistep method.

According to Figure 8, the results of HPLC analysis have shown that glycerol peaks were identified at minute 2.84 with relative areas of 96.50% for CGB sample before purification and it increases to 99.58% for CGB sample after purification with direct method. This result indicates that a higher purity of glycerol can be obtained by using the direct method. In summary, the involvement of several separation stages such as neutralization and filtration have contributed to a more quality product if compared with direct method. However, for CGB samples, the purification of crude glycerol using direct method was sufficient to achieve a good purity of glycerol.

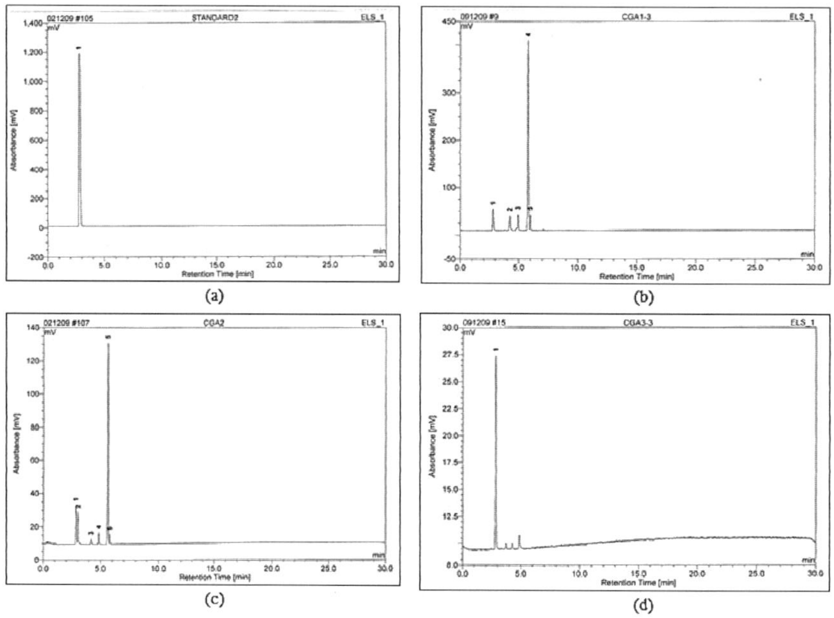

Figure 7. Result of HPLC analysis for (a) pure glycerol (b) CGA samples before purification (c) CGA samples after purification with direct method and (d) CGA samples after purification with multistep method.

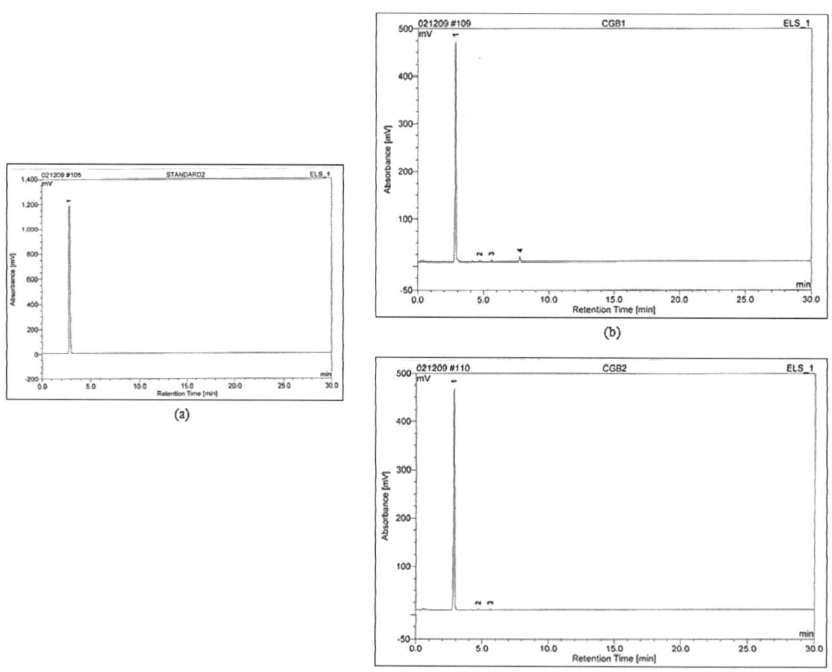

Figure 8. Result of HPLC analysis (a) pure glycerol (a) CGB samples before purification with direct method and (b) CGB samples after purification with direct method.

4. Conclusion

This study aimed to purify the glycerol from biodiesel production through direct and multistep method. Two samples, CGA and CGB were purified using both methods. This study has found that generally, the absence of glycerol and methanol after purification has been determined through GC analysis, FTIR analysis and HPLC analysis. This study has identified that purification method of CGA samples was best conducted using multistep method. While the direct purification method has found to be an effective way of purifying CGB samples. A further study could assess the effects of converting the pH value of crude glycerol before the purification process is being conducted.

Acknowledgement
The authors acknowledge the financial support from the RAGS grant vot no. R071 of Office for Research, Innovation, Commercialization, and Consultancy Management (ORICC).

References
[1] Demirbas, A. 2008. *Biodiesel*. (London: Springer).
[2] Haas, M.J.; McAloon, A.J.; Yee, W.C. and Foglia, T A 2006. A process model to estimate biodiesel production costs. *Bioresource Technology*, **94**, 671-78.
[3] FoodInfo 2015. E422: Glycerol. Retrieved July, 2015, from http://www.food-info.net/uk/e/e422.htm.
[4] Vicente, G.; Martínez, M. and Aracil, J. 2007. Optimisation of integrated biodiesel production. Part II: A study of the material balance. *Bioresource Technology*, **98**, 1754-61.
[5] Pagliaro, M.; Ciriminna, R.; Kimura, H, Rossi M, and Pina, C D. 2007. From Glycerol to Value-Added Products. *Angew. Chem. Int. Ed*, **46**, 4434-40.
[6] Buenemann, T.; Oudejans, J.C.; Gamba, P, and Rampi, A. *Process for purifying crude glycerol*. 1991, Google Patents.
[7] Ardi, M.S.; Aroua, M.K. and Hashim, N.A. 2015. Progress, prospect and challenges in glycerol purification process: A review. *Renewable and Sustainable Energy Reviews*, **42**, 1164-73.
[8] Busby, G.W. and Grosvenor, D.E. 1952. The purification of glycerin by ion-exchange. *Journal of the American Oil Chemists Society*, **29**, 318-20.
[9] Carmona, M.; Valverde, J.L.; Pérez, A, Warchol, J, and Rodriguez, J F. 2009. Purification of glycerol/water solutions from biodiesel synthesis by ion exchange: sodium removal Part I. *Journal of Chemical Technology & Biotechnology*, **84**, 738-44.
[10] Isahak, W.N.R.W.; Ismail, M.; Yarmo, M.A. Jahim J M, and Salimon J. 2010. Purification of Crude Glycerol from Transesterification RBD Palm Oil over Homogeneous and Heterogeneous Catalysts for the Biolubricant Preparation. *Journal of Applied Sciences*, **10**, 2590-95.
[11] Isahak, W.N.R.W.; Jahim, J.M.; Ismail, M, Nasir, N F, Ba-Abbad, M M, and Yarmo, M A. 2016. Purification of crude glycerol from industrial waste: Experimental and simulation studies. *Journal of Engineering Science and Technology*, **11**, 1056-72.
[12] Javani, A.; Hasheminejad, M.; Tahvildari, K, and Tabatabaei, M. 2012. High quality potassium phosphate production through step-by-step glycerol purification: A strategy to economize biodiesel production. *Bioresource Technology*, **104**, 788-90.
[13] Pavia, D.; Lampman, G. and Kriz, G. 2001. *Introduction to Spectroscopy*. Washington: Cengage Learning.
[14] Malaysia, I.P.M.K.S. 1995. *PORIM Test Methods*: Palm Oil Research Institute of Malaysia.

Experimental and numerical simulation of a moving carrier (FX63-137) through an air tunnel

W M Shibani[1], M F Zulkafli[1], B Basuno[1], M F Abdul Ghafir[1], Z Subari[1]
[1] Department of Aeronautical Engineering, Faculty of Mechanical and Manufacturing Engineering, Universiti Tun Hussein Onn Malaysia 86400 Parit Raja, Johor, Malaysia

E-mail: wanismustafa@yahoo.com

Abstract. This study investigated the movement of a carrier through an air tunnel using numerical simulation and experimental approach conducted in UTHM laboratories. The carrier was designed to have a particular trajectory moves along the air tunnel. The body of the main carrier has a shape of cube and rectangular wing of aspect ratio 2 is set in the middle of the body which has a shape of airfoil FX63-137 as its airfoil section. Experimental result of the carrier movement was investigated and compared with the simulation result obtained by Fluent's software. The experiment was conducted using a flow speed of 19.59 m/s at the inlet station. In the simulation, the flow is considered as an internal flow and has a constant temperature at the entry station. The pressure ratio between the inlet and outlet of the air tunnel is $235\ \mathrm{N/m^2}$. In addition, that the flow is considered as turbulent flow and the present study used $k-\varepsilon$ as its turbulence modelling in the numerical simulation. Comparison result between experiment and Fluent software in term of its trajectory and the speed of the carrier movement are found in a good agreement.

1. Introduction

Increasing human activities in the last two decades facilitated the significant growth of transportation of goods. Such a rapid increase in the demand for goods transport may seriously affect modern life. Present transportation systems, such as roads and railways, commonly involve route transportation. These transport systems involve problems such as traffic congestions, accidents, air and noise pollutions, and damage to roads and bridges. These issues suggest the necessity of developing an alternative way of transporting goods from one place to another. Transportation through air tunnels may provide a solution to this problem. This approach will not interfere with the existing transportation system.

The present study is to investigate feasibility of such transportation as an alternative way of delivering goods. In our approach, the goods are deliver using a carrier that float and move inside a straight air tunnel. The study of such carrier moving through an air tunnel is carried out experimentally and numerically, and their results are compared to each other. The carrier was designed as a cube-body attached with wings. In this respect, the flow problem in hand can be considered as the problem of a flow past through a wing-body configuration. As a result, their aerodynamics characteristics will be influenced by geometry parameter, such as aspect ratio, taper ratio, twist angle, wing incidence, the leading swept angle, etc.

In the numerical simulation, the movement of the carrier through air tunnel was evaluated by using Computational Fluid Dynamic (CFD) software, ANSYS Fluent. Here, the governing equations of fluid motion is treated in the form of Time Averaged Navier-Stokes Equations (TRANS). These equations were supplied by additional equation called as turbulence modelling to represent the presence of the Reynolds stress term. There is various turbulence modelling embedded in the Fluent Software such as; one equation model Spalart-Allmaras, two equation models like Standard $k-\varepsilon$, RNG $k-\varepsilon$, Realizable $-\varepsilon$, etc. The study on the use of various turbulence model can be referred to work done by Lakehal and Rodi [1], Iaccarino et. al [2] and Yakhot et. al [3]. On the other hand, the method of how the TRANS is solved numerically can be referred to Tannehill et. al [4]. The present work uses a Fluent software version 16.1 with standard k-ε as its turbulence modelling. Gosman[5] indicated that this turbulence modelling represents the most well-known turbulent model, simple and has capability for capturing turbulence transport occurrence.

The use of airfoil section FX 63-137 in the present work is adopted from work provided by Kanyako et. al [6]. Their research work found that among three types of airfoils, namely, FX 63-137, E387, and S826, the FX 63-137 airfoil provides the highest C_L/C_D ratio compared with the others. The reason is probably the fact that this particular airfoil exhibits a gradual stall compared with the other airfoils [6]. Extensive researches mostly conducted in the aircraft manufacturer industries have the intention to obtain wing configuration with higher maximum lift to drag ratio [7, and 8].

2. Description of the model

Figure 1 shows the cube shape carriers with wings attached on both of its sides. The cube are measures 0.1 m × 0.1 m × 0.1 m. The wing has a chord length of 0.05 m and it is equal to its exposed half wing span. The wings are placed in the middle of the cube with its wing incidence angle of 18°. The total mass of the carrier is 100 grams. Figure 2 shows the air tunnel used in this study. It has a length of 3 m and a circular cross section with a diameter of 0.3 m. The carrier (object) was placed initially on the leftmost side of the tunnel.

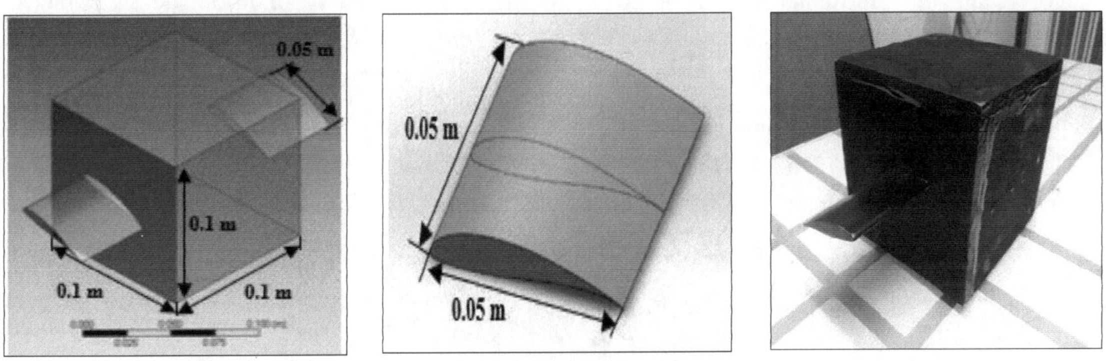

(a) (b) (c)

Figure 1. The models (a) The carrier, (b) The wing has a rectangular shape of FX63-137 airfoil section, and (c) The real carrier.

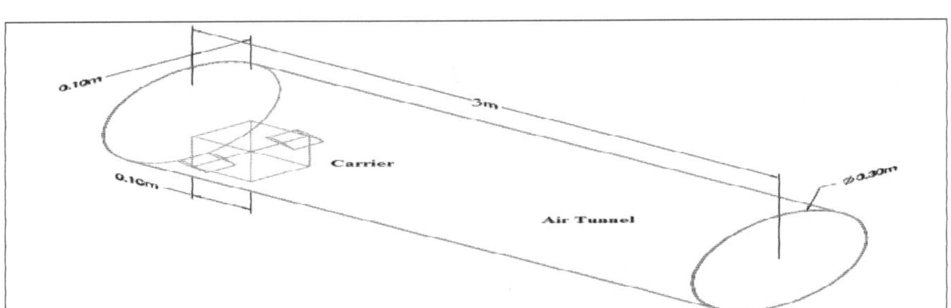

Figure 2. Three-dimensions view of the air tunnel and the carrier.

3. CFD methodology

The equations that govern the unsteady flow of an incompressible fluid are

$$\nabla . u = 0 \tag{1}$$

$$\rho \frac{D\overline{u_i}}{D_t} = F_i - \frac{\partial \bar{p}}{\partial x_i} + \mu \Delta \overline{u_i} - \rho \left(\frac{\partial \overline{u_i' u_j'}}{\partial x_j} \right) \tag{2}$$

where ρ denotes density; F is force, μ is molecular viscosity, Δ is vector operator, $\bar{u}$ is time-averaged velocity components, $\bar{p}$ is time-averaged pressure, and u' is fluctuating velocity components and index i, j refer to 1, 2, 3 indices for the three components in a Cartesian coordinate system.

Equations (1) and (2) are the well-known Time-Averaged Navier-Stokes Equations (TRANS). ANSYS Fluent [9] has been used in the present study with turbulence model $k - \varepsilon$ as proposed by Launder andSharma[10]. Guan and Park [11], &Mohammadi and Pironneau[12] have choose the $k - \varepsilon$ model after comparing with three models and they found that the $k - \varepsilon$ model are more accurate and robust. This model was also used by Chern and Wang [13] in their investigation of the flow rate of the ball valve. Such approach had been applied to study the object movement along the air tunnel as presented in [14 - 16].

The governing equation of fluid motion presenting in the form of TRANS as in Eqn. (1) and (2) can be solved using various numerical scheme such as SIMPLEC, SIMPLER, SIMPLEST, CTSSIMPLE, PISO or SIMPLE. Each has its advantages and drawbacks [17 - 19]. The present work uses a SIMPLE scheme (Semi-Implicit Method for Pressure-Linked Equations).

3.1. Boundary conditions

The physical boundary condition of these problems are as follows:

1) Each end of the air tunnel is considered as pressure inlet and outlet, while the surfaces of the carriers and the air tunnel is considered as no-slip wall boundary condition, i.e., $u = 0$. (See Figure 3).
2) "Pressure inlet" was constantly set to 235 N/m^2 higher than the atmospheric pressure. This excess dynamic pressure produces a flow speed equivalent to 19.59 m/s at the inlet.
3) The right surface of the computational domain is assigned as the outlet. The boundary condition for "pressure outlet" is implemented at the exit boundary with zero-input value of the static gauge pressure, i.e. $\frac{\partial u}{\partial x} = 0$ of the Dirichlet-type pressure boundary condition.

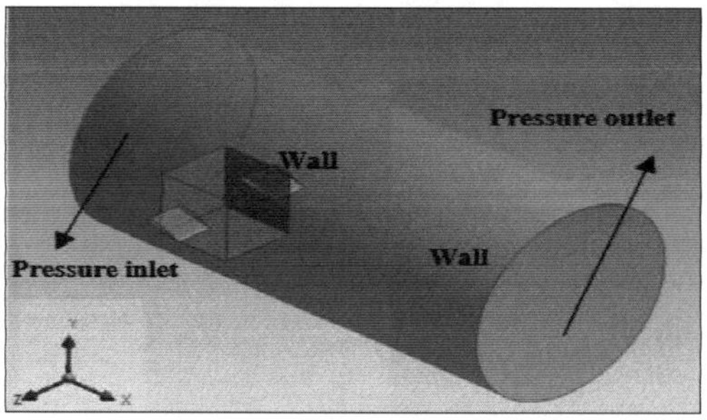

Figure 3. Boundary conditions of the air tunnel and carrier.

3.2. Mesh topologies

The carrier was designed using SOLIDWORKS software [20] and later imported into the ANSYS software [9]. A non-uniform unstructured mesh was used with a considerable fine mesh implemented in the regions of high gradients. A tetrahedron element was selected as the mesh of the carriers and the air tunnel. Tetrahedron mesh has the average orthogonal quality metric spectrum is 0.85 which is very good and the advantage of tetrahedral mesh is its flexibility. Figure 4 shows the layout of the cross-sectional plane of the mesh for the carrier. The number of mesh grids used in the simulation is 1811494.

IOP Conf. Series: Materials Science and Engineering **243** (2017) 012016 doi:10.1088/1757-899X/243/1/012016

Figure 4. Grid section of the carrier (FX63-137) and circular cross section of the air tunnel.

4. Experiment methodology

The experiment was conducted in the Aerodynamic lab of Universiti Tun Huseein Onn Malaysia (UTHM). The length of the air tunnel is 3 m as shown in Figure 5.a. The source of air was located at 2.5 m before the test section as shown in Figure 5.b. The pipes were made from clear acrylic material and connected with clear acrylic flanges. There is a small hole from the top of pipe. It was made as a door to put the carrier on the bottom of pipe. Moreover, there is a small hole in the door to pull and push the bar inside the pipe in order to hold the carrier initially while the air is blowing. The bar was made from iron and welded with small valve for better carrier holding and less air obstruction. The net linked to the end of air tunnel is to contain the carrier before fall down and to avoid damaging the carrier. The pressure difference between outlet blower and entry station of the test section is set at $235\ N/m^2$, same with condition set in the numerical simulation.

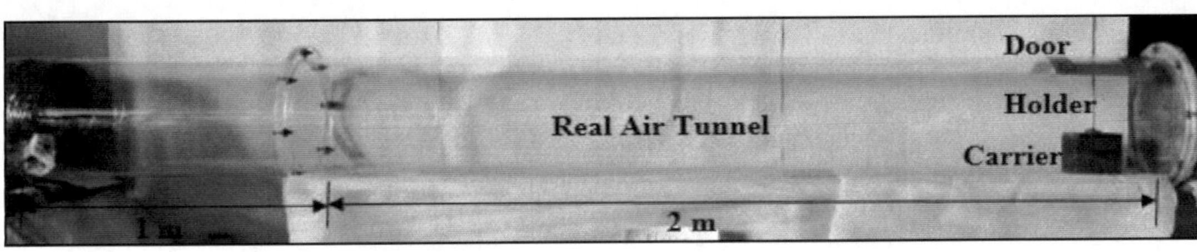

Figure 5. a. The real air tunnel and the position of the carrier.

Figure 5.b. The source of the air (blower pump) and the linked pipes (air tunnel).

5. Experimental and simulation results and discussion

Figure 6 show the movement of the carrier in the experiment taken by high speed camera. The time is written for each photo and then compared to numerical results. The dimensions shown in the photos are in meters' unit without scaling. The air tunnel showed here is 2 m in real length and diameter of 0.3 m.The results shown in the Figure 6 indicate that as the carrier floats and flew at different speed, the carrier also change in its aptitude. The presence of change of its attitude make the carrier at the time t = 0.429 s touch the ground.

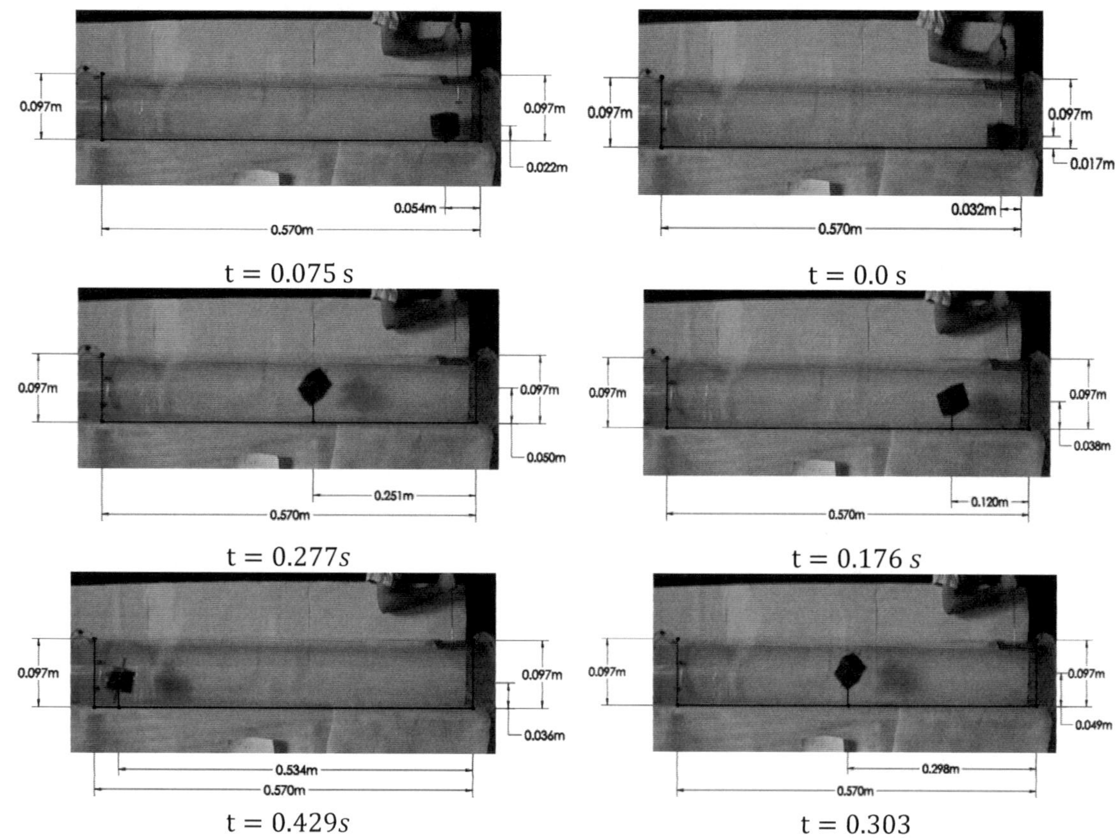

Figure 6. Experiment results for the carrier (FX63-137) captured by high speed camera.

Figure 7 shows the movement of the carrier (FX63-137) numerically. The flow pattern that surrounds the carriers at different stations is shown by the contours of pressure. The front view is shown in Figures 7, This Figure describes the flow conditions at various time steps.

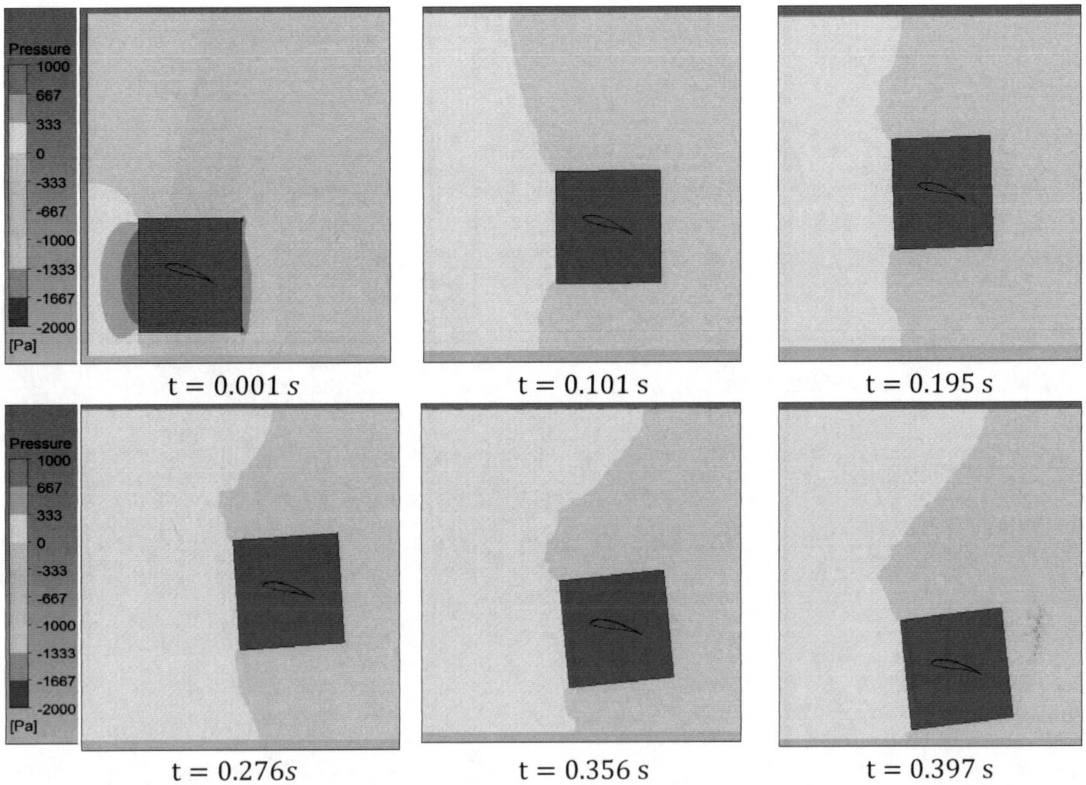

Figure 7. Pressure contours of the flow field around the carrier in various positions for XY-plane.

Based on the result shown in the Figure 7, the carrier reached speed of 0.6 m/s in 0.035 s with the lift force works on it is 5.05 N. As the time progresses, between time 0.035 s to 0.276 s, the orientation of the carrier is relatively similar to its condition at the beginning time the carrier moves. However, in the next following time, due to increasing movement speed, the aerodynamics moments work on the carrier able to rotate and change its orientation. Such condition makes the angle of attack increase gradually above stall angle, which make the lift force on the carrier less than its carrier's weight. As a result, the carrier starts to fall down. Such condition makes at time t = 0.397 s, the carrier touch the ground. At this time step, the carrier moved at speed of 4.44 m/s with the distance had been covered in the x-direction is 1.76 m.

Considering computational and experimental results indicate that the mechanism for transporting carriers through air tunnels are possible.Through the results presented in Figure 6 and Figure 7, the distance and speed experienced by the carrier can be defined. Figure 8 shows the comparison result between experimental and numerical data presented in term of velocity carrier against the distance. While Figure 9, shows their comparison in term of velocity plotted with respect to time. It can be observed from Figure 9 that the velocity magnitude of the carrier at the beginning was 0.83 m/s at time of 0.025 s for the experiment work which is higher compared to numerical result, while numerical results showed a bit high increasing after lunching of the carrier during whole journey to reach 4.44 m/s at time of 3.97 s. Figure 10 compares the time taken for the carrier to travel through the air tunnel experimentally and numerically. It indicates knowing the carrier movement between both methods, where the travelling carrier experimentally was faster than numerically at each carrier positions in x-direction.

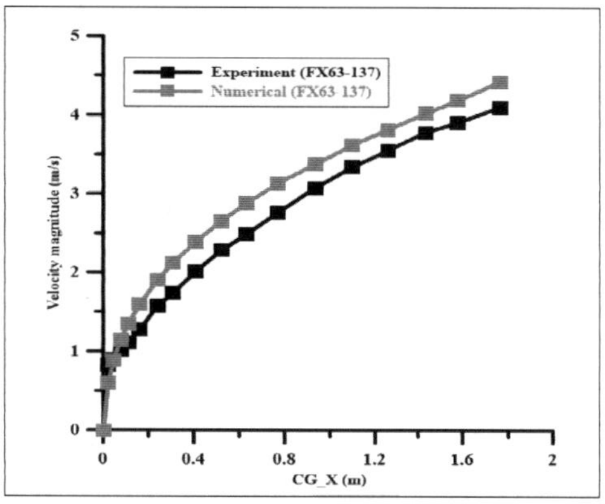

Figure 8. Distance travelled by the carrier in the X-direction versus velocity magnitude (u).

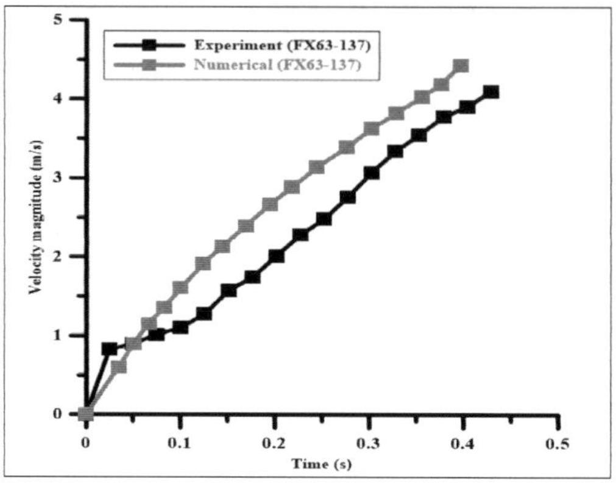

Figure 9. Velocity magnitude (u) of the carrier respect to time.

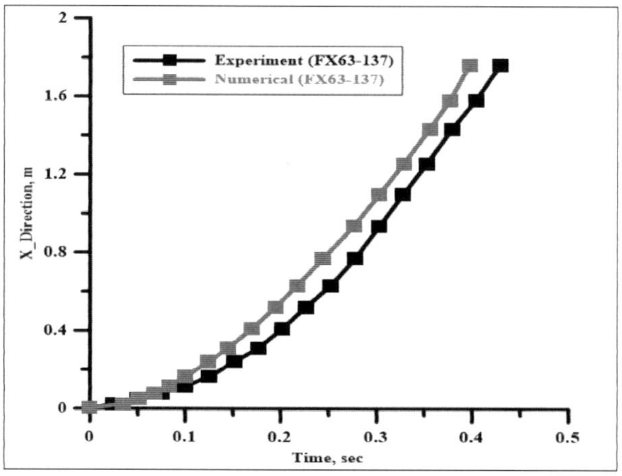

Figure 10. Distance travelled by the carrier in the X-direction versus the time taken for experiment and numerical results.

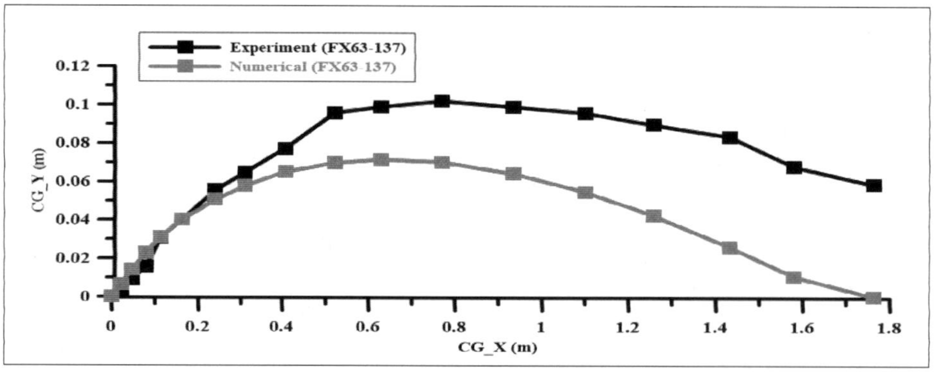

Figure 11. Carrier trajectory with characteristic Squares.

Figure 11 shows the movement comparison of the carrier along the X-direction, the motion of the carrier started from the same origin for both results. It is very clear from the graph that, good agreement with error less than 1% is obtained for both results until x = 0.4. As clear from the graph, by travelling carrier at distance of 0.53 m the error started increasing and limited to 7 % until landing position at distance of 1.76 m. At this time, new phenomena caused by the eddies started, which make the carrier roll a bit to the right side. This rolling can be seen clearly from Figure 6. Moreover, the carrier rolled to the right side and was not stable and unbalanced due to high force acted on the left side created by the continuous eddies. Furthermore, it can be seen that at a distance of 1.76 m in X-direction, the carrier has been landed probably for the numerical result, while the experimental result, the carrier has not been landed probably, because of the phenomena stated above and was very clear to see it from Figure 6, therefore, the landing was on the wing side that makes the error bigger reachedto 5.8% between both results. The distance covered by the carrier is 1.76 m from 3 m which are 58.6% of the total length of the air tunnel.

6. Conclusion

An experiment and numerical study of a moving carrier (FX63-137) have been conducted. The movement of the carrier through the air tunnel for the current model had been found experimentally and numerically to have similar behaviour. The current model allows to float up to a distance of 1.76 m which is 58.6% of the air tunnel's length. As a good agreement has been found between results, there is a positive indication for the possibility of transporting goods by this carrier through an air tunnel in the nearest future.

Acknowledgment

The authors thank Universiti Tun Hussein Onn Malaysia (Grant Code: U416) for providing financial support for this research

References

[1] Lakehal D and Rodi W 1997 Calculation of the flow past a surface-mounted cube with two-layer turbulence models *J. Wind Eng. Ind. Aerodyn.* 65-78

[2] Iaccarino G, Ooi A, Durbin P A and Behnia M 2003 Reynolds averaged simulation of unsteady separated flow *Int. J. Heat Fluid Flow* **24**(2) 147-56

[3] Yakhot A, Liu H and Nikitin N 2006 Turbulent flow around a wall-mounted cube: A direct numerical simulation *Int. J. Heat Fluid Flow* **27**(6) 994-1009

[4] Tannehill J C, Anderson D A and Pletcher R H 1997 Computational Fluid Mechanics and Heat Transfer, Series in *Computational and Physical Processes in Mechanics and Thermal Sciences* (New York: CRC Press, Taylor & Francis Group)

[5] Gosman A D 1999 Developments in CFD for industrial and environmental applications in wind engineering. *J. Wind Eng. Ind. Aerodyn.* **81**(1) 21-39

[6] Kanyako F and Janajreh I. 2013 Investigating blade performance of small horizontal axis wind turbine based on blade element momentum theory In Applications of information technology to renewable energy processes and systems (IT-DREPS) 2013 *1st International Conf. & Exhibition* on the 2013 May 29 (Amman, Jordan: IEEE) pp 85-90

[7] Hossain A, Rahman A, Iqbal A K, Ariffin M and Mazian M. 2012 Drag analysis of an aircraft wing model with and without bird feather like winglet. *Int. J. Aeros. Mech. Eng.* **6**(1) 8-13

[8] Saad M M, Mohd S B, Zulkafli M F and Shibani W M 2017 Numerical analysis for comparison of aerodynamic characteristics of six airfoils. *In AIP Conf. Proc. (Jogjakarta, 1-3 August 2016)* (Vol. 1831, No. 1) ed I Taib, N Asmuin and S A Osman (AIP Publishing) p 020004

[9] Fluent, ANSYS. ANSYS Fluent User's Guide, (Canonsburg, PA, USA, 2015) release 16.2 pp 1-2718.

[10] Launder B E and Sharma B I. 1974 Application of the energy-dissipation model of turbulence to the calculation of flow near a spinning disc. *Lett. Heat Mass Transfer* **1**(2) 131-7

[11] Guan Song X and Park Y C 2007 Numerical analysis of butterfly valve-prediction of flow coefficient and hydrodynamic torque coefficient. *In Proc. of the world congress on Engineering and computer science* pp 24-26
[12] Mohammadi. B and Pironneau. O 1993 Analysis of the K-Epsilon Turbulence Model. *Research in Applied Mathematics, John Wiley & Sons Ltd*
[13] Chern M J and Wang C C 20004 Control of volumetric flow-rate of ball valve using V-port. *Transactions-American Society of Mechanical Engineers Journal of Fluids Engineering* **126** 471-81
[14] Shibani E, Zulkafli M F, Basuno B and Ab Wahab A B 2016 Numerical Analysis of a Moving Object in the Air Tunnel. *International Journal of Mechanical & Mechatronics Engineering IJMME* **16**(03) 72-8
[15] Shibani W M, Zulkafli M F and Basuno B 2017 A comparative study on the motion of various objects inside an air tunnel. *In AIP Conf. Proc. (Jogjakarta, 1-3 August 2016)* (Vol. 1831, No. 1) ed I Taib, N Asmuin and S A Osman (AIP Publishing) p 020005
[16] Shibani W M, Zulkafli M F and Basuno B 2016 Numerical Study of a Winged Container Moving in an Air Tunnel *IOP Publishing* **160**(1) p 012041
[17] Tao Wenquan 2001 *Numerical Heat Transfer* (Xi'an: Xi'an Jiaotong University Press)
[18] Zhang Z S 2003 *Turbulence* (Beijing: National Defence Industry Press)
[19] Norberg C, Sohankar A and Davidson 1995 Numerical simulation of unsteady flow around a square two-dimensional cylinder. *In Twelfth Australian Fluid Mechanics Conf.* pp 517-520
[20] Tickoo S 2016 SolidWorks for designers (Schererville, IN, USA) 14th Edition. *Cadcim Technologies*

IOP Conf. Series: Materials Science and Engineering **243** (2017) 012017 doi:10.1088/1757-899X/243/1/012017

Numerical analysis on centrifugal compressor with membrane type dryer

M A Razali[1], M F Zulkafli[2], N Mat Isa[1] and Z Subari[2]
[1] Department of Energy and Thermofluids Engineering, Faculty of Mechanical and Manufacturing Engineering, Universiti Tun Hussein Onn Malaysia, 86400 Parit Raja, Batu Pahat, Johor, Malaysia
[2] Department of Aeronautic Engineering, Faculty of Mechanical and Manufacturing Engineering, Universiti Tun Hussein Onn Malaysia, 86400 Parit Raja, Batu Pahat, Johor, Malaysia
E-mail: fadhli@uthm.edu.my

Abstract. Moisture content is a common phenomenon in industrial processes especially in oil and gas industries. This contaminant has a lot of disadvantages which can lead to mechanical failure DEC (Deposition, Erosion & Corrosion) problems. To overcome DEC problem, this study proposed to design a centrifugal compressor with a membrane type dryer to reduce moisture content of a gas. The effectiveness of such design has been analyzed in this study using Computational Fluid Dynamics (CFD) approach. Numerical scheme based on multiphase flow technique is used in ANSYS Fluent software to evaluate the moisture content of the gas. Through this technique, two kind of centrifugal compressor, with and without membrane type dryer has been tested. The results show that the effects of pressure on dew point temperature of the gas change the composition of its moisture content, where high value lead more condensation to occur. However, with the injection of cool dry gas through membrane type dryer in the centrifugal compressor, the pressure and temperature of moisture content as well as mass fraction of H_2O in centrifugal compressor show significant reduction.

1. Introduction

Centrifugal compressors have numerous applications ranging from aircraft and process industries to oil and gas industries. This type of compressor is widely used due to high single stage pressure ratio, simple structure, long life and other favourable characteristics [1]. Centrifugal compressors are often required to maintain or increase gas flow into the pipeline system, since well flow and reservoir pressure tend to decrease over time. These happen due to depletion occur at reservoir. Thus, this type of compressor works as re-injection for the purpose of reservoir pressure maintenance. Generally, the gas extracted from the oil wells composed of many moisture contents including water vapour. However, water vapour or moisture content in gas is one of the reasons why performance of centrifugal compressor drops. Most common contaminants which lead to mechanical failure are carbon dioxide (CO_2) and hydrogen sulphide (H_2S), especially for parts that manufactured in low alloy steel. These carbon dioxide (CO2) become a dangerous contaminants when is associated with water vapour or and high pressure. Combination of carbon dioxide (CO_2), liquid water and high pressure leads to the formation of carbonic acid (H_2CO_3), which can heavily corrode the compressor component [2]. Besides, the effect of hydrates formation can lead to the compressor's leading edge blocking, can reduce the pipeline flow capacities and has potential to damage other process equipment such as filter and valves [3].

In order to control or prevent these problems, moisture content measurement in natural gas is important. The amount of moisture content in the gas can be measured by indicating the temperature of gas, called dew point temperature, where higher dew point temperature indicates high level of water vapour. Dew point temperature is the temperature where air or any gas is saturated with water and moisture content will begin to condense. All atmospheric air contains some water vapour which will begin to condense into liquid water when the gas is cooled past the saturation point, where it can hold no more water vapour. When the dew point temperature and gas temperature are same, the air is said as saturated [4]. Another factor that sensitive to dew point temperature is pressure of gas. As the gas pressure increase, the dew point temperature also increases. This happen due to the molecular of gas that compressed in compressor becomes turbulent flow. This is called dew point pressure. At this stage, the dew point pressure is higher than atmospheric pressure but the temperature of dew point always lower (or equal to) to gas temperature. These phenomenon affect the efficiency of compressor as volume production of natural gas per cycle decrease [4].

Designation of impeller is one of the solutions for DEC problem. For example, improving the blade angles of impeller [5], choosing different blade inlet and backward sweep of the impeller [6] and modification of impeller geometry in term of skew angle [7]. Another approach is to dry the gas itself. There are various methods to dry compressed air or gas such as refrigerated dryers, chemical dryers, desiccant dryers and membrane type dryers. In this study, we choose membrane type dryer as the mechanism to remove moisture content from the gas. Addition membrane type dryer on centrifugal compressor is chosen due to composition of water vapour trough continuous gas moving can be modelled numerically. In the real application, the mechanism of this kind of dryer is when the compressed air is moving through a bundle of polymer hollow fibres, the polymer hollow fibres attract the water vapour inside before it diffuse into thin selective layer until it reach outside of membrane layer [8].

2. Methodology

2.1. Multiphase flow

Multiphase flow model [9] is implemented in this work to numerically simulate the behavior of the moisture content around the compressor. Multiphase flow can be classified into four categories, gas-liquid or liquid-liquid flows, gas-solid flows, liquid-solid flows and three-phase flows. Existing moisture content or water vapor in gas indicates that this case as gas-liquid flows (presence of discrete fluid droplets in a continuous gas). There are two approaches in modelling multiphase flow; Euler-Lagrange approach and Euler-Euler approach [10]. In the Euler-Lagrange approach, fluid phase is treated as a continuum by solving the Navier-Stokes equations while Euler-Euler approach use concept of physic volume fraction which assume to be continuous functions of space and time. For the Euler-Euler approach, there are three multiphase models: volume of fluid (VOF), mixture model and Eulerian model [11]. In this study, analysis of moisture content behavior in centrifugal compressor is conducted using ANSYS Fluent with multiphase species transport model. Composition of moisture content in gas is analyzed based on pressure and temperature distribution.

2.2. Centrifugal compressor model

In order to compare the reduction of moisture content, two cases of centrifugal compressor are considered. Case 1 is the typical centrifugal compressor model and Case 2 is centrifugal compressor model with the addition of membrane type dryer. Both models were designed using SolidWorks 2014, which begins with 2D sketch, then follow with creating features and lastly combination of parts to create a complete model. The models are then saved in IGES format before imported into the ANSYS DesignModeler.

After importing geometry into the ANSYS, meshing step is required to solve the multiphase flow equations at discrete locations of the compressor's fluid domain. In this work, the author applied triangle elements to discretize the fluid domain with the addition of boundary layer mesh around the

IOP Conf. Series: Materials Science and Engineering **243** (2017) 012017 doi:10.1088/1757-899X/243/1/012017

impeller walls. Meshing model for fluid region of centrifugal compressor without and with membrane type dryer is shown in Figure 1(a) and Figure 1(b), respectively.

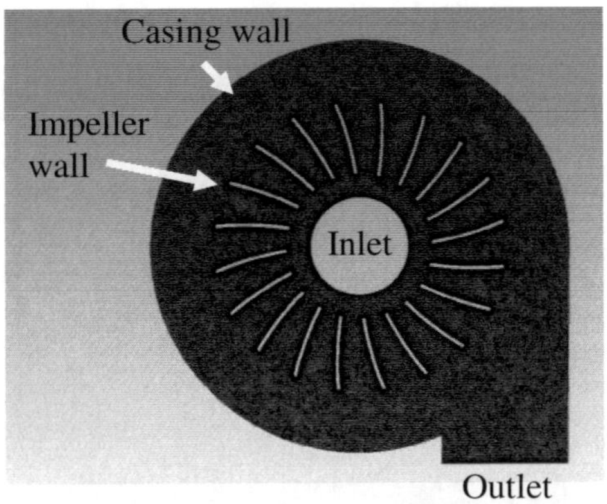

(a) Without membrane type dryer.

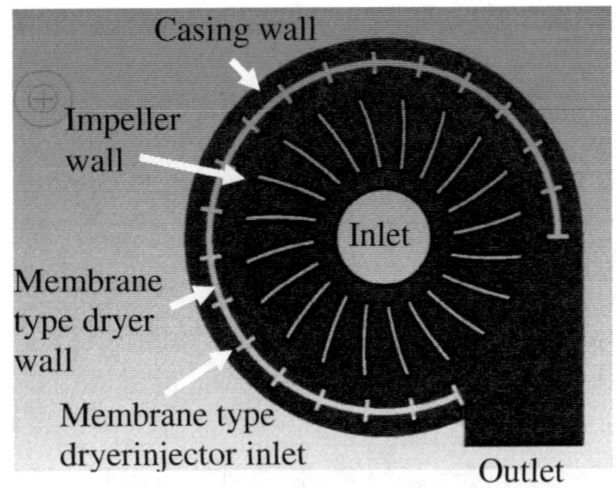

(b) With membrane type dryer.

Figure 1. Mesh of Centrifugal Compressor

2.3. Solver setup

Pressure-based solver is used in this research to solve the viscous flow and species transport of mixture content of water vapor and gas. The flow mixture content is assumed as viscous flow and k-epsilon is chosen as a turbulent model, which is considered appropriate for a flow in centrifugal compressor. As explained in the previous section, multiphase transport species is selected to obtain the changes of mixture content. The mixture-template is used to select the materials which involve mixture flow, which existing of water vapor along continuous gas.

Other than that, in order to simulate the effect of impeller rotation to the gas in the compressor, multiple rotating frames technique is used in the ANSYS Fluent. Instead of using deforming mesh to make the impeller rotate, this technique allow the impeller to be fixed while the gas rotates around it, which has the same effect as rotating impeller. The boundary conditions, as shown in figure 1(a) and 1(b), are mainly consists of Wall, Inlet and Outlet. Casing wall and membrane type dryer wall are considered as non-slip, adiabatic and stationary wall, while impeller wall is set as non-slip, adiabatic moving wall with a rotational velocity of 523.6 rad/s in anti-clockwise direction. At the compressor Inlet, gas pressure is set manually and at the Outlet, static pressure is assumed to be equal to the standard sea level atmospheric pressure. In this study, 3 cases of gas pressure is set at the compressor inlet, which is 200 kPa, 300 kPa and 400 kPa. The 40 injectors around the membrane type dryer are assumed as Inlet, where each of them inject cool gas at the speed of 30 m/s into the compressor.

3. Result and discussion

3.1. Velocity distribution

As the impeller rotating in anti-clockwise direction, Figure 2(a) and 2(b) show the mixture material moves together along the impeller before leave through the outlet. Those contour velocity magnitude indicate the mixture gas experiences highest velocity (yellow color background) at impeller region that close to the outlet. Thus, the casing region at the outlet has highest velocity compare to other region as the movement of impeller react as a re-injection to the mixture gas leaving the centrifugal compressor.

Besides that, addition of membrane type dryer into centrifugal compressor also affects the velocity of the mixture. In Figure 2(b) the mixture gas experiences high velocity at flow path from impeller region to casing region due to injection of cool gas into centrifugal compressor. Besides that, injection of cool gas through membrane type dryer forced the mixture gas moved toward the outlet region and

IOP Conf. Series: Materials Science and Engineering **243** (2017) 012017 doi:10.1088/1757-899X/243/1/012017

reduces the velocity around casing region (blue color). The changes of light blue color to red color in impeller region towards outlet show the high velocity region where the water vapor is swept away. This high velocity of mixture indicates high pressure, hence in other word, injection of gas through membrane type dryer can help to increase gas pressure when the depletion of gas occurs.

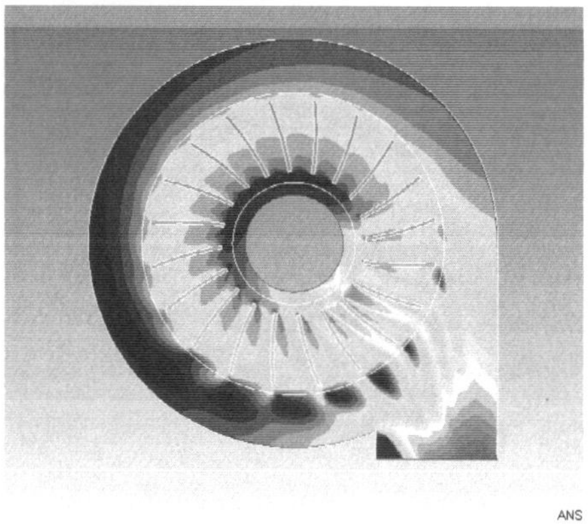

(a) Without membrane type dryer.

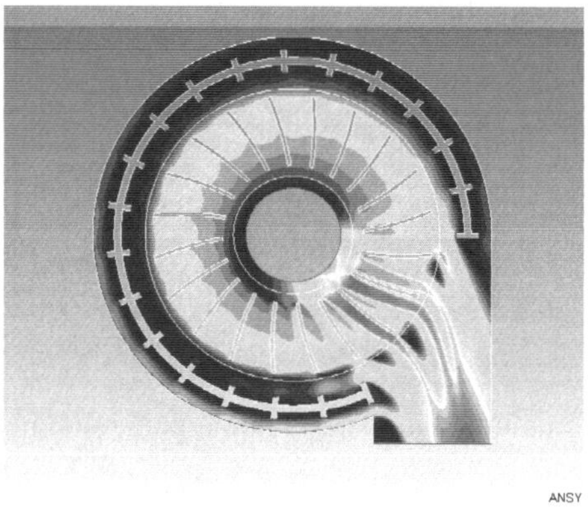

(b) With membrane type dryer.

Figure 2. Contours of velocity magnitude (compressor inlet pressure 200 kPa).

3.2. Pressure distribution

The kinetic energy of the water vapor and gas molecule increased as pressure increased in impeller region. These molecule collide each other and also cause friction to occur around impeller wall. This situation can be seen in Figure 3(a) and 3(b) where the leading edges experience high pressure compared to the trailing edge. High values of pressure indicate the gas more easily to condense and increase dew point temperature. At this stage, the dew point pressure is higher than atmospheric pressure but the temperature of dew point always lowers to gas temperature. Thus, membrane type dryer is proposed to be installed in the centrifugal compressor to control dew point temperature and pressure of the gas.

Based on Figure 3(b), there is different distribution color background for pressure of mixture. Injection of cool gas towards fluid region reduces the pressure of mixture content. The cool gas (green color) is injected with ratio of 1/10 of inlet centrifugal compressor into the fluid region act to cool the high pressure (reddish orange color) inside the centrifugal compressor. Decreasing the pressure of continuous gas with water vapor reduced the kinetic energy of molecules to collide with each other. This can be proved by the measurement of total pressure for each fluid region in the centrifugal compressor as stated in Table 1.Including the case where Inlet pressure is 200 kPa, the different value total pressure distribution for cases where Inlet pressure is 300 kPa and 400 kPa are also stated in the same table. Figure 4 illustrates the centrifugal compressor's fluid regions mentioned in the table.

IOP Conf. Series: Materials Science and Engineering **243** (2017) 012017 doi:10.1088/1757-899X/243/1/012017

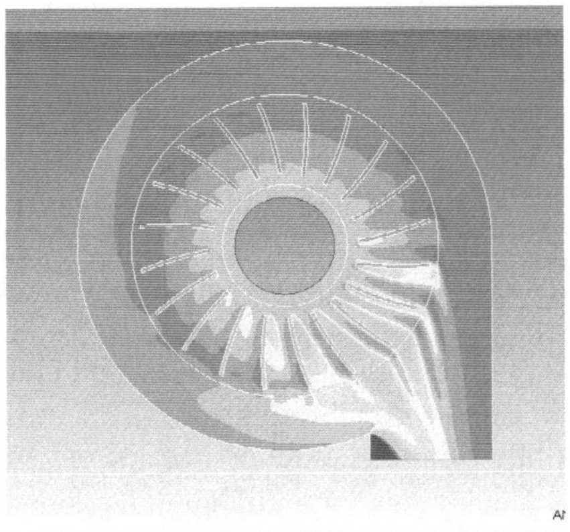

(a) Without membrane type dryer.

(b) With membrane type dryer.

Figure 3. Contours of total pressure distribution (compressor inlet pressure 200 kPa).

Table 1. Total distribution pressure for each region for two different type centrifugal compressors with different pressure at inlet of centrifugal compressor.

Pressure (kPa)	Type of centrifugal compressor	Total pressure distribution (Pa)			
		Inlet region	Impeller region	Casing region	Average
200	Case 1	200534.80	223861.89	215789.92	213099.77
	Case 2	199746.63	210154.09	171141.39	191905.44
300	Case 1	300407.32	317841.42	310803.44	308486.12
	Case 2	299644.28	305793.13	234618.04	267709.21
400	Case 1	400314.84	411525.75	405800.78	403786.28
	Case 2	399547.79	401181.74	304120.59	347240.93

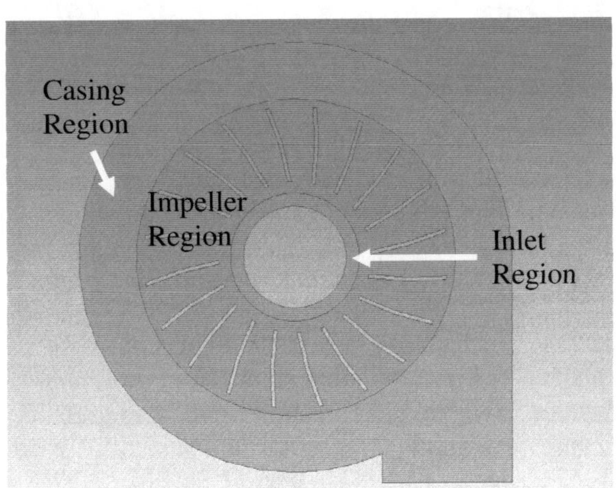

Figure 4. Fluid region of centrifugal compressor

3.3. Temperature distribution

Based on the theory of dew point temperature, the air cannot longer hold the water vapor as the temperature of air is raised to the dew point. At this point, the relative humidity is 100% if the water vapor begins to condense. Besides, higher dew point temperature indicates high level of water vapor. Thus, it is important to make the sure that the temperature of mixture content lower than dew point temperature.

The mixture content moves together along the blade of impeller when the blade is rotating. As a result, the temperature of mixture content slightly increases. The changes of green color into red color at impeller region in Figure 5(a) shows the temperature of mixture approach the dew point temperature where the water vapor inside the gas start to condensed. Furthermore, casing region near to the outlet show more in red color indicates higher temperature than other regions. This is because at this region more condensation of gas occurs to become liquid droplet. The temperature of leading edge experience high temperature compared to the trailing edge. Thus, as the temperature increase, the pressure also increases. This will lead more condensation of water vapor occur.

The contour of total temperature in centrifugal compressor for Case 1 and Case 2 show big difference value. For Case 2, the cool gas with temperature of 1/2 of inlet gas temperature is injected through membrane type dryer into centrifugal compressor to reduce the mixture temperature. Figure 5(b) shows that the cool (blue color) gas is diffused into casing region and impeller region which has high temperature initially. As a result, the continuous gas with composition of water vapor (high temperature gas) is push away to the outlet. This technique helps to cool down high temperature regions especially at impeller region and casing region. This reduction of temperature distribution help the water vapor from condense.

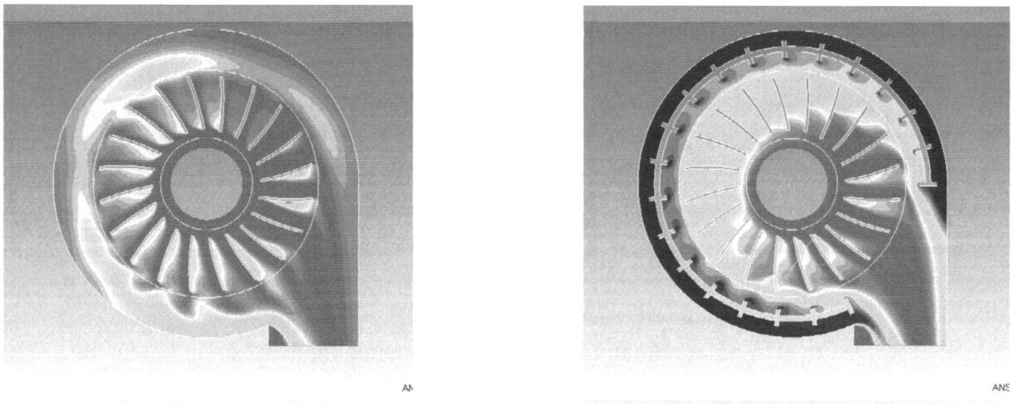

(a) Without membrane type dryer. **(b)**With membrane type dryer.

Figure 5. Contour of total temperature (compressor inlet pressure 200 kPa).

3.4. Mass fraction of H_20

Instead showing concentration water vapor in centrifugal compressor by using pressure and temperature distribution, simulation CFD software also provides multiphase species transport that can calculate mass fraction of water vapor or H_2O in centrifugal compressor. In this research, the gas extracted from the reservoir into centrifugal compressor is assumed to be composing with 50:50 ratio of water vapor to gas. Thus, the species transport model created in ANSYS Fluent has 50% air and 50% water vapor at inlet region. The composition of water vapor can be seen based on mass fraction of H_2O data inside centrifugal compressor for each region in Table 2.

Table 2. Mass fraction of H_2O for centrifugal compressor without membrane type dryer.

Pressure (kPa)	Inlet region		Impeller region		Casing region		Average	
	Case 1	**Case 2**	**Case 1**	**Case 2**	**Case 1**	**Case 2**	**Case 1**	**Case 2**
200	0.500	0.405	0.389	0.164	0.169	0.065	0.265	0.110
300	0.500	0.428	0.409	0.180	0.204	0.071	0.294	0.120
400	0.500	0.429	0.410	0.187	0.208	0.075	0.297	0.125

As shown in the table, the mass fraction of H_2O inside the centrifugal compressor slightly increase with increasing pressure. Mass fraction of H_2O represents the dew point pressure. This result prove as the pressure increase, the dew point pressure also increase. Thus, this condition leads to the increasing of dew point of temperature. Combination of water vapor with high pressure and carbon dioxide can lead the formation of carbonic acid which can corrode the centrifugal component especially impeller part. Therefore, the result is consistent with the theory. Besides proving the theory, the results of Case 2 show that mass fraction of H_2O, i.e. concentration of water vapor reduced in all compressor regions. This is due to the cool gas injected into centrifugal compressor to cool the impeller region from condensation occur and sweep the water vapor excess (extrude from reservoir) direct to the outlet

4. Conclusion

The results from the CFD were very effective and have the ability to show the behavior of multiphase flow of mixture content inside centrifugal compressor. The simulation result shows that the moisture content increase as the pressure increase. However, the addition of membrane type dryer in centrifugal compressor shows improvement in term of pressure, velocity, temperature and mass fraction of H_2O. The results show the water vapor can be reduced inside centrifugal compressor through the addition of membrane type dryer. The injection of cool gas from the membrane type dryer with moderate velocity helps to force the water vapor to exit through the outlet of the centrifugal compressor.

Acknowledgment

The authors would like to thank Universiti Tun Hussein Onn Malaysia for providing financial support to this research through IGSP Grant (Vot U416).

Reference

[1] Sivaji Ganesh C, NagpurwalaCh and Bhaskar Dixit Q H 2010 Effect of leading edge sweep on the performance of a centrifugal compressor impeller *SAS Tech J.* **9** pp 55–62
[2] Guedes AV, de Carvalho Ferreira R, et d'Audenhov F D N and da Silva Marques B 2013 CO_2 corrosion on natural gas compressors: criteria and guidelines *Proc. of the 42nd Turbomachinery Symp.*
[3] Chung D D L 2001 Materials for thermal conduction *Appl Therm Eng* **21** pp 1593–1605
[4] Kolass R and Parker C n.d.Moisture measurement in natural gas accessed 11 Nov 2016 Retrieved from: http://www.iceweb.com.au/
[5] Röyttä P, Grönman A, Jaatinen A, Turunen-Saaresti T and Backman J 2010 Effects of different blade angle distributions on centrifugal compressor performance *Int. J. Rotating Machinery* **2009** pp 1–9
[6] Tough R A, Tousi A M and Ghaffari J 2010 Improving of the micro-turbine's centrifugal impeller performance by changing the blade angles *Int. Conf. on Computational & Experimental Engineering and Sciences* **14** pp 1–22
[7] Arunachalam V C, Nagpurwala Q H, Deshpande M D and Shankapal S R 2008 Numerical studies on the effect of impeller blade skew on centrifugal compressor performance *SASTECH J.* **7** pp 33–40

[8] Fish A 2017 The importance of drying compressed air *Parker Hannifin* accessed 28 June 2017 Retrieved from: http://blog.parker.com/facts-you-need-to-know-about-drying-compressed-air
[9] Brennen C E 2005 Fundamentals of multiphase flow *Cambridge university press*
[10] Sokolichin A, Eigenberger G, Lapin A and Lübert A 1997 Dynamic numerical simulation of gas-liquid two-phase flows Euler/Euler versus Euler/Lagrange *Chemical engineering science* **52** pp 611–626
[11] Fluent Inc. 2003 Fluent 6.1 user's guide p 1093

IOP Conf. Series: Materials Science and Engineering **243** (2017) 012018 doi:10.1088/1757-899X/243/1/012018

Optimal piston crevice study in a rapid compression machine

Oku Nyong[1], Robert Woolley[2], Simon Blakey[3], Ehsan Alborzi[4]
[1-4] Low Carbon Combustion Centre, University of Sheffield, UK

E-mail: meq11oen@sheffield.ac.uk

Abstract. Multi-dimensional effects such as vortex generation and heat losses from the gas to the wall of the reactor chamber have been an issue to obtaining a reliable RCM data. This vortex initiates a flow in the relatively cold boundary layer, which may penetrate the core gas. This resulting non-uniformity of the core region could cause serious discrepancies and give unreliable experimental data. To achieve a homogenous temperature field, an optimised piston crevice was designed using CFD modelling (Ansys fluent). A 2-Dimensional computational moving mesh is assuming an axisymmetric symmetry. The model adopted for this calculation is the laminar flow model and the fluid used was nitrogen. To get the appropriate crevice volume suitable for the present design, an optimisation of the five different crevice volume was modelled which resulted to about 2-10% of the entire chamber volume. The use of creviced piston has shown to reduce the final compressed gas temperature and pressure in the reactor chamber. All the crevice volumes between 2-10% of the chamber volume adequately contained the roll up vortexes, but the crevice volume of 282 mm^3 was chosen to be the best in addition to minimising the end gas pressure and temperature drop. The final pressure trace from experiment shows a reasonable agreement with the CFD model at compression and post compression stage.

1. Introduction

Rapid Compression Machine (RCM) is an essential tool when investigating auto-ignition delay time and chemical kinetics of hydrocarbon fuels. It has an advantage of producing longer compression time of more than 20ms. Its accessibility to study combustion at conditions relevant to engines and gas turbines than other reactor facilities such as flow reactor, jet stirred reactor and shock tubes. Detailed combustion kinetic mechanism could be validated using an RCM experiments [1-5]. This help provides a better understanding of the ignition phenomena that is very pertinent to improving the performance efficiency and emission characteristic of engines[6].

One of the challenges confronting the use of RCM's designs is the complex fluid dynamics features exist in the reaction chamber. These feature result from the piston motion; a roll up vortex is formed creating a non-uniform temperature profile in the chamber [7-11]. The gases in the relatively cold boundary layer mix with the hot core gas resulting in a distribution of temperatures. This hinders the homogeneity of the core region where combustion kinetics is bound to exist. The mixing of cold boundary gases with the core gas in the chamber has a negative impact on experimental data since it causes thermal stratification within the charge making it difficult to interpret results and giving misleading results. To adequately study the combustion kinetic in RCM, it should be designed to minimise fluid motion effect that could lead to a discrepancy in data. And the effect of heat loss should also be taken into account to define the reaction chamber temperature through a simple model adequately. The model is based on a common assumption that there is an adiabatic core in the RCM, which means the temperature can be calculated. In this context, the 'Core Area' is referred to the region where the bulk gases exist in the reactor chamber and are not influenced by heat loss. The temperature distribution with the combustion chamber is typically controlled with the use of a crevice on the edge of the piston, which suppresses vortex formation [9, 11, 12]. This ideal first was initiated by Park[13] and further developed by Lee and Hochgreb[14]. The improvement was made on Park's design by changing the overall geometry of the crevice as well as increasing the original cross-sectional volume of the crevice. It was also proven that their piston crevice was adequate to suppress the roll up vortex generated in their RCM. Further computational fluid dynamics work has been used to study the physic of the roll up vortex structure, formation, temperature homogeneity[8] and also have been characterised by experiment[15] and computational investigations[12, 16-19]. The design of the crevice depends on the boundary layer thickness, which itself depends on the conditions during compression [12].

In this work, CFD study has been performed to determine the right piston crevice design that would conveniently contain the roll up vortexes and maintain a homogenous environment that would be conducive for autoignition study in the present RCM. The CFD work aimed to determine a piston crevice volume that would maintain a uniform temperature profile, minimise the compressed gas pressure and temperature in the reactor chamber.

2. Computational specification

The RCM has a cylindrical combustion chamber of 40 mm bore. The combustion chamber is oriented with the piston moving horizontally and has a stroke of 142 mm. A transient 2-D computational mesh was used, axisymmetric symmetry was assumed, and the resolution increased near the walls. Modelling was performed using Ansys Fluent. The actual machine was not truly symmetrical, because of the inlet manifold, air inlet port, and pressure transducers have modified the geometry. No slip is assumed at the cylindrical wall boundary, a uniform wall temperature of 298 K was also assumed. The compression time was approximately 30.8 ms from an initial pressure of 1bar. A mesh sensitivity study was carried out using four different angles for every time step. The highest step size (δt) was 0.25 and subsequent time step was generated by dividing by a factor of 2 to get the following 0.0125, 0.0625, 0.03125 and 0.015625. It is observed that time step greater than 0.03125 resulted in negative cell volume. To limit the computational time as well as avoiding convergence problems with the use of higher time step. A value of 0.03125 was used corresponding to the time-step size of 22.57μs for both compression stroke and post compression period was reported in all the CFD simulations. The thermal properties were generated from the NIST [20] database and specified as temperature dependent polynomials. The governing equations for the model are the conservation of energy, momentum and mass. The laminar flow model was adopted as it has been shown that it can adequately describe the experimental pressure history and velocity field inside a rapid compression machine [16, 21, 22]. The segregated implicit solver with pressure-implicit split-operator (PISO) algorithm [23] was used for pressure-velocity coupling. The pressure staggering option (PRESTO) was chosen as it prevents errors from the interpolation and pressuregradient assumptions on boundaries are prevented [24].

This scheme works better for problems with high body forces (swirl) and the second order upwind discretization for density and momentum. The second order upwind scheme changes a differential equation into an algebraic equation by Taylor series. It is more preferred than the first order scheme because it is more accurate. However, is more computationally expensive. The Navier-Stokes equation is solved along with the energy and species transport equations for all species. This evaluates a set of equations dependent on the input in fluent is represented by the following equations based on the 2D modelling.

$$\frac{\partial \boldsymbol{\rho}}{\partial \boldsymbol{t}} + \boldsymbol{\nabla}.(\boldsymbol{\rho u}) = \mathbf{0} \tag{1}$$

$$\frac{\partial \boldsymbol{u}}{\partial \boldsymbol{t}} + (\boldsymbol{u}.\boldsymbol{\nabla})\boldsymbol{u} = -\frac{\mathbf{1}}{\boldsymbol{\rho}}\boldsymbol{\nabla p} + \frac{\boldsymbol{\mu}}{\boldsymbol{\rho}}\boldsymbol{\nabla}^2\boldsymbol{u} + \boldsymbol{F} \tag{2}$$

$$\boldsymbol{\rho}\left(\frac{\partial \boldsymbol{E}}{\partial \boldsymbol{t}} + \boldsymbol{u}.\boldsymbol{\nabla E}\right) - \boldsymbol{\nabla}.(\boldsymbol{k_H}\,\boldsymbol{\nabla T}) + \boldsymbol{p\nabla}.\boldsymbol{u} = \mathbf{0} \tag{3}$$

Where u is the velocity field vector, T is the temperature, p is the pressure, E is the internal energy, ρ is the density,μ is the viscosity,K_H is the heat conduction coefficient and F is the force per unit mass. This equations 1, 2 and 3 are the continuity, momentum and, energy equation respectively. The above equation cannot be solve to give analytical result, but solved in a discretized form. The Ansys fluent operate in a fashion by developing a result using the finite volume method and integrating the above equation to give the desired result.

The velocity profile was derived from raw experimental pressure trace as employed by Mittal [25] shown in figure 1. The piston initially starts from rest and travels with a constant velocity at a point; it follows a constant speed before decelerating to rest at a constant rate. The section is divided into a three-time period of the piston motion, which is illustrated in equation 4. This equation is relevant for the numerical calculation where the parameters are used to model the compression stroke of the RCM.

$$t_{comp} = t_{accel} + t_{const} + t_{decel} \quad (4)$$

Where t_{comp} overall compression time, t_{accel} acceleration time, t_{const} constant velocity time and t_{decel} deceleration time. This profile is used as the user define function in the CFD for describing the piston trajectory.

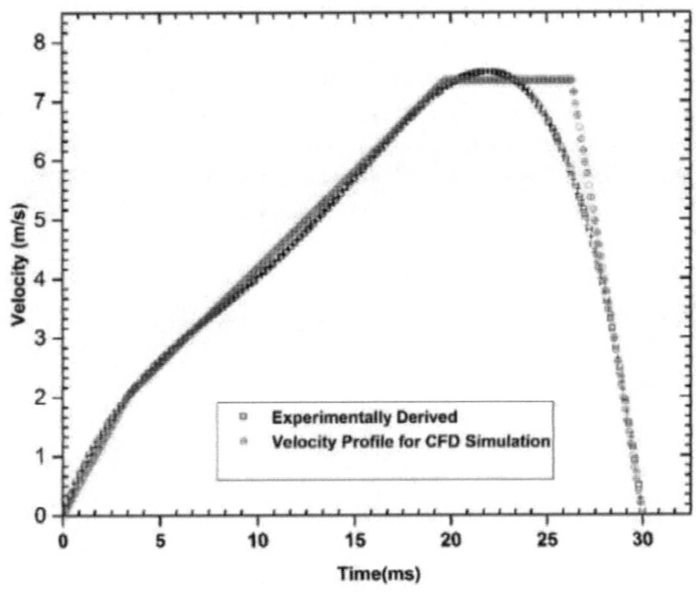

Figure 1. Piston velocity profile for 4 bar driving pressure obtained from the experimental pressure trace

Table 1 shows the dimensions of the piston crevice volume in mm.

Table 1. Dimensions of the piston crevice volume in mm

	Crevice Volume (mm^3)	Clearance between piston and wall	Inclined Angle.			Height of Crevice
			a	b	c	d
1	282	0.2	15°	2.53	4	14
2	564	0.2	15°	3.58	4	14
3	846	0.2	15°	4.39	4	14
4	1128	0.2	15°	5.06	4	14
5	1143	0.2	15°	5.66	4	14

Figure 2, shows a typical computational mesh at the end of the stroke for flat and crevice piston. The mesh shown to consist of 20,042 and 26,042 cells respectively, fine grids were used on the walls and piston face and coarse meshes in between the piston and the top dead centre (TDC) to adequately capture the fluid flow taking place in the chambers. Table 1, shows five different crevice volume that was considered in this study.

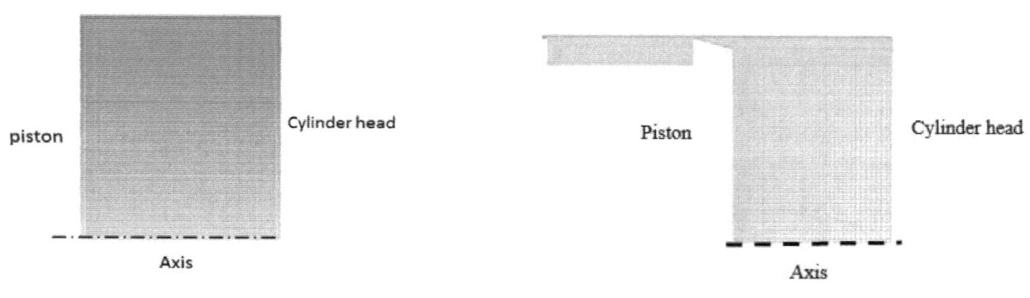

(a) (b)

Figure 2. 2D computational grid for Flat and Crevice piston.

3. Results

3.1 Flat piston

Figure 3, shows the compressed gas temperature contour and velocity profile at the end of compression and up to 17 ms post compression time. The use of flat piston produces a vortex which is demonstrated in the figure below. At the post-compression stage where ignition delay is bound to occur the vortex brought to settle in the core region of the chamber and found to be severe with time distorting the homogeneity of the temperature field in the chamber.

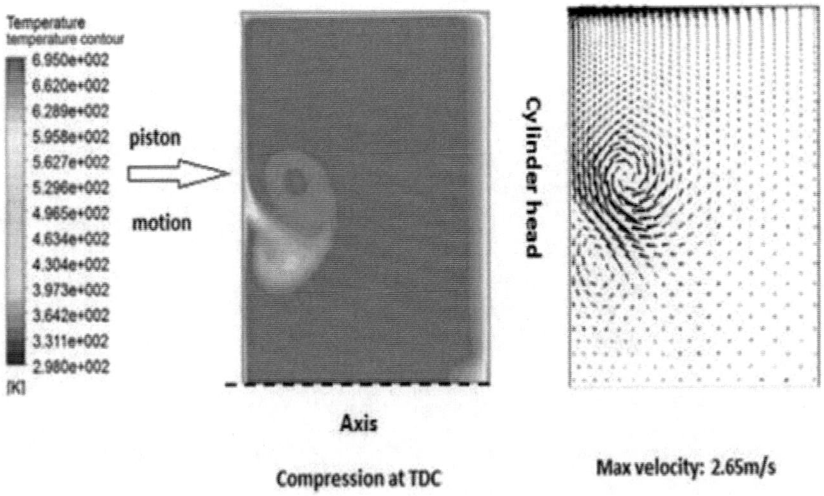
Temperature
temperature contour
6.950e+002
6.620e+002
6.289e+002
5.958e+002
5.627e+002
5.296e+002
4.965e+002
4.634e+002
4.304e+002
3.973e+002
3.642e+002
3.311e+002
2.980e+002
[K]
piston
motion
Cylinder head
Axis
Compression at TDC
Max velocity: 2.65m/s

(a)

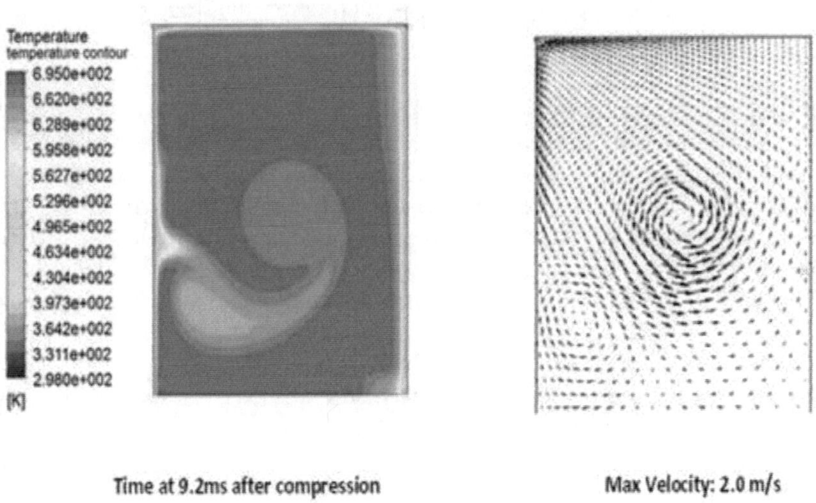
Temperature
temperature contour
6.950e+002
6.620e+002
6.289e+002
5.958e+002
5.627e+002
5.296e+002
4.965e+002
4.634e+002
4.304e+002
3.973e+002
3.642e+002
3.311e+002
2.980e+002
[K]
Time at 9.2ms after compression
Max Velocity: 2.0 m/s

(b)

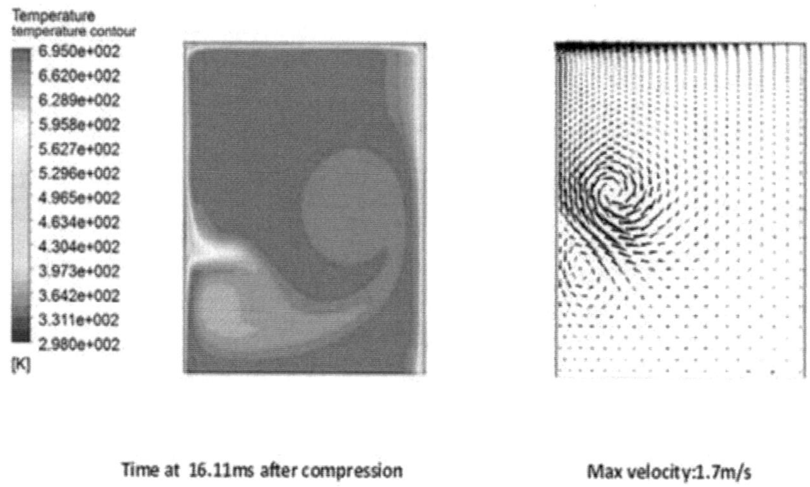

(c)

Figure 3. Contour of velocity and temperature profile for Flat piston head at Ti = 298K, Pi = 0.7 bar. (a). Compression time at TDC (b). 9.2 ms after post compression time (c). 16.11 ms after post compression time.

3.2 Crevice piston

The piston crevice with 2% of the entire chamber volume, $282mm^3$, was the best at suppressing the boundary layer and providing uniform temperature field with a minimum drop in compressed gas temperature and pressure. Figure 4, presents the contour temperature profile for different crevice volume from (a-e). All the crevice volumes used contained the vortex but tends to have a reduced compressed gas temperature/pressure as the crevice volume is increased from 282 – 1410 mm^3. Figure 5, shows the isometric view of the crevice piston that was finally machined for the present RCM.

Figure 6(a), shows the final compressed gas pressure at the current condition and in comparison with the experimental pressure trace. There seems to be an agreement between the CFD model and the experiment. Matching pressure traces gives an estimated compressed gas temperature as shown in Figure 6(b).

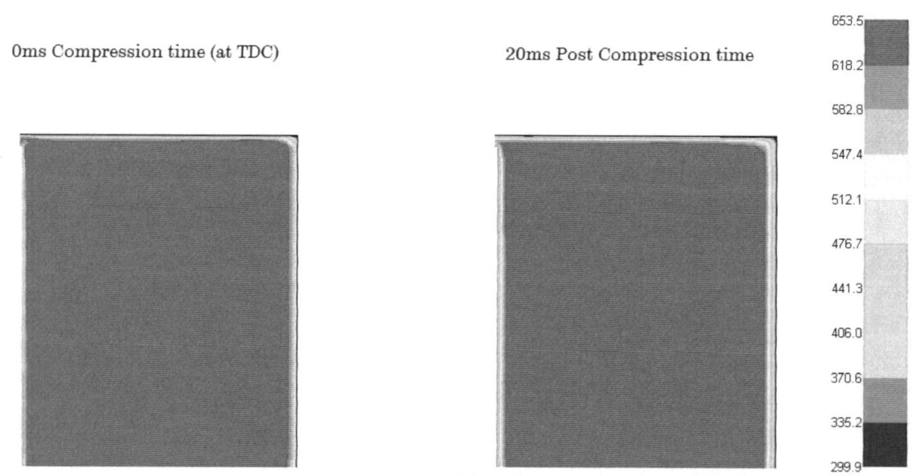

(a)

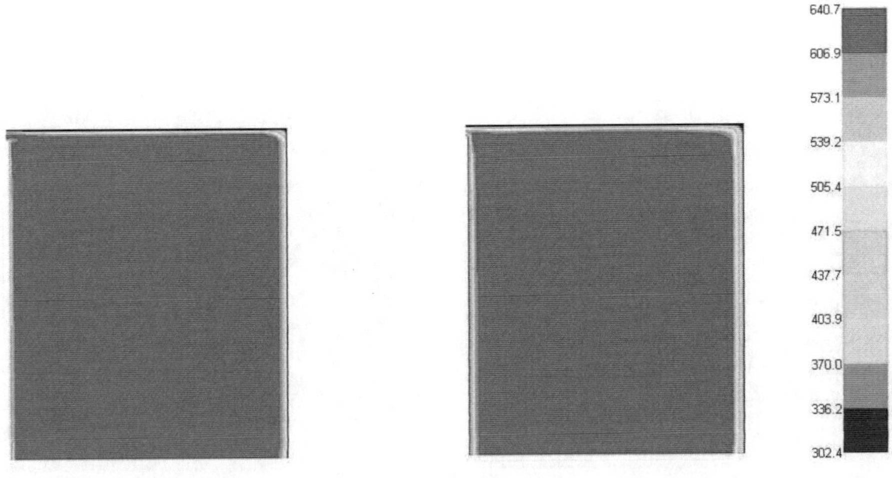
640.7
606.9
573.1
539.2
505.4
471.5
437.7
403.9
370.0
336.2
302.4

(b)

632.3
599.2
566.2
533.2
500.2
467.2
434.2
401.2
368.1
335.1
302.1

(c)

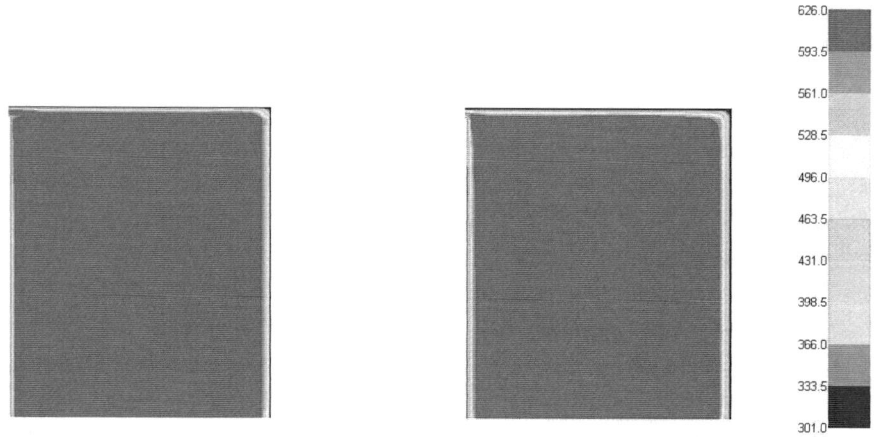

(d)

(e)

Figure 4. Shows the contour of the compressed gas temperature profile for piston crevice head with angled channel. (a) volume = 282 mm^3 (b) volume = 564 mm^3 (c) Volume = 846 mm^3 (d) Volume = 1128 mm^3 (e) volume 1410 mm^3

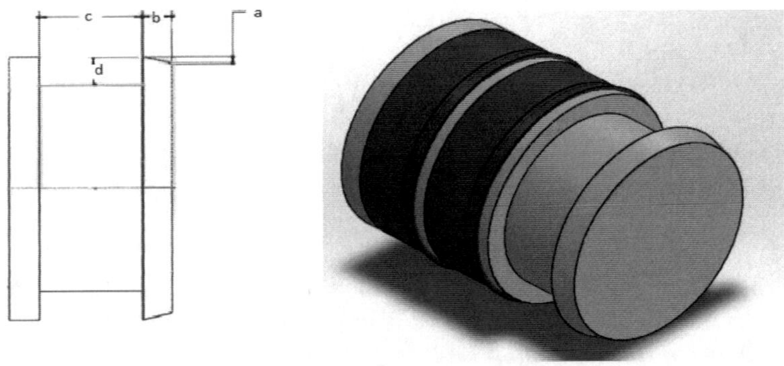

Figure 5. shows (a). The diagram of the crevice region of the piston (b). An isometric view of the angle channel design showing the crevice volume.

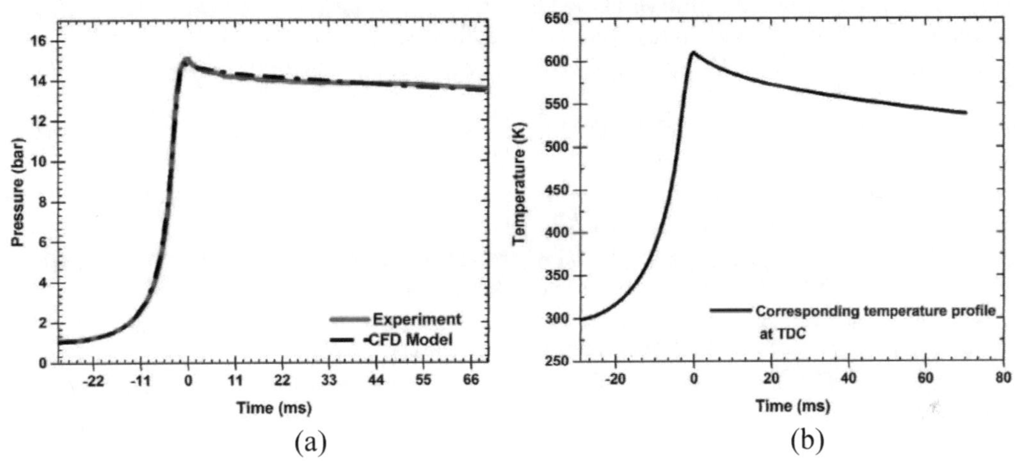

Figure 6. (a) Comparison of Experimental pressure trace with the CFD Model (b) End of compressed gas temperature corresponding to P_C =15 bar, T_C = 615 K.

4. Conclusion

The CFD modelling demonstrated that the use of flat piston generates a roll-up vortex, which could distort the temperature uniformity of the chamber. The results showed that proper machining a crevice on the peripheral surface of the piston could control the vortex build up from the walls of the chamber and maintain the temperature homogeneity of the chamber. The simulation has shown that piston crevice with about 2% of the entire reaction chamber volume was the best for the present RCM. The model result from CFD was machined and tested experimentally. At both compression and post compression period, the model was in agreement with the experimental pressure trace.

References

[1] Westbrook C K, Pitz W J and Herbinet O 2009 A comprehensive detailed chemical kinetic reaction mechanism for combustion of n-alkane hydrocarbons from n-octane to n-hexadecane *Combustion and Flame* **156** 181-99

[2] Curran H J, Gaffuri P and Pitz W 2002 A comprehensive modeling study of iso-octane oxidation *Combustion and flame* **129** 253-80

[3] Dooley S, Curran H J and Simmie J M 2008 Autoignition measurements and a validated kinetic model for the biodiesel surrogate, methyl butanoate *Combustion and Flame* **153** 2-32
[4] Mittal G, Burke S M and Davies V A 2014 Autoignition of ethanol in a rapid compression machine *Combustion and Flame* **161** 1164-71
[5] Mittal G and Sung C-J 2009 Autoignition of methylcyclohexane at elevated pressures *Combustion and Flame* **156** 1852-55
[6] Cowart J S, Keck J C and Heywood J B 1990 Engine knock predictions using a fully detailed and a reduced chemical kinetic mechanism *Proc. Combust. Inst.* **23** 1055-62
[7] Griffiths J F, Jiao Q and Schreiber M 1992 Development of thermokinetic models for autoignition in a CFD Code: Experimental validation and application of the results to rapid compression studies *Symposium (International) on Combustion* **24** 1809-15
[8] Desgroux P, Gasnot L and Sochet L R 1995 Instantaneous temperature measurement in a rapid-compression machine using laser Rayleigh scattering *Applied Physics B Laser and Optics* **61** 69-72
[9] Lee D and Hochgreb S 1998 Rapid compression machines: Heat transfer and suppression of corner vortex *Combustion and Flame* **114** 531-45
[10] Griffiths J F, MacNamara, J P and Mohamed C 2001 Temperature fields during the development of autoignition in a rapid compression machine *Faraday Discussions* **119** 287-303
[11] Mittal G and Sung C J 2006 Aerodynamics inside a rapid compression machine, *Combustion and Flame* **145** 160-80
[12] Würmel J and Simmie J M 2005 CFD studies of a twin-piston rapid compression machine *Combustion and Flame* **141** 417-30
[13] Park P and Keck J C 1990 Rapid compression machine measurements of ignition delays for primary reference fuels *SAE Technical Paper*
[14] Park P 1990 *Rapid compression machine measurements of ignition delays for primary reference fuels* (Cambridge:Massachusetts Institute of Technology)
[15] Mittal G 2006 A Rapid Compression Machine – Design, Characterization, and Autoignition Investigations (Cleveland:Case Western Reserve University)
[16] Mittal G and Sung C-J 2006 Aerodynamics inside a rapid compression machine *Combustion and Flame* **145** 160-80
[17] Brett L, MacNamara J and Musch P 2001 Simulation of methane autoignition in a rapid compression machine with creviced pistons *Combustion and flame* **124** 326-29
[18] Lee D and Hochgreb S 1998 Rapid Compression Machines: Heat Transfer and Suppression of Corner Vortex *Combustion and Flame* **114** 531-45
[19] Griffiths J F, Piazzesi R and Sazhina E M 2012 CFD modelling of cyclohexane auto-ignition in an RCM *Fuel* **96** 192-203
[20] Lemmon E, McLinden M and Friend D 2011 Thermophysical Properties of Fluid Systems in NIST Chemistry WebBook *NIST Standard Reference Database Number 69*, Ed Linstrom P J and Mallard WG (Gaithersburg:National Institute of Standards and Technology)
[21] Goldsborough S S and Potokar C J 2007 The influence of crevice flows and blow-by on the charge motion and temperature profiles within a Rapid Compression Expansion Machine used for chemical kinetic (HCCI) studies *SAE Paper* 01-0169
[22] Mittal G and Chomier M 2013 Interpretation of experimental data from rapid compression machines without creviced pistons *Combustion and Flame*
[23] Issa R I, Gosman A D and Watkins A P 1986 The computation of compressible and incompressible recirculating flows by a non-iterative implicit scheme *Journal of Computational Physics* **62** 66-82
[24] Anderson J D 1995 *Computational fluid dynamics* vol 206 (New York: McGraw-Hill)
[25] Mittal G, Raju M P and Sung C-J 2008 Computational fluid dynamics modeling of hydrogen ignition in a rapid compression machine *Combustion and Flame* **155** 417-28

IOP Conf. Series: Materials Science and Engineering **243** (2017) 012019 doi:10.1088/1757-899X/243/1/012019

Computation of tabs and wing generator modification on temperature distribution and flow of free jet

I Piya[1], W Palasai[1] and C Nuntadusit[2]
[1] Department of Mechanical Engineering, Princess of Naradhiwas University, Naradhiwas, Thailand
[2] Department of Mechanical Engineering, Faculty of Engineering, Prince of Songkla University, Songkhla, Thailand

E-mail: piya.ibroheng@gmail.com

Abstract. The mixing process is important improve the performance of jet-engineering equipment. The purpose research, aim to study temperature distribution might be simulate mixing layer of jet flow with surrounding fluid.In this study measure the temperature at the position the jet exit. Z =1D, 2D, 4D and 6D. Calculate the temperature coefficient and a computational fluid dynamic (CFD) using ANSYS, Fluent (V.15.0) was selected to this computation. The investigation model was jets discharging from pipe nozzle installing wing generator having an inner diameter d=28.15 mm and wall fitted to the outlet of the jet. Promoters were attached at the nozzle exit. 4 types of turbulence promoters were triangular tab with tip angle of 45° and 90° and vortex generators with attack angle of 45° and 60°. In addition, the conventional pipe nozzle was studied to comparing as base results and modifications. Installing at 2 and 4 positions give the measuring temperature. The Reynolds number of test fluid was constant at Re=29,500. Increase the strength of jet mixing was found the case of vortex generator installed with 4 positions, attack angle of 60° gives the strongest mixing in jet when comparing with the other cases.

1. Introduction

Jet mixing normally are used in many industrial and engineering applications, because their common devices and mixing efficient are require for good performance in all mixing applications. Primary applications of jet mixing include combustion industries air distribution system industrial mixing and cooling of electronic packages

The purpose of passive techniques to enhance the mixing process for achieving the optimum performance of jet mixing. Many techniques were used surface or geometrical modifications to the flow channel by incorporating inserts or additional devices. These techniques not require any direct input of external power source to improve the entrainment.

Indeed, jet flow have been extensively studied in previous work with demand increase the performance and how interaction of the media between the jet fluid and surrounding fluid [1,2]. The Parameter Which affects the characteristics of independent jets[3,4].When the Reynolds number is higher, the flow pattern become turbulent which eddy or swirl occurs reverse current in the liquid better mixture and faster. However, increase the turbulent is require pumping power.

The insert tab or wing vortex generator were affects to the mixing of layers characteristics. Then indicates for mixing jet flow and surrounding [5,6].

The previous research show decrease of temperature or flux the jet mass flow that indicates higher induction of the mixture[7]. In addition, research has measuring intensity fluid in position per time and define volume flow rate of plane [8, 9].

The work presents investigates the effects of tab enhancement and vortex generators region exit nozzle. This research to compare experimental data and numerically investigate flow and reduction of media temperature which indicate mixing flow layers. The results will be compared to the case of conventional jet under the same of mass flow rate.

2. Experimental setup and method

2.1. Experimental model and parameters

The experimental setup of the jet discharging from the round pipe and vortex generator arrangement is shown in figure 1. An origin of the Cartesian coordinates was located on the probe measure velocity as shows in Fig.1. The X-, Y-axes are normal to the axial jet, and Z-axis is identical direction to the axial jet.

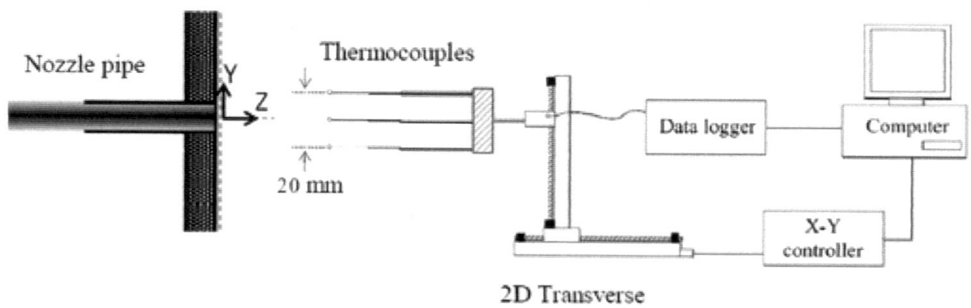

Figure 1. Computational model and boundary conditions.

2.2. Experimental setup

The experiment was conducted for jet from round pipe with installing tabs and vortex generator at jet outlet. The 4 types of configuration and orientation at jet outlet are shown in Figure . 2. The opposite positions of delta tabs insert with 90° and 45° and 4 positions of delta tabs with 45°.

The conventional pipe without tab was also studied to compare the results of the jet with delta tabs.The comparison of mixing layer characteristics on the constant jet mass flow rate (Re=29,500). During the pipe with delta test run, the jet mass flow rate was controlled in based on the identical with the conventional round pipe for the required Reynolds number. The inner diameter of round pipe was D=28.15 mm, and measuring distance at jet outlet were H=1D, 2D, 4D and 6D.

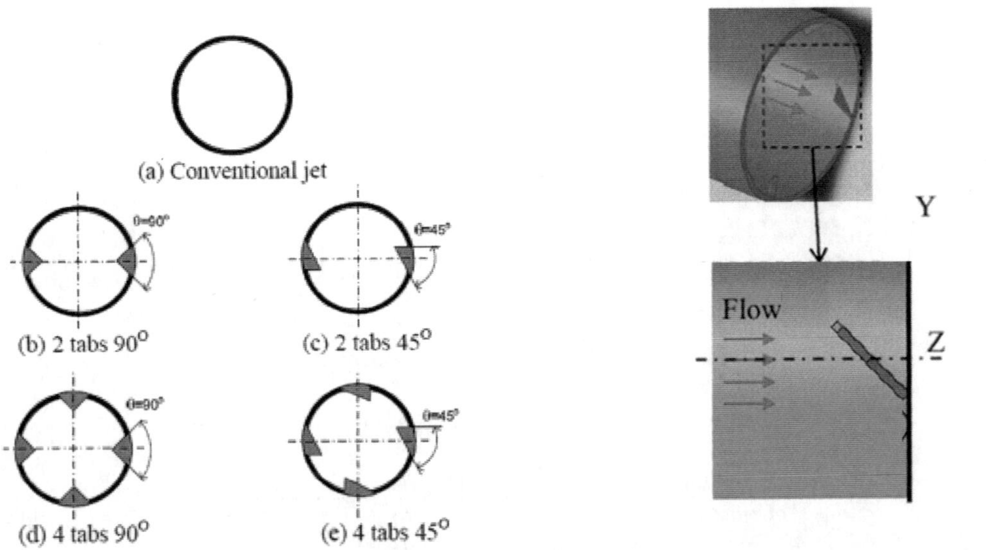

(a) Tab arrangement (b) vortex generator arrangement

Figure 2. Computational model and boundary conditions.

2.3. Data reduction

For average temperature (T), each point can be calculated from the following equation.

$$\bar{T} = \frac{\sum_{i=1}^{N} T_i}{N} \tag{1}$$

Temperature coefficient (ŋ) can be calculated from the following relationship

$$\eta = \frac{\bar{T} - T_\alpha}{T_j - T_\alpha} \tag{2}$$

where T_i is the temperature measured at any time

N is the amount of data

T_α is the ambient temperature

T_j is the jet temperature

3. Numerical Simulation

3.1. Computational model and boundary conditions

ANSYS FLUENT (version 15.0) was adopted in the numerical simulation programming. Computational model were divided in three sections and boundary conditions as shown in Figure 3. The first section were included the main pipe nozzles diameter (d) =28.15 mm. The air-induced ducts were assembled at the end of main pipe nozzle in the second section.

IOP Conf. Series: Materials Science and Engineering **243** (2017) 012019 doi:10.1088/1757-899X/243/1/012019

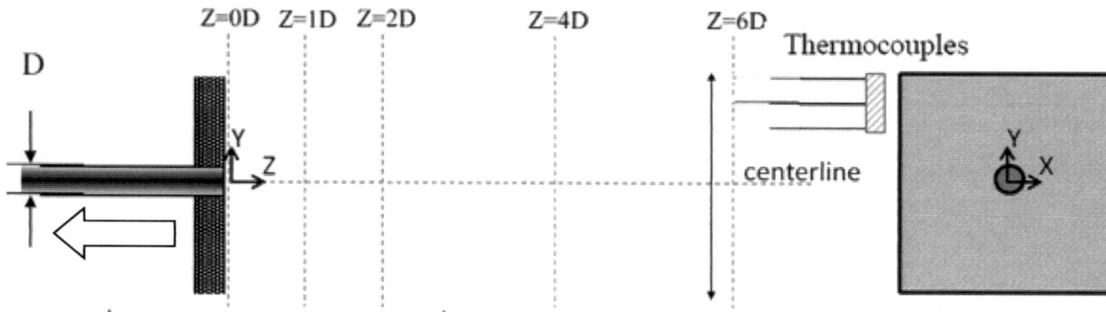

Figure 3. The diagram of experimental setup

Table 1. The details of boundary conditions

Boundary condition	Define
Constant velocity inlet	16 ms^{-1} (Re=29,500)
Front and side surface	Pressure Outlet
Top and bottom surfaces	Pressure Outlet

The length (L), inner diameter (D) of air-induced duct and the measuring probe distance (H) were fixed at H=1D, 2D, 4D and 6D, respectively. The details of boundary conditions were summarized in Table 1.

3.2. Grid generation and grid dependency

The rectangular grid was mainly applied in this numerical model. The grids in region having high velocity gradient as near pipe nozzles surface were finely controlled. The generating grid on Z-X plane at center of nozzle is shown in Figure 4. The number of generated grid was varied to achieve an accurate solution by considering the dimensionless wall distances (y^+) of the first node less than 1, calculated from

$$y^+ = \frac{y_1 u_\tau}{v} \tag{3}$$

Where y^+ was the distance of the first node to the wall, u_τ was shear velocity, y_l is the distance to the nearest wall and v is the kinematic viscosity of air.

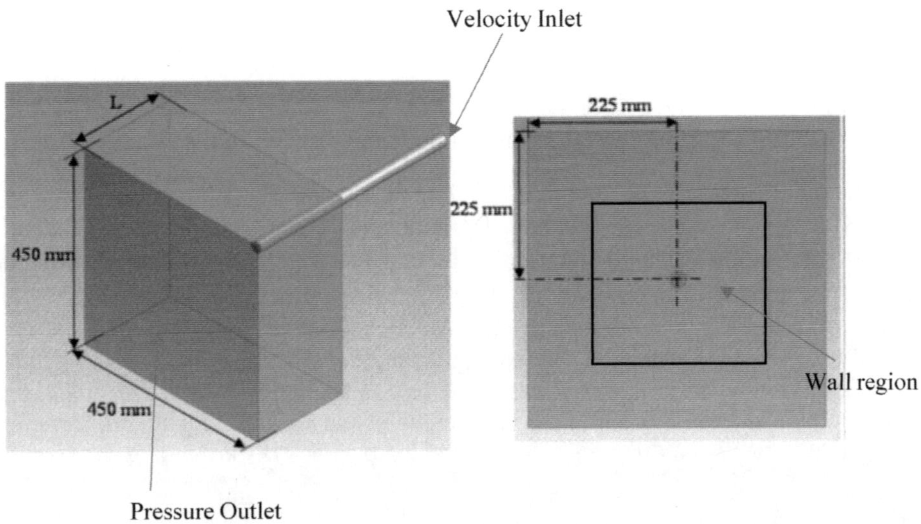

Figure 4. Computational model and boundary conditions.

The effects of grid dependency on y^+ distribution on the target surface passing. In this model, the number of grids at 767,365 elements was selected to numerical simulation. Because of the grids can be processed as similar square grid of 1,085,907 nodes is shown in Figure 5

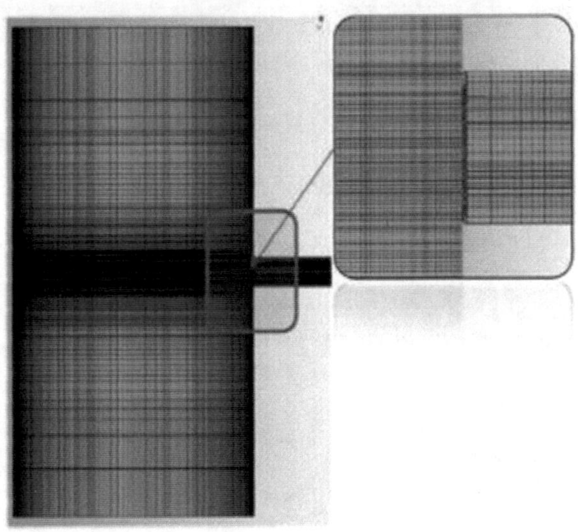

Figure 5. Generated rectangular grid for the numerical model.

3.3. Calculation method and algorithm
Computations were conducted by solving Reynolds averaged continuity and Navier-Stokes equations under existing boundary conditions. The SST k-ω turbulence model was selected for applying in this simulation. Since, this turbulence model has been adopted in solving many numerical simulations of jet problems. More accurately predicted the solutions of jet mixing problems with moderate computation cost. The SIMPLE algorithm was used with second order upwind scheme for all spatial discretization. The convergence of iterative solution was insured when the residual of all the

IOP Conf. Series: Materials Science and Engineering **243** (2017) 012019 doi:10.1088/1757-899X/243/1/012019

variables was less than the specified values. The specified value was 1×10^{-5} for continuity and momentum equations.

4. Results and discussion

This part presents the results of flow characteristics and temperature distribution characteristics.

4.1. Flow characteristics

Figure 6 shows the distribution of the velocity vector of the conventional jet found that distribution was quite symmetrical for all case. The velocity for 2 position flow direct position tab. The potential core has decrees area when installation triangular tab. However, triangular tab tip 90° has increase velocity and potential core region is distorted. Increasing triangular tab tip to 4 number as shown in Figure 7, it was found that the potential core are distorted cross shape and velocity flow of triangular tab tip 45° increasing. When adjust attack angle are 45° and 90° effect surrounding air is induce near of vortex generator position.

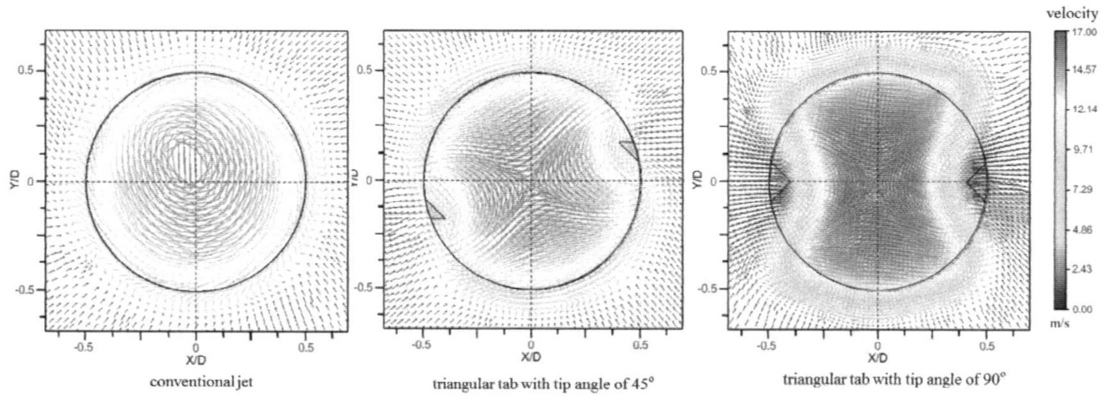

Figure 6. Vector of conventional and modification of jet nozzle on X-Y plane of position Z=1D ,Vj =16 m/s) (CFD results, Re=29,500).

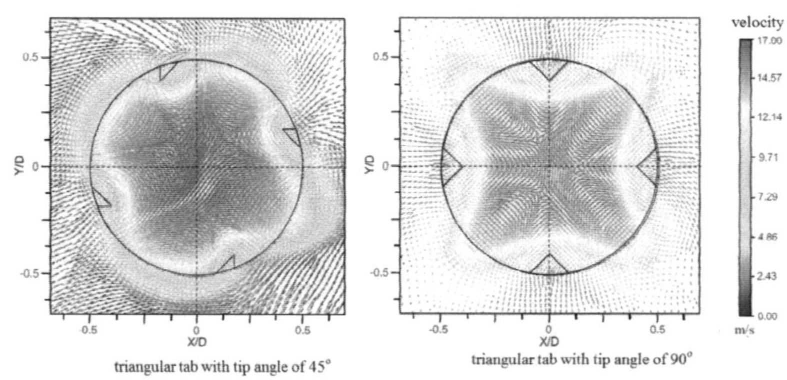

Figure 7. Vector of modification of jet nozzle at 4 position on X-Y plane of position Z=1D,Vj=16 m/s (CFD results, Re=29,500.

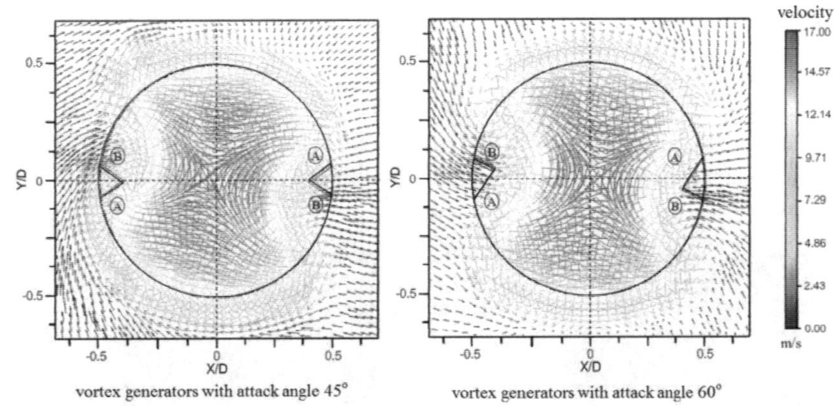

Figure 8. Vector of modification of jet nozzle with attack angle on X-Y plane of position Z=1D, Vj=16 m/s (CFD results, Re=29,500).

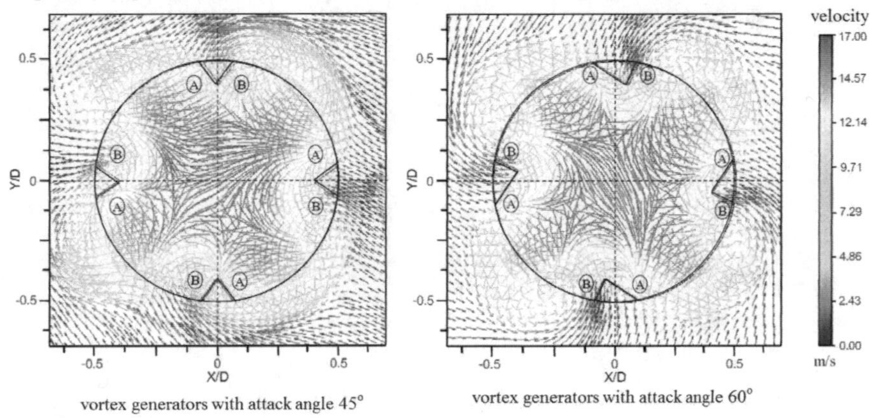

Figure 9. Vector of modification of jet nozzle at 4 position on X-Y plane of position Z=1D, Vj=16 m/s (CFD results, Re=29,500).

4.2. Temperature distribution characteristics

Distribution of temperature coefficients of free conventional jets was measured by the thermocouple with various points and input to calculated using equations (1 and 2).

The area the temperature coefficient is high represents the potential core of the jet and the low temperature coefficient represents the area the jet mixing with air surrounding. In the experiments, the Reynolds number of fixed to constant jet mass flow rate (Re=29,500). The jet temperature and ambient air temperature were Tj and T_{α} respectively on the XY plane. Each image shows the distribution of temperature coefficients at the distance from the outlet jet to Z =1D, 2D ,4D and 6D
In figure 10 to 18 show temperature distribution characteristics with various tab and attack angle.

For all Temperature distribution characteristics divided into 3 regions, first the temperature coefficient is higher (ŋ> 0.85) occurs at the center of the jet (Potential core). Second, the temperature coefficients is medium (0.85≤ŋ≤0.45), main reason shear layer effect on jet flow mixing with air ambient then reduce the temperature coefficient. Third, the range of temperature coefficients is low (ŋ≤0.45). The area around the jet mixing with the surrounding air, so that the temperature coefficient of the jet near the surrounding air.

Conventional jet case distance from exit jet to Z=6D has potential core (ŋ> 0.9). For other installation Tab and wing vortex generator has distorted potential core at position Z=2D.The

IOP Conf. Series: Materials Science and Engineering **243** (2017) 012019 doi:10.1088/1757-899X/243/1/012019

phenomenon of jet mixing with ambient air has disappeared. When distance to Z = 6D the distribution coefficient was similar conventional jet case.

The distribution temperature coefficient wide region distance from exit nozzle to Z = 4D for the attack angles 60° at 2 position. When Installation 4 position for wing generator has attack angle 60° increase the mixing layer. However. The effect of installing attack at the outlet causes a mixture between the jet and the surrounding fluid. More than similar conventional jet.

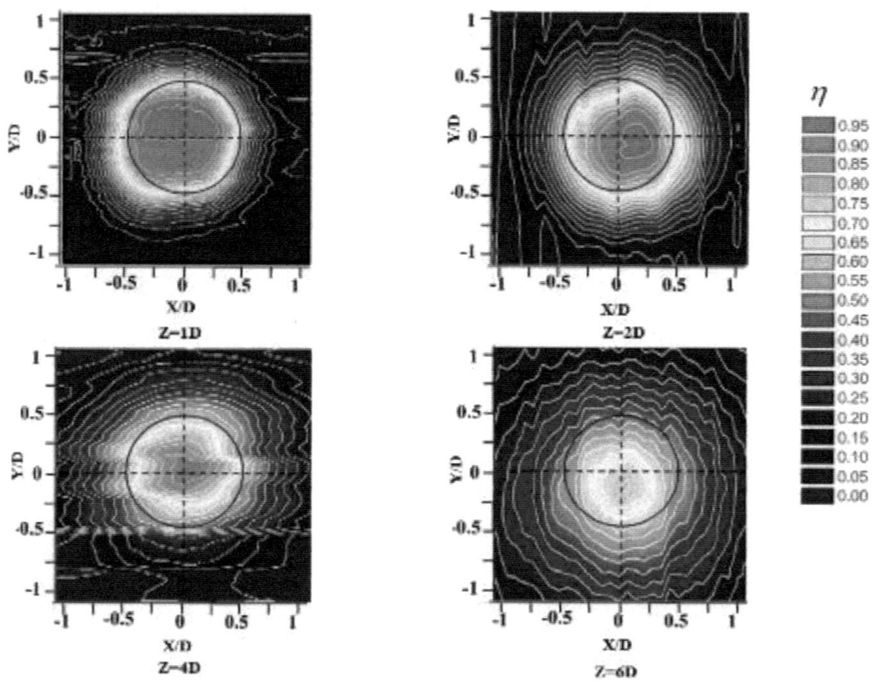

Figure 10. The temperature distribution characteristics (η) of the free jet conventional. On the XY plane at various z positions.

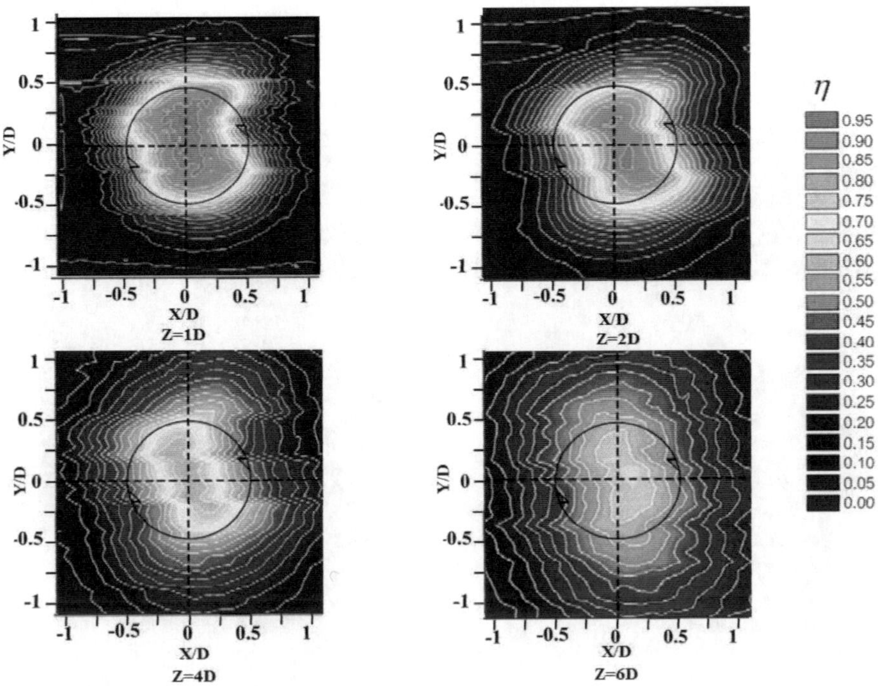

Figure 11. The temperature distribution characteristics (ŋ) of the delta tabs generator with 45° at 4 position. On the XY plane at various z positions.

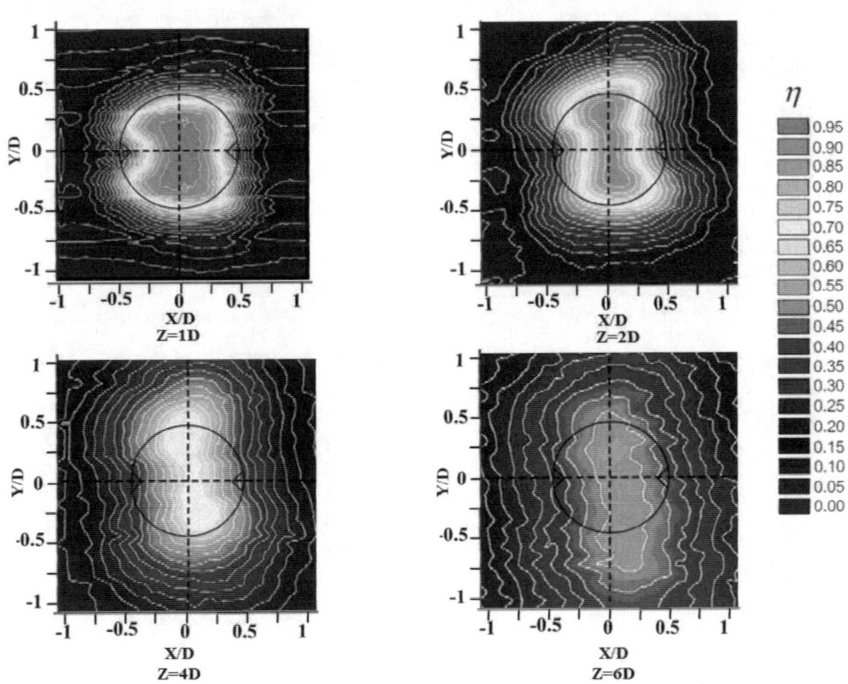

Figure 12. The temperature distribution characteristics (ŋ) of the delta tabs generator with 90°. On the XY plane at various z positions.

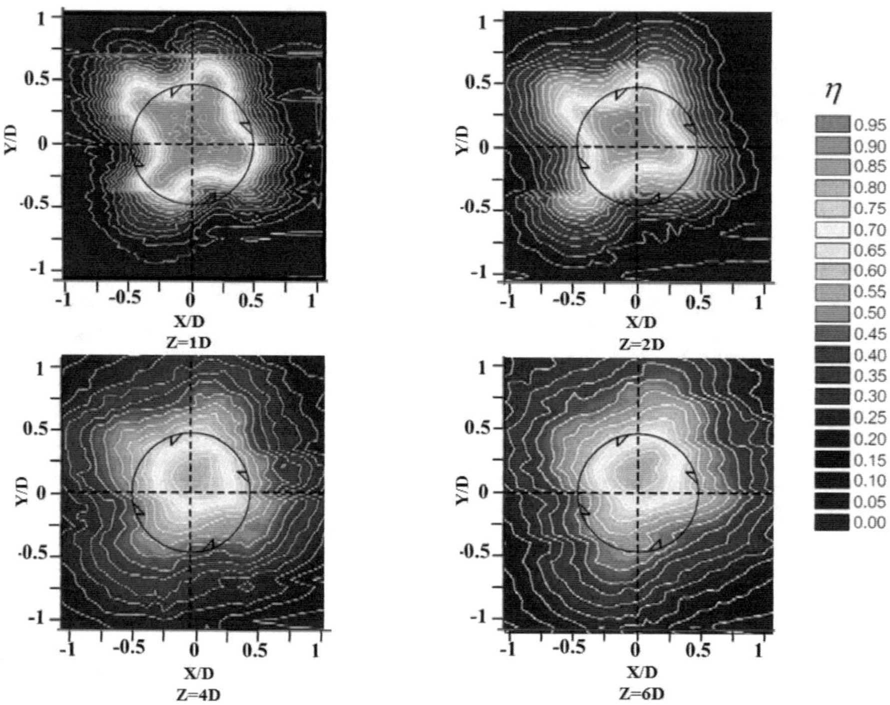

Figure 13. The temperature distribution characteristics (η) of the delta tabs generator with 45°. On the XY plane at various z positions.

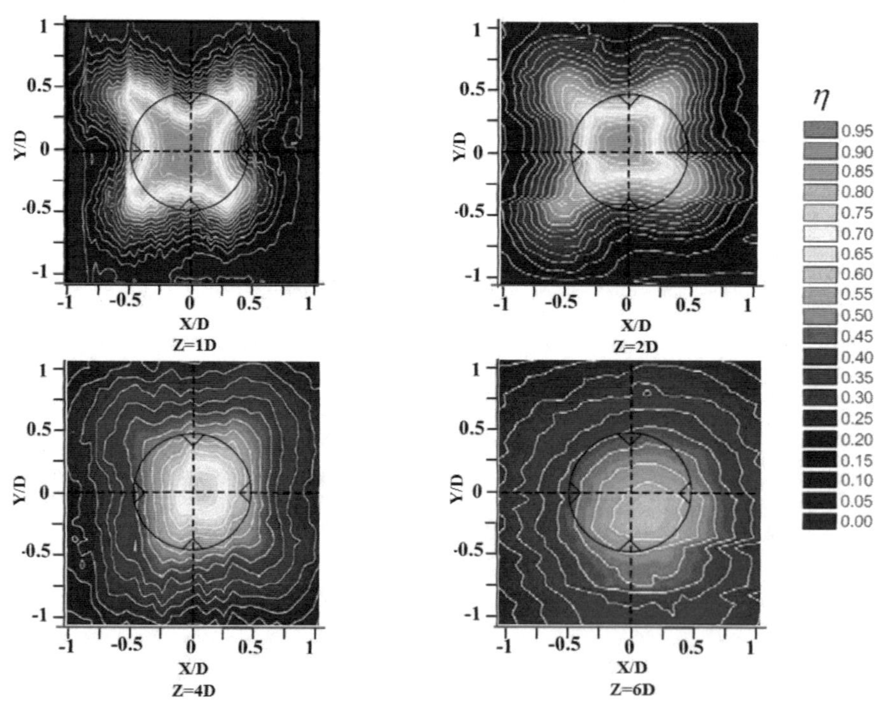

Figure 14. The temperature distribution characteristics (η) of the delta tabs generator with 90°. On the XY plane at various z positions.

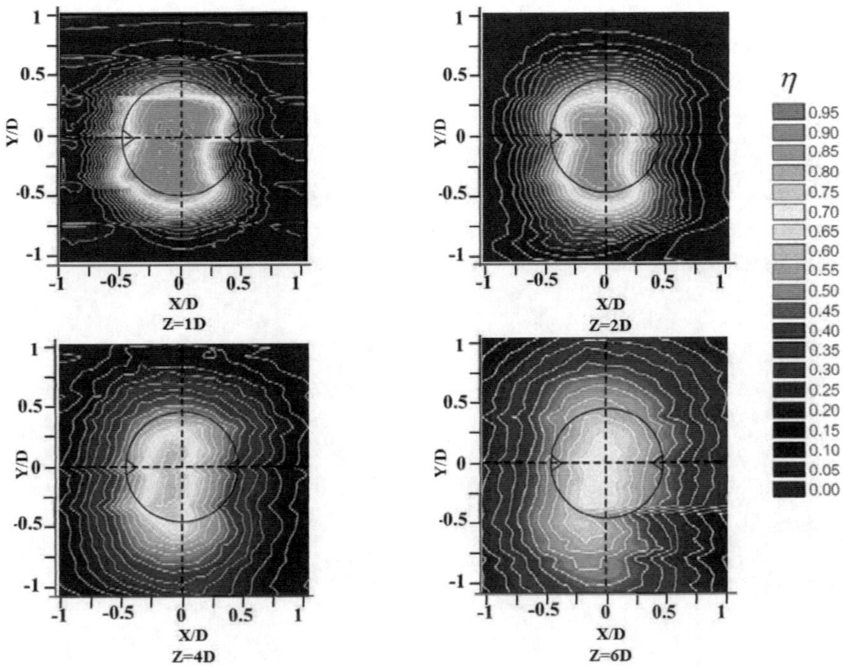

Figure 15. The temperature distribution characteristics (ŋ) of the wing generator attack angle 45°. On the XY plane at various z positions.

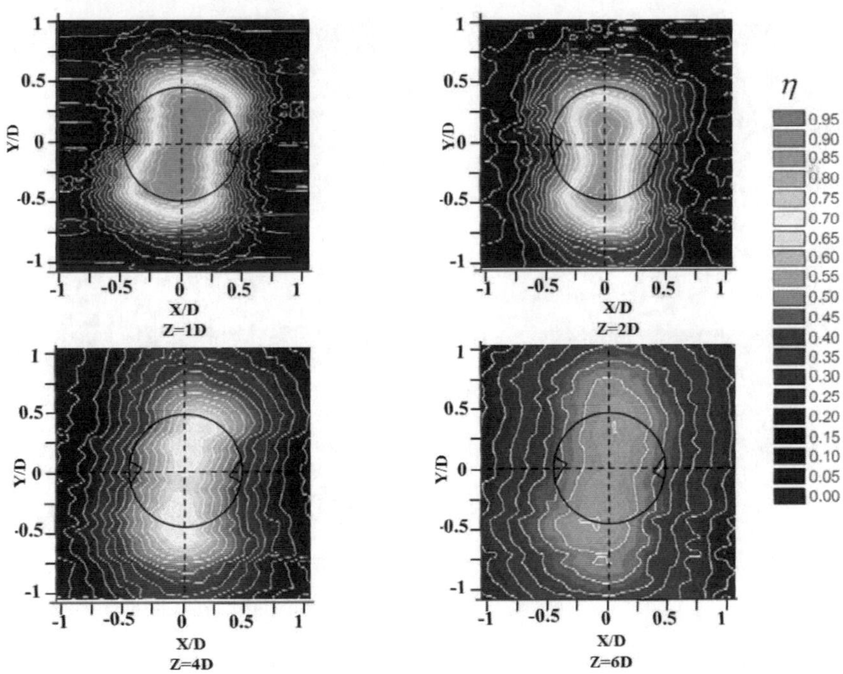

Figure 16. The temperature distribution characteristics (ŋ) of the wing generator attack angle 60°. On the XY plane at various z positions.

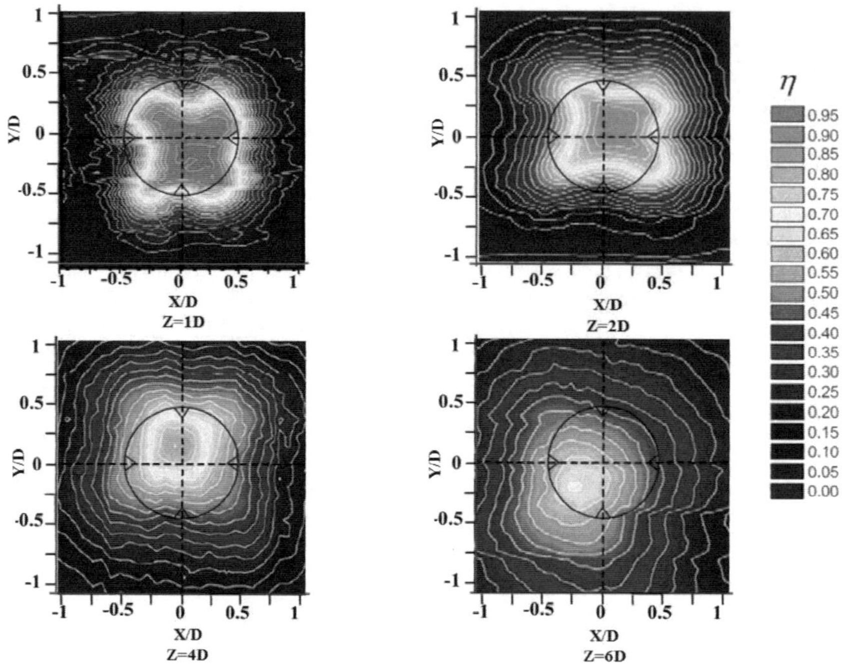

Figure 17. The temperature distribution characteristics (ŋ) of the wing generator attack angle 45°. On the XY plane at various z positions.

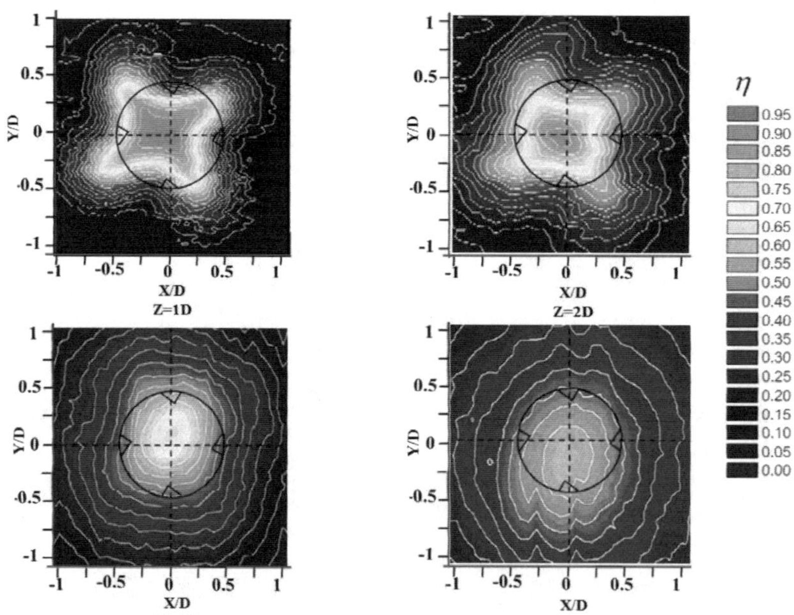

Figure 18. The temperature distribution characteristics (ŋ) of the wing generator attack angle 60°. On the XY plane at various z position.

IOP Conf. Series: Materials Science and Engineering **243** (2017) 012019 doi:10.1088/1757-899X/243/1/012019

5. Conclusions

In this paper, flow and Distribution of temperature coefficients characteristics of modification jet nozzle were studied experimentally and numerically. The results concluded as follow:

1) Tabs and wing generator attack are improve the flow cross section high intensity flow velocity and direction than conventional jet.
2) Tabs insert has high distribution of temperature coefficients than that of generator attack. The high temperature distribution is delta tabs generator with 90° due to high turbulent intensity on the flow area .
3) Attack angle of 60° have highest of temperature distribution then improve the mixing jet when compared with the other cases. Due to large of flow intensity.

For future work would be simulation study on effect of counter rotating vortex at near field region with temperature distribution.

References

[1] Gutmark E J and Grinstein F F 1999 Flow control with noncircular jets *Annual Review of Fluid Mechanics* **31** 239-272

[2] Seok J K 2005 Reynolds number effects on the behavior of a nonbuoyant round jet *Experiments in Fluids* **38** 801-812

[3] Brown G L and Roshko A 1974 On density effect and large structure in turbulent mixing layers *J. Fluid Mech* **64** 775-816

[4] Bernal L P and Roshko A 1986 Streamwise vortex structure in plane mixing layers *J. Fluid Mech.* **170** 499-525

[5] Ashforth-Frost S and Jambunathan K 1996 Effect of nozzle geometry and semi-confinement on the potential core of a turbulent axisymmetric free jet *Int. C. Heat Mass Transfer* **23** 155-162

[6] Bradbury L J S and Khadem A H 1975 The distortion of a jet by tabs *J. Fluid Mech.* **70** 801-813

[7] Reeder M F and Samimy 1996 The evolution of a jet with vortex generating tab: real time visualization and quantitative measurements *J. Fluid Mech.* **311** 73-118

[8] Kamotani Y and Greber I 1972 Experiment on a turbulent jet in cross flow *The American Institute of Aeronautics and Astronautics* **11** 1425-1429

[9] Maruyama T Ban Y and Mizushina T 1981 Jet mixing of fluids in tanks *Journal of Chemical Engineering of Japan* **15** 342–348

IOP Conf. Series: Materials Science and Engineering **243** (2017) 012020 doi:10.1088/1757-899X/243/1/012020

A CFD study on turbulent forced convection flow of Al_2O_3-water nanofluid in semi-circular corrugated channel

R K Ajeel[1] and W S I W Salim[1]
[1] Faculty of Mechanical and Manufacturing Engineering, Universiti Tun Hussein Onn Malaysia, 86400 Parit Raja, Batu Pahat
Johor, Malaysia

E-mail: malisya.ra75@gmail.com, wsaiful@uthm.edu.my

Abstract. The performance of heat exchangers especially for single phase flows can be enhanced by many augmentation techniques. One of the most popular method used is a passive heat transfer technique. Researchers have been quite active in the search of novel ways on heat transfer augmentation techniques using various types of passive techniques to increase heat transfer performances of heat exchanger. Computational Fluid Dynamics (CFD) simulations of heat transfer and friction factor analysis in a turbulent flow regime in semi-circle corrugated channels with Al_2O_3-water nanofluid is presented in this paper. Simulations are carried out at Reynolds number range of 10000-30000, with nanoparticle volume fractions 0-6% and constant heat flux condition. The results for corrugated channels are examined and compared to those for straight channels. Results show that the Nusselt number increased with the increase of nanoparticle volume fraction and Reynolds number. The Nusselt number was found to increase as the nanoparticle diameter decreased. Maximum Nusselt number enhancement ratio 2.07 at Reynolds number 30,000 and volume fraction 6%.

1. Introduction

The enhancement of convection heat transfer is a very interesting topic for different kinds of industrial and engineering applications and can be improved passively by changing the flow geometry and boundary conditions or by enhancing the thermophysical properties of the fluid. The ability of a fluid medium to transfer a large amount of heat across a small temperature gradient enhances the efficiency of energy conversion, as well as improves the design and performance of heat exchangers. Therefore, research on enhancement technique in such channels has become very prominent. For this purpose, using nanofluids as a cooling fluids in corrugated channels instead of traditional fluids can enhance thermal conductivity of the base fluids and thereby a further improvement in thermal performance of heat exchangers with a more compact design.

Various numerical and experimental studies on the forced convection flow of conventional fluids or nanofluids in channels exist in the literature [1-12]. Sunden and Skoldheden [1] experimentally studied the heat transfer and pressure drop in corrugated channels and smooth tubes. Sawyers [2] numerically and experimentally studied the effect of three dimensional hydrodynamics on the enhancement of heat transfer in corrugated channels for Reynolds numbers in the range of 9–149. Fabbri [3] studied laminar convective heat transfer in a channel composed of smooth and corrugated walls. Vasudevaiah and Balamurugan [4] studied theoretically the convective heat transfer in a corrugated microchannel. The transport equations of continuum theory were adopted under incompressible flow conditions. Wang and Chen [5] applied a simple coordinate transformation method and the spline alternating direction implicit method for determining the heat transfer rates for flow through a sinusoidally curved converging–diverging channel. Gradeck et al. [6] experimentally studied the effects of the hydrodynamic conditions on the enhancement of heat transfer for single phase flow in corrugated channels. Naphon [7–10] conducted numerical and experimental study of forced convection heat transfer and flow developments in a channel with V-corrugated upper and lower plates in which all configuration peaks lie in an in-phase arrangement. The results show that wavy angle and channel height have significant effect on the temperature distribution and flow development. It was found that the sharp edge of the wavy plate (V-shaped) has a significant effect on the enhancement of heat transfer. The turbulent flow of nanofluids with different volume fractions of nanoparticles flowing through a two-dimensional duct under constant heat flux condition was simulated by Rostamani et al. [11]. Ahmed et al. [12] investigated numerically of the turbulent forced convection of nanofluid flow in triangular-corrugated channel over Reynolds number ranges of 1000-5000. It is found that the average Nusselt number, pressure drop, heat transfer enhancement, thermal–hydraulic performance increase with increasing in the volume fraction of nanoparticles and with decreasing in the diameter of nanoparticles.

In this paper, forced convection of nanofluids in semi-circular corrugated channels is numerically studied using finite volume method. The effects of Reynolds number, nanoparticles volume fraction (ϕ) and nanoparticle diameter (dp) on heat transfer and friction factor are presented and analysed. The aim of this research is to determine the optimum corrugation profile for heat transfer augmentation and to serve as reference for experimental work in the future.

2. Computational Model

Figure 1 shows the geometry of the corrugated channel which will be denoted here as the test section while Figure 2 shows the computational model. The total length of the channel is L_{total}=700mm. The length of the test section is L_2=200mm, with an upstream rectangular section of L_1=400mm upstream to ensure a fully developed flow at the leading edge. The downstream section has a length of L_3=100mm which is used to prevent the occurrence of adverse pressure effects caused by reversed flow which might at the trailing section. The channel height (H) is 10mm while the channel width (W) is 50mm. The semi-circular corrugated height h= 2.5mm with fixed pitch (p= 1.5H). The geometry configuration was achieved by using SolidWorks. The fluid medium is Al_2O_3-water nanofluid. It is assumed that the flow is steady and incompressible.

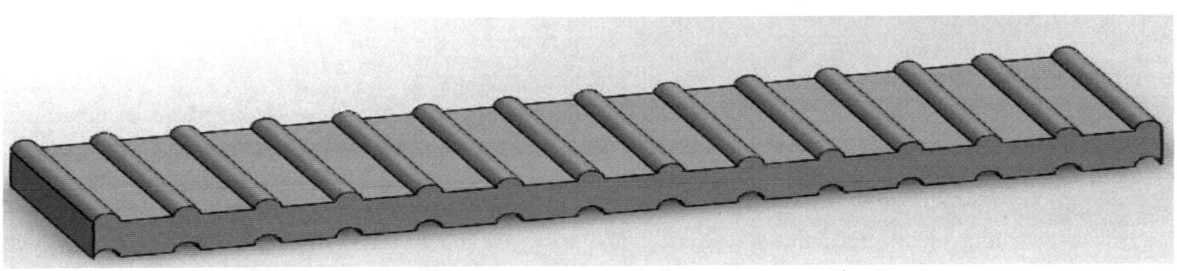

Figure 1. Semi-circular corrugated channel (test section).

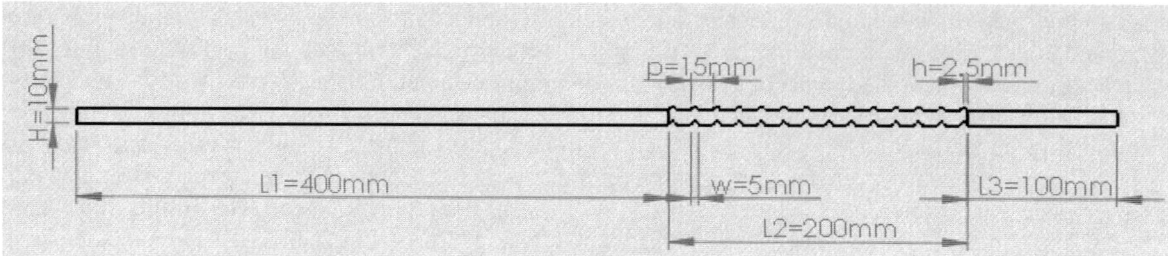

Figure 2. Physical model of the present study.

Five sets of grid distributions are tested on the semi-circular corrugated channel using water as working fluid to analyze the effects of grid size on the results. They are 162632, 267696, 329668, 413520 and 544235 elements, respectively. By comparing the third, fourth and fifth mesh configurations, in terms of average Nusselt number, the corresponding percentage relative errors are 0.021%. Therefore, the fourth grid case has been adopted to obtain an acceptable compromise between the computational time and the results accuracy.

3. Numerical procedure

This simulation utilizes FLUENT (V16.1) as the solver to the governing equations for turbulent three-dimensional incompressible steady–state Reynolds-averaged Navier-Stokes equations shown below:

Conservation of mass:

$$\frac{\partial}{\partial x_i}(\rho u_i) = 0 \tag{1}$$

Conservation of momentum:

$$\frac{\partial}{\partial x_j}(\rho u_i u_j) = -\frac{\partial p}{\partial x_i} + \frac{\partial y}{\partial x_j}\left[\mu\left(\frac{\partial u_i}{\partial x_j} + \frac{\partial u_j}{\partial x_i}\right) - \rho \bar{u}_i \acute{u}_j\right] - \frac{\partial \rho \overline{\acute{u}_i \acute{u}_j}}{\partial x_j} + \rho g_i \tag{2}$$

Conservation of energy:

$$\frac{\partial}{\partial x_j}(\rho u_j T) = \frac{\partial}{\partial X_j}\left[\left(\frac{\mu}{Pr} + \frac{\mu_t}{\sigma_t}\right)\frac{\partial T}{\partial x_j}\right] \tag{3}$$

The governing equations are discretized by finite volume method and solved in steady-state implicit format. The SIMPLE algorithm is used to couple the velocity and pressure fields. The second-order upwind scheme is applied and standard *k-ε* turbulent model with standard wall function is selected.

The thermophysical properties of the nanofluid used in present study are the ones used by Ahmed et al. [12] as shown in Table [1]. At the inlet, the water velocity can be set for different values according to the Reynolds number. The flow is simulated at five Reynolds number values which are 10000, 15000, 20000, 25000 and 30000. Top and bottom walls for the test section are subjected to a constant heat flux of $10kW/m^2$ while adiabatic boundary condition is applied to the remaining walls. The inlet temperature of the working fluid is T_{in}=300k. Al_2O_3-water nanofluid is set to flow steadily over the test section at Reynolds numbers stated above.

Table 1. Thermophysical properties of nanoparticle and base fluid at T=300 K

Material	Density ρ (kg/m^3)	Dynamic viscosity μ (N s/m^2)	Thermal conductivity K (W/m.k)	Specific heat C_P (J/kg.k)
Water	998.2	1.00E-03	0.6	4182
Al_2O_3	3600	0	36	765

4. Results and discussion

The effect of Reynolds number, nanoparticles volume fraction and nanoparticles diameter on heat transfer and friction factor are analysed and discussed in this section.

4.1 Validation of straight channel simulation results

For validation, the results obtained by simulation for straight channel which included Nusselt number (Nu) and friction factor (*f*) are compared with empirical correlations of Dittus–Boelter and Petukov, respectively [13]:

$$Nu = 0.023Re^{0.8}pr^{0.4} \tag{4}$$

$$f = (0.79 \ln(Re) - 1.64)^{-2} \tag{5}$$

Comparisons of Nusselt number and friction factor which shows a very good agreement demonstrated in figure 3 and 4.

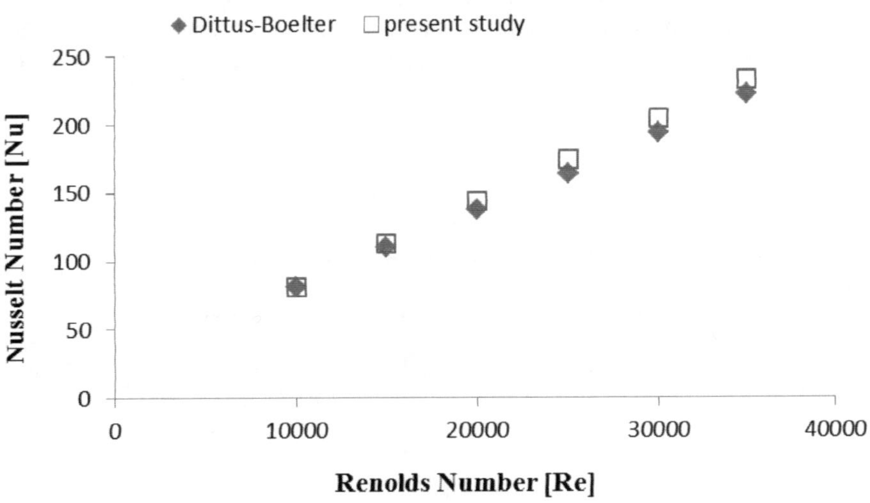

Figure 3. Smooth duct simulated Nusselt number vs. Reynolds number

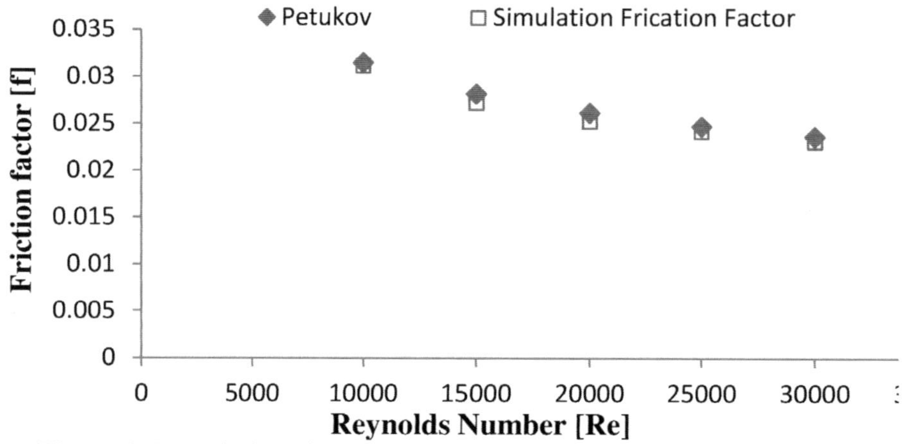

Figure 4. Smooth duct simulated frication factor vs. Reynolds number

4.2 Effect of nanoparticles volume fraction on heat transfer and friction factor

The effect of nanoparticles volume fractions on Nu at dp=20 nm is displayed in Figure 5. As expected, Nu increases with ϕ due to improve thermal conductivity of nanofluid. This enhancement is due to the high thermal conductivity of nanoparticles and the role of Brownian motion of nanoparticles on the enhancement of thermal conductivity, which is due to the large surface area of nanoparticles for molecular collisions. Previous studies [11] and [12] reported a similar result. Figure 6 displays the friction factor for different values of ϕ . It is observed that the friction factor increases due to the increase in the viscosity of the nanofluid. Also, with an increase in nanoparticles volume fraction, the possibility and number of contacts between walls and corrugated surfaces increase, which leads to increase in friction factor. The result shows the higher Nusselt number enhancement ratio is 2.07 at Reynolds number 30000 and volume fraction 6%.

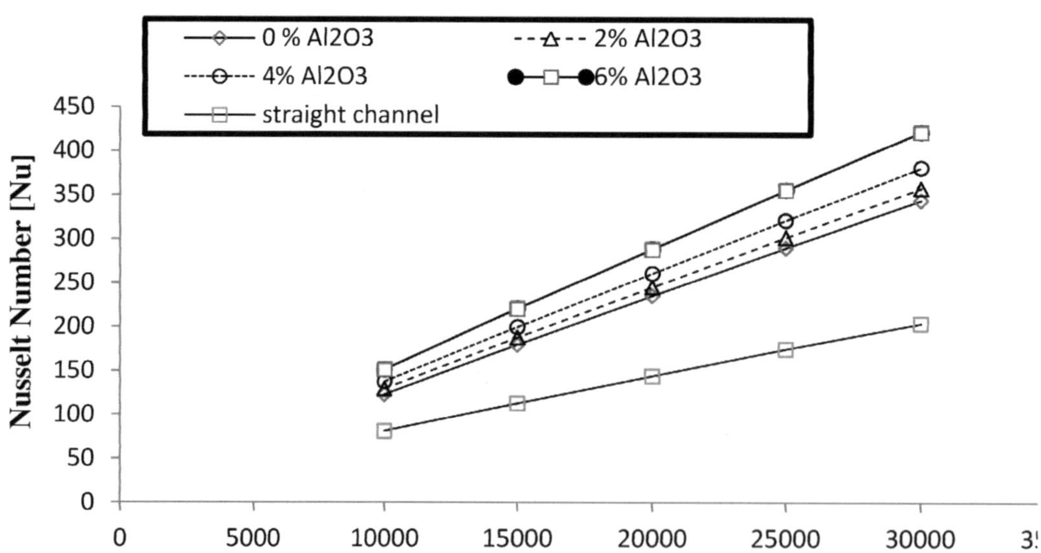

Figure 5. Nusselt number vs. Reynolds number for different ϕ at dp= 20 nm

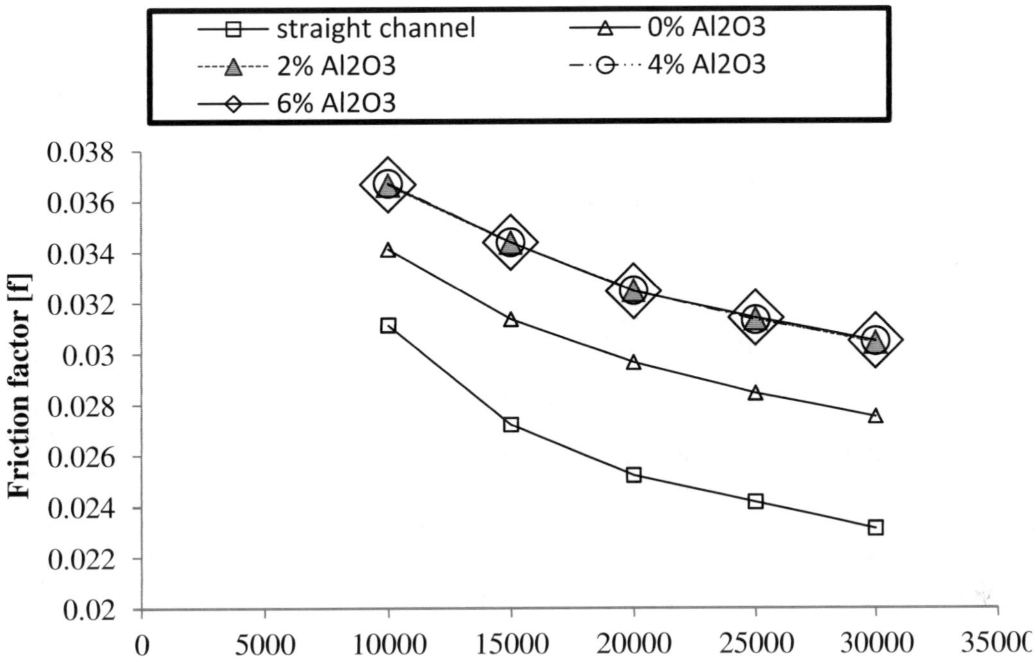

Figure 6. Nusselt number vs. Friction factor for different ϕ at dp=20 nm

4.3 Effect of nanoparticles diameters on heat transfer and friction factor

The effect of nanoparticle diameter on Nusselt number in the range of 20 to 60 nm with constant volume fraction of 6% is shown in figure 7. The results reveals that the nanofluid with smaller particle diameter gives higher Nusselt number as seen for Al_2O_3–water nanofluid. This is due to stronger Brownian motion at smaller nanoparticle diameters, which leads to higher thermal conductivity of nanofluid. The effect of particle size may be attributed mainly to two reasons which are the high specific surface area of the nanoparticles and the Brownian motion. As the particle size reduces, the surface area per unit volume increases. Since the heat transfer is dependent on the surface area, the effectiveness of nanoparticles in transferring heat to the base liquid increases. However, reducing the particle size means increasing the Brownian motion velocity, which again adds up to the contribution by the nanoparticles to the total heat transfer by continuously creating additional paths for the heat flow in the fluid. As presented in this figure the nanofluid with 20 nm nanoparticle diameter has the highest Nusselt number, whereas, the nanoparticle with a diameter of 60 nm has the lowest Nusselt number.

This is consistent with the numerical studies of Ahmed et al. [12]. The comparison of friction factor for Al_2O_3–water nanofluid with different nanoparticle diameters is presented in Figure 8 at various Reynolds numbers. It is noted that there is a slight change in the friction factor when nanoparticle diameters of Al_2O_3 nanofluid are varied. So, for high heat transfer enhancement, the Al_2O_3-water with 6% volume fraction and 20nm particle can be the best choice.

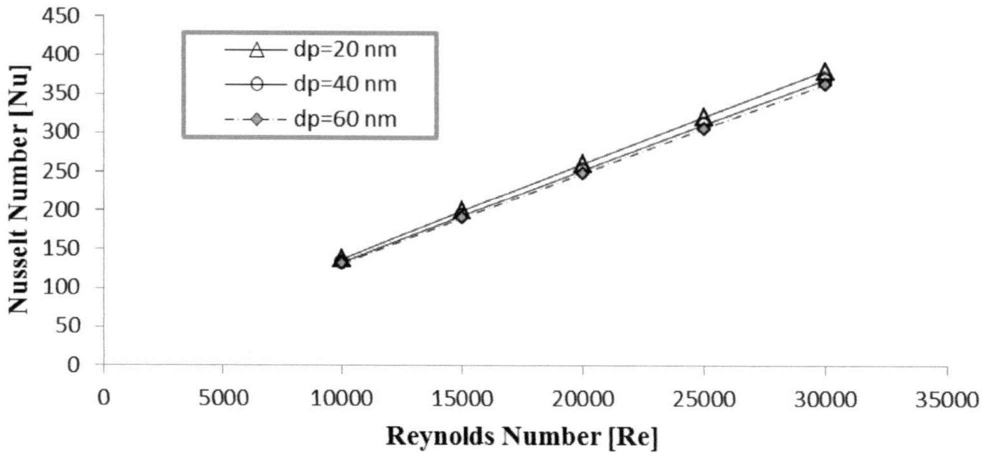

Figure 7. Nusselt number vs Reynolds number for different dp at ϕ =6%

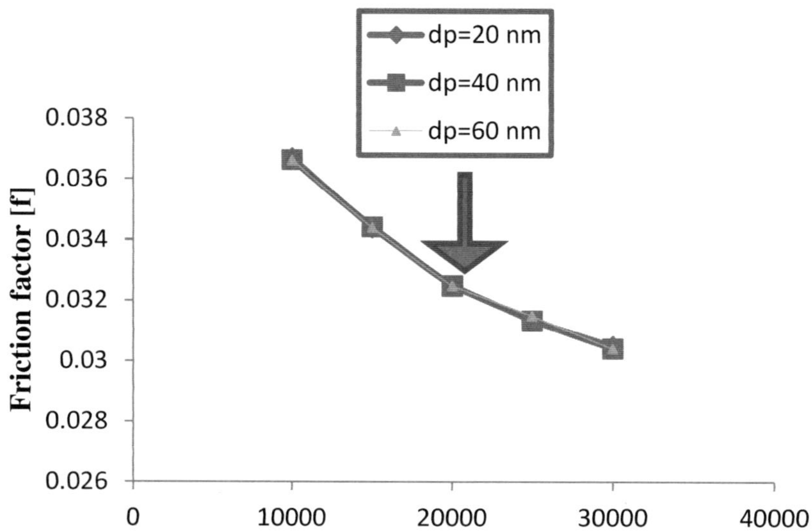

Figure 8. Friction factor vs Reynolds number at ϕ =6%

5. Conclusion

Numerical simulations of turbulent forced convection heat transfer in a semi-circular corrugated channel subjected to uniform heat flux were carried out. The computations were performed for a symmetrical semi-circular corrugated channel with varying Reynolds numbers ($10000 \le Re \le 30000$), volume fractions ($0\% \le \phi \le 6\%$) and nanoparticle diameters (20nm ≤ dp ≤60 nm).The results of numerical solution showed that Nu and f increase with increasing the Re, and nanoparticles volume fraction . In addition, by decrease of nanoparticle diameter, the Nusselt number increased. The results of the present study are consistent with the results presented by [7], [8], [9], [10], [11] and [12]. Finally, higher Nusselt number enhancement ratio which indicates the optimum configuration is 2.07 at Reynolds number 30000 and volume fraction 6%. Based on the above results, the use of nanofluids in semi-circular corrugated channel is a suitable method to achieve a good enhancement in the performance of many thermal devices as a passive method.

Acknowledgement
The authors wish to acknowledge that this paper and the work was funded by the Fundamental Research Grant Scheme (FRGS 1589), Universiti Tun Hussein Onn Malaysia.

References

[1] Sundén B, Sköldheden T. Heat transfer and pressure drop in a new type of corrugated channels. International communications in heat and mass transfer. 1985 Sep 1;12(5):559-66.
[2] Sawyers DR, Sen M, Chang HC. Heat transfer enhancement in three-dimensional corrugated channel flow. International Journal of Heat and Mass Transfer. 1998 Nov 1;41(22):3559-73.
[3] Fabbri G. Heat transfer optimization in corrugated wall channels. International Journal of Heat and Mass Transfer. 2000 Dec 1;43(23):4299-310.
[4] Vasudevaiah M, Balamurugan K. Heat transfer of rarefied gases in a corrugated microchannel. International journal of thermal sciences. 2001 May 1;40(5):454-68.
[5] Wang CC, Chen CK. Forced convection in a wavy-wall channel. International Journal of Heat and Mass Transfer. 2002 Jun 30;45(12):2587-95.
[6] Gradeck M, Hoareau B, Lebouche M. Local analysis of heat transfer inside corrugated channel. International journal of heat and mass transfer. 2005 May 31;48(10):1909-15.
[7] Naphon P. Laminar convective heat transfer and pressure drop in the corrugated channels. International communications in heat and mass transfer. 2007 Jan 31;34(1):62-71.
[8] Naphon P. Heat transfer characteristics and pressure drop in channel with V corrugated upper and lower plates. Energy conversion and management. 2007 May 31;48(5):1516-24.
[9] Naphon P. Effect of corrugated plates in an in-phase arrangement on the heat transfer and flow developments. International Journal of Heat and Mass Transfer. 2008 Jul 15;51(15):3963-71.
[10] Naphon P. Effect of wavy plate geometry configurations on the temperature and flow distributions. International Communications in Heat and Mass Transfer. 2009 Nov 30;36(9):942-6.
[11] Rostamani M, Hosseinizadeh SF, Gorji M, Khodadadi JM. Numerical study of turbulent forced convection flow of nanofluids in a long horizontal duct considering variable properties. International Communications in Heat and Mass Transfer. 2010 Dec 31;37(10):1426-31.
[12] Ahmed MA, Yusoff MZ, Ng KC, Shuaib NH. Numerical investigations on the turbulent forced convection of nanofluids flow in a triangular-corrugated channel. Case Studies in Thermal Engineering. 2015 Sep 30;6:212-25.
[13] Jaisankar S, Radhakrishnan TK, Sheeba KN. Experimental studies on heat transfer and friction factor characteristics of forced circulation solar water heater system fitted with helical twisted tapes. Solar Energy. 2009 Nov 30;83(11):1943-52.

IOP Conf. Series: Materials Science and Engineering **243** (2017) 012021 doi:10.1088/1757-899X/243/1/012021

Computational modelling of flow and tip variations of aortic cannulae in cardiopulmonary bypass procedure

Siti A Thomas[1], Shirly Empaling[1], Nofrizalidris Darlis[2], Kahar Osman[3,4], Jeswant Dillon[5], Ishkrizat Taib[6], Ahmad Zahran Md Khudzari[1,4]

[1]Faculty of Biosciences & Medical Engineering, Universiti Teknologi Malaysia, 81310, UTM Johor Bahru, Johor, Malaysia
[2]Faculty of Engineering Technology, Universiti Tun Hussein Onn Malaysia, , 86400, Parit Raja, Johor, Malaysia
[3]Faculty of Mechanical Engineering, Universiti Teknologi Malaysia, 81310, UTM Johor Bahru, Johor, Malaysia
[4]IJN-UTM Cardiovascular Engineering Centre, Institute of Human Centered Engineering, Universiti Teknologi Malaysia, 81310, UTM Johor Bahru, Johor, Malaysia
[5]Institut Jantung Negara, Jalan Tun Razak, 50400, Kuala Lumpur, Malaysia
[6]Flow Analysis Simulation and Turbulence Research Group, Faculty of Mechanical and Manufacturing, Universiti Tun Hussein Onn, 86400, Parit Raja, Johor, Malaysia
Email: iszat@uthm.edu.my[6]

Abstract. Aortic cannulation has been the gold standard for maintaining cardiovascular function during open heart surgery while being connected onto the heart lung machine. These cannulation produces high velocity outflow which may lead to adverse effect on patient condition, especially sandblasting effect on aorta wall and blood cells damage. This paper reports a novel design that was able to decrease high velocity outflow. There were three design factors of that was investigated. The design factors consist of the cannula type, the flow rate, and the cannula tip design which result in 12 variations. The cannulae type used were the spiral flow inducing cannula and the standard cannula. The flow rates are varied from three to five litres per minute (lpm). Parameters for each cannula variation included maximum velocity within the aorta, pressure drop, wall shear stress (WSS) area exceeding 15 Pa, and impinging velocity on the aorta wall were evaluated. Based on the result, spiral flow inducing cannulae is proposed as a better alternatives due to its ability to reduce outflow velocity. Meanwhile, the pressure drop of all variations are less than the limit of 100 mmHg, although standard cannulae yielded better result. All cannulae show low reading of wall shear stress which decrease the possibilities for atherogenesis formation. In conclusion, as far as velocity is concerned, spiral flow is better compared to standard flow across all cannulae variations.

1. Introduction

Over recent years, there has been an explosive growth of interest in the development and improvement of the aortic cannula. Aortic cannula is a surgical equipment that was inserted into the ascending aorta in conjunction with heart lung machine to drain the blood away from heart during the open heart surgery [1]. Establishing an arterial cannulation is an important events in cardiopulmonary bypass procedure as it maintain a normal circulatory flow of the blood within our body [2]. However, uses of this medical device may results in many possible complication and even death [3]. This occurs mainly as the blood exit from the cannula with high jet velocity results in the sandblasting effect; which may lead to erosion of blood wall vessel, particularly dangerous for patient in which arteries is hardening followed by deposition of plague [4, 5]. These plague deposition has been implicated in the occurrence of postoperative stroke up to 25% [6].

Numerous research has been done to investigate the best design of aortic cannula used during the cardiac surgery [7]. Reduction in velocity and blood damage became their main objective. Uses of the existing cannula such as standard end-hole cannula, the soft-flow cannula, and the dispersion cannula in cardiac surgery however still results in event of thromboembolism and atherosclerosis development [8]. Studies have demonstrated that flow ejected into the aorta from the left ventricle is swirling around the central axis i.e. a spiral/helical flow. It is physiological and has many benefits in maintaining blood elements balance [9]. Few researches have tried to emulate spiral flow into the blood vessel flow with varying degrees [10]. Despite several researches conducted, only Meedos' X-Flow Cannulae™ is in market. Although spiral flow cannula has been shown to offer better alternative, there are still some improvements that needs to be done especially on the range of flowrate and type of cannula tip that spiral flow are effective.

In this study, we proposed a straight and curved tip design of cannula in conjunction with the flow characteristics; standard flow (forward flow) and the spiral flow. The study aim was to determine which variations is the best in reducing the velocity, low pressure drop, wall shear stress and the impinging velocity onto the wall of aorta.

2. Numerical modelling

2.1. Geometry: Characteristics of Models

There were three different design factors: 1) cannula tip (straight and curved), 2) flow induced (straight flow and spiral flow), and 3) flowrate out of the cannula. These 3 design factors lead to a full factorial design of experiments. Eight cannulae of different tips were designed using SolidWorks; straight tip cannula, curved tip cannula, straight and curved tip cannula with spiral inducer as illustrated in Figure 1. Geometrical properties of each cannula are listed in Table 1. Each cannula was connected to patient specific aorta model as shown in Figure 2. The patient specific aorta was modelled from a healthy male from Insitut Jantung Negara (IJN). The original image was a Computed Tomography (CT) scan which has undergone image processing using a commercial software (Mimics, Materialise Inc., Leuven, Belgium).

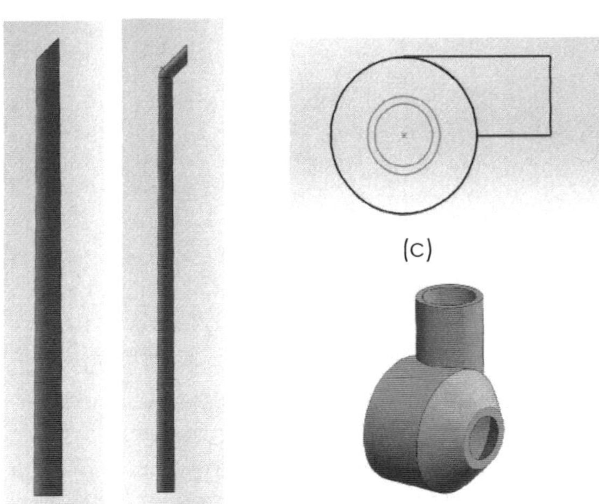

(a) (b (c) (d)

Figure 1. (a) straight tip cannula; (b) curved tip cannula; (c) spiral inducer; (d) spiral flow inducer in 3D.

Table 1. Geometrical properties of each cannula designed

Cannula	Geometrical properties	
	Normal cannula	**Spiral inducing Cannula**
Straight tip	Cannula length : 270mm Inlet diameter : 10mm Outlet diameter: 8mm	Cannula length : 170mm Inlet diameter : 10mm Outlet diameter: 8mm
Curved tip	Cannula length : 270mm Inlet diameter : 10mm Outlet diameter: 8mm	Cannula length : 170mm Inlet diameter : 10mm Outlet diameter : 8mm

For the design factor of flowrate, three levels were chosen: 3, 4 and 5 liter per minute (lpm). Overall, there were 12 variations (2 × 2 × 3) from which all were tested numerically. Since the simulations were done using computer, no replications was carried out.

Figure 2. Patient specific aorta model

2.2. Computational Models

Flow characteristics in cannulae was simulated by using commercial code: ANSYS CFX 14.0 software (Canonsburg, PA, USA). This solver using a vertex centered finite volume method governing equations on a spatially rectangular computational mesh designed in the Cartesian coordinate system with the planes orthogonal to its axes. Additional refining was completed in specified blood regions, at the arterial surfaces during calculation. The time derivatives are approximated with an implicit second-order accurate in both space and time. Both flow rates inlet and pressure outlet are computed to solve the continuity and Navier-Stokes equations. Hence, the physical laws describing the problem of aorta and cannulation are the conservation of mass and the conservation of momentum. The flow is governed by the following incompressible Navier-Stokes equation:

$$\nabla . u = 0 \tag{1}$$

$$\rho\left(\frac{\partial u}{\partial t} + u.\nabla u - f\right) = \mu\nabla^2 u - \nabla p \tag{2}$$

where u is the flow vector, ρ is the density of the fluid used in the flow, p is the pressure and μ is the fluid viscosity.

Newtonian fluid was considered in this simulation due to the shear rate of blood in aorta was calculated more than 100/s. Thus, the turbulence model was imposed in predicting flow characteristics during cannulation procedure. The turbulence model used in simulation was the shear stress transport (SST). Grid independence test was carried out which showed that 3 million nodes was sufficient to carry out a well-balanced numerical study on the model. Computational simulation proceeded after completing the meshing process with more than 3 million of nodes for all designs as shown in Figure 3.

2.3. Boundary Condition

The drafted cannulae were then imported into ANSYS CFX software to be simulated. Newtonian blood flow with constant haemodynamic properties of density = 1056.4 kg/m^3 and viscosity = 0.0036 Pa s was simulated with no-slip wall boundary condition which was assumed rigid and impermeable [11]. The flow rates used is 3L/min, 4L/min and 5L/min and maximum iteration of 10000. The resulting Navier-Stokes equation of this modelling is solved within the framework of the CFX Software. The turbulence model used in the simulation was shear stress transport (SST). Grid independence test was carried out which showed that 3 million nodes was sufficient to carry out a well-balanced numerical study on the model. Computational simulation proceeded after completing the meshing process with more than 3 million of nodes for all designs as shown in Figure 3.

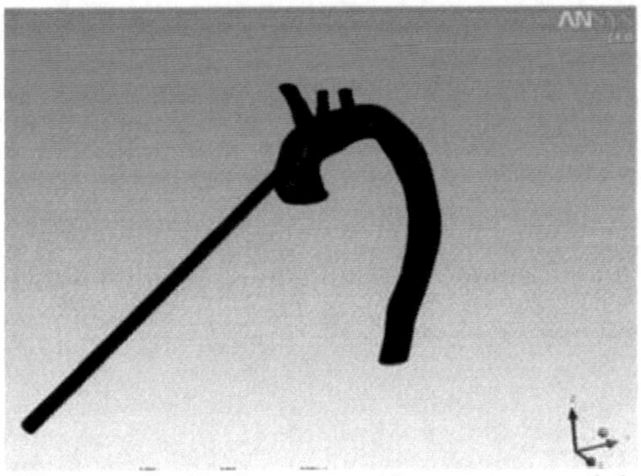

Figure 3. Patient specific aorta model when finished meshing process

2.4. Output Measurement

Several output parameter of haemodynamic has been acquired based on the boundary condition set. Those haemodynamic parameters are: maximum velocity inside the aorta (termed as flow output), the impinging velocity on the aorta wall, pressure drop and the percentage of wall shear stress that exceeded 15 Pa have been obtained. Wall shear that exceeded 15 Pa has higher chances of causing haemolysis and endothelial lesion [7]. The impinging velocity was taken at a point at the end of a straight line from cannulae tip until the aorta wall. The overall process of the system can be summarized as shown in Figure 4.

IOP Conf. Series: Materials Science and Engineering **243** (2017) 012021 doi:10.1088/1757-899X/243/1/012021

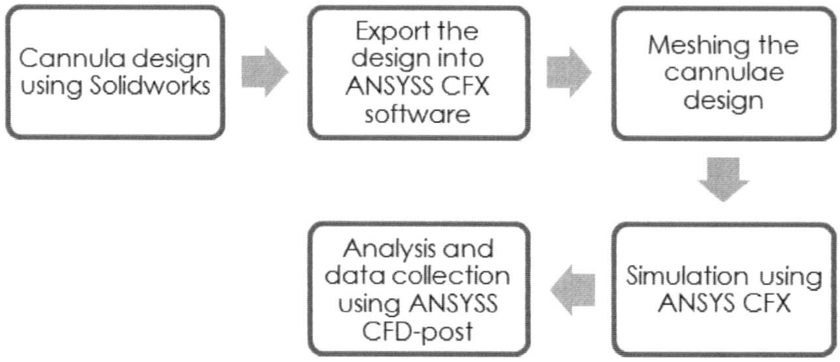

Figure 4. Block diagram of the processes of the system

2.5. Selection of Best Cannulae Group

All results will then be used for determining the best group of cannulae. The value from each parameters will then be ranked, where the smaller the value is the better result. A weighted scale will then be applied to the rank number of each group, this is called the scoring process. It is decided that the weighted scale for each parameters are: 1) flow output: 40%, 2) impinging velocity: 30%, 3) WSS > 15 Pa (%): 15%, and 4) pressure drop: 15%. These scale is justified due to the importance of velocity on the haemolysis as well as damage on aorta blood wall.

3. Results and Discussion

Although the flow rates were varied into 3 lpm, 4 lpm and 5 lpm, the output results were then averaged to yield new results for easier analysis. Table 2 is the original data for all 12 variations, while the averaged data within the same group is presented in Table 3. The maximum velocity inside the aorta for straight cannula shows a lower reading compared to curved tip cannula independent to their flow type. Straight tip cannula with spiral inducer showed the lowest velocity which is 1.61 m/s followed by straight tip cannula, curved tip cannula with spiral inducer and curved tip cannula with velocity of 1.68 m/s, 1.73 m/s and 1.88 m/s respectively. Figure 5 shows the velocity streamline for each cannula.

Table 2. Flow output for all variations. The unit is in (m/s)

FLOW OUTPUT (m/s)		FLOW RATE (lpm)	TIP	
			STRAIGHT	CURVE
FLOW TYPE	STANDARD	3	1.30	1.46
		4	1.66	1.85
		5	2.09	2.32
	SPIRAL	3	1.22	1.29
		4	1.58	1.71
		5	2.03	2.18

Table 3. Averaged flow output from cannulae. The unit is in (m/s)

FLOW OUTPUT (m/s)	STRAIGHT	CURVED
STANDARD	1.68	1.88
SPIRAL	1.61	1.73

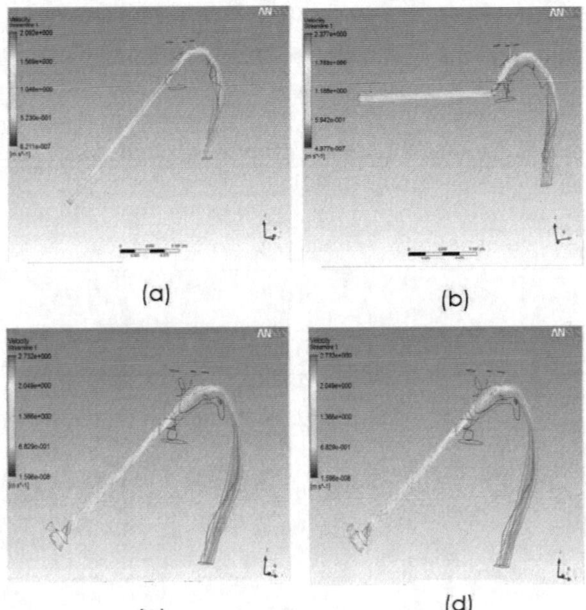

Figure 5. Velocity streamline of cannula: (a) straight tip; (b) curved tip; (c) straight tip with spiral inducer; (d) curved tip with spiral inducer. These images are from simulation of 5 lpm.

Although all instances in Figure 5 show some degree of spiral flow (even from straight cannula tip), it is much more noticeable at the descending aorta. However, overall, the spiral flow is much more pronounced for cannulae with spiral flow inducing chamber (Figure 5(c) and 5(d)).

Table 4 shows the complete result for impinging velocity. For impinging velocity, both curved and straight cannula with spiral inducer showed lower reading compared to the others with 0.38 m/s and 0.43 m/s respectively. It was the followed by curved tip cannula and straight tip cannula of standard flow with slightly higher velocity reading.

Next, the pressure drop was measured by taking the differences between static pressure at the inlet and outlet of the cannula. High pressure can cause adverse effect. The optimum pressure drop should be below 100 mmHg [12]. All cannula yielded results below 100 mmHg, with straight tip cannula having the lowest pressure drop of 10.21 Pa, followed by curved tip cannula, straight tip

Table 4. Impinging velocity of cannulae groups

IMPINGING VELOCITY (m/s)	STRAIGHT	CURVED
STANDARD	0.52	0.43
SPIRAL	0.43	0.38

cannula with spiral inducer and curved tip cannula with spiral inducer having 12.75 Pa, 31.41 Pa and 33.31 Pa respectively.

Figure 6 shows the wall shear stress of the cannula and aorta for cannulae groups at 5 lpm. Based on previous study, wall shear stress in the aorta must not exceed 15 Pa, to prevent vascular damage [13]. It can be seen from Figure 6 that all simulations show similar trend, whereas most of the aorta wall has low WSS reading, while the maximum WSS value can be found at the ascending aorta, especially within the surrounding of three arterial branches. Site directly distal to the left subclavian artery also show a concentrated area of elevated WSS value compared to its surrounding, suggesting that would be the point where flow is focused i.e. the sandblasting area. Percentage of WSS exceeding 15 Pa has been calculated and presented in Table 6. Curved cannula tip with spiral inducer showed the highest percentage among all with 1.75 %.

Table 5. Averaged pressure drop for all four conditions

PRESSURE DROP (Pa)	STRAIGHT	CURVED
STANDARD	10.21	12.75
SPIRAL	31.41	33.31

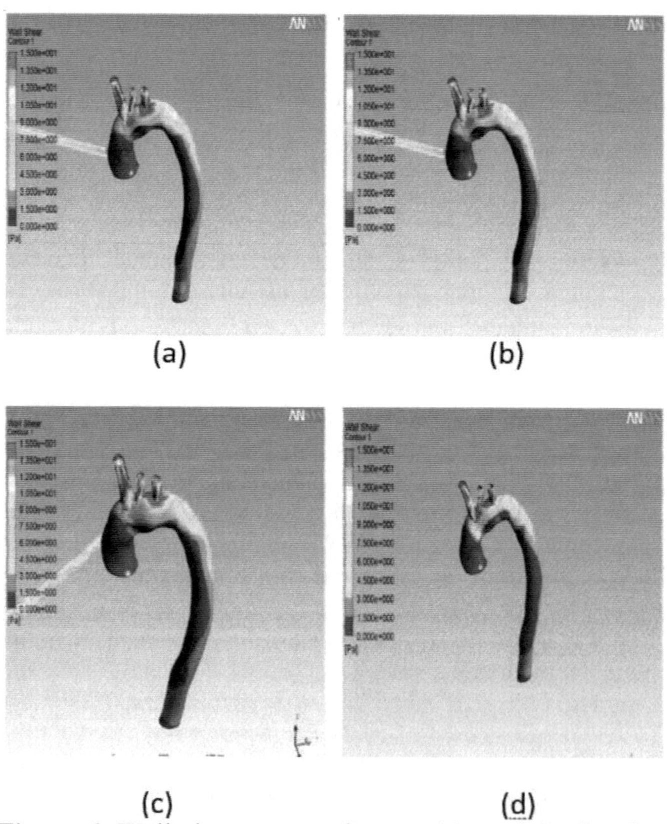

Figure 6. Wall shear stress of aorta: (a) straight tip; (b) curved tip; (c) straight tip with spiral inducer; (d) curved tip with spiral inducer. These images are from simulation of 5 lpm.

Table 6. Percentage of WSS more than 15 Pa.

WSS > 15 Pa (%)	STRAIGHT	CURVED
STANDARD	1.41	0.91
SPIRAL	1.47	1.75

3.1. Selection of best cannula group

All results were then ranked in each category in an ascending manner, where the best one was ranked the first as shown in Table 7. After that each rank value is multiplied by the weighted percentage set earlier. The resulting value for each category and for each group were summed up, resulting in the final ranking of best cannula group. As tabulated in Table 7, the cannula with straight tip with spiral inducer was ranked at number one, with accumulated mark of 1.9. This can be attributed to the group being first in the velocity output which carried higher weightage than the rest. Although in the other category (WSS > 15 Pa, and Pressure Drop), cannulae group with spiral flow inducer, group 3 has not fared well, as well as group 4, the most important part that contributed to the net value was from the velocities (velocity output and impinging velocity).

In this study, three different design factors were investigated. The effect on haemodynamic within the aorta by varying different levels of each design factors revealed interesting findings. The sandblasting effect due to high jet flow from cannula tip has been shown to affect healthy aorta wall in animal testing. It is imperative that jet flow and associated adverse effect on aorta wall as well as on blood be minimised. For easier analysis, the results from variations of input velocity into the cannulae were averaged. By averaging the results for three, four and five lpm, it gave a general idea of output from just variations of cannulae tip and type of flow. As far as velocity from the cannula is concerned, cannulae with spiral flow showed a much reduced flow velocity output. This coincide with results by Pekkan et al.

Table 7. Selection of the best cannula group

		Velocity Output		Impinging Velocity		WSS > 15 Pa		Pressure Drop		SUM	RANK
		Rank	Weighted 40%	Rank	Weighted 30%	Rank	Weighted 15%	Rank	Weighted 15%		
1	Straight tip	2	0.8	4	1.2	2	0.3	1	0.15	2.45	2
2	Curved tip	4	1.6	2	0.6	1	0.15	2	0.3	2.65	3
3	Straight tip with inducer	1	0.4	2	0.6	3	0.45	3	0.45	1.9	1
4	Curved tip with inducer	3	1.2	1	0.3	4	0.6	4	0.6	2.7	4

4. Conclusion

This paper focus are on the cannula design variations of three design factors for cannula and its haemodynamic effect. All cannulae were implanted at the ascending aorta to simulate flow condition during cardiopulmonary bypass. Based on the results, spiral flow was definitely induced for cannula with the spiral flow inducer. Cannulae with spiral flow inducer generally were better in term of velocity factors, while cannulae without spiral flow inducer were better in other haemodynamic parameters. After the scoring process, it emerged that the cannulae of straight tip with spiral flow inducer was the best group compared to others.

Acknowledgement
This research has been funded in part using Ministry of Higher Education Fundamental Research Grant Scheme : R.J1300000.7809.4F588. The authors would also like to express their appreciation to Hairul Lail Ismail from MJIIT, UTM for his technical assistance with the preparation of prototype model.

References
[1] Blauth C I, Cosgrove D M, Webb B W, Ratliff N B, Boylan M, Piedmonte M R, Lytle B W and Loop F D 1992 Atheroembolism from the ascending aorta. An emerging problem in cardiac surgery *The Journal of thoracic and cardiovascular surgery* **103**(6) 1104-11 discussion 1111-2
[2] Kaufmann T A S, Schlanstein P, Moritz A and Steinseifer U 2014 Development of a Hemodynamically Optimized Outflow Cannula for Cardiopulmonary Bypass *Artificial Organs* **38**(11) 972-978
[3] Magner J B 1971 Complications of aortic cannulation for open-heart surgery *Thorax* **26**(2) 172-173
[4] Minakawa M, Fukuda I, Yamazaki J, Fukui K, Yanaoka and Inamura T 2007 Effect of cannula shape on aortic wall and flow turbulence: hydrodynamic study during extracorporeal circulation in mock thoracic aorta *Artificial organs* **31**(12) 880-886
[5] Goto T, Inamura T, Shirota M, Fukuda W, Fukuda W, Daitoku K, Minakawa M and Ito K 2016 Hydrodynamic evaluation of a new dispersive aortic cannula (Stealthflow) *Journal of Artificial Organs* 1-7
[6] Katz E S, Tunick P A, Rusinek H, Ribakove G, Spencer F C and Kronzon I 1992 Protruding aortic atheromas predict stroke in elderly patients undergoing cardiopulmonary bypass: experience with intraoperative transesophageal echocardiograph *Journal of the American College of Cardiology* **20**(1) 70-77
[7] Menon P G, Antaki J F, Undar A and Pekkan K 2013 Aortic outflow cannula tip design and orientation impacts cerebral perfusion during pediatric cardiopulmonary bypass procedures *Annals of biomedical engineering* **41**(12) 2588-2602
[8] Grooters R K, Steeg D A V, Stewart M J, Thieman K C and Schneider R F 2003 Echocardiographic comparison of the standard end-hole cannula, the soft-flow cannula, and the dispersion cannula during perfusion into the aortic arch *The Annals of thoracic surgery* **75**(6) 1919-1923
[9] Stonebridge P A, Hoskins P R, Allan P L and Belch F F 1996 Spiral laminar flow in vivo *Clinical Science* **91**(1) 17-21
[10] Shafii N S, Darlis N, Dillon J, Osman K and Khudzari A Z M 2014 The Effect of Internal Profile on Cannula Outflow *Revolution* **30** 3
[11] Neidlin M, Sonntag S J, Schmitz-Rode T, Steinseifer U and Kaufmann T A S 2016 Investigation of hemodynamics during cardiopulmonary bypass: A multiscale multiphysics fluid–structure-interaction study *Medical engineering & physics* **38**(4) 380-390
[12] Kira Y, Kochel P J, Gordon E E and Morgan H E 1984 Aortic perfusion pressure as a determinant of cardiac protein synthesis *American Journal of Physiology-Cell Physiology* **246**(3) C247-C258
[13] Avrahami I, Dilmoney B, Hirshorn O, Brand M, Cohen O, Shani L, Nir R and Bolotin G 2013 Numerical investigation of a novel aortic cannula aimed at reducing cerebral embolism during cardiovascular bypass surgery *Journal of biomechanics* **46**(2) 354-361

Effect of ground proximity on the flow over STOL CH750 multi-element airfoil

Aslam Abdullah[1], **Muhamad 'Ammar Kamsani**[2], **Kamil Abdullah**[3]
[1,2]Department of Aeronautical Engineering
[3]Department of Energy and Thermofluids Engineering
[1,2,3]Faculty of Mechanical and Manufacturing Engineering, Universiti Tun Hussein Onn Malaysia, Johor

E-mail: aslam@uthm.edu.my

Abstract. The paper presents the influence of ground distance on aerodynamic characteristics of the flow over short take-off and landing STOL CH750 aircraft multi-element airfoil at various flap angle of deflection. The angle of attack is kept constant throughout the computations. ANSYS is used for the grid generation and the computational calculation. Further investigation suggests the best range of deflection angle where the aerodynamic performance of the airfoil increases in the presence of ground. Furthermore, a cushion of high-pressure air region between the airfoil pressure side and the ground surface also emerges.

1. Introduction

The phenomenon of ground effect becomes apparent in the proximity of the runway, and is beneficial in the control of fixed wing aircrafts during take-off and landing. When an aircraft is flown at about the equivalent to one wing span above the runway, the interaction between the airflow around the airfoil and the ground surface modifies the fluid velocity's vertical component and changes its normal pattern which consequently affects the lift-to-drag ratio L/D [1,2,3]. In particular, the corresponding rotation of the resultant force vector gives change to the component of lift and drag forces [4]. A well trained pilot takes the phenomena into account in order to enhance the safety prior to airborne and touchdown [5].

The paper evaluates the influence of ground on the multi-element airfoil during landing. The problem is analysed with regard to the aerodynamic characteristics of the flow over NACA 4412 airfoils which form the cross sections of STOL CH750 light sport aircraft (LSA) wing with constant chord length [6,7]. The evaluation focuses on fixed angle of attack, but the distance between the airfoil and the ground varies. The main airfoil is equipped with a junkers flap with various angle of deflection (AoD), namely 0°, 10°, 25°, 40°. Such AoD is crucial in achieving touchdown point accuracy and preventing ballooning especially if the aircraft is close to the runway during flare; the recovery from such problem would require emergency procedures [8]. The junkers flap is slotted below the trailing edge of the main airfoil. The design allows the airflow to pass between the main airfoil and the flap, even when the latter is retracted. Because of the increase in airflow, a significant increase in lift is added to the airfoils, even at low speeds. The flap is employed nowadays in many modern ultralights [9]. A steady-state RANS investigation is performed using the commercial CFD code ANSYS. The relationship between AoD and the nature of ground effect at constant landing speed is observed. The effect is also considered for different distances between the airfoil and the ground.

2. Governing equations and modelling assumptions

A finite volume numerical method based on solving RANS equations describing the case under consideration is used for calculating the flow around the airfoil. In the case of incompressible, steady, 2-D viscous flow, the equations are given by:

$$\rho\left(u_1\frac{\partial u_1}{\partial x_1}+u_2\frac{\partial u_1}{\partial x_2}\right)=F_1-\frac{\partial P}{\partial x_1}+\mu\left(\frac{\partial^2 u_1}{\partial x_1^2}+\frac{\partial^2 u_1}{\partial x_2^2}\right)-\rho\left(\frac{\partial < u_1'u_1' >}{\partial x_1}+\frac{\partial < u_1'u_2' >}{\partial x_2}\right), \qquad (2.1)$$

and

$$\rho\left(u_1\frac{\partial u_2}{\partial x_1}+u_2\frac{\partial u_2}{\partial x_2}\right)=F_2-\frac{\partial P}{\partial x_2}+\mu\left(\frac{\partial^2 u_2}{\partial x_1^2}+\frac{\partial^2 u_2}{\partial x_2^2}\right)-\rho\left(\frac{\partial < u_1'u_2' >}{\partial x_1}+\frac{\partial < u_2'u_2' >}{\partial x_2}\right), \qquad (2.2)$$

where u_1, u_2 are time-averaged velocity vectors, P is the pressure, μ is the dynamic viscosity, u_1', u_2' are fluctuating velocity vectors, and F_1, F_2 are body forces. It is necessary that the continuity equation,

$$\frac{\partial u_1}{\partial x_1}+\frac{\partial u_2}{\partial x_2}=0, \qquad (2.3)$$

is satisfied by (2.1) and (2.2).

Since the Reynolds averaging is involved in the derivation of (2.1) and (2.2), then a set of new unknowns called the Reynolds stresses need to be modeled to ensure that the equations can be closed. In particular, the k-ε turbulence model which gives a general description of turbulence by means of two transport equations is applied.

The method is applied with second-order upwind discretization schemes. The algorithm called SIMPLE is employed for the velocity-pressure coupling. The airfoil velocity and the Reynolds number are 25 m/s and 1.7 x 10^6, respectively. The values are within the range of CH750 landing and take-off conditions. The Mach number of flow is equal to 0.07, so then the compressibility effects could be neglected since they should be taken into account for a Mach number greater than 0.3. The investigated AoD and heights of airfoil flights represented by the relative ratio of the height-to-airfoil chord length (h/c) are introduced.

3. Geometry, grid and computational domain

The flow is studied for the NACA 4412 for both the main airfoil and the junkers flap. The airfoil has a flat shape of the pressing (bottom) side. The geometry of the airfoil is given in Fig. 1.

For discretization of the computational domain, an unstructured type of grid with quad elements is selected. The airfoil model moving above the ground and the grid used for calculations are given in Fig. 2.

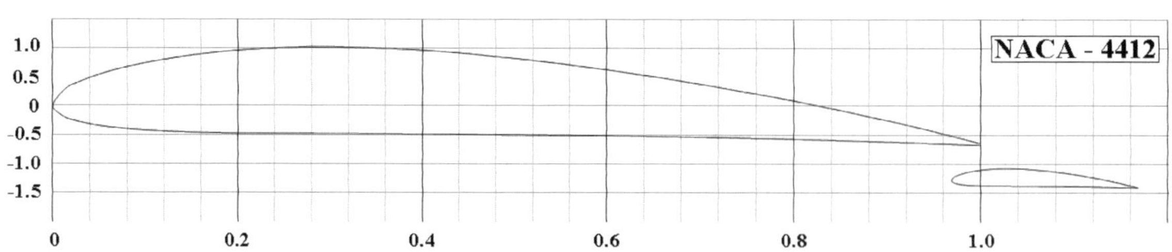

Figure 1. Geometry of NACA 4412 airfoils (main airfoil with junkers flap)

Inlet and outlet boundary conditions are specified on the outer sides of computational domain with necessary turbulence and flow parameters. The boundary condition on the airfoil is a no-slip condition with zero relative speed enforced. This should produce better representation of the reversing flow assumption.

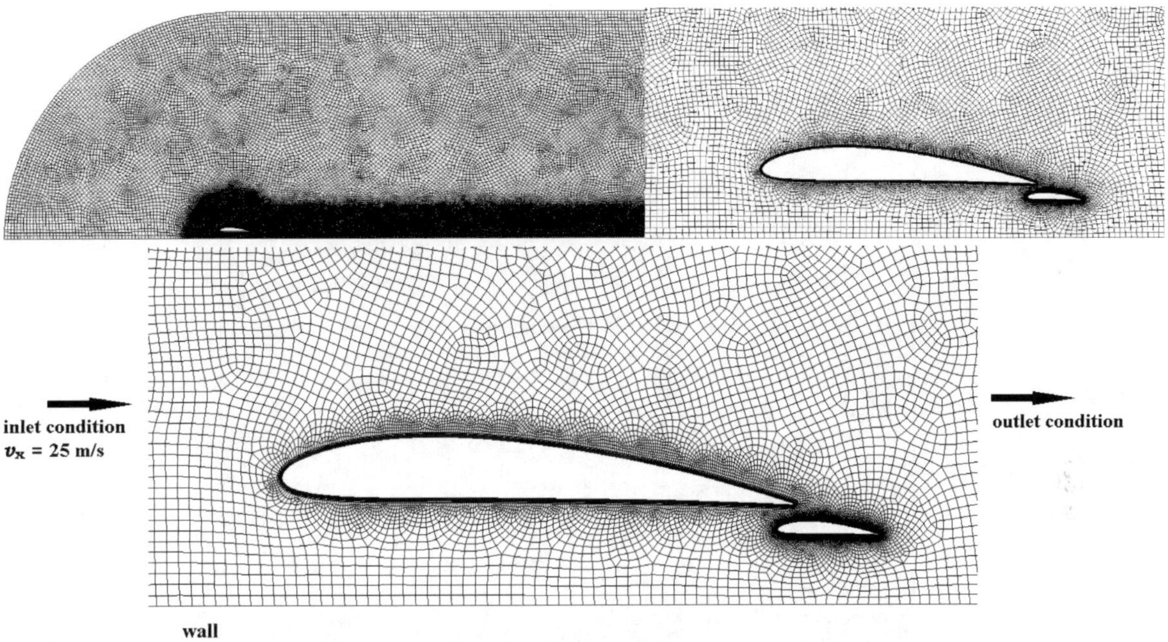

Figure 2. Grid applied for airfoil moving close to the ground surface

4. Results of calculations

The velocity field around the NACA 4412 airfoil in clean configuration is shown in Fig. 3. Note that the airfoil is in ground proximity. It is obvious that the velocity is relatively low in particular at the pressure side when the airfoil approaches the ground, and thus gives direct influence to *L/D* a moment after the roundoff. This indicates that the airfoil floats on a cushion of high-pressure air region above the ground surface. Furthermore, the stagnation point shifts to the lower side of the airfoil due to the ground effect.

The ground effect on the velocity field around the airfoil in dirty configuration is shown in Fig. 4. Clearly the velocity drop is acute at the pressure side when AoD changes from 10° to 25°, and consequently 40°, giving therefore great influence to *L/D* during the final stage of the approach until touchdown. Apart from the shift of the stagnation point to the lower airfoil side, the flow separation occurs near the flap at AoD 25° and 40°.

Although the AoD increase has a positive influence on the lift, in general its impact on the drag is greater such that *L/D* decreases. The percentage of *L/D* reduction in the presence of the ground is shown in table 1. In this case the basic aerodynamic characteristics are therefore in agreement with those for unbounded airflow. The next question is whether the *L/D* can be much improved by the ground effect.

IOP Conf. Series: Materials Science and Engineering **243** (2017) 012022 doi:10.1088/1757-899X/243/1/012022

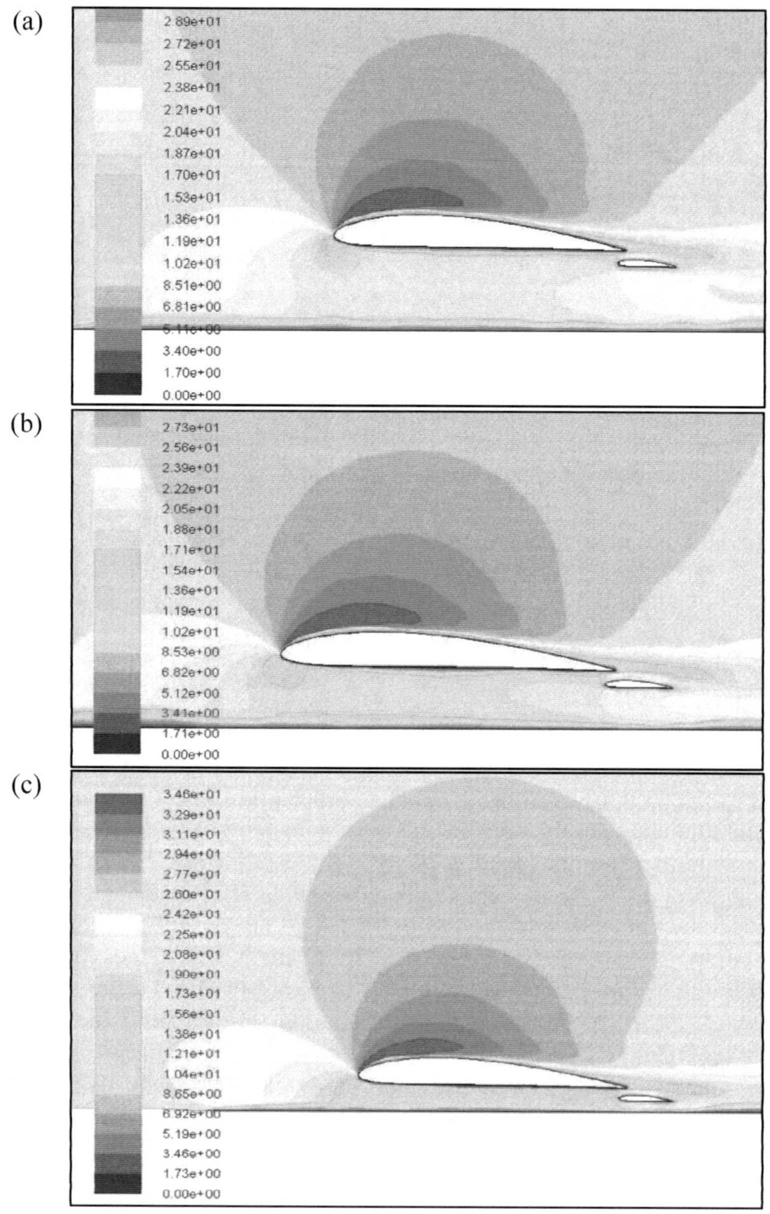

Figure 3. Velocity contour for the airfoil in clean configuration
(a) $h/c = 0.25$ (b) $h/c = 0.15$ (c) $h/c = 0.05$

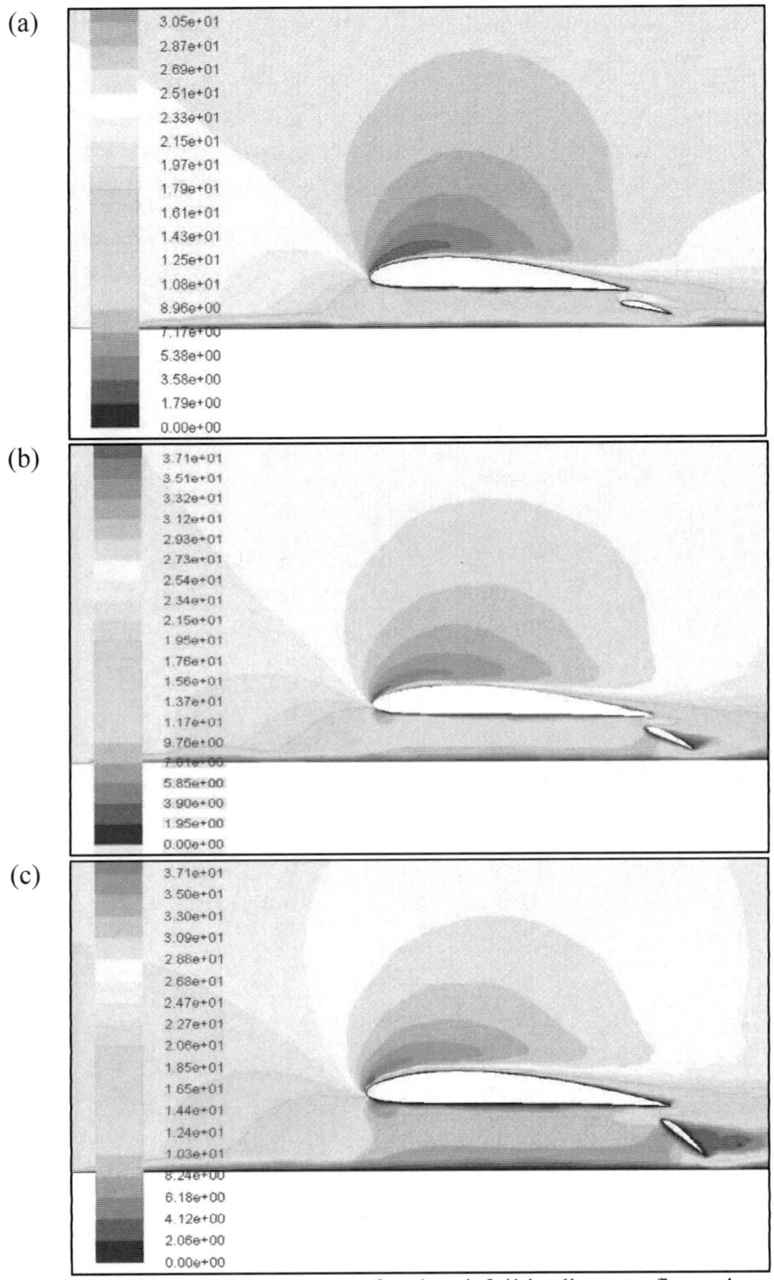

Figure 4. Velocity contour for the airfoil in dirty configuration (a) AoD = 10° (b) AoD = 25° (c) AoD = 40°

Table 1. Influence of AoD on *L/D* at *h/c* = 0.15

AoD	C_l	C_d	*L/D*	$\Delta\, C_l$ (%)	$\Delta\, C_d$ (%)	Δ *L/D* (%)
0°	2.891E-02	8.330E-04	34.73			
10°	3.781E-02	1.074E-03	35.20	31	29	1
25°	4.269E-02	2.081E-03	20.51	48	150	-41
40°	4.921E-02	3.114E-03	15.80	70	274	-55

IOP Conf. Series: Materials Science and Engineering **243** (2017) 012022 doi:10.1088/1757-899X/243/1/012022

The computation results based on different settings of the distance between the airfoil and the ground are tabulated in table 2. Lift and drag are the forces which act on the whole geometry consisting of main airfoil and flap. The coefficients C_l and C_d profile shows that the drag, in contrast to the lift is lower for any AoD when the airfoil approaches the ground. This unconditionally contributes to the increase of *L/D* in the vicinity of the ground, and keeps the aircraft to 'float' above the runway relatively longer before the touchdown. It is also worth to note that the airfoil best performs at AoD 10° where the average lift-to-drag ratio increment $(\Delta\, L/D)_{av}$ with respect to *h/c* is approximately 44%. While at AoD 25° and 40°, the overall performance is relatively low since the flow separation compromises the flap lift (see figure 4).

Table 2. Influence of *h/c* on *L/D* at various AoD

AoD	*h/c*	C_l	C_d	*L/D*	$\Delta\, C_l$ (%)	$\Delta\, C_d$ (%)	$\Delta\, L/D$ (%)
	0.25	2.697E-02	8.900E-04	30.28			
	0.20	2.797E-02	8.340E-04	33.53	3.7	-6.3	11
0°	0.15	2.891E-02	8.330E-04	34.73	7.2	-6.4	15
	0.10	3.024E-02	8.260E-04	36.60	12.1	-7.2	21
	0.05	3.288E-02	8.220E-04	40.00	21.9	-7.6	32
	0.25	3.677E-02	1.479E-03	24.86	--	--	--
	0.20	3.679E-02	1.110E-03	33.15	0.1	-24.9	33
10°	0.15	3.781E-02	1.074E-03	35.20	2.8	-27.4	42
	0.10	3.841E-02	1.051E-03	36.56	4.5	-28.9	47
	0.05	3.867E-02	1.021E-03	37.89	5.2	-31.0	52
	0.25	4.243E-02	2.197E-03	19.31	--	--	--
	0.20	4.262E-02	2.172E-03	19.62	0.4	-1.1	2
25°	0.15	4.269E-02	2.081E-03	20.51	0.6	-5.3	6
	0.10	4.403E-02	2.028E-03	21.71	3.8	-7.7	12
	0.05	4.522E-02	1.989E-03	22.73	6.6	-9.5	18
	0.25	4.596E-02	3.174E-03	14.48	--	--	--
	0.20	4.818E-02	3.168E-03	15.21	4.8	-0.2	5
40°	0.15	4.921E-02	3.114E-03	15.80	7.1	-1.9	9
	0.10	4.930E-02	3.090E-03	15.95	7.3	-2.6	10
	0.05	4.943E-02	3.084E-03	16.03	7.6	-2.8	11

5. Conclusion

The study focuses on the effect of ground proximity on the flow over STOL CH750 multi-element airfoil. From the study, several conclusion can be made:

(i) AoD significantly influence the near ground air and cause higher pressure at the pressure side of the airfoil thus slowing down the touch down of the airplane. Higher AoD will result in increase of lift.

(ii) However, higher AoD also contribute to the increase of drag which helps the airplane to slow down at faster rate.

(iii) The distance between the ground and the airfoil also records similar increasing pattern of lift as the airfoil approaching the ground but records decrease of drag force.

It is important to take into consideration the AoD and *h/c* in the operational of an airplane during landing.

Acknowledgement
The author would like to thank both Universiti Tun Hussien Onn Malaysia (UTHM) and Ministry of Higher Education of Malaysia for the financial support through Fundamentals Research Grant Scheme (Vot 1544).

References
[1] Mugunthan M, Pravalika D, Babu KMK, and Kumar SS 2015 Flow field investigation of flat bottom aerofoil under ground effect *OSR-JMCE* **12** 83-88
[2] Mobassher M T, Adi M, Yasser MA, Saeed J, Agoes P, and Rahimuddin 2014 Experimental Investigation of a wing-in-ground effect craft *The Scientific World Journal* **14** 1-7
[3] Amir MAU, Maimun A, Mat S and Saad MR 2016 Computational analysis of aerodynamic characteristics for wing in ground effect craft in lateral stability *IOP Conf. Series: Materials Science and Engineering* **152** 1-9
[4] Luo SC and Chen YS 2012 Ground effect on flow past a wing with a NACA0015 cross-section *Experimental Thermal and Fluid Science* **40** 18-28
[5] Kirill VR 2006 Wing-in-ground effect vehicles *Progress in Aerospace Sciences* **42** 211–283
[6] AMD 2009 STOL 750LS-SLSA with O-200 Airplane and systems description *Engine Engine Pilot Operating Handbook* **2** (Eastman, Georgia) 0-7
[7] Zenair 2017 *Light Sport Utility Aircraft STOL CH750 Specifications and Performance* (Missouri, USA) 1-2
[8] AERO'Connor Aviation – Flight Training 2014 Emergency procedures *Private Pilot Syllabus* (Frederick, USA) Lesson 5 p 5
[9] Andrea DM, Marco RC and Ernesto B 2012 A retrospective of high-lift device technology *International Journal of Mechanical, Aerospace, Industrial, Mechatronic and Manufacturing Engineering* **6** 2561-2566

IOP Conf. Series: Materials Science and Engineering **243** (2017) 012023 doi:10.1088/1757-899X/243/1/012023

Potential reduction of energy consumption in public university library

Z Noranai [1] **and ADF Azman** [2]
Centre for Energy and Industrial Environment Studies (CEIES),
Faculty of Mechanical & Manufacturing Engineering,
University Tun Hussein Onn, Batu Pahat, 86400, Johor Darul Takzim, Malaysia
E-mail: zamrin@uthm.edu.my [1], azraidanial557@yahoo.com [2]

Abstract. Efficient electrical energy usage has been recognized as one of the important factor to reduce cost of electrical energy consumption. Various parties have been emphasized about the importance of using electrical energy efficiently. Inefficient usage of electrical energy usage lead to biggest factor increasing of administration cost in Universiti Tun Hussein Onn Malaysia. With this in view, a project the investigate potential reduction electrical energy consumption in Universiti Tun Hussein Onn Malaysia was carried out. In this project, a case study involving electrical energy consumption of Perpustakaan Tunku Tun Aminah was conducted. The scopes of this project are to identify energy consumption in selected building and to find the factors that contributing to wastage of electrical energy. The MS1525:2001, Malaysian Standard - Code of practice on energy efficiency and use of renewable energy for non-residential buildings was used as reference. From the result, 4 saving measure had been proposed which is change type of the lamp, install sensor, decrease the number of lamp and improve shading coefficient on glass. This saving measure is suggested to improve the efficiency of electrical energy consumption. Improve of human behaviour toward saving energy measure can reduce 10% from the total of saving cost while on building technical measure can reduce 90% from total saving cost.

1. Introduction

Energy sources are among one of the basic requirements that have been used since the existence of human on earth. Rapid development of human civilization demands huge usage of energy resources of the non-renewable type. According to the International Energy Agency, energy demand is growing at a faster rate of 36% between 2008 and 2035[1]. Efficient usage of energy is one approach that is being studied and implemented. As Malaysia moves towards the status of a developed nation in 2020, our energy requirement will become more intensive energy consumption are divided into several main sector which is building, commercial, industrial and transportation [2]. Building sector is the among the major energy consumers in this country. Rapid growth of University Tun Hussein Onn Malaysia (UTHM) leads to an increase of electrical energy usage. It is expected that the university will expand six times greater than the current condition. Hence, it is crucial to study on the potential reduction of energy consumption. The objective of this study is to investigate the potential reduction of energy consumption, to investigate the cause of wastage electrical energy and to propose steps or opportunities for energy saving consumption.

2. Case Study

Tunku Tun Aminah Library building was selected as a case study in this research. The library building located in main campus of UTHM, Parit Raja, Batu Pahat, Johor, at 2° toward North and 103° toward East. The library building block is five stories building. The total floor area of the building is 16,000 m^2 and capable to occupied 4000 student capacity. This library was design to storage 300,000 pieces of books. The typical equipments used in this library building are including air conditioner units, lamps, computers, laptop, photocopy machine, printers and other small equipments. The building is oriented in a way that the longitudinal axis of the building is aligned with North West – South East axis. UTHM Library is shaded from North East and South West exposure due to presence of adjacent buildings. Table 1 shows the details of general information of the library building

Table 1. General information of UTHM Library

No	Building Specification	Description
1	Area	16,000 meter square
2	Storey	5 storey building
3	Year Built	May 2010
4	Student Capacity	4000 people
5	Storage Capacity	300,000 books
6	Amenities	2 x seminar rooms, 100 x carrel rooms 40 x discussion rooms, 1 x auditorium 2 x 24 hours reading room

The library building operation can be categories into three main purposes. First purpose is to provide study area to the university students. Second purpose is to provide teaching and learning activity, such as seminar rooms and auditorium. Third purpose is function as administration office for top management, discussion room and meeting room.

3. Analyses

3.1. Electrical energy consumption pattern

Electricity consumption in Tunku Tun Aminah Library was studied by analyzing the actual electricity consumption of the electric bills issued by TNB. Electric tariff bills are issued in the early of every month to UTHM through Pejabat Pentadbiran Hartabina. Table 2 shows the detail of electric bill 2016 for Tunku Tun Aminah Library Building. Based on that table, the trends of electricity consumption from January 2016 to December 2016 are not consistent. In August 2016, the electricity consumption was only at 230,552kWh. It is the lowest electricity consumption because mostly students are not in UTHM due to semester break. However in March 2016 the electricity consumption reach up to 417,980kWh, highest compare to another month. Air conditioning contributes to a major portion of the electricity bill. The electricity energy consumption for air conditioning includes air conditioning for office rooms, class rooms, and lecture hall. Electricity consumption for lighting consists of office room, passageway, walkway, lobby, building and landscape lighting.

Based on research conducted by Z Noranai (2008), the breakdown of electricity consumption for air conditioning is 75% followed by electricity for lighting and other equipments is 25% [3].

Table 2. Detail electric bill year 2016 for Tunku Tun Aminah Library

Month	Electrical Consumption (kWh)
January	349,270
February	292,400
March	417,980
April	393,740
May	401,420
June	370,690
July	307,164
August	230,552
September	264,217
October	277,071
November	303,641
December	306,516
Total	**3,914,664**

3.2. Cause of wastage electrical energy

Tunku Tun Aminah Library is one of the buildings that represent a very high energy consumption percentage compare to another UTHM building. In this study, several factors cause of wastage electrical energy is identified:

3.2.1. Usage of lighting at inappropriate area. Based on the observation conducted, it is found lamps at corridor and stairs way area are still on even thought that area already receive enough brightness from the sun. Lux meter equipment are been used for measuring brightness or luminance. The reading of lux meter at corridor area is 650 lux, exceed the lighting requirement of MS 1525:2007 which is 300 – 500 lux [4]. At stairs way area, the reading of lux meter also exceeds the requirement lighting needed for stairs which is 100 lux. The table 3 shows the amount of energy wastage due to using light at inappropriate area.

Table 3. Amount of energy waste with type of waste

No	Type of Waste	Amount of energy waste per month (kWh)	Amount of energy waste per year (kWh)
1	Usage of lighting at inappropriate area – Corridor	168.48	2021.76
2	Usage of lighting at inappropriate area – Stairs way	339.36	4720.32

IOP Conf. Series: Materials Science and Engineering **243** (2017) 012023 doi:10.1088/1757-899X/243/1/012023

3.2.2. Airflows. Air flows or air leakage is one of the cooling load components. Air leakage occurs when outside air enter and conditioned air leaves uncontrollably through cracks and openings [5]. This factor will cause the increasing of cooling load and electricity consumption. After the observation is conducted, some of the causes of air leak was identified which is toilet doors at floor 4 and windows at locker area are left open. Table 4 below show the energy loss due to air flows.

Table 4. Amount of energy waste with type of waste

No	Type of Waste	Amount of energy waste per month (kWh)	Amount of energy waste per year (kWh)
1	Airflow – Toilet door	62	744
2	Airflow – Windows	28	336

3.2.3. Usage of air conditioning in unused room. The use of air conditioning in unused room will contribute to wastage electrical energy. Based on observation conducted, it is found Permata Hikmah Library using air conditioning for 13 hours even though that library only operates for 3 hours per day. This library is supposed to using split unit air conditioning to control the room temperature inside. The table 5 below show the amount of energy wastage due to this factor.

Table 5. Amount of energy waste with type of waste

No	Type of Waste	Amount of energy waste per month (kWh)	Amount of energy waste per year (kWh)
1	Usage of air conditioning in unused room	4100	49200

3.2.4. Misuse of electric equipment. Ground floor contain 24 hours study room that operating every day. There are several facilities provided such as fan, air conditioner, table and chair. Based on observation conducted, some of users still switch on the fan even though the air conditioner is already operating. This situation will contribute to the waste of electricity. The table 6 below show the amount of energy wastage due to this factor.

Table 6. Amount of energy waste with type of waste

No	Type of Waste	Amount of energy waste per month (kWh)	Amount of energy waste per year (kWh)
1	Misuse of electric equipment	540	6480

3.3. Energy saving opportunity
Key figures for energy consumption and energy saving potential for Tunku Tun Aminah Library were presented based on observation, review of relevant literature and calculation conducted. As a result, it can be concluded that the main energy consumption is the air conditioning system followed by lighting system. There are several energy saving approaches including minimize the lamp number, minimize the lamp usage, put a sensor and change the lamp type. Table 7 illustrated the amount of energy saving with its type of saving.

Table 7. Amount of energy waste with type of waste

No	Saving action	Amount of energy saving per month (kWh)	Amount of energy saving per year (kWh)
1	Change the fluorescent type of lamp in toilet into LED tube lamp type.	1,950	23,400
2	Change the fluorescent type of lamp in closed space into LED tube lamp type	3089	37,068
3	Installing the Sensor	2,238	26,856
4	Reduce the number of lights in the Open reading space	17,355	208,260
5	Enhancing Layer Window Films	32,160	385,920
	Total	**56,792**	**681,504**

As an example potential energy saving calculation method

Fluorescent lamp Energy	=	Number of lamps x power x hours
	=	500 pieces x 32 watts x 2600 hours
	=	4160 kWj
LED lamp energy	=	Number of lamps x power x hours
	=	500 pieces x 17 watts x 2600 hours
	=	2210 kWj
Potential energy saving	=	Fluorescent lamp Energy - LED lamp energy
	=	4160 – 2210 kWj
	=	1950 kWj per month.

4. Conclusion and recommendation
There are two types of potential cost saving; with less or no costs and with costs whether medium costs or high costs. Rescheduling of air conditioner system and lighting operation is consider as no costs potential saving per year. Retrofitting specific lighting system typically would bring great saving up to RM8990 per month, but this measure is required medium initial investment costs. Increasing the

IOP Conf. Series: Materials Science and Engineering **243** (2017) 012023 doi:10.1088/1757-899X/243/1/012023

shading coefficient of window film will save up to RM11680 per month. As conclusion, if all the recommendation made are implemented, it is estimated possible to save up to RM20279 per month. It is suggested for further economical analysis to find out payback period and return of investment.

Reference

[1] Zaharah E A E S 2013 *Potensi Penjimatan Penggunaan Tenaga oleh Isi Rumah di Malaysia* Perkem VIII pg 1004 - 1010 ISSN: 2231-962X

[2] Dong Z L, Zhang Z, Jianmin Z and Keru W 2004 *Development of thermal energy storage concrete* Cement and Concrete Research vol.34

[3] Noranai Z 2014 *Energy Saving Measures for University Public Library: A Case Study of UTHM Library* Applied Mechanics and Materials (Trans Tech Publ)

[4] MS1525 2007 *Code of practice on energy efficiency and use of renewable energy for non-residential buildings*, Department of Standards Malaysia, Cyberjaya Selangor

[5] Noranai Z 2012 *Development of Cooling Load Demand Program for Building in Malaysia.* International Journal of Mechanical and Industrial Engineering

Effect of piston profile on performance and emission characteristics of a GDI engine with split injection strategy – A CFD study

Y Karaya[1] and J M Mallikarjuna[2]
[1] MS Research Scholar, Indian Institute of Technology Madras, Chennai, India
[2]Associate Professor, Indian Institute of Technology Madras, Chennai, India

E-mail: yashas.1993@gmail.com[1], jmmallik@iitm.ac.in[2]

Abstract. Gasoline direct injection (GDI) engines have gained popularity in the recent times because of lower fuel consumption and exhaust emissions. But in these engines, the mixture preparation plays an important role which affects combustion, performance and emission characteristics. The mixture preparation in turn depends mainly upon combustion chamber geometry. Therefore, in this study, an attempt has been made to understand the effect of piston profile on the performance and emission characteristics in a GDI engine. The analysis is carried out on a four-stroke wall guided GDI engine using computational fluid dynamics (CFD). The spray breakup model used is validated with the available experimental results from the literature to the extent possible. The analysis is carried out for four piston profiles viz., offset pentroof with offset bowl (OPOB), flat piston with offset bowl (FPOB), offset pentroof with offset scoop (OPOS) and inclined piston with offset bowl (IPOB) fitted in an engine equipped with a six-hole injector with the split injection ratio of 30:70. All the CFD simulations are carried out at the engine speed of 2000 rev/min., with the overall equivalence ratio of about 0.65±0.05. The performance and emission characteristics of the engine are compared while using the above piston profiles. It is found that, the OPOB piston is preferred compared to that of the other pistons because it has better in-cylinder flow, IMEP and lower HC emissions compared to that of other pistons.

1. Introduction

Developing engines with higher fuel economy and lower exhaust emissions have been a key research area in recent times [1]. With poor part-load fuel efficiency and higher emissions, port fuel injection (PFI) engines are facing difficulty in fulfilling the emission norms. On the other hand, GDI engines, because of lean mode of operation and mixture stratification, are capable of achieving better fuel economy and lower emissions. In GDI engines, fuel is injected directly into the engine cylinder which helps increasing volumetric efficiency and compression ratio. In a wall guided GDI engine, piston is profiled in such a way that the fuel-air mixture is directed towards the spark plug to form a combustible mixture at the time of ignition. Piston with a bowl gives better air-fuel mixing, higher tumble ratio and turbulent kinetic energy compared to that of the flat piston [1]. It is also beneficial in maintaining in-cylinder turbulent intensity and promotes fuel evaporation and reduces soot emissions [2]. Also, piston with a smoother top surface accelerates combustion and increases in-cylinder temperature and pressure [3]. Stoichiometric mixture around the spark plug results in higher heat release and NO_x emissions, compared to that of the lean mixture [4]. With split injection ratio of

IOP Conf. Series: Materials Science and Engineering **243** (2017) 012024 doi:10.1088/1757-899X/243/1/012024

30:70, mixture is slightly homogenous which results in better ignitable mixture around the spark plug which will lead to a better combustion [5].

In the present study, the effect of piston top profile on the performance and emission characteristics of a four-valve, wall guided GDI engine is studied. The engine is operated in the split injection mode with the split injection ratio of 30:70. Four piston top profiles are considered for the analysis.

2. CFD Methodology

2.1. Engine Geometry and Meshing

The CFD analysis is carried out on a four-stroke, four-valve, wall guided GDI engine whose specifications are taken from Costa et al., [6] and are shown in Table 1. Figure 1 shows the computational model of the engine. Table 2 gives the details of fuel injection parameters. The grid embedding for this study is done as per details given in *Krishna and* Mallikarjuna [7], which are not shown here.

2.2. Piston Profile

In this study, four piston profiles are considered viz., offset piston with offset bowl (OPOB), flat piston with offset bowl (FPOB), offset piston with offset scoop (OPOS) and inclined piston with offset bowl (IPOB). Figure 2 shows various piston profiles used.

Figure 1. Computational model of the engine

Table 1. Detailed specifications of the engine

Parameters	Specification
Displacement volume	398.5 cm3
Stroke	81.3 mm
Bore	79 mm
Connecting rod	143 mm
Compression ratio	10.6:1
Speed (rpm)	2000
Number of valves	4
Exhaust valve opening	27° before BDC
Exhaust valve closure	0° after TDC
Inlet valve closure	36° after BDC
Inlet valve opening	3° before TDC

(*360 CAD is suction at TDC)

Table 2. Details of fuel injection parameters

No. of holes	Six
Fuel	Gasoline
Start of Pilot Fuel Injection	210^0 BTDC
Start of Main Fuel Injection	90^0 BTDC
Pilot fuel injection duration	6.9 CAD
Main fuel injection duration	16.2 CAD
Fuel injection pressure	120 bar
Spray cone angle	15 degree
Mass of fuel injected per	22 mg

*CAD- Crank Angle Degrees

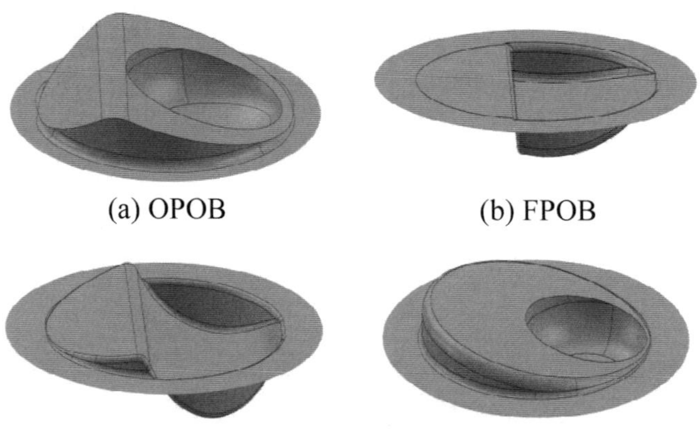

(a) OPOB (b) FPOB

(c) OPOS (d) IPOB

Figure 2. Various piston profiles used in the study

2.3. CFD Modeling

The governing equations used, in the study, to generalize the in-cylinder flows are given below [8].
Conservation of mass:

$$\frac{\partial \rho}{\partial t}+\frac{\partial(\rho u_j)}{\partial x_j}=s_m \tag{1}$$

General transport equation:

$$\frac{\partial(\rho\varphi)}{\partial t}+div(\rho\varphi u)=div(\Gamma grad\varphi)+S_\varphi \tag{2}$$

Where, ϕ is a general variable of conservative form of all fluid flow equations including equations for scalar quantities such as temperature and pollutant concentration, etc.

The CFD simulation is carried out using the CONVERGE. Mass, momentum and energy are calculated at each node of the grid using unsteady Navier-Stokes equations which is implicitly discretized based on finite volume method over the Cartesian grid [8]. RNG k-ε turbulence model is used to analyze the turbulence characteristics [9]. The KH-RT model is used to capture fuel spray break-up [7]. The pressure implicit for splitting of operator (PISO) algorithm is used to solve the pressure-velocity coupling [10]. The O'Rourke and Amsden model is used for wall heat transfer calculations [11]. For the analysis of droplet evaporation, the FROSSLING model is used [12]. For the analysis of film splash and fuel turbulent dispersion, the O'Rourke model is used [12]. The droplet collision has been analyzed by the NTC collision model [12] and initial droplet distribution is analyzed using the blob injection model [12]. The inlet and exhaust boundaries are maintained at atmospheric conditions [11,14]. In order to obtain the initial conditions in various regions of the computational domain, full-cycle simulations are carried out successively four times. The flow characteristics, at the end of the fourth cycle, are used as the initial conditions for all the simulations. The constants used for the KH-RT spray model and k-ε turbulence model are taken from the literature [13]. In this study, all the CFD simulations are carried out from the inlet valve opening (IVO) to the exhaust valve opening (EVO) at the engine speed of 2000 rev/min., with the overall equivalence ratio of 0.65±0.05. The time step varies between 10^{-8} and 10^{-5} seconds depending on the grid CFL criteria. The pilot injection is maintained at 510 crank angle degree (CAD), the main injection is maintained at 630 CAD and the spark timing is maintained at 710 CAD. The convergence tolerance for momentum and pressure are taken as 10^{-5} and 10^{-8} respectively. Whereas, for the density, the energy and the species, the convergence tolerance is taken as 10^{-4}.

2.4. CFD Model Validation

The CFD models have already been validated by Krishna et al., [12] in their work. Validation was done based on the experimental results obtained from Costa et al., [6] on the same engine. Comparison of in-cylinder pressure with crank angles obtained from both the CFD and experimental results are shown in Figure 3. The comparison of ER distribution at the time of ignition for both CFD and experimental results are shown in Figure 4. From Figures 3 and 4, it is evident that; the two results are in good agreement. Therefore, the selected CFD models are used for further analysis.

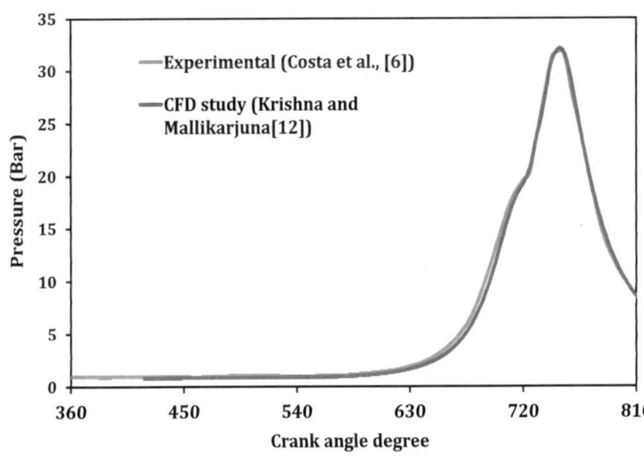

Figure 3. Comparison of in-cylinder pressure with crank angle degree

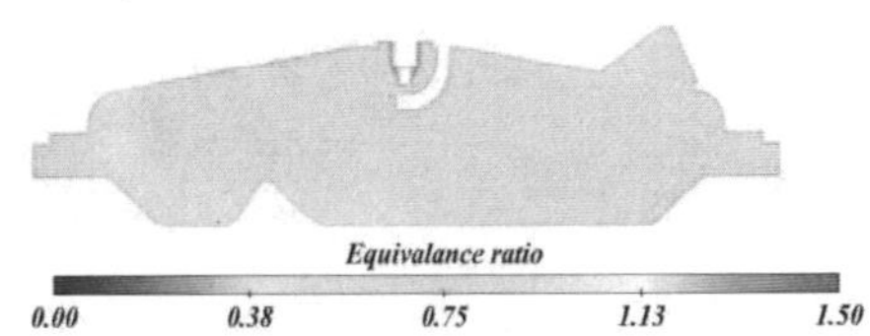

(a) From CFD analysis of Krishna and Mallikarjuna [6]

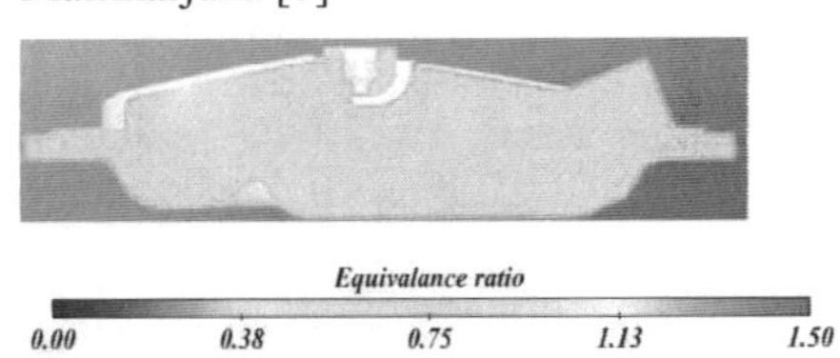

(b) From CFD analysis of Costa et al., [16]

Figure 4. Comparison of ER distribution on central plane at spark timing

3. Results and Discussions

3.1. Effect of Piston Profile on In-cylinder flow

Figure 5 shows the comparison of in-cylinder flows for various piston profiles at various CADs. From figure 5, it is seen that, at 640 CAD, the piston profiles considered have minimum impact on in-the cylinder flows. Whereas, at 675 CAD, for the OPOB and the FPOB pistons, a strong vortex occurs near the spark plug which leads to better mixture stratification compared to that of the other two pistons. At the end of the compression stroke (700 CAD), all piston profiles show vortices in the combustion chamber. But, for the OPOB piston, the intensity of the vortex is observed to be higher than that of the other pistons which in turn help enhanced combustion and performance.

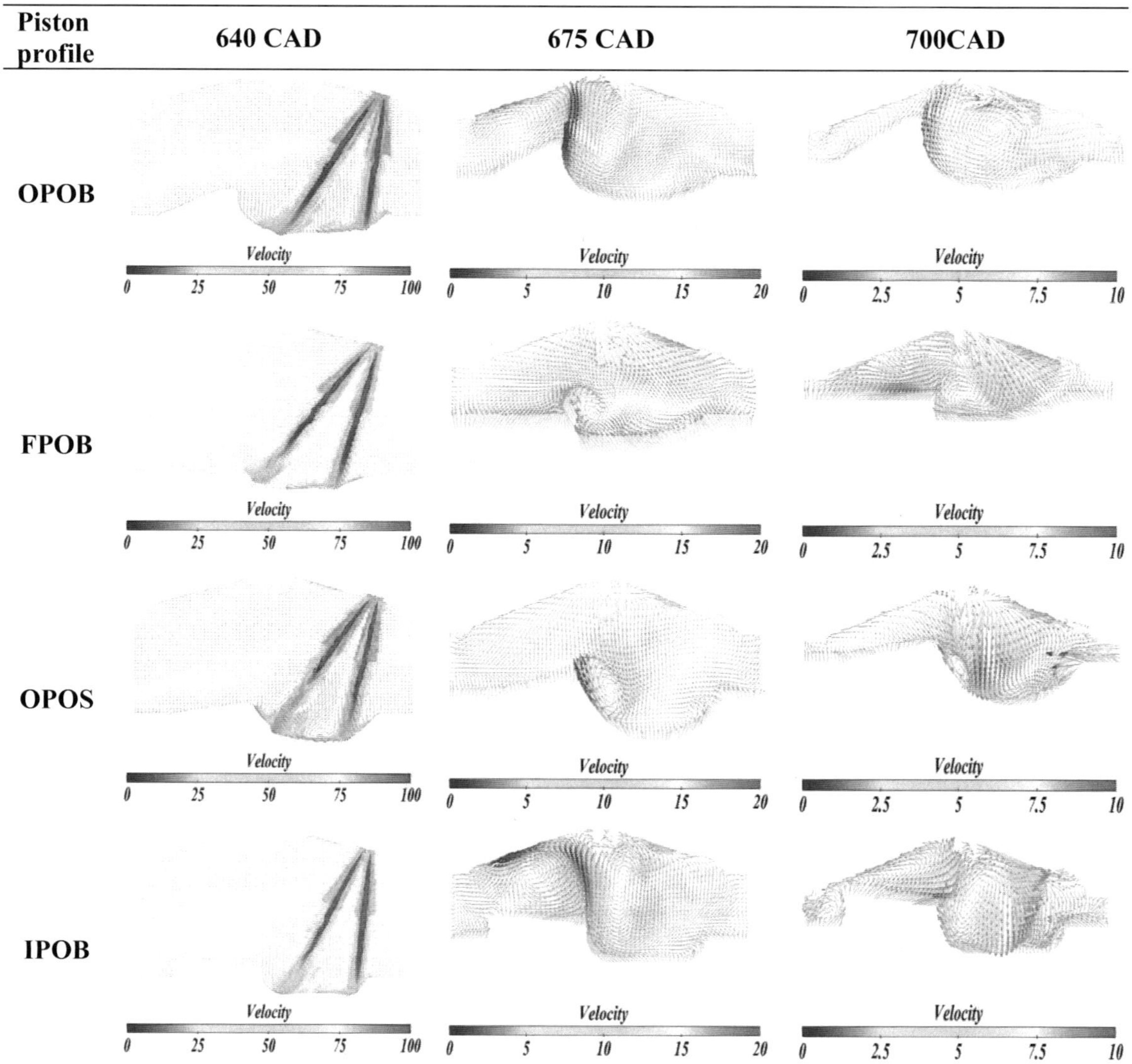

Figure 5. Comparison of in-cylinder flows for various piston profiles at various CAD

3.2. Effect of Piston Profile on Equivalence Ratio(ER) Distribution

Figure 6 shows the ER distribution for various piston profiles. From figure 6, it is seen that, the ER near the spark plug location for OPOB, FPOB, OPOS and IPOB piston profiles are about 1.12, 0.85, 0.6 and 1.45 respectively. For OPOB and FPOB, mixture is almost stoichiometric near the spark plug location because of better in-cylinder flow. Whereas for OPOS the mixture is very lean and for IPOB, the mixture is very rich near the spark plug location.

3.3. Effect of Piston Profile on In-cylinder Pressure

Figure 7 shows the comparison of in-cylinder pressure for various piston profiles. From figure 7, it is seen that, for the OPOB piston profile, the peak in-cylinder pressure is about 48 bar which occurs at about 740 CAD. Whereas, for the FPOB, OPOS and IPOB pistons, the peak in-cylinder pressures occur at about 740 CAD, which are lower by about 27, 38 and 27% compared to that of the OPOB piston. Since, the ER is nearer to stoichiometric around the spark plug in the case of OPOB piston, the in-cylinder pressure is higher compared to the other pistons considered.

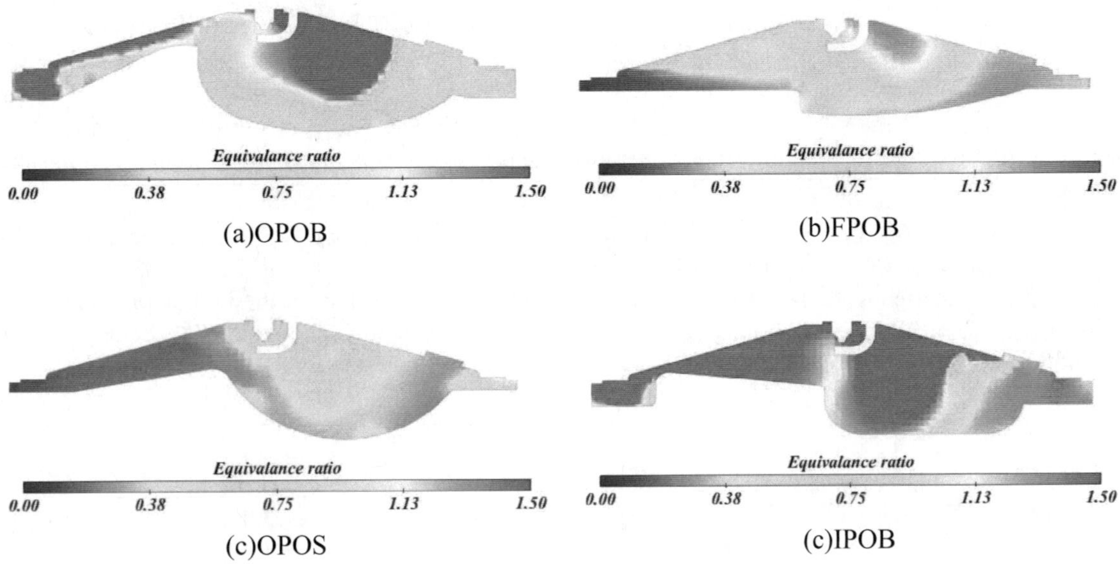

Figure 6. Comparison of ER for various piston profiles

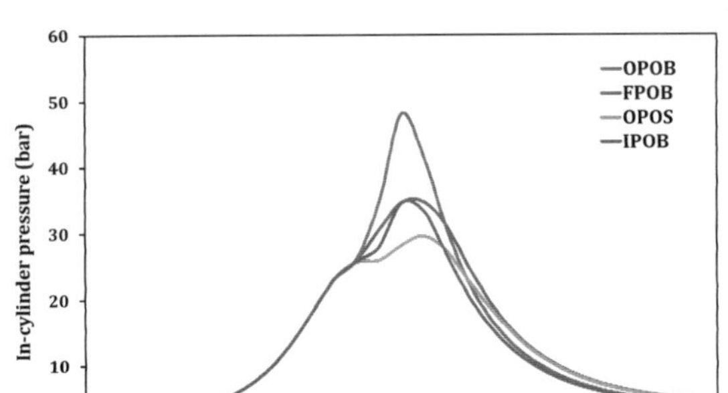

Figure 7. Comparison of in-cylinder pressure for various piston profiles

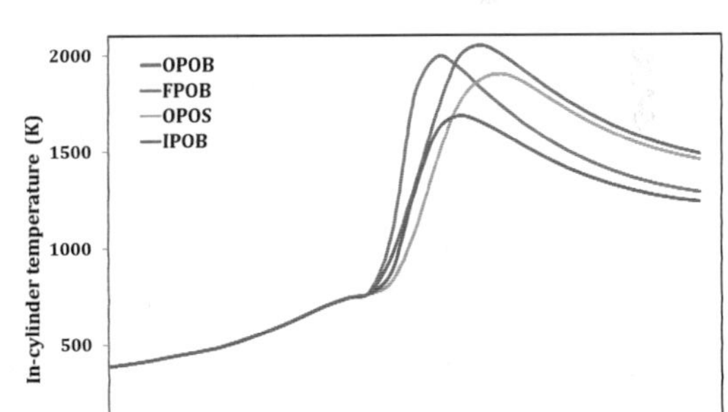

Figure 8. Comparison of in-cylinder temperature for various piston profiles

3.4. Effect of Piston Profile on In-cylinder Temperature

Figure 8 shows the comparison of in-cylinder temperature for various piston profiles. From figure 8, it is seen that, for the OPOB piston, the peak in-cylinder temperature is about 1987 K which occurs at about 750 CAD. Whereas, for the FPOB, the peak in-cylinder temperature is about 3% higher compared to that of OPOB piston. But, for the OPOS and the IPOB, the peak in-cylinder temperatures are about 5 and 15% lower compared to that of the OPOB piston. It means that, the combustion is slower in the FPOB compared to that of the OPOB, as the peak temperature is occurring closer to the TDC in the OPOB. However, because of slower combustion in the FPOB, more fuel is burnt away from the TDC, which gives higher in-cylinder temperature, in the case of the FPOB compared to that of the OPOB piston.

3.5. Effect of Piston Profile on Heat Release Rate

Figure 9 shows the comparison of heat release rate (HRR) for various piston profiles. The HRR is calculated in CONVERGE by using equation 3

$$\frac{dQ}{d\theta}=\frac{\gamma}{\gamma-1}\left(P\frac{dV}{d\theta}\right)+\frac{1}{\gamma-1}\left(V\frac{dP}{d\theta}\right)+\frac{dQ_W}{d\theta} \tag{3}$$

Where, Q is the heat released on combustion of fuel, Q_w is the heat transfer out of the cylinder, P is instantaneous in-cylinder pressure and γ is ratio of specific heat at constant pressure to constant volume.

From figure 9, it is seen that, for the OPOB piston, the peak HRR is about 31 J/CAD. Whereas, for the FPOB, the OPOS and the IPOB pistons, the HRRs are lesser by about 21.2, 28.6 and 8.4% respectively, compared to that of the OPOB piston profile. It is seen that, in the OPOB piston, the in-cylinder flows and tumble ratio are better than that of other pistons. This results in better mixture preparation inside the combustion chamber which leads to faster combustion.

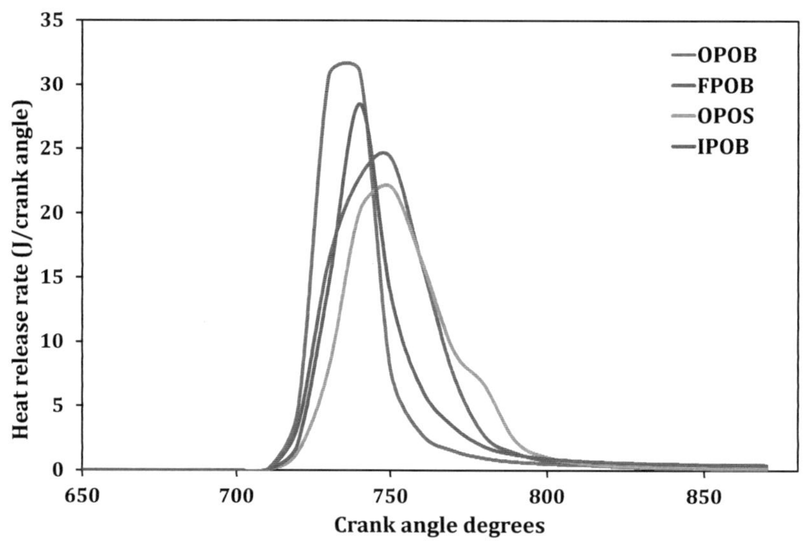

Figure 9. Comparison of heat release rate for various piston profiles

3.6. Effect of Piston Profile on Percentage Fuel Burnt [PFB]

In this study, the combustion is considered to be completed when 90% of the injected fuel is burnt. Figure 10 shows the comparison of the crank angle position at which 10, 50 and 90% of fuel is burnt. From figure 10, it is seen that, for the OPOB piston, the 90% of the fuel is burned at about 755.7 CAD. However, for the FPOB, the OPOS and the IPOS pistons, the 90% of the fuel is burned at about 769.7, 778.4 and 775.9 CADs respectively. Hence, it is seen that, the fastest combustion takes place with the OPOB piston, compared to that of the other pistons. This is because of better ER distribution in the combustion chamber for the OPOB as mentioned in Section 3.2. In addition, the slowest combustion takes place with the OPOS, because of too lean mixture around the spark plug due to which the flame propagation becomes difficult. For the IPOB piston, the 50% fuel burnt is faster as compared to that of the FPOB piston, which implies that more fuel is available in the case of the IPOB piston at 10-50% PFB phase. However, with the FPOB piston, more fuel is available during 50-90% of the PFB phase. It also resulted in higher HRR with the IPOB piston compared to that of the FPOB piston as seen from figure 9.

3.7. Effect of Piston Profile on IMEP

Figure 11 shows the comparison of IMEP for various piston profiles. From figure 11, it is seen that, for the OPOB piston, the IMEP is about 9.56 bar. With the FPOB, the OPOS and the IPOB pistons, the IMEP is lesser by about 4, 10 and 20% compared to that of the OPOB piston. Higher IMEP, in the case of the OPOB, is mainly because of the better mixture formation at the time of ignition and higher peak in-cylinder pressure.

3.8. Effect of Piston Profile on HC Formation

Figure 12 shows the comparison of HC formation for various piston profiles. From figure 12, it is seen that, the HC formation, at the EVO (at 873 CAD), is negligible for the OPOB, FPOB and OPOS pistons. This is because of better mixture preparation which resulted in almost complete combustion with these pistons. Even though, for the OPOS piston, there is a very lean mixture around the spark plug, at the time of ignition, due to better in-cylinder motion, most of the mixture is burnt at a later stage of the cycle. Therefore, there are very less HC emissions, at the EVO, with the OPOS piston. However, for the IPOB piston, there is a very rich mixture around the spark plug at the time of ignition because of which the combustion becomes slower. Therefore, with the IPOB piston, the HC emissions, at the EVO, are almost doubled compared to that of the OPOB piston.

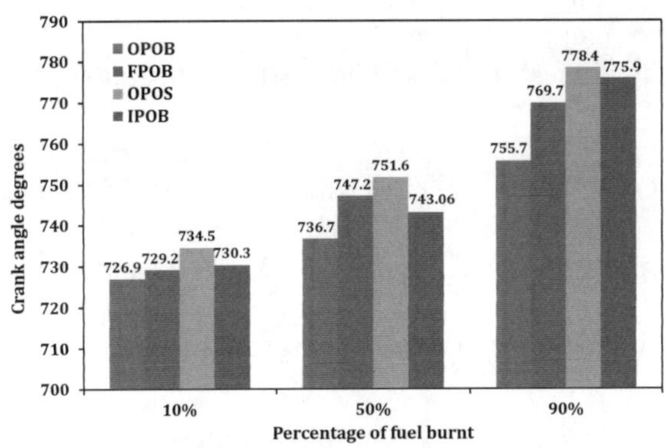

Figure 10. Comparison of percentage of fuel burnt for various piston profiles

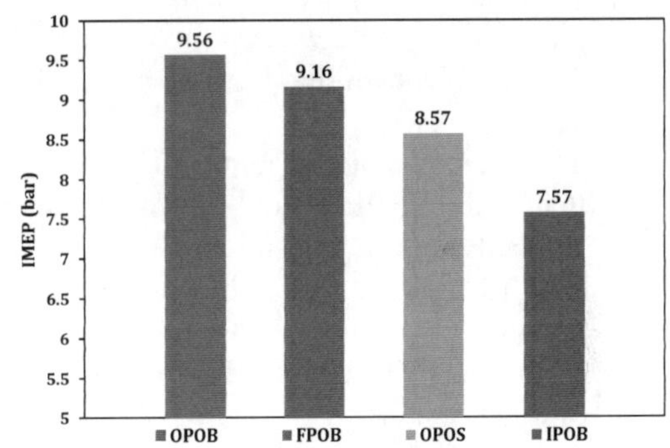

Figure 11. Comparison of IMEP for various piston profiles

3.9. Effect of Piston Profile on NO_X Formation

Figure 13 shows the comparison of NO_X formation for various piston profiles. From figure 13, it is seen that, the NO_X formation at the time of the EVO is the highest for the OPOB piston. Whereas, with the FPOB, OPOS and IPOB pistons, the NO_X emissions are lower by about 15, 50 and 64% compared to that of the OPOB piston. The NO_X emissions, with the OPOB piston, are higher because of better combustion resulting in higher in-cylinder temperature compared to that of the other pistons. Whereas, in the OPOS and IPOB pistons, the mixture at the spark plug location is far from the stoichiometric value. Therefore, there is a poor combustion resulting in lower in-cylinder temperature. Hence, lower NOX emissions are observed in the OPOS and IPOB pistons.

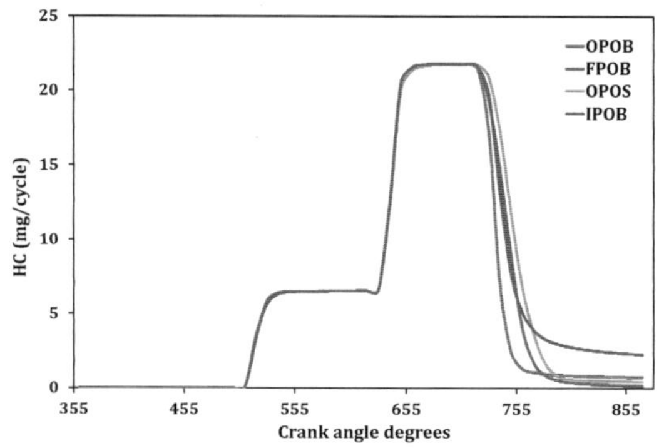

Figure 12. Comparison of HC formation for various piston profiles

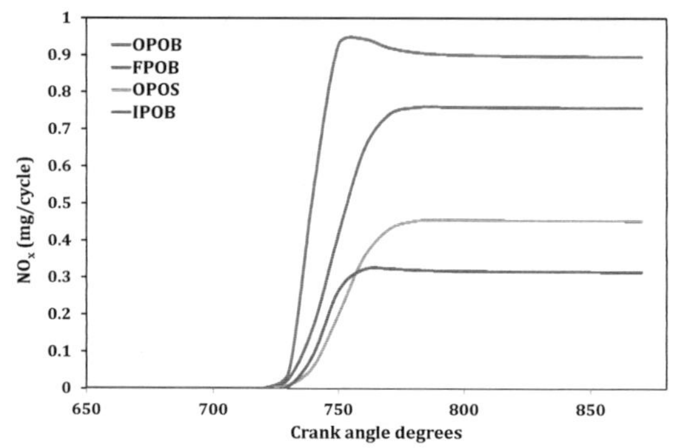

Figure 13. Comparison of NO_X temperature for various piston profiles

4. Conclusion

This study presents the effect of piston profile on the combustion, performance and emission characteristics of a wall-guided GDI engine. Four piston profiles are considered for the analysis. From the results, the following conclusions are drawn:

- Better in-cylinder flows are observed in the OPOB piston compared to that of other pistons.
- In the OPOB and the FPOB pistons, mixture is almost stoichiometric around the spark plug, whereas for the OPOS piston, there is very a lean mixture near the spark plug and for the IPOB piston, there is very a rich mixture near the spark plug at the time of ignition.
- The peak in-cylinder pressure and IMEP are the highest for the OPOB piston compared to that of other pistons.
- The peak in-cylinder temperature is lower by about 3%, in the case of the OPOB piston, than that of the FPOB piston. But, higher than the OPOS and the IPOB pistons by about 5 and 15% respectively.
- The heat release rate is higher for the OPOB piston followed by the IPOB, FPOB and OPOS pistons.
- With the OPOB piston, HC emissions are lower and NO_x emissions are higher compared to that of other pistons.

Finally, it is concluded that, the piston profile plays an important role on the combustion, performance and emission characteristics of a GDI engine. Even though with the OPOB piston, the NO_X emissions are higher, it is preferred, compared to that of others because of better in-cylinder flow characteristics, higher IMEP and lower HC emissions. Also, in the future, it is planned to apply particle image velocimetry (PIV) and planar laser induced florescence (PLIF) techniques using an optical engine for different piston profiles to study the mixture formation and flame propagation under different operating conditions.

5. Acknowledgments

Authors are grateful to the high performance computing facility (HPCF) at Indian Institute of Technology Madras, Chennai, which is used to carry out the CFD simulations presented in this work. Also I would like to thank my Co-scholars, Mr. Krishna Addepalli S., Priyanka D Jadhav, Om Prakash Saw and Ankit Raut for their support during this study.

References

[1] Harshavandhan B and J M Mallikarjuna 2014 Effect of piston profile on in-cylinder flows and air fuel interaction in a direct injection spark ignition engine-A CFD analysis Energy **81** 361-372

[2] Zhaolei Z, Xuefeng T and Xiaoyu Z 2015 Effects of split injection proportion and the second injection time on the mixture formation in a GDI engine under catalyst heating mode using stratified charge strategy *Applied Thermal Engineering* **84** 237-245

[3] Zheng Z, Liu C, Tian X and Zhang X 2015 Numerical study of the effect of piston top contour on GDI engine performance under catalyst heating mode *Fuel* **157** 64–72

[4] Costa M, Catapano F, Sementaa P, Sorge U and Vaglieco B M 2016 Mixture preparation and combustion in a GDI engine under stoichiometric or lean charge: an experimental and numerical study on an optically accessible engine *Applied Energy* **180** 86–103

[5] Zhaolei Z, Chuntao L, Xuefeng T and Xiaoyu Z 2015 Numerical study of the effect of piston top contour on GDI engine performance under catalyst heating mode *Fuel* **157** 64–72

[6] Costa M, Marchitto L, Merola S S and Sorge U 2014 Study of mixture formation and early flame development in a research GDI (gasoline direct injection) engine through numerical simulation and UV-digital imaging *Energy* **77** 88-96

[7] Krishna A S and J M Mallikarjuna 2016 Parametric analysis of a 4-stroke GDI engine using CFD *Alexandria Engineering Journal*

[8] Versteeg H K and W Malalasekera 1995 An Introduction to Computational Fluid Dynamics *Longman scientific and technical publication* 11-25

[9] Krishna A S, Mallikarjuna J M, Kumar D and Babu Y R 2013 In-Cylinder Flow Analysis in a Two- Stroke Engine –A Comparison of Different Turbulence Models Using CFD *SAE Technical Paper 2013-01-1085* doi. 10.4271/2013-01-1085.

[10] Krishna A S, Mallikarjuna J M and Kumar D 2016 Effect of engine parameters on in-cylinder flows in a two-stroke gasoline direct injection engine *Applied Energy* **176** 282–294

[11] Krishna A S and Mallikarjuna J M 2017 Effect of Engine Parameters on Mixture Stratification in a Wall-Guided GDI Engine - A Quantitative CFD Analysis *SAE International Journal of Commercial Vehicles* **10**(2)

[12] 2015 Converge v2.2.0, Theory Manual, Convergent Science Inc

[13] Saw O P and Mallikarjuna J M 2016 Comparison of in-cylinder flows in a GDI engine with and without fuel injection *International Conference on Development of Smart Cities. Interface, Governance and Technology, September 9-10, 2016*, Dr. Ambedkar Institute of Technology, Bengaluru India

[14] Krishna A S, Saw O P and Mallikarjuna J M 2017 Effect of Mixture Distribution on Combustion and Emission Characteristics in a GDI Engine – A CFD Analysis *SAE Technical Paper 2017-24-0036*

Abbreviations

CFD	Computational Fluid Dynamics
GDI	Gasoline Direct Injection
PFI	Port Fuel Injection
IMEP	Indicated Mean Effective Pressure
PIV	Particle Image Velocimetry
PLIF	Planar Laser-Induced Fluorescence
IVO	Inlet valve Opening
EVO	Exhaust Valve Opening
ER	Equivalence Ratio
CAD	Crank Angle Degree
RNG	Re-Normalized Group
KH-RT	Kelvin-Helmholtz-Rayleigh-Taylor
ISO	Pressure Implicit with Splitting of Operators

IOP Conf. Series: Materials Science and Engineering **243** (2017) 012025 doi:10.1088/1757-899X/243/1/012025

Effect of spark plug and fuel injector location on mixture stratification in a GDI engine - A CFD analysis

O P Saw[1] and J M Mallikarjuna[2]
[1] MS Research Scholar, Indian institute of technology Madras, Chennai, India
[2] Associate Professor, Indian institute of technology Madras, Chennai, India

E-mail: opsawiitm@gmail.com[1], jmmallik@iitm.ac.in[2]

Abstract: The mixture preparation in gasoline direct injection (GDI) engines operating at stratified condition plays an important role in deciding the combustion, performance and emission characteristics of the engine. In a wall-guided GDI engine, with a late fuel injection strategy, piston top surface is designed in such a way that the injected fuel is directed towards the spark plug to form a combustible mixture at the time of ignition. In addition, in these engines, location of spark-plug and fuel injector, fuel injection pressure and timing are also important to create a combustible mixture near the spark plug. Therefore, understanding the mixture formation under the influence of the location of spark plug and fuel injector is very essential for the optimization of the engine parameters. In this study, an attempt is made to understand the effect of spark plug and fuel injector location on the mixture preparation in a four-stroke, four-valve and wall-guided GDI engine operating under a stratified condition by using computational fluid dynamics (CFD) analysis. All the CFD simulations are carried out at an engine speed of 2000 rev/min., and compression ratio of 10.6, at an overall equivalence ratio (ER) of about 0.65. The fuel injection and spark timings are maintained at 605 and 710 CADs respectively. Finally, it is concluded that, combination of central spark plug and side fuel injector results in better combustion and performance.

1. Introduction

The GDI engine technology has received considerable attention over the last few years as a way to improve fuel efficiency along with meeting future emission norms in spark ignition (SI) engines [1]. In a wall guided GDI engine, fuel is injected directly into the combustion chamber at a high pressure [2]. In these engines, air-fuel mixing in a short duration poses a major challenge. In order to assist mixing of air with fuel, piston top profile is shaped to enhance swirl and tumble flows [3]. However, intake port angle, fuel injector orientation, fuel spray foot print etc., play major role in the mixture formation [4-5].

However, the mixture stratification mainly depends on the engine operating conditions viz., intake port design, combustion chamber geometry, engine speed and compression ratio, position of fuel injector etc. Whereas, the combustion characteristics of the engine depend upon in-cylinder flows, location of spark-plug, mixture stratification inside the combustion chamber etc. Among these, the location of fuel injector and spark plug together impact largely on fuel-air mixing, combustion, performance and emission characteristics.

Previously Church et al., [2] conducted experiments, using a pulsed laser and high-speed imager for particle tracking to quantify the effect of change in intake port geometry on in-cylinder motions in a modular engine with replaceable intake port blocks. They used three different intake ports and found that the in-cylinder flows exhibited large-scale motions up to 90° before TDC, but they disappeared at TDC. The smaller intake port angles had higher tumble ratios at BDC and at 90° before TDC. While, the larger port angles had significantly lower tumble ratios at BDC.

Costa et al., [7] studied mixture preparation and combustion in a GDI engine under stoichiometric and lean conditions. Experiments were carried out in an optically accessed turbocharged GDI engine at the engine speed of 1500 rev/min. The engine performance and emissions were studied for the fuel injection pressures of 15 and 6 MPa under the stoichiometric and lean conditions respectively. Their results showed that the stoichiometric mixture resulted in higher cumulative heat release compared to that of the lean stratified mixture. They reported that the higher injection pressure resulted in improved combustion in the case of lean stratified mixture. The NO emissions were higher in the case of stoichiometric mixture, compared to that of lean mixture. Soot emission trends were opposite to that of NO emissions. They also reported that increasing fuel injection pressure and advancing the ignition timing improved power output and lowered the cyclic variation. Also, CO and soot emissions reduced, but increased NO emissions.

Li et al., [8] investigated the effect of split fuel injection with various dwells and mass ratios on stratified charge formation in a GDI engine. The experiments were performed using a swirl type injector in a container filled with the nitrogen gas. The laser absorption and scattering (LAS) technique was used to measure the concentration of liquid and vapour phase sprays. The laser induced fluorescence-particle image velocimetry (LIF-PIV) technique was used for analysing the spray and ambient air flows. The tests were carried out for ambient pressures of 1 and 0.6 MPa, for two ambient temperatures of 500 and 300 K. The fuel injector pressures of 5 and 4.6 MPa; and total fuel quantity of 9.4 and 14 mg were used. They reported that an optimum combustion in a GDI engine could be achieved by using two fuel injection pulses. They found that the penetration of liquid fuel decreased with increased dwell period or percentage of fuel mass in the first pulse, but with a marginal reduction in the penetration of the vapour fuel.

Costa et al., [9] analysed mixture formation and early flame development in a GDI engine through numerical simulations and ultraviolet digital imaging. They investigated the effect of injection and ignition timing on flame propagation development. The tests were carried out at full-throttle operation, compression ratio (CR) of 10.5 and at the engine speed of 2000 rev/min., in a four-valve, single-cylinder wall-guided GDI engine. The fuel injection pressure of 100 bar and equivalence ratio of 1.1±1% were used. The flame growth was detected by an intensified CCD camera and the AVL-FIRE was used to perform simulations. They reported that the early fuel injection resulted in the formation of homogeneous mixture which led to stable combustion. They also noted that the stratified mixture increased cycle-by-cycle variations.

From the above discussion, it is understood that, the mixture formation, in a GDI engine, is influenced significantly by many engine parameters and conditions. Also, in a wall guided GDI engine, it mainly depends upon piston shape, engine speed, fuel injection strategy, etc. However, studies on the evaluation of the effects of fuel injector and spark plug location on mixture formation are rarely reported in the literature. Therefore, the present CFD analysis investigates the effect of fuel injector and spark plug location on mixture formation in a wall guided GDI engine.

2. CFD Methodology

2.1. Engine geometry and meshing

The CFD analysis has been done on a single-cylinder, four-stroke, four-valve, wall guided GDI engine which has a pentroof cylinder head. The spark plug is considered to be placed at the centre of cylinder head and the fuel injector is placed between the two intake ports as shown in Figures 2(a) and (b). The piston has a pentroof shape with an offset bowl as shown in Figure 2 (c). The engine specifications are given in the Table 1 [9]. A single fuel injection system has been used and the details about fuel injection events are given in Table 2 [9].

Table 1. Engine specifications

Parameter	Specifications
Displaced volume	398.5 cm^3
Stroke	81.3 mm
Bore	79 mm
Connecting rod	143 mm
Compression ratio	10.6 : 1
Speed (rpm)	2000
Charge induction system	Naturally aspirated
Spark timing	10^0 before TDC
Number of valves	4
Exhaust valve opening	27° before BDC
Exhaust valve closure	0° after TDC
Inlet valve closure	36° after BDC
Inlet valve opening	3° before TDC
Intake valve lift	6.5 mm
Exhaust valve lift	6.5 mm

(*360 CAD is suction at TDC)

Table 2. Fuel injection system parameters

Parameter	Detail
No. of holes	Six
Fuel	Gasoline (C_8H_{18})
Start of fuel injection	115 before TDC
Fuel injection duration	23.1 CAD
Fuel injection pressure	120 bar
Spray cone angle	15 degree
Mass of fuel injected per cycle	22 mg

2.2. Cases considered in this study

In the present study, four cases have been studied with different location of the spark plug (SP) and fuel injector (FI). Figure 1 shows the top views of the engine, for all the cases.

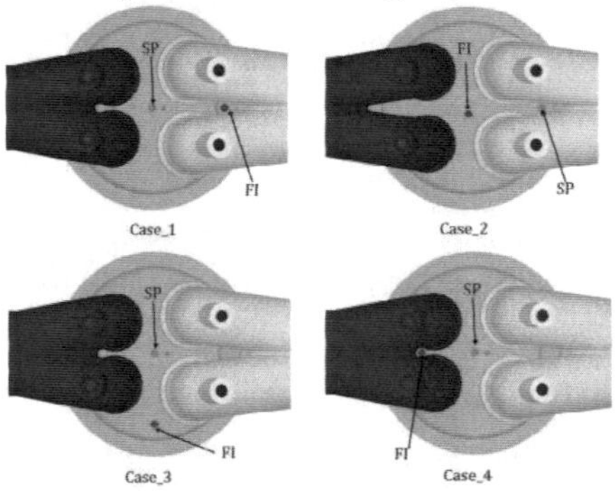

Figure 1. SP and FI location in all four cases considered

2.3. CFD Modelling and Boundary conditions

In the present study, the CFD analysis has been carried out using the CONVERGE and the post processing of the 3D output is done by using EnSight 10.1. Here, the SAGE model is used for the analysis of combustion, this model is based on the detailed chemical kinetics with modifications of Givler et al., [10]. This model considers 152 reactions along with 48 species. For modelling the ignition event, a high temperature source between the spark plug electrodes is introduced at the time of ignition. Modeling of wall heat transfer is done by the O'Rourke and Amsden models [11]. The droplet collision and initial droplet distribution has been analyzed by the NTC collision model [12] and the blob injection model [12] respectively. Film splash and fuel turbulent dispersion is analysed by the O'Rourke model [13]. The FROSSLING model is used for the analysis of droplet evaporation [14]. The turbulence is analysed using the RNG k-ε model [15]. Whereas, the pressure-velocity coupling is done by using the PISO algorithm [16]. The KH-RT model is used to study the spray breakup [17]. The calculation of embedded cell size, shown in figure 2 (e), is done by equation 1.

$$Cell\ size\ after\ embedding = \frac{Cell\ size\ before\ embedding}{2^{embed\ scale}} \tag{1}$$

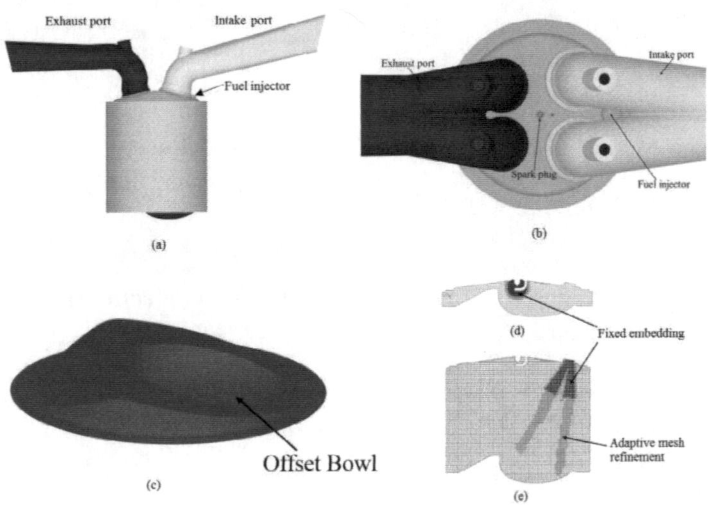

Figure 2. (a) Computational domain of the engine (b) Top view of engine (c) Piston with offset piston (d) Embedding at Spark (e) Adaptive mesh refinement at fuel injection

2.4. Governing equations

To generalize the in-cylinder flows, the following governing equations are used *[20]*:
Conservation of mass:

$$\frac{\partial \rho}{\partial t} + \frac{\partial(\rho u_j)}{\partial x_j} = s_m \tag{2}$$

General transport equations:

$$\frac{\partial(\rho\varphi)}{\partial t} + div(\rho\varphi u) = div(\Gamma grad\varphi) + S_\varphi \tag{3}$$

where, ϕ is a general variable of conservative form of all fluid flow equations including equations for scalar quantities such as temperature and pollutant concentration, etc.

3. Validation of the CFD Models

The validation of the CFD models for the same engine has been already done by Krishna et al., [13] in their previous work. They did the validation by using experimental results available in the literature [9]. Some of the validation results of them are shown here [13]. Figure 3 shows the comparison of in-cylinder pressures with crank angles, obtained from the experimental and CFD results [13]. Figure 4 shows the comparison of the equivalence ratio (ER) distribution between the CFD and experimental results of Costa et al., [9]. From Figures 3 and 4, it can be seen that, the two results are in good agreement. Therefore, these models can be used with confidence for the further analysis.

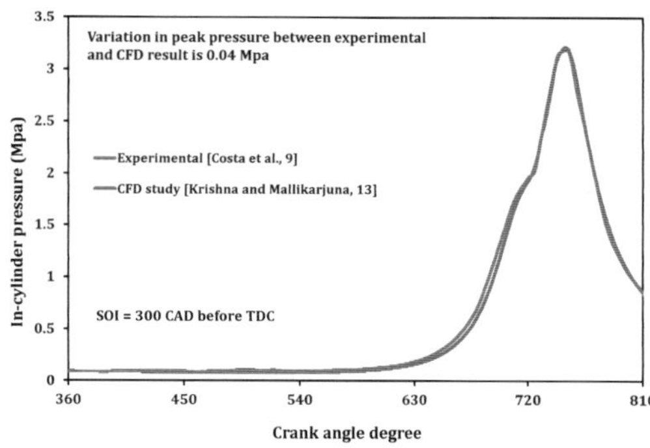

Figure 3. Comparison of in-cylinder pressure with crank angle degree

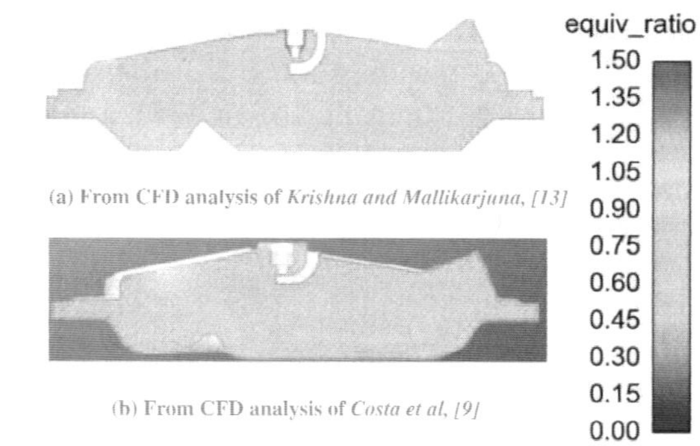

Figure 4. Comparison of ER distribution on central plane at spark timing

4. Result and discussions

In this section, the effect of the spark plug (SP) and the fuel injector (FI) location on various parameters has been discussed.

4.1. Effects of SP and FI location on Turbulent kinetic energy

Figure 5 shows the effect of the SP and FI location on turbulent kinetic energy (TKE) distribution for all the four cases considered at various crank angles. From Figure 5 it is seen that at 630 CAD the TKE is almost similar for cases 1, 2 and 3 whereas for case 4 the TKE is lower than that of the other cases. But at 650 CAD the TKE is maximum for case 1 followed by case 3, 2 and 4, this improves the fuel-air interaction for case 1 and hence better ER distribution. Also similar trend as of 650 CAD is also observed at 670 CAD. From Figure 5 it is observed that case 1 produces highest TKE.

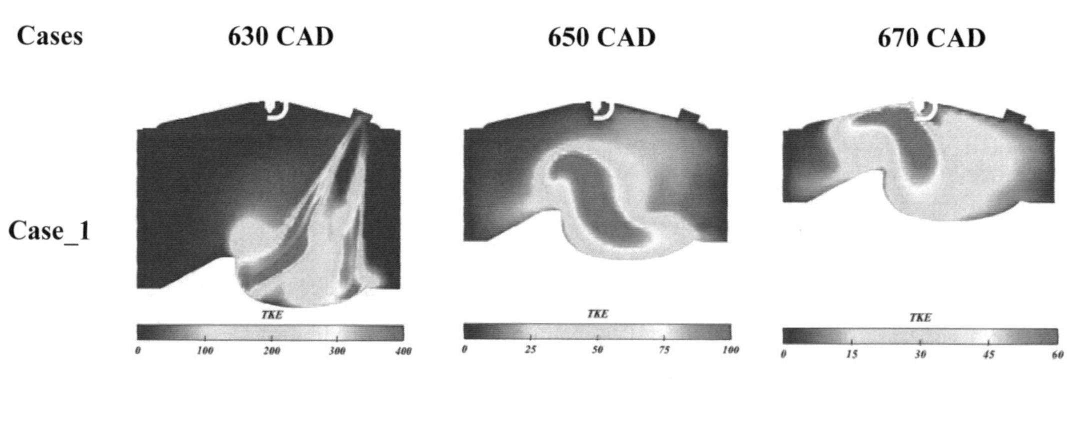

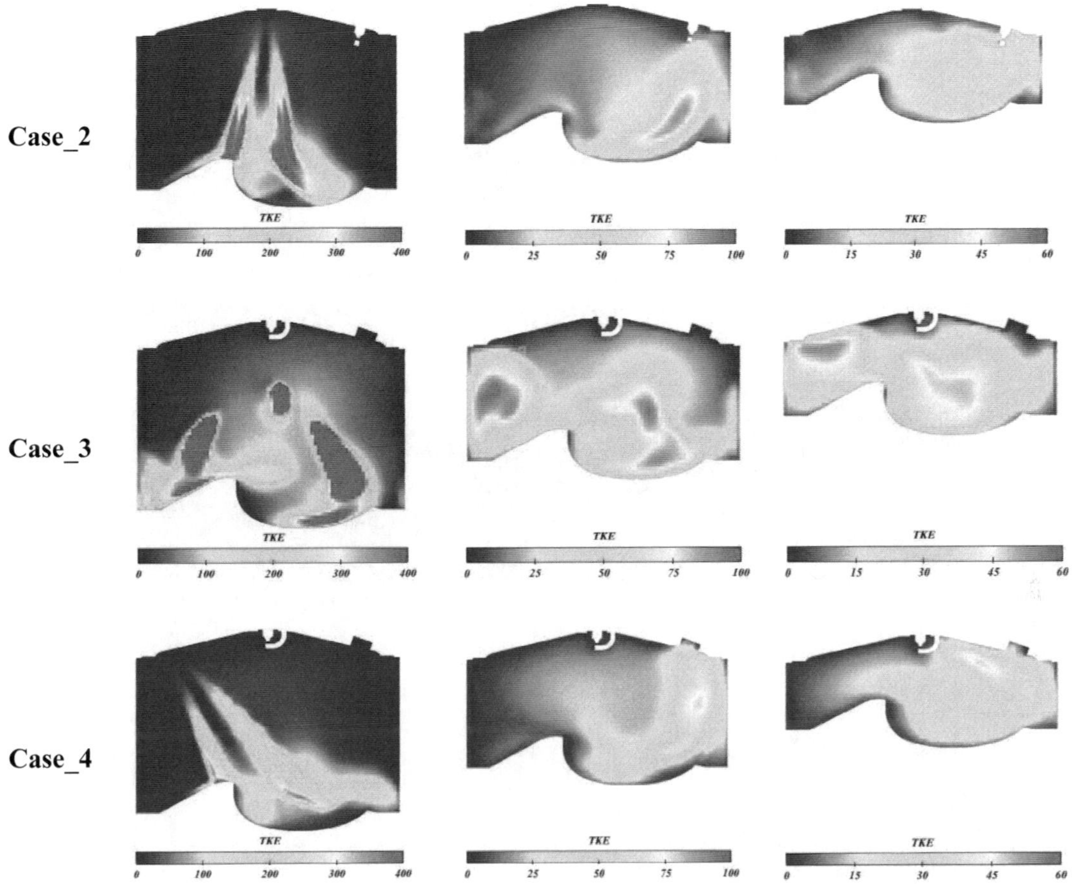

Figure 5. Comparison of TKE on the central plane at various CADs

4.2. Effects of SP and FI location on ER distribution

Figure 6 shows the effect of the SP and FI location on ER distribution for all the four cases considered. The ER around the spark plug, at the time of spark, for the cases 1, 2, 3 and 4 are about 0.99, 1.89, 1.18 and 1.77. Also, it is observed that, the distribution of the mixture is most favourable in the case 1 followed by the case 3, case 4 and case 2. Better ER distribution in the case 1 and case 3 results in better flame propagation which in turn results in better combustion. Whereas, in the case 2 and case 4, the fuel-air mixture is too rich around the spark plug, which leads to quenching of flame and then it becomes difficult for flame propagation. This is because of the higher TKE inside the combustion chamber for case 1 and 3 as discussed in Section 1.5.

4.3. Effects of SP and FI location on Mixture Stratification

In this study, the mixture stratification is characterized by a parameter called "stratification index" (SI) developed by Krishna and Mallikarjuna [13]. The SI is calculated based on the ER distribution in the hemispherical zones with spark plug as the center, as shown in Figure 7. In any combustion chamber, the value of SI varies between 0 and 1. The SI value of 0 indicates an ideally homogenous mixture (the ER is constant throughout the combustion chamber) and the SI value of 1 indicates an ideally stratified mixture (the ER varies continuously from higher to lower value away from the spark plug location). A negative value of the SI indicates a mal-distributed mixture (the ER varies continuously from a lower to higher value away from the spark plug location) in the combustion chamber. However, in a GDI engine, the SI value of 0.5 to 1 is desirable and closer to 1 is better.

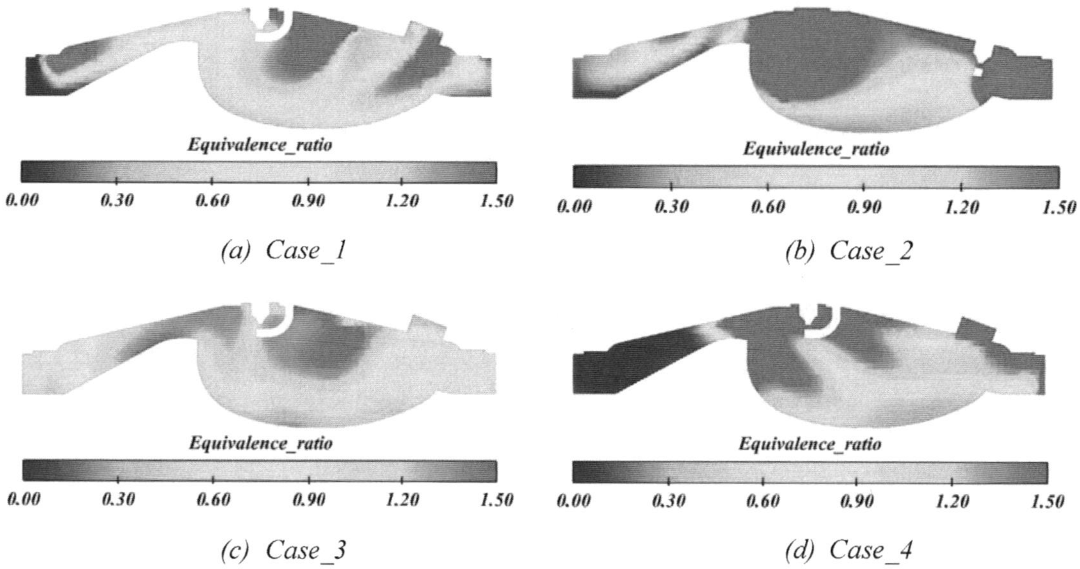

Figure 6. Comparison of ER distribution on the central plane at the spark timing

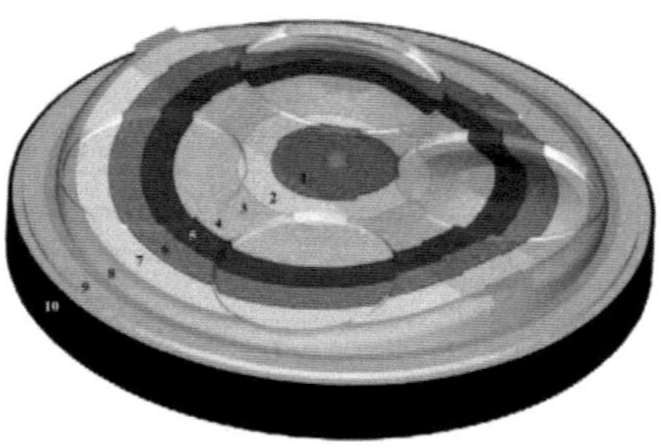

Figure 7. Hemispherical zones in the combustion chamber with spark plug as centre

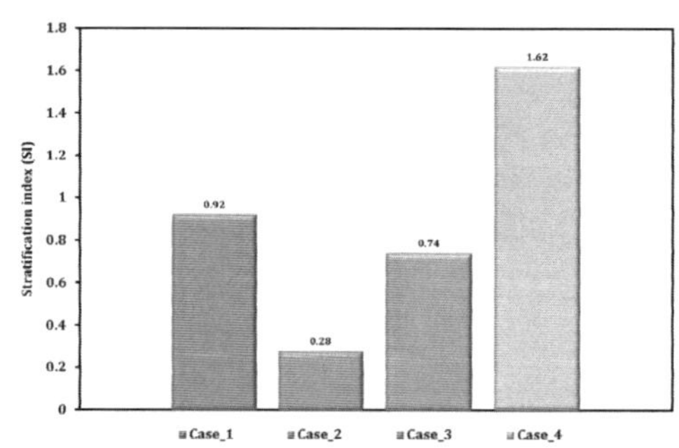

Figure 8. Comparison of SI for various cases of SP and FI location

4.4. Effects of SP and FI Location on Stratification index

Figure 8 shows the comparison of SIs, for various cases considered. From Figure 7, it is seen that, for case 1, 2, 3 and 4, the Sis are about 0.92, 0.28, 0.74 and 1.62. It indicates that, for the case 1, the mixture is almost ideally stratified. Whereas, for the case 2, the stratification is poor. The case 3 also has fairly stratified mixture, but in the case 4, a very poor stratification is seen because of too rich mixture around the spark plug.

4.5. Effects of SP and FI location on In-cylinder pressure and Temperature

Figure 9 shows the comparison of in-cylinder pressures for various SP and FI locations. From Figure 8, it is found that, for the case 1, the peak in-cylinder pressure is about 39 bar, which occurs at about 740 CAD. Whereas, for the cases 2, 3 and 4, the peak in-cylinder pressures are lower by about 32,

24.3 and 27.4% respectively, compared to that of the case 1. However, it is also observed that, after 800 CAD, all the in-cylinder pressure traces follow the same trend.

Figure 10 shows the comparison of in-cylinder temperatures for various cases of SP and FI locations. From Figure 10, it is also found that, for the case 1, the peak in-cylinder temperature is about 2078 K. Whereas, for the cases 2, 3 and 4, the peak in-cylinder temperatures are lower by about 14.3, 13.7 and 22.1% respectively, compared to that of the case 1.

Hence, it is observed that, the in-cylinder pressure and temperature are following the same trend as observed in Figures 6 and 8. The FI location in the case 1 is in such a way that, the interaction between intake air and fuel injected happens in a better way. This in turn results in higher in-cylinder pressure and temperature.

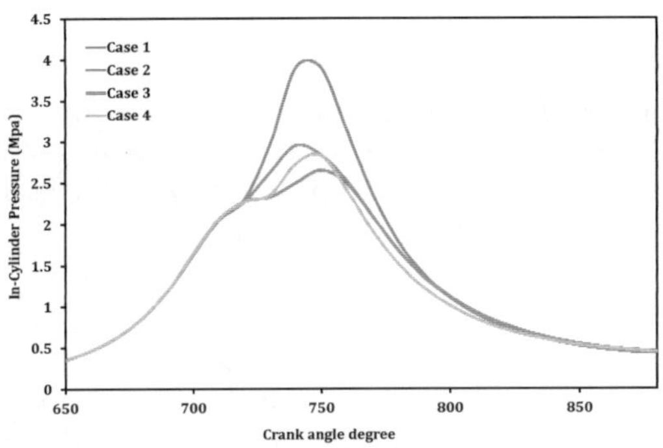

Figure 9. Comparison of in-cylinder pressure for various cases of SP and FI location

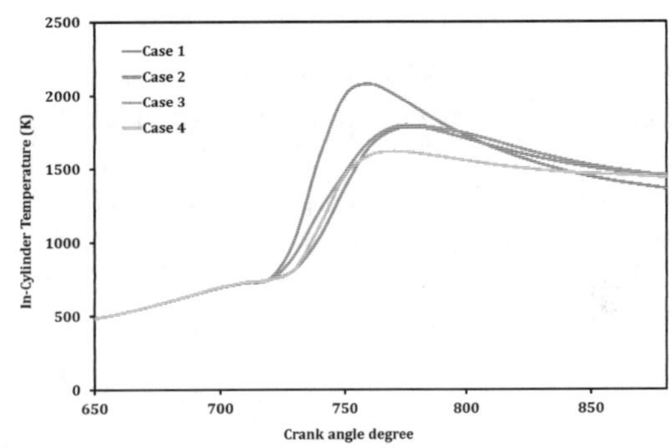

Figure 10. Comparison of in-cylinder temperature for various cases of SP and FI location

4.6. Effects of SP and FI location on Heat release rate and Mass of fuel burnt

Figure 11 shows the comparison of HRR for various cases. From Figure 11, it is seen that, for the case 1, the peak heat release rate is about 34.3 J/CAD. Whereas, for the cases 2, 3 and 4, the peak HRR is lesser by about 38.8, 47.8 and 30.6% respectively, compared to that of the case 1. This is because of better mixture preparation in the case 1 which leads to faster combustion.

Figure 12 shows the variation of the percentage of fuel burnt with crank angle for the various cases. From Figure 12, it is seen that, for the case 1, the center of combustion (COC or CA50) is at about 740 CAD, which indicates a faster and improved combustion. Whereas, for the cases 2, 3 and 4, the COC's are at about 753, 751 and 750 CADs which are quite closer to each other. From Figure 11, it is also seen that, for the case 1, 90% fuel is burned at about 759 CAD. Whereas, for the case 2, 3 and 4, 90% of the fuel is burned at about 791, 790 and 817 CAD respectively. Hence, it is seen that, the faster combustion takes place with the case 1, compared to that of the other cases. This is because of better ER distribution in the combustion chamber for the case 1 as mentioned in Section 4.2. From Figure 12, it is also observed that, the slower combustion takes place with the case 4, because of too rich mixture near the spark plug due to which the spark gets quenched and then it becomes difficult for flame propagation.

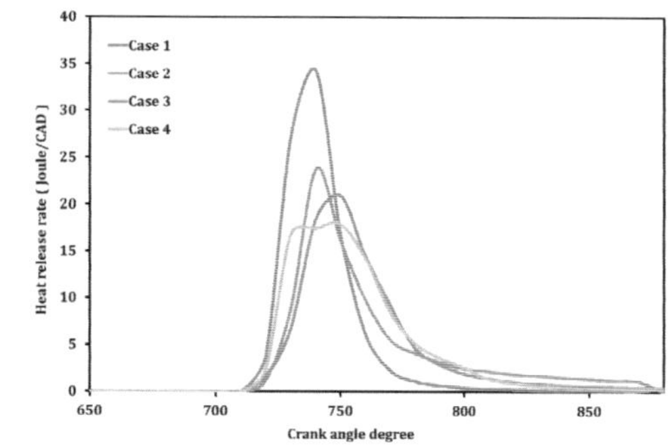

Figure 11. Comparison of heat release rate for various cases of SP and FI location

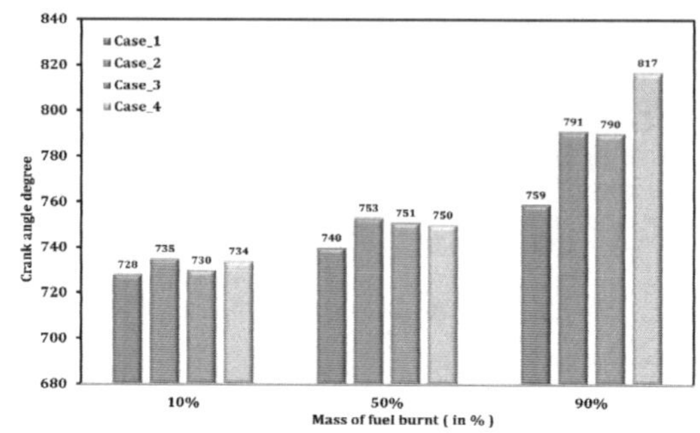

Figure 12. Comparison of mass fuel burnt for various cases of SP and FI location

4.7. Effects of SP and FI location on Indicated thermal efficiency (ITE)

Figure 13 shows the comparison of indicated thermal efficiency for the various cases. From Figure 13, it is seen that, for the case 1, the ITE is about 36.1%. Whereas, for the cases 2, 3 and 4, it is lower by 9.42, 4.8 and 8.2% than that of the case 1. This is because of the higher IMEP and power output, which is obtained by faster combustion in the case 1.

4.8. Effects of SP and FI location on NO_X emission

Figure 14 shows the comparison of NOx emissions at the EVO (873 CAD) for the various cases. From Figure 14, it is seen that, with the faster and enhanced combustion, the NOx emissions increase. Also it is found that, for the case 1, the NOx emissions are about 1 mg/cycle. For the cases 2, 3 and 4, they are lesser by about 63, 52 and 74% respectively, compared to that of the case 1.

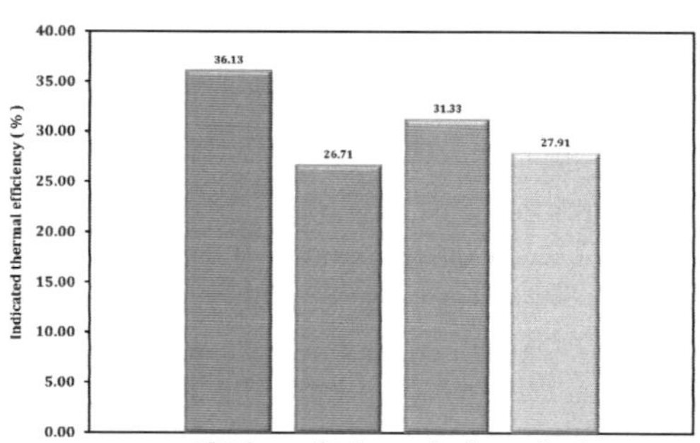

Figure 13. Comparison of indicated thermal efficiency for various cases of SP and FI location

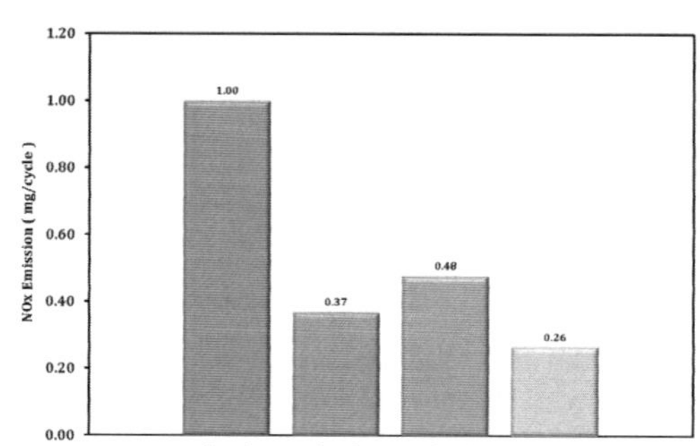

Figure 14. Comparison of NOx emission at EVO for various cases of SP and FI location

4.9. Effects of SP and FI location on Soot and CO emission

Figure 15 shows the comparison of soot emissions at the EVO (at 873 CAD), for various cases considered. From Figure 15, it is seen that, with the faster and enhanced combustion, the soot emissions decrease. Also it is found that, for the case 1, the soot emissions are about 0.017 mg/cycle. For the cases 2, 3 and 4, they are higher by about 123.3, 23.5 and 100% respectively, compared to that of the case 1.

Figure 16 shows the comparison of CO emissions at the EVO, for the various cases considered. From Figure 16, it is seen that, with the faster and enhanced combustion, the CO emissions decrease. This is because, with the better combustion, more carbon will be converted into CO_2. Also, it is found that, for the case 1, the CO emissions are about 1.66 mg/cycle. For the cases 2, 3 and 4, they are higher by about 133.3, 24.7 and 196.3% respectively, compared to that of the case 1. Hence, the case 1 shows the least soot and CO emissions as compared to that of the other cases.

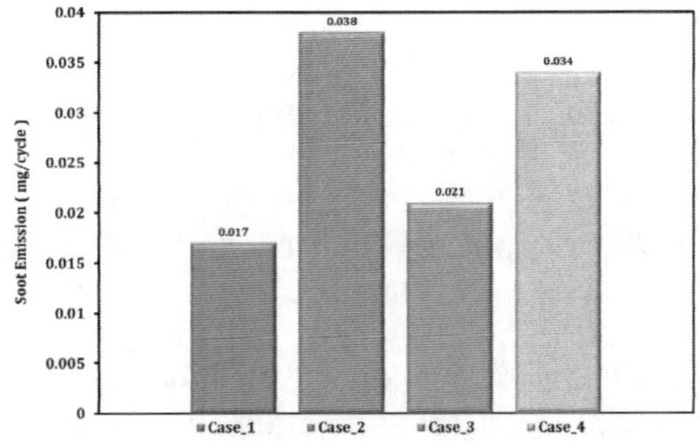

Figure 15. Comparison of soot emission at EVO for various cases of SP and FI location

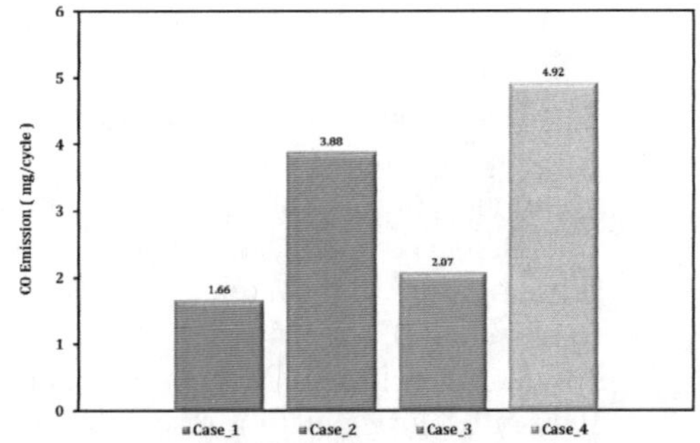

Figure 16. Comparison of CO emission at EVO for various cases of SP and FI location

5. Conclusions

In this study, a CFD simulation study is conducted on a four-stroke four-valve GDI engine to evaluate the effect of SP and FI locations on the mixture stratification, combustion and emission characteristics of the engine. From the analysis of results, the following conclusions are drawn:

- The mixture stratification is found best for the case1 followed by the cases 3, 2 and 4.
- The case 1 shows the higher in-cylinder pressure and temperature, heat release rate and faster burning compared to that of the other cases.
- The case 1 gives the ITE of about 36.1%, whereas for the cases 2, 3 and 4, it is lower by about 9.42, 4.8 and 8.2%.
- The NOx emissions are the highest for the case 1 followed by the cases 3, 2 and 4. The soot and CO emissions are the least for the case 1.

Finally, it is concluded that, the case 1 results in the better mixture stratification, which leads to better combustion and performance. However, the NOx emissions are higher, but the soot and CO emissions are least. Also, in this study, the extent to which location of SP and FI can be changed is limited by the space available in the engine head and other geometrical parameters.

Acknowledgments

Authors are grateful to the High-Performance Computing Facility (HPCF) at Indian Institute of Technology Madras, Chennai, which is used to carry out the CFD simulations presented in this work. Also, we would like to thank my co-scholars, Mr. Krishna Addepalli S., Priyanka D. Jadhav, Yashas K and Ankit Raut for their support during this study.

References

[1] Krishna S, Mallikarjuna J M and Davinder K 2016 Effect of Engine Parameters on In-Cylinder Flows in a Two-Stroke Gasoline Direct Injection Engine *Applied Energy* **176** 282-294

[2] Zhao F Q, Lai M C and Harrington D L 1999 Automotive Spark Ignited Direct-Injection Gasoline Engines *Progress in Energy and Combustion Science* **25** 437–562

[3] Krishna S and Mallikarjuna J M 2017 Effect of Engine Parameters on Mixture Stratification in a Wall-Guided GDI Engine – A Quantitative CFD Analysis *SAE Paper* No. 2017-01-0570

[4] Krishna S and Mallikarjuna J M 2016 Parametric analysis of a 4-stroke GDI engine using "CFD" (In press) Alexandria Engineering Journal

[5] Costa M U S and Allocca L 2011 Numerical study of the mixture formation process in a four-stroke GDI engine for two-wheel applications *Simulation Modelling Practice and Theory* **19** (2011) 1212–1226
[6] Church W and Farell P V 1998 Effects of intake port geometry on large scale in cylinder flows *SAE Technical Paper* 98484 doi: 10.4271/98484
[7] Costa M, Catapano F, Sementaa P, Sorge U and Vaglieco B M 2016 Mixture preparation and combustion in a GDI engine under stoichiometric or lean charge: an experimental and numerical study on an optically accessible engine *Applied Energy 180* (2016) 86–103
[8] Li T, Nishida K, Zhang Y and Hiroyasu H 2007 Effect of split injection on stratified charge formation of direct injection spark ignition engines. *International Journal of Engine Research* 2007; 8:205e18
[9] Costa M, Marchitto L, Merola S S and Sorge U 2014 Study of mixture formation and early flame development in a research GDI engine through numerical simulation and UV-digital imaging Energy **77** (2014) 88-96
[10] Givler S, Raju D, Pomraning E M, Senecal P K, Salman N and Reese R 2013 Gasoline Combustion Modeling of Direct and Port Fuel Injected Engines using a Reduced Chemical Mechanism *SAE Technical Paper* No. 2013-01-1098
[11] O'Rourke P J and Amsden A 1987 A The TAB Method for Numerical Calculation of Spray Droplet Breakup *SAE Technical Paper No. 872089*
[12] Ishiyama T, Shioji M, Ihara T and Katsuura A 2003 Modeling and Experiments on Ignition of Fuel Sprays Considering the Interaction Between Fuel-Air Mixing and Chemical Reactions *SAE Paper* No. 2003-01-1071
[13] Krishna A S and Mallikarjuna J M 2017 Effect of Engine Parameters on Mixture Stratification in a Wall-Guided GDI Engine– A Quantitative CFD Analysis *SAE International Journal Commercial Vehicle* **10**(2) doi: 10.4271/2017-01-0570
[14] Amsden A A, O'Rourke P J and Butler T D 1989 KIVA-II: A Computer Program for Chemically Reactive Flows with Sprays *Los Alamos National Laboratory Report No. LA-11560-MS* 1989
[15] Wang B, Miles P C, Reitz R D, Z Han and Petersen B 2011 Assessment of RNG Turbulence Modelling and the Development of a Generalized RNG Closure Model *SAE Paper No. 2011-01-0829*
[16] Krishna A S, Mallikarjuna J M and Kumar D 2016 Effect of engine parameters on in-cylinder flows in a two-stroke gasoline direct injection engine *Applied Energy* **176** (2016) 282–294
[17] Larmi M and J T 2003 Diesel 2003 Spray Simulation and KH-RT Wave Model *SAE Paper No. 2003-10-27*
[18] Sagayaraj A R G, Mallikarjuna J M and Ganesan V 2013 Energy efficient piston configuration for effective air motion – A CFD study *Applied Energy* **102** (2013) 347–354
[19] Imaoka Y, Shouji K, Inoue T and Noda T 2015 A Study of a Multistage Injection Mechanism for Improving the Combustion of Direct-Injection Gasoline Engines *SAE Technical Paper 2015-01-0883*
[20] 2015 Converge v2.2.0, Theory Manual, Convergent Science Inc
[21] Krishna A S, Saw O P and Mallikarjuna J M 2017 Effect of Mixture Distribution on Combustion and Emission Characteristics in a GDI Engine – A CFD Analysis *SAE Technical Paper 2017-24-0036*

Abbreviations

SP	Spark Plug
FI	Fuel Injector
TDC	Top dead center
ER	Equivalence ratio
HRR	Heat release rate
ITE	Indicated thermal efficiency
GDI	Gasoline direct injection
CFD	Computational fluid dynamics
CA50	The crank angle at which 50% of the fuel has burnt
COC	Center of combustion
RNG	Re-normalized group
KH-RT	Kelvin-Helmholtz-Rayleigh-Taylor
PISO	Pressure implicit with splitting of operators
CFL	Courant Fredrick and Lewy Number

Flame spread behavior over combustible thick solid of paper, bagasse and mixed paper/bagasse

Mohd Azahari Razali[1], Sofian Mohd[2], Azwan Sapit[1], Akmal Nizam Mohammed[1], Ahmad Husaini Ismail[2], Mohd Faisal Hushim[3], Norrizam Jaat[3] and Amir Khalid[3]
[1]Centre for Energy and Industrial Environment Studies (CEIES), Universiti Tun Hussein Onn Malaysia, 86400 Pt. Raja, Batu Pahat, Johor, Malaysia
[2] Faculty of Mechanical & Manufacturing Engineering, Universiti Tun Hussein Onn Malaysia, 86400 Pt. Raja, Batu Pahat, Johor, Malaysia
[3]Advanced Technology Centre (ATC), Faculty of Engineering Technology, Universiti Tun Hussein Onn Malaysia, 86400 Pt. Raja, Batu Pahat, Johor, Malaysia

Email: azahari@uthm.edu.my

Abstract. Flame spread behavior on combustible solid is one of important research related to Fire Safety Engineering. Now, there are a lot of combustible solid composed from mixed materials. In this study, experiments have been conducted to investigate flame spread behavior over combustible solid composed by paper, bagasse and mixed paper/bagasse. Experimental data is captured by using video recording and examined flame spread shape and rate. From the results obtained, shows that the different materials produce different flame spread shape and rate. Different flame shape is seen between all types of samples. Flame spread rate of 100% paper is faster than the one of 100% bagasse. Based on the result, it is also inferred that the material composition can be influenced on the flame spread shape and flame spread rate of mixed paper/bagasse.

1. Introduction

Fire safety engineering is an application of science to protect human and their environments from the destructive effects of the fire. Researches on the fire safety engineering have been done for many years and it is beginning to focus on the technical capability for fire preventions in complex issues. In order to improve the technical capability for fire preventions, the fundamental approach is required which can be applied at the design stage. Such approach requires detailed understanding of combustion process from an engineering standpoint.

The combustion involves the flame spread phenomenon along the surface of combustible solid [1]. A lot of researches have been done in order to discover the mechanism [1-5]. An understanding of factors affecting the flame spread mechanism is indispensable to discover the mechanism. In other to elucidate the complex mechanism, it is necessary to know about the flame spread mechanism.

There are several researches are conducted to examine the effect of material on flame spread behavior of combustible solid [6-8]. However, all these researches are focused only on the flame spread over fabric. Instead of fabric, the paper is also one of the important factor influences on fire destruction. Nowadays, there are several materials such as oil palm leaf, bagasse or pineapple leaf are used as alternative source in paper fabrication. Some product also fabricated by composed these alternative sources with the paper.

Before flame spread behavior over mixed paper is examined, it would be indispensable to study the phenomenon for pure material since this phenomenon has not been studied in detail. Thus, in this study, the flame spread behavior over paper, bagasse and mixed paper/bagasse is examined.

2. Experimental setup

In this research, the experiments have been conducted for pure paper, pure bagasse and mixed paper/bagasse. The specimen size is 8 cm x 7.5 cm.

Figure 1 shows experimental setup in this research. Before the burning test, the sample is dried in a desiccator for more than 48 hours and the humidity in the desiccator is controlled to below 40%. The sample is clamped to the holder and a burner is used for igniting the sample at a point on its top edge. The surface of the sample is videotaped to observe the flame spread behavior.

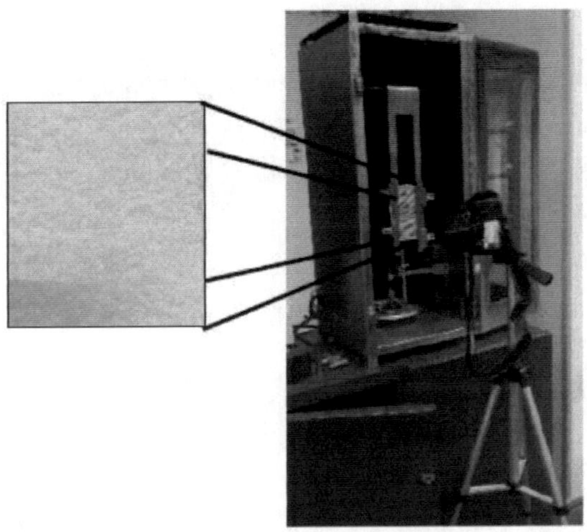

Figure 1. Experimental setup.

3. Results and discussion

3.1. Flame spread shape

In this experiment, the sample is ignited at a point on its top edge. The flame spreads in the downward direction. The sample is videotaped to observe the flame spread behavior. Figure 2 shows shape of flame front for pure paper, pure bagasse and mixed paper/bagasse. It is seen that from Figure 2 (a) to (c), all samples exhibit similar shape of flame front, which is 'U' type. However, size of the shape shrinks from pure paper, mixed paper/bagasse to pure bagasse.

Figure 3 shows shape of flame front for every 30s. This figure shows detail propagation of flame shape during the experiments. It is seen from Figure 3 (a), the shape of flame front is in 'V' shape at the beginning of combustion. After few minutes, the shape changes from 'V' to 'U' shape which is remained until the end of sample.

From Figure 3 (b) and (c), significant different of the shape is seen for other samples. It is seen that for paper/bagasse, the shape is in 'U' at the beginning. The flame is propagated after the ignition; however, the propagation is limited to the central region only and the flame is retarded after few minutes.

For the case of pure bagasse, the flame is spread after the ignition. However, the flame is retarded in the very short period that the one of paper/bagasse, as seen in Figure 3 (c). Results infer that the type of material and material composition may influence on the flame spread shape.

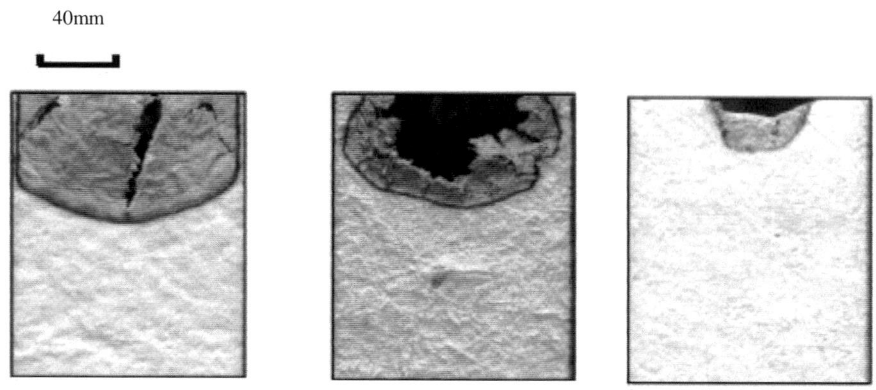

(a) pure paper (b) paper/bagasse (c) pure bagasse

Figure 2. Shape of flame front for different type of sample.

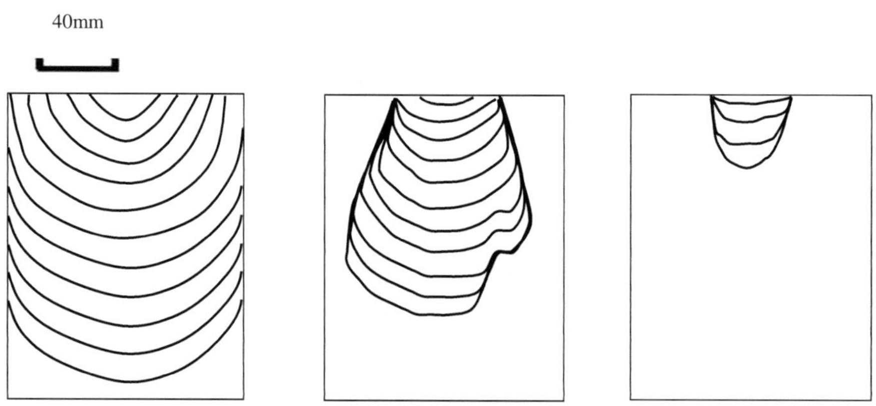

(a) pure paper (b) paper/bagasse (c) pure bagasse

Figure 3. Shape of flame front for every 30s.

3.2. Flame spread rate

Flame spread rates are obtained by measuring the position of the most preceding point of the burning front at each time. In order to avoid the influence of the initial transition process after the ignition, the position is measured from $x = 10$ mm, where x is the vertical distance from the top edge of the sample in downward direction.

Figure 4 shows the flame spread rate for pure paper, paper/bagasse, and pure bagasse. Result shows that the flame spread rate for pure paper is highest, which is 0.111 mm/s. It is followed by paper/bagasse and pure bagasse; the former and the latter are 0.093 mm/s and 0.086 mm/s, respectively. From this research, it is inferred that the material has significant influences on not only the flame spread shape, but also on the flame spread rate. However, the additional results are needed in order to examine about the influence of the material composition. This result will be useful to validate simulation works in the future.

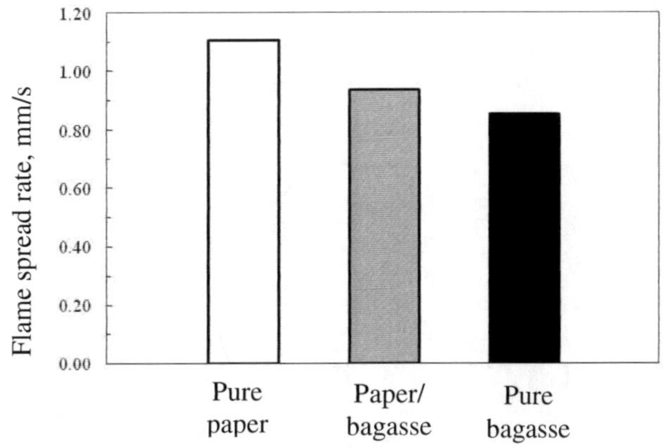

Figure 4. Flame spread rate for different type of sample.

4. Conclusion

In this study, flame spread over pure paper, pure bagasse, and paper/bagasse is examined and following results are obtained:

1. Shapes of flame front differ between samples. Pure paper burns and flame spreads in 'U' shape. However, flame is distinguished when pure bagasse is ignited. For paper/bagasse, the sample burns and the flame is only spread in the central region during burning process.
2. Flame spread rate for pure paper is highest, followed by the values of paper/bagasse and pure bagasse. Results infer the material has significant influences on not only the flame spread shape, but also on the flame spread rate.

Acknowledgment

The authors would like to thank Universiti Tun Hussein Onn Malaysia (UTHM) and Ministry of Higher Education Malaysia (MOHE) for their financial support of the present work through Exploratory Research Grant Scheme (ERGS-E034) and Fundamental Research Grant Scheme (FRGS-1465).

References

[1] Zhang Y, Ji J, Li J, Sun J, Wang Q and Huang X 2012 Effect of altitude and sample width on the characteristic flame spread over white sheets *Fire Safety* **51** 120-125

[2] Pizzo Y, Consalvi J L, Querre P, Coutin M and Porterie B 2009 Width effect on the early stage of upward flame spread over PMMA slabs: Experimental observations **44**(3) 407-414

[3] Watanabe Y, Hiroyuki T and Ito A 2011 Flame spread along a thin solid randomly distributed combustible and non-combustible areas *Proceeding of Combustion Institute* **33**(2) 2449-2455

[4] Sato K, Suzuki K, Sakai Y and Sega S 1994 Effect of flame retardant on the behavior and temperature profile of flame spreading over paper *Fire Safety Science* **4**(1) 503-514

[5] Quintiere J G 2001 The effects of angular orientation on flame spread over thin materials *Fire Safety Journal* **36** 291-312

[6] Mohd Azahari B R, Azwan B S, Mohd Faisal B H, Amir B K, Suzuki M and Masuda W 2016 Flame spread behavior over combined fabric of cotton/polyester *ARPN Journal of Engineering and Applied Sciences* **11**(12) 7553-7557

IOP Conf. Series: Materials Science and Engineering **243** (2017) 012026 doi:10.1088/1757-899X/243/1/012026

[7] Mohd Azahari B R, Suzuki M, Masuda W, Hamidon B S, Azwan B S, Akmal Nizam B M, Mohd Faisal B H and Amir B K 2017 Effect of synthetic material on angle dependency of flame spread behavior over combined fabric *AIP Conference Proceedings* **1831**(1) 020014

[8] Mohd Azahari B R, Azwan B S, Akmal Nizam B M, Mohd Faisal, Hamidon B S and Hazahir B P 2017 Effect of thread angle on flame spread behaviour over combined fabric of kenaf/polyester *AIP Conference Proceedings* **1831**(1) 020016

Thread angle dependency on flame spread shape over kenaf/polyester combined fabric

Mohd Azahari Razali[1], Azwan Sapit[1], Akmal Nizam Mohammed[1], Md Nor Anuar Mohamad[1], Normayati Nordin[1], Azmahani Sadikin[1], Mohd Faisal Hushim[2], Norrizam Jaat[2] and Amir Khalid[2]

[1] Centre for Energy and Industrial Environment Studies (CEIES), Faculty of Mechanical & Manufacturing Engineering, Universiti Tun Hussein Onn Malaysia, 86400 Pt. Raja, Batu Pahat, Johor, Malaysia

[2] Advanced Technology Centre (ATC), Faculty of Engineering Technology, Universiti Tun Hussein Onn Malaysia, 86400 Pt. Raja, Batu Pahat, Johor, Malaysia

E-mail: azahari@uthm.edu.my

Abstract. Understanding flame spread behavior is crucial to Fire Safety Engineering. It is noted that the natural fiber exhibits different flame spread behavior than the one of the synthetic fiber. This different may influences the flame spread behavior over combined fabric. There is a research has been done to examined the flame spread behavior over kenaf/polyester fabric. It is seen that the flame spread shape is dependent on the thread angle dependency. However, the explanation of this phenomenon is not described in detail in that research. In this study, explanation about this phenomenon is given in detail. Results show that the flame spread shape is dependent on the position of synthetic thread. For thread angle, $\theta = 0°$, the polyester thread is breaking when the flame approach to the thread and the kenaf thread tends to move to the breaking direction. This behavior produces flame to be 'V' shape. However, for thread angle, $\theta = 90°$, the polyester thread melts while the kenaf thread decomposed and burned. At this angle, the distance between kenaf threads remains constant as flame approaches.

1. Introduction

Fire safety engineering is an essential science for improving the safety knowledge from the fire destructive. Now, the research starts to focus on the technical capability for fire preventions in complex issues. For improving the technical capability, the fundamental approach in the design stage is essential. Such approach requires detailed understanding of combustion process from an engineering viewpoint.

One of the important technical areas in fire safety engineering is combustion process of combustible solid such as fabric. There are several researches have been done related to the flame spread rate. Bhattacharjee et al. [1] have made an apparatus for flame spread study. The spread rate can be measured by tracking the position of visible edge of flame through video recording. There are several factors influence on flame spread rate. Fernandez-Pello et al. [2] found that the flame spread rate can be affected by ambient and fuel conditions. Ding et al. [3] simulated the fire spread. Results show that spread characteristics have major relation with factors such as material of solid, width, thickness, placed angle and atmospheric pressure. It is also discovered that the increment in oxygen concentration will cause the external radiant flux required for flame spread to decrease. Flame spread characteristics of many thin materials will be varying with affect from material thickness, external heat flux, oxygen concentration, pressure and forced flow velocity [4]. Higher heat flux results in higher heat release rate and peak mass loss rate and the fabrics will burns more violently. The density of the structure and moisture content also has effects on burning behavior of fabrics. The arrangement order of types of fabrics in increasing fire risk: wollen fabrics > sponge fabrics > cotton fabrics > linen fabrics [5]. It is stated that cellulosic fabrics burn with a yellow flame, light smoke, and have glowing embers. Synthetic fabrics catch fire quickly or shrink from the flame initially. Nylon, lastol, olefin, polyester, and spandex burn slowly and melting. It also can melt and pull away from small flames without igniting. The residue is hot molten and difficult to get rid. The burning of these fabrics can self-extinguish [6].

Nowadays, fabric is always fabricated by using two different type of thread; natural/natural threads, natural/synthetic threads or synthetic/synthetic threads. The natural thread has different flame spread characteristic than the one of synthetic thread; which may have some influence on flame spread of combined fabric. There are several researches have been done to examine the flame spread rate over combined fabric. Mohd Azahari et al. [7] have conducted experiment to study the flame spread behavior over combined fabric of cotton/polyester. Experiment is conducted for several weft trade angles from 0° to 90°. Results show a significant difference in the shape of burning front between θ = 0° and 90°. It is also found that the flame spread rate decreases as the angle increases. Experiments also have been conducted to study the flame spread behaviour over combined fabric of cotton/polyester and cotton/nylon [8]. For θ = 0°, a significant difference is seen between these two combined fabrics in the shape of burning front. However, the shape of the burning front is the same for both fabrics at θ = 90°. From the research, it is also seen that both fabrics have similar angle dependency; the flame spread rate decreases as the angle increases.

Recently, several industrial have used kenaf as natural thread. Mohd Azahari et al. [9] have been examined the effect of thread angle on flame spread behavior over kenaf/polyester fabric. From the results obtained, it is seen that thread angle have some influence on the flame spread shape of fabrics. However, there is no detail explanation about the influence of the angle on flame spread shape. The data of detail behavior of flame spread is essential to describe this influence. Thus, in this study, the detail flame spread behavior over kenaf/polyester combined fabric is explored.

2. Experimental setup

Fabric samples are made by means of a weaving machine. Plain weave is chosen as the structure of the fabric. The plain weave has warp threads perpendicular to weft threads. The fabric composed of different materials is referred to as a combined fabric. For combined fabric, kenaf is used as the warp thread and polyester as the weft thread. **Table 1** shows the characteristic values of these threads.

Table 1. Size of threads.

	Diameter [mm]
Kenaf thread	0.7
Polyester thread	0.5

Thread distance of all samples have been controlled for approximately 3.0 mm. Samples are cut with different weft thread angle of 0°, 45° and 90°. The weft thread angle is referred to the angle between polyester thread and the horizontal line as shown in **Figure 1**.

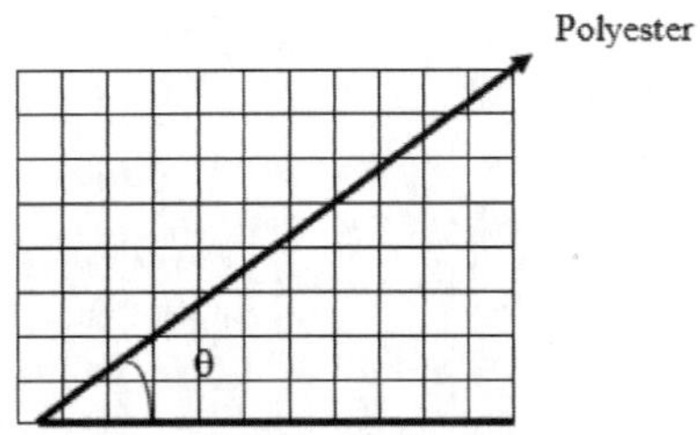

Figure 1. Illustration of fabric angle.

Figure 2 shows the experimental setup. Before the burning test, the fabric sample is dried in a desiccator for more than 48 hours and the humidity in the desiccator is controlled to below 40%. The fabric sample is clamped to the holder and its area of 100 mm x 200 mm is exposed to burn. A burner is used for igniting the fabric at a point on its top edge. The surface of the sample is videotaped to observe the flame spread behavior.

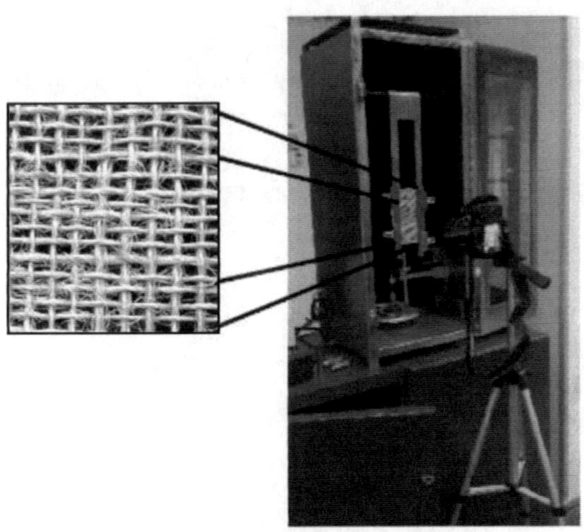

Figure 2. Experimental setup.

3. Result and discussion

This experiment was carried out with a fire lit at the top end of the experiment. This sample combustion is carried out facing down. The fire behavior during the experiment recorded for the purposes of observation and analysis of the data collected.

Figure 3 (a) to **(c)** show shapes of flame front for θ = 0°, 45°, and 90°. From **Figure 3(a)**, the flame spreading with slightly 'V' shape for θ = 0°. However, the shapes started to change into curvy flat line shape for θ = 45° and 90° as seen in **Figure 3(b)** to **(c)**.

IOP Conf. Series: Materials Science and Engineering **243** (2017) 012027 doi:10.1088/1757-899X/243/1/012027

(a) 0°

(b) 45°

(c) 90°

Figure 3. Shape of flame front of different weft thread angle.

Detail observation of flame front for every 5 s is shown in **Figure 4 (a)** to **(c)**. The flame front is determined at the front edge of discoloration when it decomposes. For $\theta = 0°$, it is seen that the flame front starts with semicircle shape as shown in **Figure 4(a)**. After few second, the shape changes to 'V' shape. The 'V' shape remains as time increases until the fabric completely burn. **Figure 4(b)** shows the shape of flame front of the fabric at $\theta = 45°$. It is seen, the flame front is in the semicircular shape at the beginning of the combustion. Then, the semicircular shape turns into 'V' shape. After that, the flame front becoming flat line until the fabric completely burn. This similar phenomenon is also seen for $\theta = 90°$.

IOP Conf. Series: Materials Science and Engineering **243** (2017) 012027 doi:10.1088/1757-899X/243/1/012027

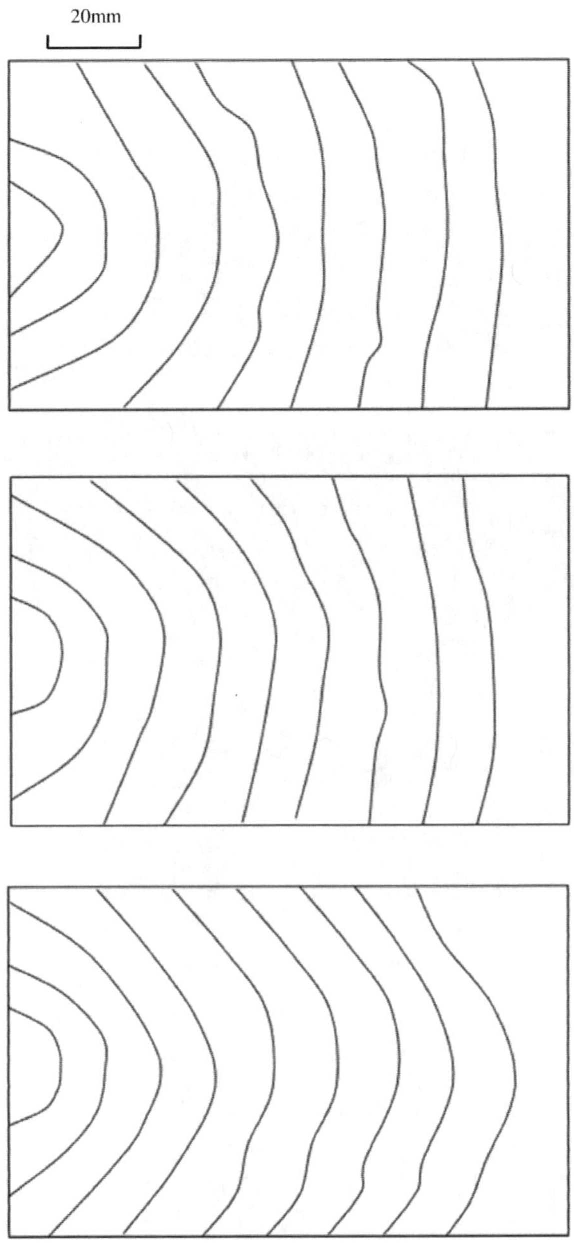

(a) 0° (b) 45° (c) 90°

Figure 4. Shape of flame front for every 5 s.

This different of flame spread shape between the weft thread angle may influenced by the decomposition behavior at leading point of combustion. **Figures 5** and **6** show the flame spread behavior at leading edge for $\theta = 0$ °and 90°, respectively.

Figures 5 (a) to **(d)** show the flame spread behavior at leading edge in every 0.30 s as shown in 'O'. As seen in figure, the polyester thread melts when the edge of discoloration on the cotton thread approaches. The polyester thread breaks into two and it pulls the kenaf thread when it shrinks.

The case is different when the thread angle is 90°. As shown in **Figure 6 (a)** to **(d)**, shows the different shrinking behavior of polyester thread at $\theta = 90°$. The polyester thread melts gradually in the direction of the thread and attaches to the kenaf thread. However, the distance between kenaf threads remains constant as flame approaches.

IOP Conf. Series: Materials Science and Engineering **243** (2017) 012027 doi:10.1088/1757-899X/243/1/012027

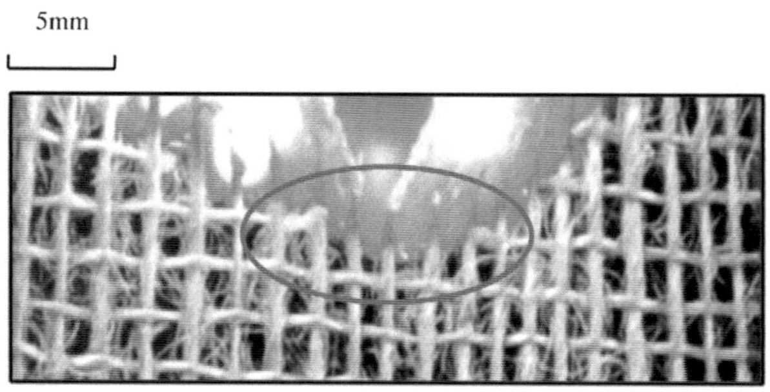

(a) t = t_0

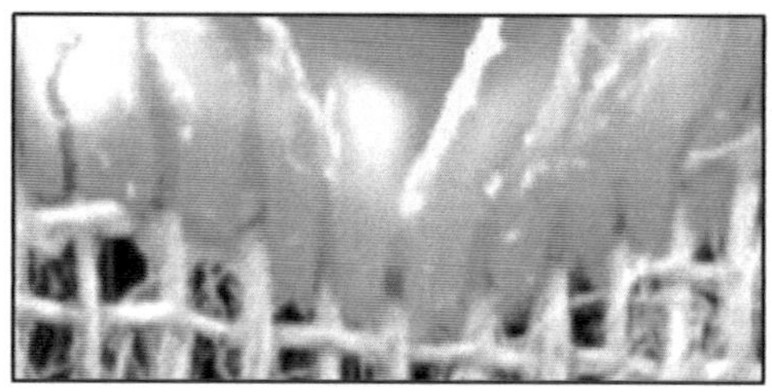

(b) t = t_0 + 30s

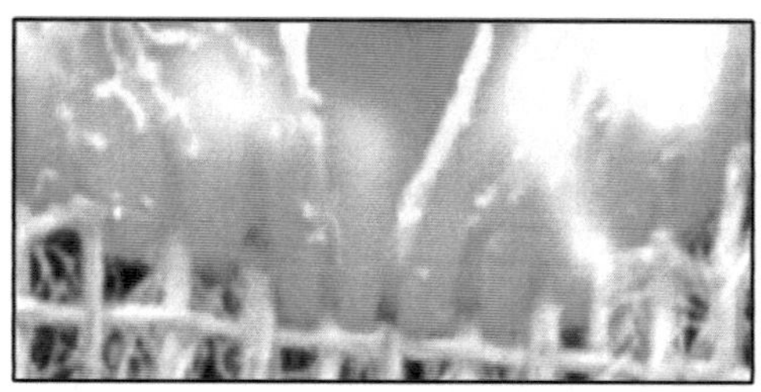

(c) t = t_0 + 60s

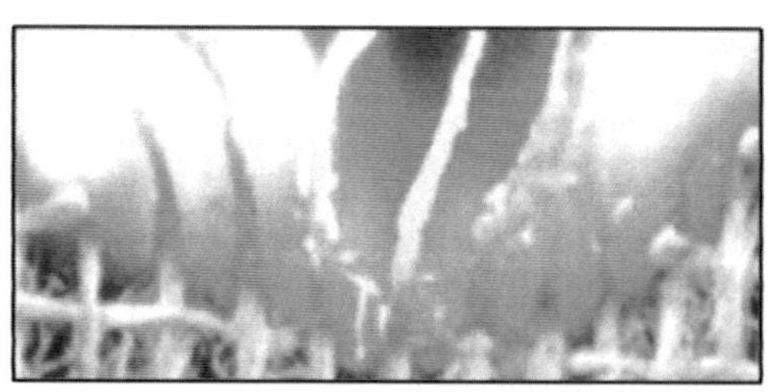

(d) t = t_0 + 90s

Figure 5. Flame spread behavior at leading edge for $\theta = 0°$.

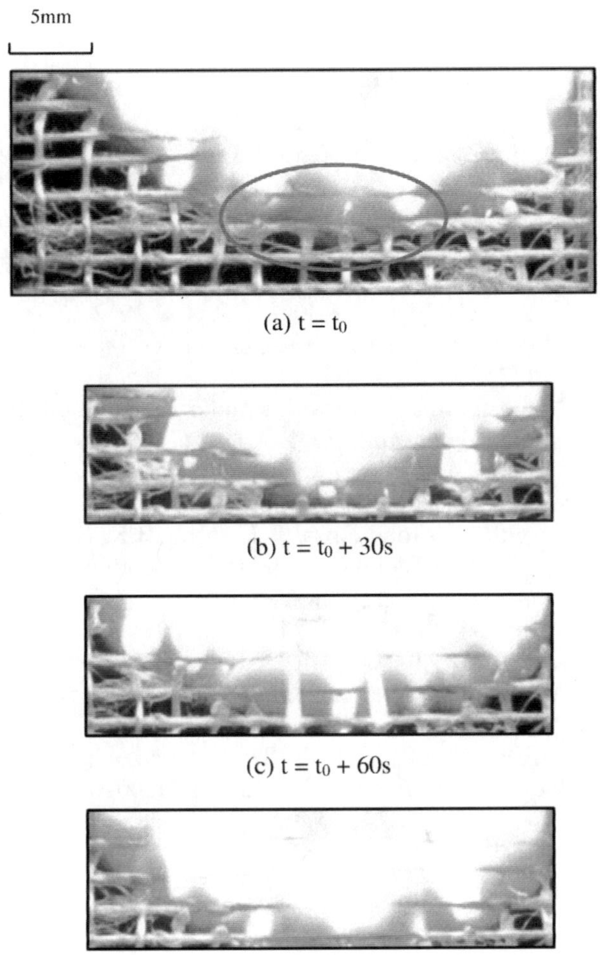

(a) $t = t_0$

(b) $t = t_0 + 30s$

(c) $t = t_0 + 60s$

(d) $t = t_0 + 90s$

Figure 6. Flame spread behavior at leading edge for $\theta = 90°$.

4. Conclusion

In this study, thread angle dependency on flame spread shape over kenaf/polyester fabrics is examined and following results are obtained:

1. Flame spread shape is differed for combined fabric at $\theta = 0°$ and 90°. For $\theta = 0°$, the shape turns to be in 'V' shape but the flame is in flat shape when $\theta = 90°$.
2. Flame spread shape is influenced by the flame spread behavior at leading point. For $\theta = 0°$, the polyester thread melts when the edge of discoloration on the cotton thread approaches. The polyester thread breaks into two and it pulls the kenaf thread when it shrinks. For $\theta = 90°$, the polyester thread melts gradually in the direction of the thread and attaches to the kenaf thread. The distance between kenaf threads remains constant as flame approaches.

Acknowledgment
The authors would like to thank Universiti Tun Hussein Onn Malaysia (UTHM) and Ministry of Higher Education Malaysia (MOHE) for their financial support of the present work through Fundamental Research Grant Scheme (FRGS-1465).

References
[1] Bhattacharjee S, Bundy M, Paolini C, Patel G and Tran W 2012 A novel apparatus for flame spread study *Proc. of the Combustion Int.* (San Diego, USA: San Diego State University) pp 2513-2521
[2] Fernandez-Pello A C, Ray S R and Glassman I 1981 *Proc. Combustion Int.* pp 579-589
[3] Ding Y, Wang C and Lu S 2014 Large Eddy Simulation of Fire Spread *Procedia Engineering* (Heifei, China: University of Science and Technology of China) pp 537-543
[4] Osorio A F, Fernandez-Pello C, Urban D L and Ruff G A 2012 Limiting Conditions for Flame spread In Fire Resistant Fabrics *Proc. of the Combustion Int.* (Berkeley, USA: University of California) pp 2691-2697
[5] Bei P, Liwei C and Chang Lu 2012 An Experimental Study on the Burning Behavior of Fabric used Indoor *Proc. of the Combustion Int.* (San Diego, USA: San Diego State University) pp 2513-2521
[6] *Facts about Fabric Flammability* 2003 (USA: North Central Regional Extension Publication)
[7] Mohd Azahari B R, Azwan B S, Mohd Faisal B H, Amir B K, Suzuki M and Masuda W 2016 Flame spread behavior over combined fabric of cotton/polyester *ARPN Journal of Engineering and Applied Sciences* **11**(12) 7553-7557
[8] Mohd Azahari B R, Suzuki M, Masuda W, Hamidon B S, Azwan B S, Akmal Nizam B M, Mohd Faisal B H and Amir B K 2017 Effect of synthetic material on angle dependency of flame spread behavior over combined fabric *AIP Conference Proceedings* **1831**(1) 020014
[9] Mohd Azahari B R, Azwan B S, Akmal Nizam B M, Mohd Faisal B H, Hamidon B S and Hazahir B P 2017 Effect of thread angle on flame spread behaviour over combined fabric of kenaf/polyester *AIP Conference Proceedings* **1831**(1) 020016

IOP Conf. Series: Materials Science and Engineering **243** (2017) 012028 doi:10.1088/1757-899X/243/1/012028

Numerical investigation of a thick plate restriction orifice on the pressure drop performance

K H Yau[1*], E C Kua[2] and S Balvinder[3]
[1] Dunham-Bush Industries Sdn Bhd, 43000 Kajang, Malaysia
[2] School of Engineering, Monash University Malaysia, 47500 Selangor, Malaysia
[3] Faculty of Engineering and the Built Environment, SEGI University, 47810 Selangor, Malaysia

Email: yaukh@dunham-bush.com

Abstract. This paper presents a numerical study on the thick plate restriction orifice on the pressure drop performance due to various orifice ratio, β. The restriction orifice was investigated using commercial software package namely, ANSYS. The restriction orifice was modelled using built-in modeler and simulated using Fluent module. The orifice ratio, β was varied in the range of 0.5 to 0.75. Various flow velocities were applied from 50 to 420 m/s. The fluid flow in the constriction was hydrocarbon in vapour phase. The preliminary results of discharge coefficients were compared with literature and theoretical values between a sharp-edged and thick plate orifice to show a consistent trend. The results yielded that as the Reynolds number, *Re* increases, the pressure drop performance increases exponentially. This is more prominent at high *Re* of 8.75×10^6 where the pressure drop increases by 63% from the baseline of *Re* number, 1.08×10^6 based on orifice ratio of 0.75. For a hydrocarbon with low rheological properties and high Re number, the orifice ratio, β is best at higher range of 0.75 which show a pressure drop of 0.18 MPa.

1. Introduction

A restriction orifice is commonly used as the fundamental of orifice meters for many decades, and until recently, the restriction orifice is utilized as pressure-reducing device to reduce the acoustic induced-vibration in a piping system. Typically, the piping system in a high pressure letdown scenario of a typical process industry is usually installed with a pressure control valve for gas or vapour flow. The downstream flow after the pressure control valve has very high pressure which causes vibration in the pipe and subsequently, induce an undesirable high frequency noise. In addition to noise generation, the vibration eventually causes dynamic failure in the pipe over a short period of time.

Some of the restriction orifices [1-7] are: (a) sharp-edged, (b) square-edged (thick plate), (c) quadrant-edged and (d) conical-edged orifice as shown in figure 1, respectively. The shape of the orifice affects the flow behaviour at the constriction and subsequently, affects the local pressure drop and velocity at the exit of the orifice which creates a vena contracta shown in figure 1. The orifice ratio, β is defined as the diameter of the orifice, d to the inner diameter of the pipe, D also shown in figure 1.

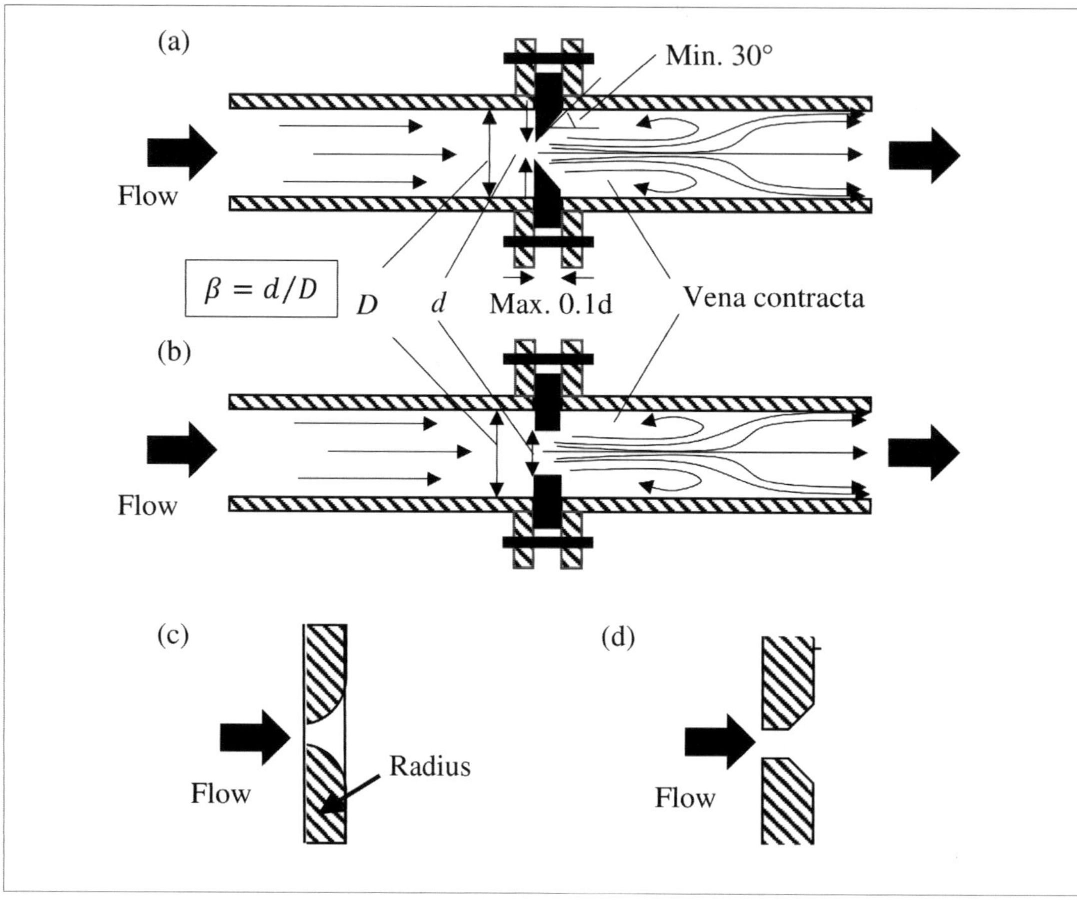

Figure 1. Restriction orifice for (a) sharp-edged, (b) square-edged, (c) quadrant-edged and (d) conical-edged [4,5,7].

There are two measurements of the pressure drop performance [7] for a restriction orifice which is defined at three pressure locations, conveniently labelled as P_1, P_2 and P_3 as shown in figure 2. These locations, P_1, P_2 and P_3 are called upstream pressure, throat pressure and final recovery pressure, respectively. The pressure drop between P_1 and P_2 is called the orifice pressure drop, ΔP_{orif}. On the other hand, the pressure drop between P_1 and P_3 is called the permanent pressure drop, ΔP_{perm}.

The hydraulic performance of the orifice is defined by the discharge coefficient, C_D which is shown in equation (1) is based on [7]. The discharge coefficient is dependent on the volume flowrate, orifice ratio, orifice area and the pressure drop [4,5].

Discharge coefficient, C_D

$$C_D = \frac{Q_v\sqrt{1-\beta^4}}{A_O\varepsilon\sqrt{2\Delta P_{orif}/\rho}} \tag{1}$$

where C_D is the discharge coefficient (-), Q_v is the volumetric flowrate (m^3/s), ε is the expansibility factor (1 for incompressible fluid) (-), ΔP_{orif} is the pressure drop across orifice ($P_1 - P_2$) (Pa), ρ is the fluid density (kg/m^3), β is the orifice diameter to inner pipe diameter (-) and A_o is the cross-sectional area of the orifice (m^2).

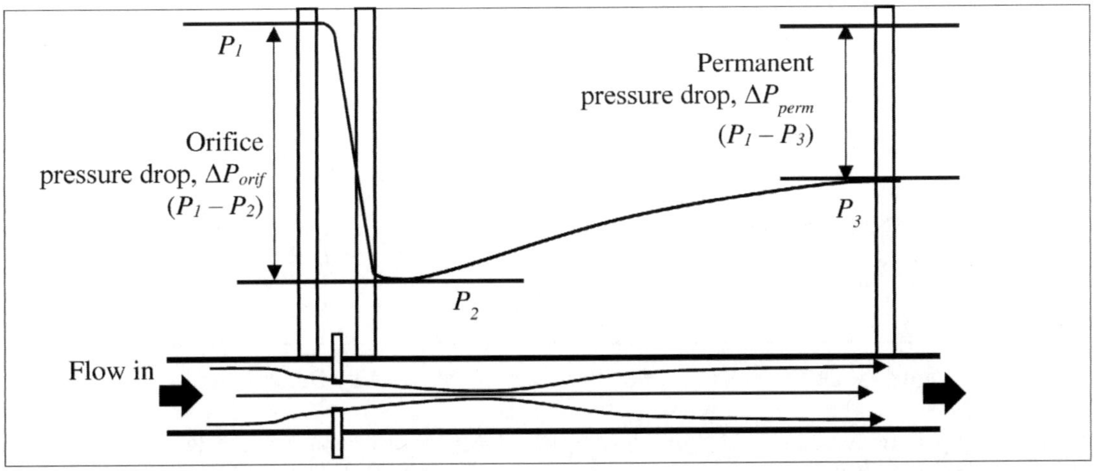

Figure 2. Orifice pressure drop ($P_1 - P_2$) and permanent pressure drop ($P_1 - P_3$) [7].

In these papers from Kumar [3] and Huang [7], they studied the effect of various orifice design on the pressure drop performance based on discharge coefficient, C_D and compared the results to a standard flat orifice using air as single-phase gas fluid. In Kumar's study [3], they investigated numerically the effect of slotted orifice on the discharge coefficient, C_D by using air as two-phase flow, liquid-vapor medium. They varied the aspect ratios of slotted orifices from 1.5 to 3.0 and fixed at orifice ratio, β of 0.4. They found that for a slotted orifice, a low orifice ratio is more sensitive to liquid presence in the stream and hence, is preferable for wet gas metering.

On the other hand, Huang [7] studied experimentally on a perforated orifice which shown a higher discharge coefficient, C_D of 22.5 to 25.6% than a standard orifice using water as single phase, liquid medium. Their perforated orifice had at least 6 orifices in a diameter as compared to a single orifice. They varied different orifice ratios, β for perforated orifices from 0.338 to 0.668 and also varied mass flow rates from 0 to 4000 kg/hr.

In another paper by Buker [8], they studied the Reynolds number dependence on a sharp-edged orifice and compared with two different designs of orifice with attached protrusion on the bottom and top as shown in figure 3. They found that the Reynolds number for a sharp-edge orifice have insignificant effect on the discharge coefficient, C_D at lower range of orifice ratio from 0.1 to 0.3. An inflection point on the discharge coefficient, C_D can be observed when the orifice ratios are increased from 0.5 to 0.65 for different Reynolds number.

There is one paper by Geng [9] also studied the slotted orifice plate and a standard flat orifice using numerical and experimental method for a gas flow. The standard flat orifice was simulated under one single orifice ratio for different mass flow rates. The slotted orifice shown a higher drag coefficient by about 5% as compared to a standard flat orifice. The results were compared experimentally to show a relative error within ± 10% for a slotted orifice.

A review on recent papers on orifice plates shows that most of the researchers are keen on irregular shapes such as slotted [10] and concentric shape [11] where both papers used tomography method and numerical method to observe the flow in the orifice qualitatively, respectively. The tomography method [10] was performed on a two-phase flow in the slotted orifice to show an optimum homogeneous flow at x/D of 1.5 and 2.5. The numerical results of concentric shape [11] was compared with theoretical discharge coefficients and found that the relative errors agreed within ± 10.5%. Flow separation was observed at the bottom part of the upstream and downstream of concentric plate, and not present for an eccentric orifice plate.

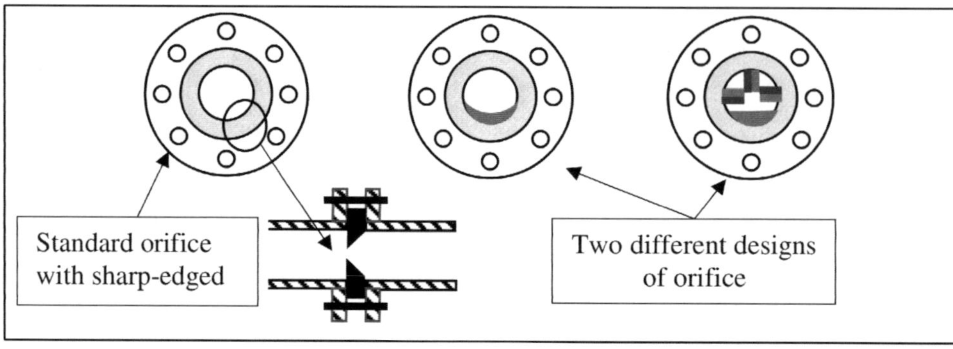

Figure 3. Different shape of orifice used in the study by Buker et al. [8]

All of the papers reviewed [3-11] show that the standard flat orifice used were at limited range and variations which were reported at a single parameters of geometrical design, constriction ratio, flow rate and fluid, i.e. water or gas. Other information pertaining to the standard orifice data is very scarce and difficult to obtain due to confidentiality and proprietary from the manufacturer. Millers [6] coefficient of discharge, C_D are widely used in the theoretical calculations for orifice meters, flow nozzles and Venturi meters. Thus, in this paper, the aim is to study the effect of various Reynolds number, *Re* and orifice ratio, β on the pressure drop performance and drag coefficients of a thick plate restriction orifice.

2. Methodology

The numerical software package by Fluent, ANSYS (Version 17) was utilized to solve the fluid complex equations of continuity, momentum and energy as shown in equation (2), (3) and (4), respectively. A square-edged restriction orifice was modelled in 3-dimensional in a cylindrical pipe. In the setup of the model, some assumptions made were: (a) steady-state flow, (b) isothermal condition across the flow boundary, (c) no-slip boundary condition, (d) no penetration from the pipe wall region to the gas flow margin, (e) no wall roughness, and (f) incompressible flow.

Continuity equation:

$$\frac{\partial \rho}{\partial t} + \nabla \cdot (\rho \vec{u}) = 0 \qquad (2)$$

Momentum equation:

$$\frac{\partial}{\partial t}(\rho \vec{u}) + \nabla \cdot (\rho \overrightarrow{uu}) = -\nabla p + \nabla \tau \qquad (3)$$

Energy equation:

$$\frac{\partial}{\partial t}(\rho \vec{E}) + \nabla \cdot \left(\vec{u}(\rho E + P)\right) = 0 \qquad (4)$$

where ρ is the density, u is the velocity, P is the pressure, τ is the viscous stress tensor, E is the energy.

All cases were simulated in turbulence flow regime (Re > 1×10^5) using standard k-ε model based on the recommendation by Kumar et al. [3]. The equations for k and ε are as follows:

$$\frac{\partial}{\partial t}(\rho k) + \frac{\partial}{\partial t}(\rho k u_i) = \frac{\partial}{\partial x_j}\left[\left(\mu + \frac{\mu_t}{\sigma_k}\right)\frac{\partial k}{\partial x_j}\right] + G_k - \rho\varepsilon \qquad (5)$$

$$\frac{\partial}{\partial t}(\rho \varepsilon) + \frac{\partial}{\partial t}(\rho \varepsilon u_i) = \frac{\partial}{\partial x_j}\left[\left(\mu + \frac{\mu_t}{\sigma_\varepsilon}\right)\frac{\partial \varepsilon}{\partial x_j}\right] + C_{1\varepsilon}\frac{\varepsilon}{k}G_k - C_{2\varepsilon}\rho\frac{\varepsilon^2}{k} \qquad (6)$$

where G_k is the generation of turbulence kinetic energy due to mean velocity gradients and $C_{1\varepsilon}$, $C_{2\varepsilon}$, σ_k, σ_ε are the standard k-ε model constants, and μ_t is the turbulent viscosity $\left(\rho C_\mu \frac{k^2}{\varepsilon}\right)$.

IOP Conf. Series: Materials Science and Engineering **243** (2017) 012028 doi:10.1088/1757-899X/243/1/012028

The geometrical dimensions for restriction orifice followed from design standard BS1042, and manufacturer technical sheet from Fisher (6011, US) as shown in figure 4. The entrance and exit length from the orifice was based on the recommendation from Kumar [3] which was 5*D* before and 10*D* after the orifice to ensure the flow was fully developed. The diameter of the pipe, *D* was at 0.3 m. Hence, the length of the pipe was 1.5 m and 3.0 m for entrance length and exit length, respectively.

The fluid used throughout the investigation was a hydrocarbon in gas phase in a process line where the properties of the process gas are summarized in table 1. The variable parameters for simulation are summarized in table 2. The velocities and orifice ratio selected were arbitrarily and within the recommended range as proposed by an industry.

The mass flow rate, Q_v can be calculated using standard orifice meter shown in equation (7) based on the re-arrangement of equation (1). The calculated theoretical flow rate as a function of Reynolds number, *Re* is based on pipe diameter, D given in equation (8).

$$Q_v = \frac{C_D \cdot A_O \cdot \varepsilon \sqrt{2\Delta P_{orif}/\rho}}{\sqrt{1-\beta^4}} \tag{7}$$

$$Re_D = \frac{\rho v D}{\mu} \tag{8}$$

where ρ is the fluid density (kg/m^3), v is the fluid velocity (m/s), D is the pipe diameter (m) and μ is the fluid dynamic viscosity (Pa.s).

Pressure drop, ΔP of an internal flow due fluid viscous effect can be calculated in equation (9) based on the Darcy friction factor.

$$\Delta P = f \frac{L}{D} \frac{\rho v^2}{2} \tag{9}$$

where f is the Darcy friction factor (-), L is the pipe length (m), D is the pipe diameter (m) and $\rho v^2/2$ is the dynamic pressure (Pa)

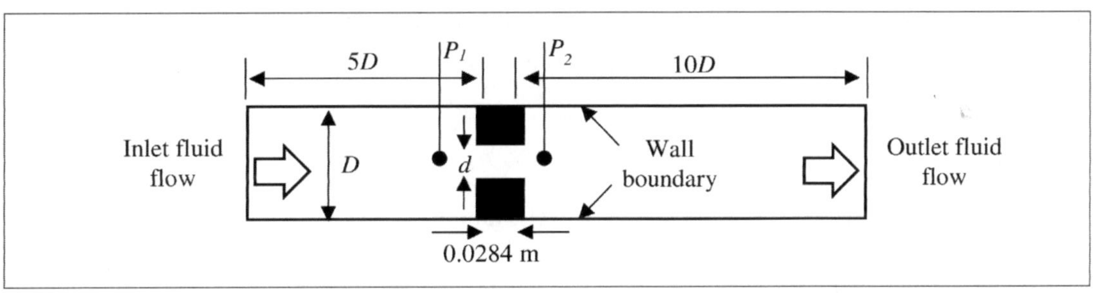

Figure 4. 3-D model of orifice with boundary conditions.

Table 1. Fluid properties of the hydrocarbon gas.

Properties	Value
Density (kg/m^3)	9.07×10^{-1}
Viscosity (cP)	1.30×10^{-2}
Molecular Weight (g/mol)	21.3

Table 2. Variable parameters for simulation.

Parameter	Value
Velocity, v (m/s)	50, 100, 200, 300, 420
Orifice ratio, β (-)	0.5, 0.7, 0.75

IOP Conf. Series: Materials Science and Engineering **243** (2017) 012028 doi:10.1088/1757-899X/243/1/012028

The downstream velocity, v_2 can be calculated using theoretical formula based on the orifice ratio as shown in equation (10).

$$v_2 = \sqrt{\frac{2\Delta P_{perm}}{\rho(1-\beta^4)}} \tag{10}$$

where ΔP_{perm} is the permanent pressure drop ($P_1 - P_3$) (Pa), ρ is the fluid density (kg/m3) and β is the orifice ratio (-).

2.1. Mesh Generation

The model was generated with unstructured tetrahedral mesh for simple boundaries. For complex regions such as the constriction area in the restriction orifice were generated with structured hexahedral with at least mesh size of 0.001 mm as shown in the cross-section in figure 5.

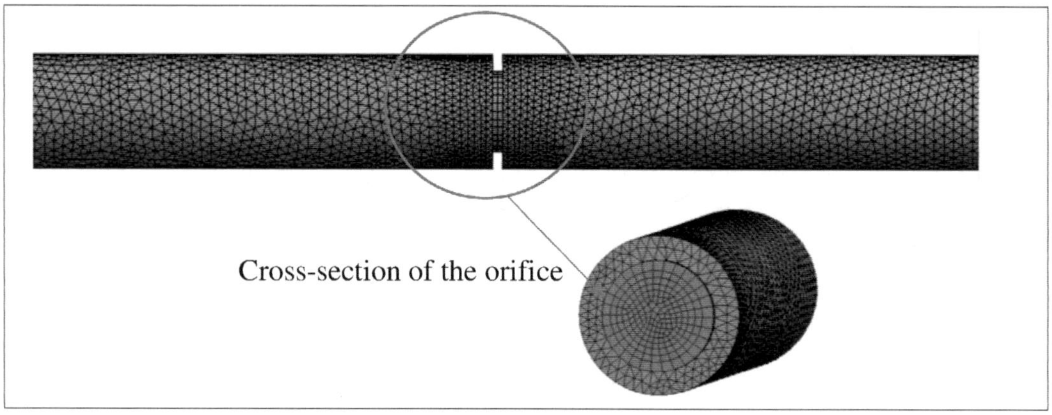

Figure 5. Mesh generation of restriction orifice.

2.2. Convergence Criterion

The convergence criterion of the simulation was based on the residual value of calculated variables such as mass, velocity components, turbulent kinetic energy, k, and turbulent kinetic energy dissipation rate, ε. The simulation was considered as converged when all other variables residual achieved a value of less than 10^{-4} when the plots reached a minimum plateau.

3. Grid Independence Study

The number of mesh generated was from 0.56×10^6 to 1.10×10^6 and the acceptable mesh size was chosen at 0.96×10^6 based on the trend that plateau after that mesh size. The result of the grid independence study is plotted in figure 6. Based on this selected mesh size, all other models due to different orifice ratios, β were mesh generated within this range.

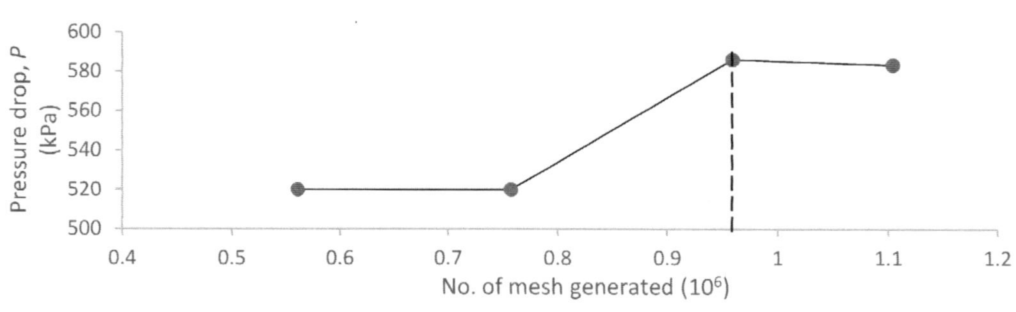

Figure 6. Grid independence study on the restriction orifice.

4. Results and Discussions

The results from the numerical study such as drag coefficient, C_D, pressure distribution, P, orifice pressure drop, ΔP_{orif} velocity contour, and pressure contour for different orifice ratio, β are discussed. As a direct comparison with similar setup and conditions are difficult as aforementioned in the literature review, the closest possible range of conditions, fluid and geometry are compared with Kumar [3], Miller [6] and Buker [9]. The drag coefficients, C_D are shown in figure 7, 8 and 9. A theoretical comparison is also demonstrated using internal flow equation of volume flowrate from the pressure drop performance is shown in figure 9. The plots of pressure distribution against axial distance, x from the pipe entrance to the exit where total length is 4.5 m, are plotted in figure 10.

Shown in figure 7, the shape of the orifice affects the discharge coefficient by about 64% as compared between a sharp-edged and a thick plate orifice. The thick plate orifice has lower discharge coefficient than the sharp-edged orifice. Similarly, both have an increasing trend when the orifice ratio increase. However, based on the theoretical results by Buker [9], the discharge coefficient, C_D for a sharp-edged orifice reaches to a peak at orifice ratio, β of 0.63 and C_D reduces beyond that orifice ratio. An extrapolation of the theoretical results by Buker [9] shows that an intersection at orifice ratio of 0.71 between thick plate and sharp-edged orifice.

The results of discharge coefficients, C_D for different Reynolds number, Re are plotted in Figure 8 where a comparison with Buker [9] for orifice ratio of 0.5 for sharp-edged orifice. Buker [9] results were of higher margin than the current study due to the shape of the orifice. A thick plate orifice has lower discharge coefficients due to larger drag area than the sharp-edge orifice, and also due to the rheological properties of the hydrocarbon fluid where density and viscosity is lower than air by about 26.3% and 27.3%, respectively. The flow performance of downstream velocities were calculated based on the permanent pressure drop, ΔP_{perm} from simulation using equation (9) and compared with theoretical equation (10) which shown a deviation of 6.63% to 20.2% in figure 9. This is expected due to the absence of discharge coefficient, C_D in the theoretical flow equation, and a similar error margin of 10.5% was also demonstrated in paper [11].

Based on figure 10, the pressure distributions show a similar trend with Kumar [3] where a direct comparison is not possible due to the different fluid and velocities being used. The model was extended to orifice ratio, β of 0.70 and 0.75 for velocities of 50 to 420 m/s to observe the trend. The pressure distribution can be observed that as the velocity increases the pressure decreases by about 460%, at the highest velocity of 420 m/s for orifice ratio of 0.5. This is expected as the orifice is so narrow that the passage will incur a high pressure drop. On the other hand, the orifice ratio of 0.70 and 0.75 yields about 87% and 48% in pressure reduction, respectively.

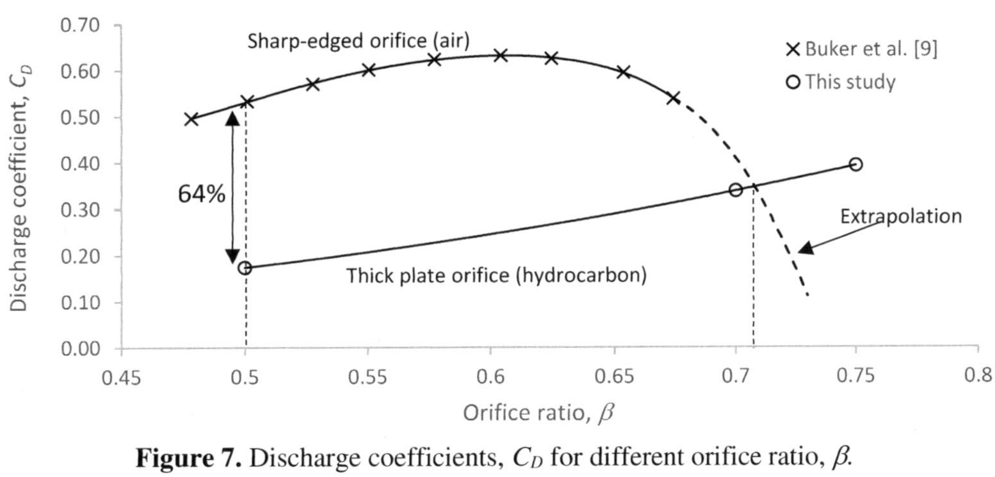

Figure 7. Discharge coefficients, C_D for different orifice ratio, β.

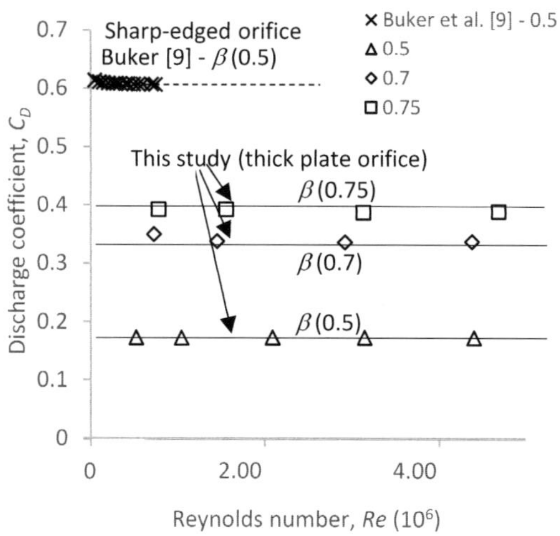

Figure 8. Discharge coefficient, C_D versus Reynolds number, *Re* for different orifice ratio, *β*.

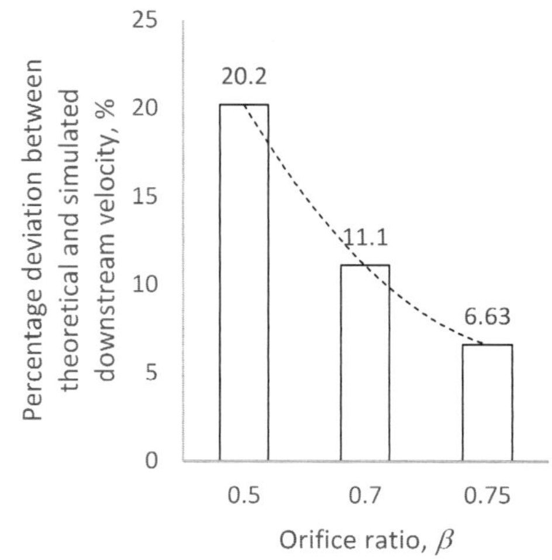

Figure 9. A comparison of flow performance between simulation and theoretical values

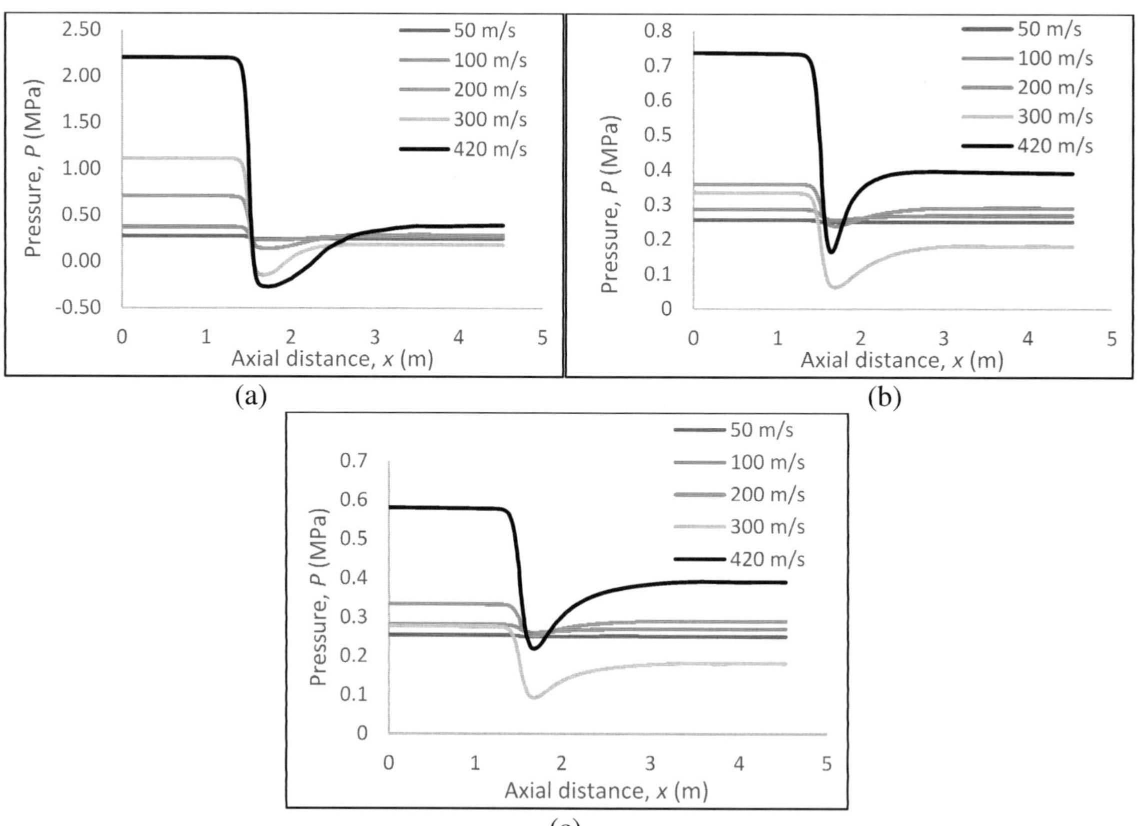

Figure 10. Pressure profiles along the distance, x for different orifice ratios, β of (a) 0.5, (b) 0.7 and (c) 0.75.

The permanent pressure drop, P_{perm} ($P_1 - P_3$) for different orifice ratios, β and different Reynolds number, *Re* are plotted in figure 11. As the orifice ratio increases, the pressure drop also increases exponentially from *Re* of 1.04×10^6 to 8.75×10^6 as shown in the trend lines. All trend lines are plotted with best fit, R^2 of 0.99, on average. The increment in pressure drop is about 63% at orifice ratio of 0.75 from the same range of Reynolds number. A tremendous amount of penalty in pressure drop is observed when the orifice ratio reduces from 0.75 to 0.50 which yields to about 837% at the highest *Re* of 8.75×10^6.

The velocity and pressure contour for different orifice ratios, β is shown in figure 12 and figure 13, respectively. As orifice ratio of 0.70 is an intermediate result between ratio of 0.5 and 0.75, only two sets of results are shown here for orifice ratio of 0.50 and 0.75.

Based on figure 12 (a), the velocity contour at orifice ratio of 0.50 shows that the velocity vectors are widely scattered along axial distance demonstrating a trend of non-uniformity. On the other hand, orifice ratio of 0.75 shows velocity vector of more uniform distribution shown in figure 12 (b). This is similar to the pressure distribution shown in figure 13 (a) and (b) where orifice ratio of 0.75 is more uniform than 0.50. This can be seen in the darker blue regimes between orifice ratio of 0.75 and 0.50 for Reynolds number from 1.04×10^6 to 8.75×10^6. It is desirable to have uniform pressure distribution in a piping system than a non-uniform distribution for different Reynolds number.

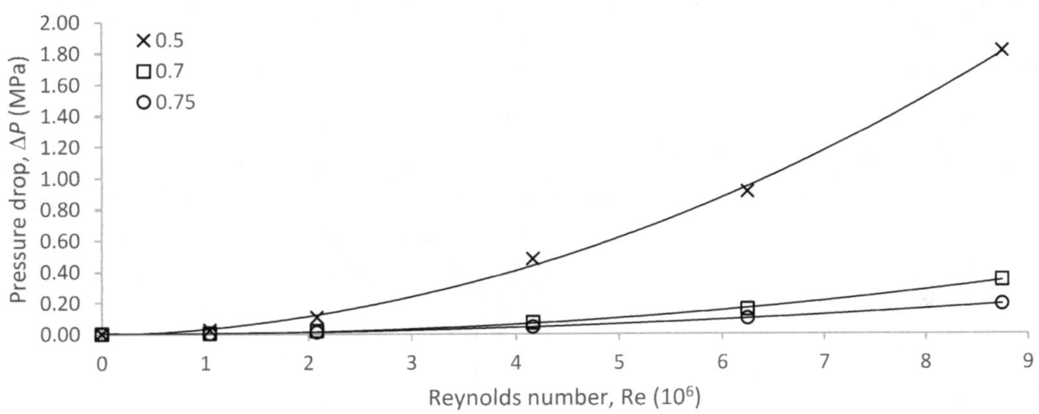

Figure 11. Pressure drop, ΔP_{perm} vs. Reynolds number, *Re* for different orifice ratios, β.

IOP Conf. Series: Materials Science and Engineering **243** (2017) 012028 doi:10.1088/1757-899X/243/1/012028

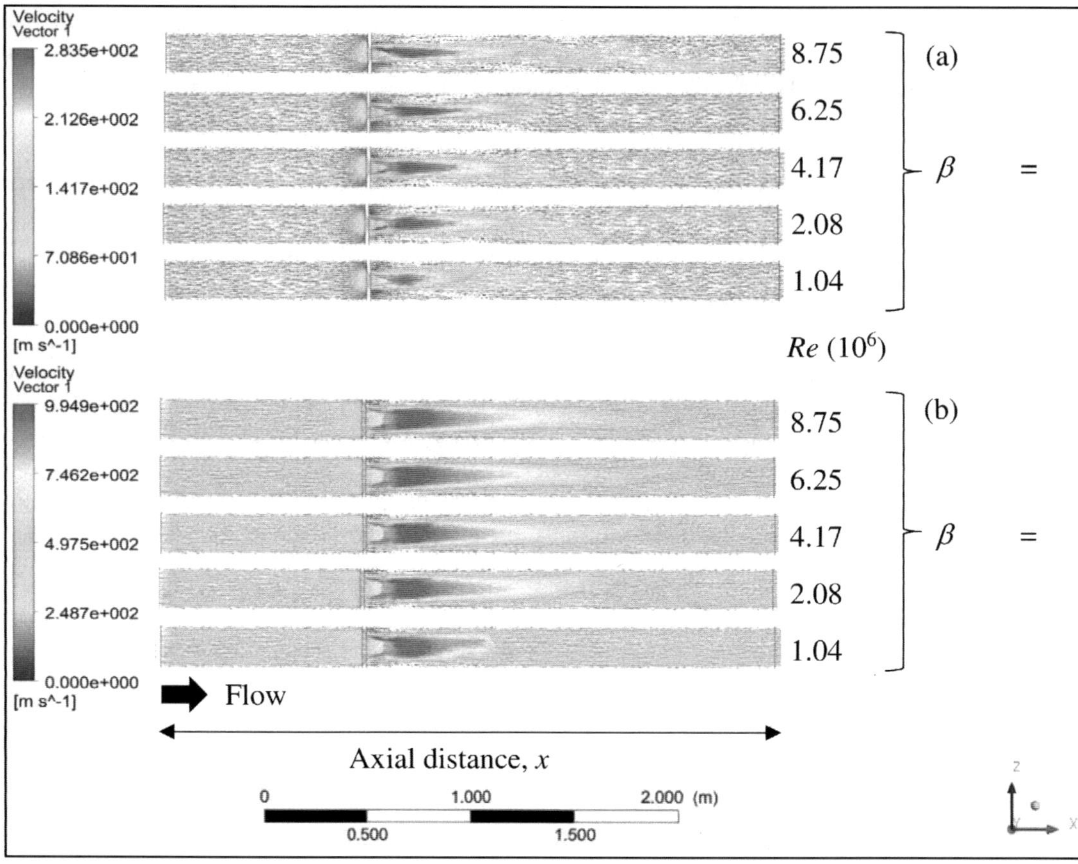

Figure 12. Velocity contour for orifice ratio of (a) 0.50 and (b) 0.75.

IOP Conf. Series: Materials Science and Engineering **243** (2017) 012028 doi:10.1088/1757-899X/243/1/012028

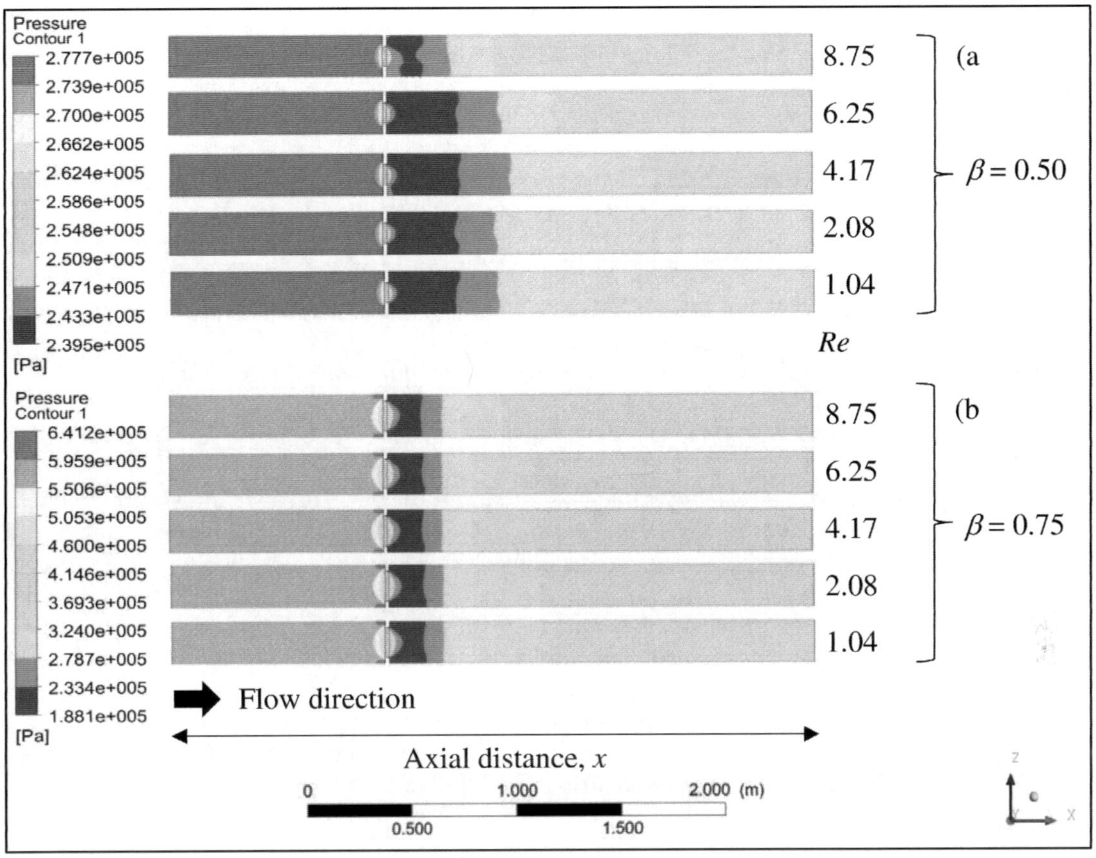

Figure 13. Pressure contour for orifice ratio of (a) 0.50 and (b) 0.75.

5. Conclusions

In this study, the velocity and pressure distribution due to different orifice ratios, β from 0.50, 0.70 and 0.75 are compiled and analyzed for its pressure drop performance. Based on both contours, velocity and pressure, some conclusions are made:

a. Restriction orifice with flat protrusion has lower discharge coefficients that sharp-edged protrusion.
b. Thick plate orifice of 0.50 provides a tremendous penalty in pressure drop that yields to about 837%.
c. Thick plate orifice of 0.75 sufficiently provides a pressure drop of 63% at the highest Reynolds number.
d. Velocity and pressure contours shows that the thick plate orifice of 0.75 provides uniform distribution than orifice ratio of 0.50.

6. Recommendation and Future Studies

For future studies, it is recommended that: (a) an experimental study to verify the thick plate shape; (b) a numerical study of a thick plate in a piping system consisting of various components, and (c) a numerical study on the thick plate in comparison with various orifice shapes on the pressure drop performance.

Acknowledgement
The authors would like to appreciate Dunham-Bush Industries Sdn Bhd and Engineering, Monash University Malaysia for supporting this study.

References
[1] Shrouf R D 2003 *Pressure and flow characteristics of restrictive Flow orifice devices* (United Sates: Sandia National Laboratories Report, SAND2003-1874)
[2] Geng Y F, Zheng J W and Shi T M 2006 Study on the metering characteristics of a slotted orifice for wet gas flow *Flow Meas. Instrum.* **17** 123-8
[3] Kumar P and Wong M B 2011 A CFD study of low pressure wet gas metering using slotted orifice meters *Flow Meas. Instrum.* **22** 33-42
[4] Munson B R, Young D F, Okiishi T H and Wade W 2009 *Fundamentals of Fluid Mechanics* (New York: John Wiley)
[5] Perry R H and Green D W 2008 *Perry's Chemical Engineers' Handbook* (New York: McGraw-Hill)
[6] Miller R 1996 *Flow Measurement Engineering Handbook* (New York: McGraw-Hill)
[7] Coulson J M and Richardson J F 1996 *Chemical Engineering: Vol. 1* (London: Butterworth-Heinemann)
[8] Huang S, Ma T, Wang D and Lin Z 2013 Study on discharge coefficient of perforated orifices as a new kind of flow meter *Exp. Therm. Fluid Sci.* **46** 74-83
[9] Buker O, Lau P, Tawackolian K 2013 Reynolds number dependence of an orifice plate *Flow Meas. Instrum.* **30** 123-132
[10] Annamalai G, Pirouzpanah S, Gidgopuram S R and Morrison G L 2016 Characterization of flow homogeneity downstream of a slotted orifice plate in a two-phase flow using electrical resistance tomography *Flow Meas. Instrum.* **50** 209-215.
[11] Zahariea D 2016 Numerical analysis of eccentric orifice plate using ANSYS Fluent software *20th Innovative Manufacturing Engineering and Energy Conf.* **161** 012041

Pressure recovery performance of 2-D turning diffuser by varying area ratios and inflow Reynolds numbers

M S Ruzaini[1, a], **N Nordin**[1, b], **A Sadikin**[1], **A N Mohammed**[1], **A Sapit**[1], **K Abdullah**[1], **Y Ramli**[1], **A F Idris**[1]

[1] Flow Analysis, Simulation and Turbulence Research Group, Center for Energy and Industrial Environment Studies, Faculty of Mechanical and Manufacturing Engineering, Universiti Tun Hussein Onn Malaysia, Johor, Malaysia

E-mail: syahirruzaini@gmail.com [a], mayati@uthm.edu.my[b]

Abstract. The paper aims to investigate the effects of varying area ratio, AR = 1.2 and 4.0 and inflow Reynolds number, Re_{in} = 5.478 x 10^4 - 1.547 x10^5 on the performance of 90° two-dimensional turning diffuser. The optimum configuration area ratio and Re_{in} to produce good pressure recovery is determined. The rig was developed to produce fully developed entrance flow by adopting arrangement of mesh net and sufficient hydrodynamic entrance length, $L_{h,turb}$= 28 D_h. Digital manometer was used to measure the inlet and outlet static pressures and Particle Image Velocimetry (PIV) to visualize the flow structure. The present results were compared with empirical solution of Asymptotic Computational Fluid Dynamics (ACFD) results to give acceptable deviation of ±7.4%. The AR=4.0 produces pressure recovery 20% more than AR=1.2 when applies low Re_{in}<1.40 x 10^5. However, it is subjected to severe flow separation and circulation at Re_{in}>1.40 x 10^5 that considerably disturbs the recovery. Therefore, turning diffuser of AR=1.2 is optimum applied for high Re_{in}>1.40 x 10^5.

1. Introduction

Diffuser is a common engineering device which has the simplest design of an expanding area in the flow direction. It is applied widely in applications such as heating, ventilation and air conditioning (HVAC), wind-tunnel, gas turbine cycle and aircraft engine as an adapter to join the conduits of different cross-sectional areas or an ejector to decelerate the flow and raise the static pressure before discharging to the atmosphere. Diffuser is also be used to suck more water from a standard spigot which has a great impact in the overall performance of the system. There would be losses whenever a diffuser is interrupted by the inclusion of pipe fitting such as bend [1] and it is all dependent on the geometrical and operating parameters applied.

Fox and Kline [2] have suggested that the area ratio (AR) of a sharp 90° turning diffuser should be within 1.2-2.0 in order to avoid severe stall. The stall is undesirable as it would decrease the core flow area, induce the presence of secondary flow vortices, increase the form drag and ultimately affect the pressure recovery [3 - 5]. Since early 1980s, the variation of operating conditions, Re_{in} have been taken into consideration to affect the performance of diffusers such as 55° 2-D turning diffuser (Re_{in} = 7.8 x 10^5 – 1.29 x 10^6) [6], annular diffuser (Re_{in} = 6.0 x10^3 – 6.0 x 10^5) [7] and combined 90° bend diffuser (Re_{in} = 8.8 x 10^4 – 1.94 x 10^5) [1]. Nordin [8] has recently established an empirical correlation integrating the effects of both geometrical and operating parameters on the performance of

90° 2-D turning diffusers using Asymptotic Computational Fluid Dynamics (ACFD) technique. This correlation will be used to verify the current work:

$$C_{p\,ACFD} = 0.200 + \left(\left[\frac{Re_{in}}{Re_{in\,ref}}\right]^2 - 1\right)0.0059 - \left(\left[\frac{L_{in}/W_1}{L_{in}/W_{1ref}}\right]^1 - 1\right)0.6569 - \left(\left[\frac{W_2/W_1}{W_2/W_{1ref}}\right]^3 - 1\right)0.0288 \quad (1)$$

where,

$C_{p\ \mathrm{ACFD}}$ = Outlet pressure recovery from ACFD
$\mathrm{Re_{in}}$ = Inflow Reynold number
L_{in} = Inner wall length (m)
W_1 = Inlet width (m)
W_2 = Outlet width (m)
$\mathrm{Re_{in\ ref}}$ = 6.382 x 10^4
L_{in}/W_{1ref}= 4.37
W_2/W_{1ref} =2.16

Upon verification, the effect of varying AR = 1.2 and 4.0 and $\mathrm{Re_{in}}$ = 5.478 x 10^4-1.547 x10^5 on the pressure recovery of 90° 2-D turning diffuser will be discussed and the optimum configuration to produce good pressure recovery will be decided.

2. Methodology

2.1. Design and development of rig

The experimental rig as shown in Figure 1 was designed to incorporate flow treatment consisting of an arrangement of mesh net and sufficient hydrodynamic entrance length, $L_{h,turb}$ = 28 D_h [9]. The turning diffusers of AR 1.2 and 4.0 with identical inlet dimension were designed to follow Chong et al [5] procedures and fabricated from acrylic of 1.2 mm thickness (Figure 2). The rig was proven to produce fully developed entrance flow at specified $\mathrm{Re_{in}}$ as depicted in Figure 3.

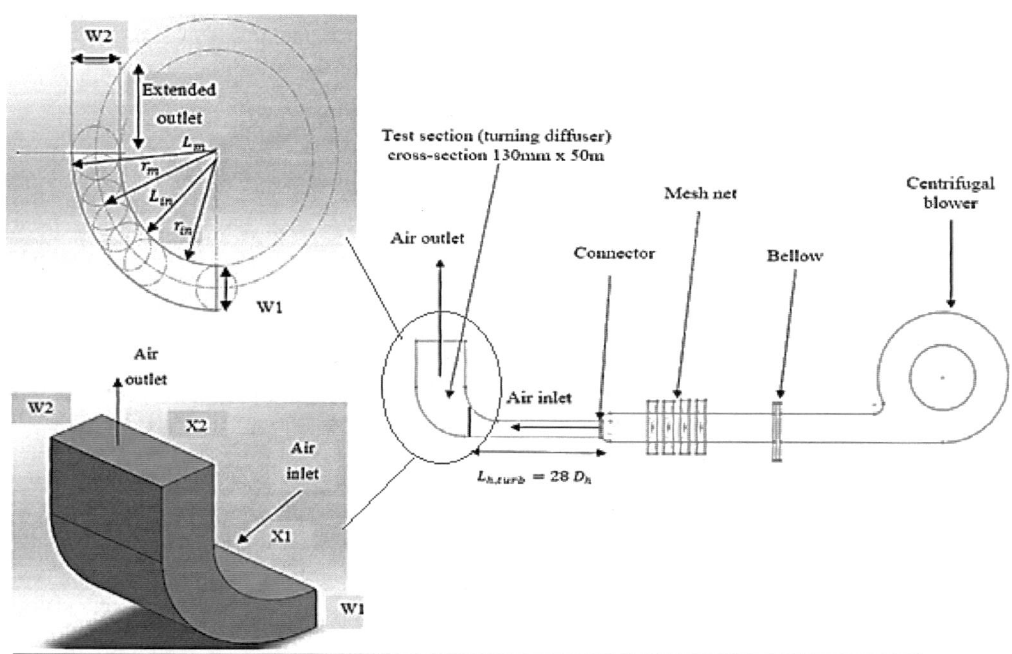

AR	W_1(mm)	X_1(mm)	W_2(mm)	X_2(mm)	r_{in}(mm)	L_{in}(mm)	r_m(mm)	L_m(mm)
1.2	50	130	60	130	120	173.28	150	251.05
4.0	50	130	200	130	120	228.97	220	403.02

Figure 1. Experimental rig

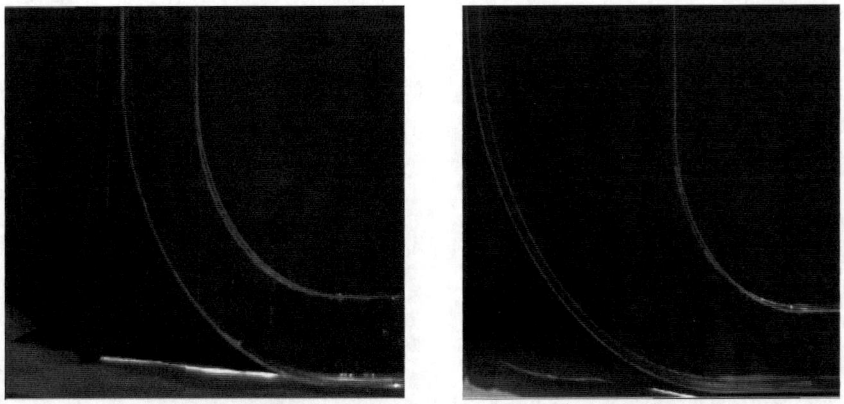

(a) (b)

Figure 2. 90° 2-D turning diffuser (a) AR=1.2 (b) AR=4.0

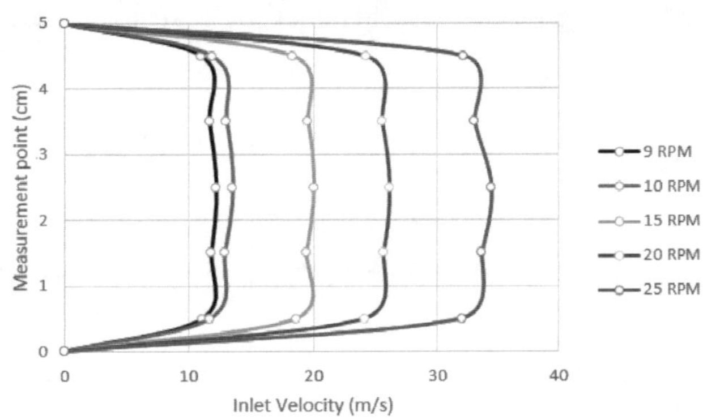

Figure 3. Fully developed entrance flow at specified $Re_{in} = 5.478 \times 10^4 - 1.547 \times 10^5$(9 – 25 RPM)

2.2. Measurement parameters and setup

Pressure recovery coefficient, C_p was used to evaluate the performance of turning diffusers. Equation (2) was applied in the present work to find the value of C_p. The outlet, P_{out} and inlet, P_{in} static pressures were measured using a digital manometer of resolution 0.1 Pa connected to pressure tappings as shown in Figure 4. In addition, the mean inlet velocity (V_{inlet}) was calculated using one-seventh power law velocity, with the maximum inlet velocity (V_{max}) measured using air flow meter at the center diffuser inlet [4].

$$C_p = \frac{2(P_{out} - P_{in})}{\rho V_{in}^{2}} \tag{2}$$

where,

P_{out}= outlet average static pressure (Pa)
P_{in}= inlet average static pressure (Pa)
ρ= air density (kg/m^3)
V_{in}= mean inlet air velocity ($0.817V_{max}$, m/s)

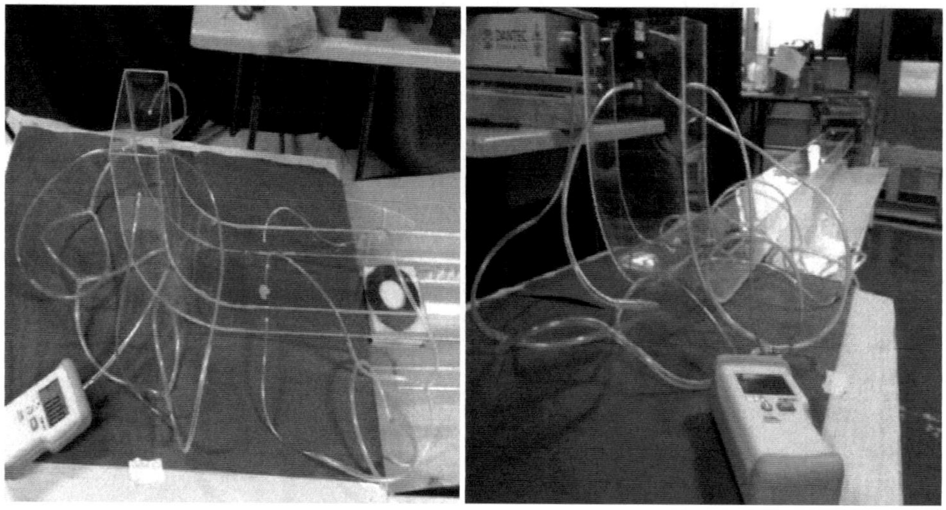

Figure 4. Average inlet (P_{in}) and outlet (P_{out}) static pressure measurement

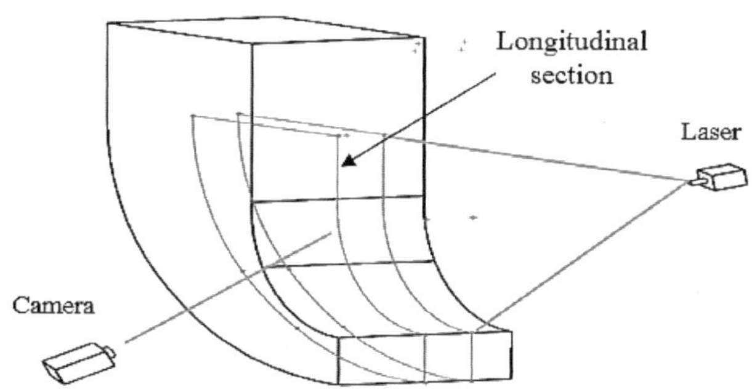

Figure 5. PIV setup

Particle image velocimetry (PIV) was used to visualize the flow structure within centre longitudinal section of turning diffuser. Following is the PIV measurement procedures:

1. The camera was arranged to be perpendicular to the target plane as illustrated in Figure 5.
2. The calibration was performed by adopting IMF: DLT.
3. The laser was calibrated to get better aligned to the target plane with thickness of 2mm.
4. The blower was set to speed of 9 RPM. Seeding particle was then injected to the blower.
5. Measurements were taken after the seeding particle and air completely mixed.
6. The camera was in double frame mode and the optimum time between pulses was applied.
7. The images obtained were analyzed using cross correlation technique and the bad vectors were masked.
8. Steps 4 – 8 were repeated by increasing the blower speed to 10, 15, 20 and 25 RPM.

3. Results analysis and discussion

The results obtained by present work were compared with ACFD results calculated using Equation (1) [8]. As depicted in Table 1, there is relatively huge deviation in the case of AR=1.2 of average 19.1% as the ACFD correlation applied was not specifically developed to solve small AR of less than 2.16 [8]. On the other hand, the present and ACFD results for AR=4.0 satisfy each other well with deviation of 7.4%.

Table 1. Comparison of present work with ACFD solution

Re_{in}	AR=1.2			AR=4.0		
	$C_{p\ present\ work}$	$C_{p\ ACFD}$	Dev (%)	$C_{p\ present\ work}$	$C_{p\ ACFD}$	Dev (%)
5.478x10^4	0.295	0.296	0.3	0.354	0.321	9.3
6.072x10^4	0.298	0.306	2.6	0.359	0.322	10.3
8.974x10^4	0.296	0.369	19.8	0.357	0.328	8.1
1.174x10^5	0.305	0.453	32.7	0.351	0.337	3.9
1.547x10^5	0.360	0.599	39.9	0.332	0.351	5.4
			19.1			7.4

Figure 6 shows the effects of varying AR and Re_{in} on C_p. The C_p increases with the increase of Re_{in} for AR=1.2 to maximum recovery of 0.360 at Re_{in}=1.547 x 10^5. Although producing higher recovery than AR=1.2, AR=4.0 shows an opposite trend of reducing C_p while increasing the Re_{in}. In fact, the C_p of AR=4.0 is considerably affected to produce C_p less than AR=1.2 at maximum Re_{in}. This is due to unstable flow happen within the turning diffuser as marked in Figure 7.

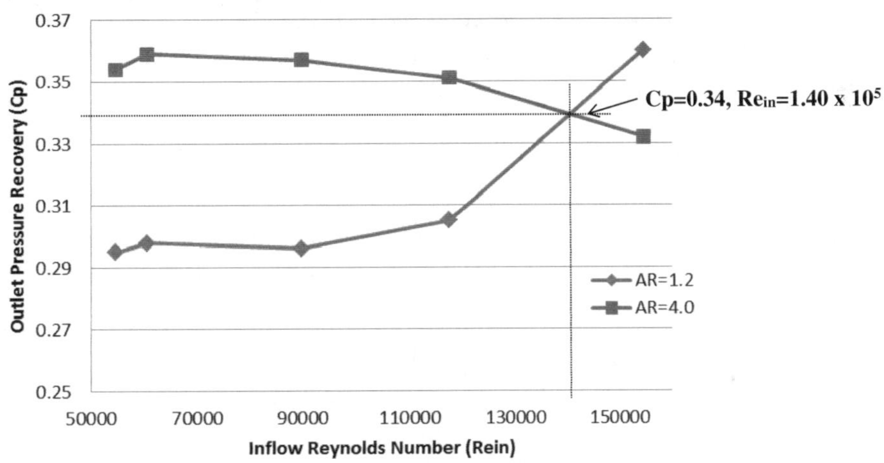

Figure 6. Effect of varying AR and Re_{in} on outlet pressure recovery

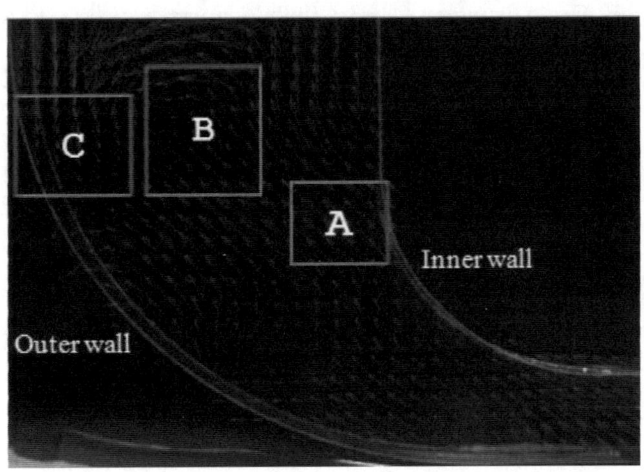

Figure 7. Flow structure of AR=4.0 at Re_{in}=1.547 x 10^5

The flow near to the inner wall (Region A) is subjected to curvature induced effect, where under a strong adverse pressure gradient the flow tend to detach from the wall and deflect towards the outer wall. Consequently, reverse backflow is formed to produce flow vortices as seen in Region B and C. The flow separation and vortices cause form drag that ultimately disturbs the C_p.

Overall, it is promising to apply 90° 2-D turning diffuser of AR=4.0 when $Re_{in} < 1.40 \times 10^5$ is applied, to give the optimum recovery 20% more than the AR=1.2. On the other hand, when $Re_{in} > 1.40 \times 10^5$, the AR=1.2 should be opted as the AR=4.0 is subjected to highly distorted flow that inherently affects the overall performance of 90° 2-D turning diffuser.

4. Conclusion

In conclusion, the current work manages to investigate the effects of varying AR and Re_{in} on C_p of 90° 2-D turning diffuser. The C_p increases with the increase of Re_{in} for AR=1.2 while otherwise happens in the case of AR=4.0. In order to get optimum pressure recovery it is promising to apply 90° 2-D turning diffuser of AR=1.2 for high $Re_{in} > 1.40 \times 10^5$ and AR=4.0 for low $Re_{in} < 1.40 \times 10^5$.

Acknowledgements

This work was financially supported by the Faculty of Mechanical and Manufacturing Engineering (FKMP), Universiti Tun Hussein Onn Malaysia (UTHM). Immeasurable appreciation is extended to Head of Center for Energy and Industrial Environment Studies, Dr. Mohd Faizal Mohideen Batcha for all provided research supports. A sense of gratitude is extended to all Flow Analysis Simulation and Turbulence Research Group (FAST) members for the insight sharing. Also thanks to Mr. Zainal Abidin Alias (Assistant Engineer of PIV Laboratory, UTHM) for the technical-lab assist.

References

[1] El-Askary W A and Nasr M 2009 Performance of a bend diffuser system: Experimental and numerical studies *Computer & Fluids* **38** 160-170
[2] Fox R W and Kline S J 1962 Flow regime data and design methods for curved subsonic diffusers *J. Basic Eng. ASME* **84** 303-312
[3] Gan G and Riffat S B 1996 Measurement and computational fluid dynamics prediction of diffuser pressure-loss coefficient *Applied Energy* **54**(2) 181-195
[4] Nordin N, Karim Z A A, Othman S and Raghavan V R 2014 The performance of turning diffusers at various inlet conditions *Applied Mechanics and Materials* **465-466** 597-602
[5] Chong T P, Joseph P F and Davies P O A L 2008 A parametric study of passive flow controlfor a short, high area ratio 90 deg curved diffuser *J. Fluids Eng.* **130**
[6] Sullerey R K, Chandra B and Muralidhar V 1983 Performance comparison of straight and curved diffusers *J. of Def. Sci.* **33** 195-203
[7] Shuja S Z and Habib M A 1996 Fluid flow and heat transfer characteristics in axisymmetric annular diffusers *Computer & Fluids* **25** 133-150
[8] Nordin N 2016 Performance investigation of turning diffusers at various geometrical and operating parameters *PhD Thesis Department of Mechanical Engineering* (Bandar Seri Iskandar: Universiti Teknologi PETRONAS)
[9] Nordin N, Karim Z A A, Othman S and Raghavan V R 2013 Design & development of low subsonic wind tunnel for turning diffuser application *Advanced Materials Research* **614-615** 586-591

IOP Conf. Series: Materials Science and Engineering **243** (2017) 012030 doi:10.1088/1757-899X/243/1/012030

Study on airflow characteristics of rear wing of F1 car

A R S Azmi[1], A Sapit[1], A N Mohammed[1], M A Razali[1], A Sadikin[1], N Nordin[1]
[1]Flow Analysis, Simulation and Turbulence Research Group, Faculty of Mechanical and Manufacturing Engineering, Universiti Tun Hussein Onn Malaysia, Johor, Malaysia

E-mail: ar.salami94@gmail.com

Abstract. The paper aims to investigate CFD simulation is carried out to investigate the airflow along the rear wing of F1 car with Reynold number of 3 x 10^6 and velocity, u = 43.82204 m/s. The analysis was done using 2-D model consists of main plane and flap wing, combined together to form rear wing module. Both of the aerofoil is placed inside a box of 350mm long and 220mm height according to regulation set up by FIA. The parameters for this study is the thickness and the chord length of the flap wing aerofoil. The simulations were performed by using FLUENT solver and k-kl-omega model. The wind speed is set up to 43 m/s that is the average speed of F1 car when cornering. This study uses NACA 2408, 2412, and 2415 for the flap wing and BE50 for the main plane. Each cases being simulated with a gap between the aerofoil of 10mm and 50mm when the DRS is activated. Grid independence test and validation was conduct to make sure the result obtained is acceptable. The goal of this study is to investigate aerodynamic behavior of airflow around the rear wing as well as to see how the thickness and the chord length of flap wing influence the airflow at the rear wing. The results show that increasing in thickness of the flap wing aerofoil will decreases the downforce. The results also show that although the short flap wing generate lower downforce than the big flap wing, but the drag force can be significantly reduced as the short flap wing has more change in angle of attack when it is activated. Therefore, the type of aerofoil for the rear wing should be decided according to the circuit track so that it can be fully optimized.

1. Introduction

Aerofoil in motorsports are called wings where it can generate high downforce by having a high angle of attack, therefore increasing the drag of the aerofoil. Daniel Bernoulli deduced a formula proving that the total energy in steadily flowing fluid system is a constant along the flow path. An increase on the fluid's speed will resulting in decrease in its pressure. Adding up the pressure variation times the area around the entire body determines the aerodynamic force on the body [1].

Aerofoilsare mostly designed with more thickness on the lower side. As downforce generating, the lower airstream is slightly reduced in surface, hence increasing the flow speed and decreasing the pressure. On top of the wing, the airspeed is lower, thus the pressure difference will generate a downward force on the wing. More specifically, the shape of the wing will turn air upwards and change its velocity [1].

Drag Reduction System (DRS) is a new development that have been introduced in F1 in the year of 2011. The DRS is a system where the flap wing will be opened up when it is activated for overtaking aid. Drivers are free to activate the DRS as they wish within the designated DRS zones during practice and qualifying. However, if the DRS is disabled at any point during Q1, Q2 or Q3, it will remain disabled for the remainder of the relevant period. The rule for activating the DRS is the driver only may activate it when they are within one second of the car in front at the DRS detection point. The DRS will automatically be disabled that is resetting the rear wing flap to its original position the first time the driver uses the brakes after activation [2].

Computational Fluid Dynamic (CFD) is a sophisticated computationally-based design and analysis technique that uses computational technology that enables researcher to study the dynamics of things that flow. Using CFD, a computational model that represents a system or device can be build. Then the fluid flow physics and chemistry can be applied to this virtual prototype and the software will output a prediction of the fluid dynamics and related physical phenomena. Besides, CFD software can give the power to simulate flows of gases and liquids, heat and mass transfer, moving bodies, multiphase physics, chemical reaction, fluid-structure interaction and acoustics through computer modelling. A lot of time can be save that normally need to use for building prototype by using CFD [3].

CFD is broadly used in the racing industry to predict the downforce and the drag of race cars that would experience at high velocities. CFD provides numerical solutions to the governing equations of fluid dynamics throughout the desired flow region. It takes into account for complex problems to be solved without losing the integrity of the problem due to over-simplification. It is this ability to solve large problems that makes CFD an efficient tool for the automotive industry. CFD allows engineers to examine the airflow over an automobile or a particular part such as a wing or hood, and see the aerodynamic effect of changing the geometry of any particular area of the vehicle [4].

Therefore, simulation has been done to see how the thickness and chord length of the flap wing influence the drag force, downforce, lift and drag coefficient of the rear wing.

2. Methodology

2.1. Rear wing design

There are a total of 12 models that were simulated for three different thickness of NACA 2408,2412, and 2415 for the flap wing when it is close at 10mm gap and 50mm when it is activated. The difference in length of the flap wing also being tested namely big and short flap wing. The simulations has been ran using FLUENT solver with initially compared with three different types of turbulences model. The geometry of all 12 models were modeled in SolidWork software before imported into ANSYS. The geometry of the models for big flap is shown in the Table 1 while the Table 2 shows the one with short flap wing. In the case of big flap, the chord length for the main plane is 250mm while for the flap wing is 175mm, and for the case of short flap, the chord length of the main plane is 290mm while the flap wing is 120mm.

IOP Conf. Series: Materials Science and Engineering **243** (2017) 012030 doi:10.1088/1757-899X/243/1/012030

Table 1. Rear Wing Model with Big Flap Wing

NACA	Model	
	10mm	50mm
2408	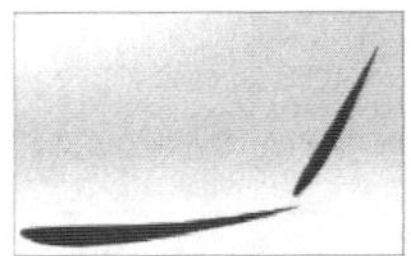	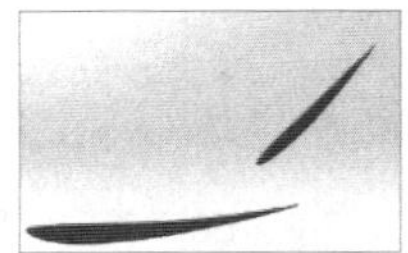
2412		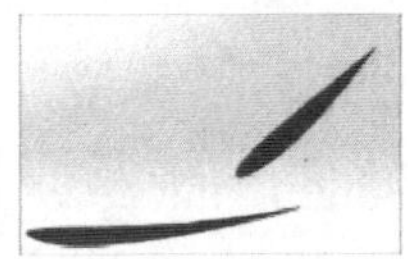
2415	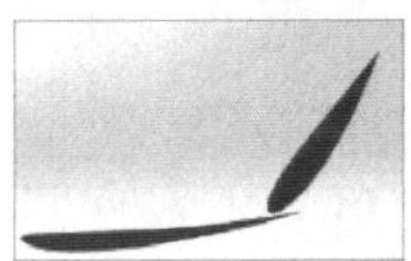	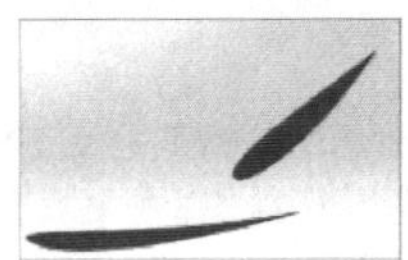

Table 2. Rear Wing Model with Short Flap Wing

NACA	Model	
	10mm	50mm
2408	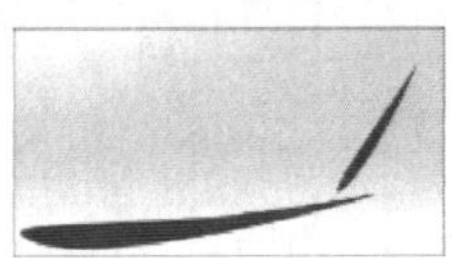	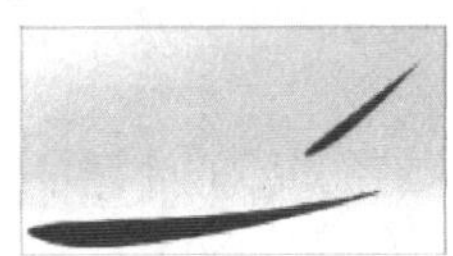
2412	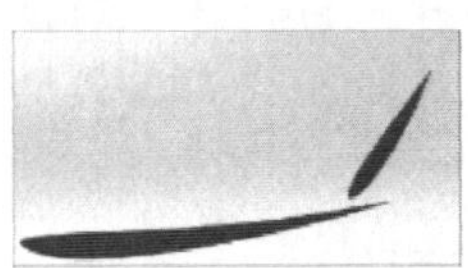	
2415		

IOP Conf. Series: Materials Science and Engineering **243** (2017) 012030 doi:10.1088/1757-899X/243/1/012030

2.2. Grid independence test (GIT)

Grid independence test has been conducted to see the accuracy of the mesh and the result obtained. The test has been done with three grids which started with coarse, medium, and then fine meshing. The medium is the refinement of coarse meshing and the fine is the refinement of the medium meshing. The element, nodes, and the result of lift coefficient value of those three grids are presented in the Table 3 below.

Table 3. NACA 2415 GIT

Grid type	Nodes	Elements	Iteration to converged	Lift coefficient, cl
Coarse	22899	22595	174	0.188771
Medium	46238	45760	211	0.195898
Fine	98762	98184	436	0.204577

2.3. Validation

In term of validation of the result, the single aerofoil of NACA 2415 have been chosen to compare the numerical result that have been ran using CFD FLUENT with the experimental result. The reason of validation with only single aerofoil was because there is hardly any experimental data for the rear wing of F1 cars as they are a closely guarded secret. Therefore, a single aerofoilhas been chosen for validation before conducting with the real cases. The validation of the data has been conducted to make sure that the data of this numerical research is on the right track by having small margin error when compared to the experimental result. The experimental data that had been compared with the numerical data was obtained from a book produced by the director of NASA [5]. The experimental data of the NACA 2415 is shown in the Figure 1 and Table 4 below.

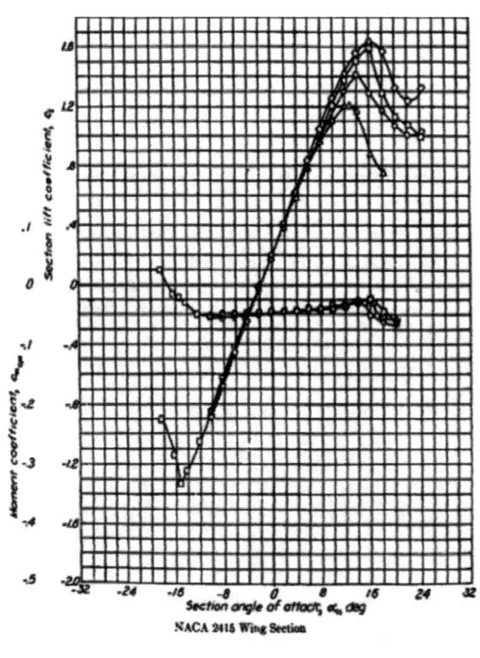

Figure 1. NACA 2415 experimental result

Table 4. Experimental data of NACA 2415 at Re=3x10^6

Experimental lift coefficient, cl	Experimental angle of attack, α
0.09193992713054255	-1.0179211469534124
0.1616102372700614	-0.3154121863799304
0.200682992	0
0.23620367901892836	0.28673835125447766
0.2860037323380431	0.8888888888888857

The Table 4 shows the experimental data of NACA 2415 with Reynolds number of 3X10^6. The numerical simulation of the NACA 2415 has been conducted at an angle of attack of 0 degree with three different types of turbulent models which are k-epsilon, k-omega, and k-kl-omega transition. The result was then compared with the 0 degree angle of attack experimental data which is 0.200682992. The comparison of the data is shown in the Table 5.

Table 5.Comparison of numerical and experimental data

Turbulence model	Experimental lift coefficient, cl	Numerical lift coefficient, cl	Error percentage
K-epsilon	0.200682992	0.18822	6.21
K-omega	0.200682992	0.21998	9.62
K-Kl- ω	0.200682992	0.20458	1.94

From the Table 5, k-kl-ωturbulence model has the smallest error percentage of 1.94For airfoils operating in the Re range of 10^6, the adverse pressure gradient is eliminated by turbulent flow at transition thus preventing separation. An increase in Re induces turbulence in the boundary layer, imparting high energy to oppose separation. From the result it can be seen that k-kl-ω is more reliable for predicting separation bubble formation, growth and reattachment for this cases. Therefore, k-kl-ω has been used to calculate all the cases.

3. Results analysis and discussion

3.1. Contour results

The pressure and velocity contour for rear wing of F1 car have been extracted from the post-processing for all models. The pressure and the velocity contour of both when the wings is at 10mm and 50mm are presented in the Table 6 and Table 7.

IOP Conf. Series: Materials Science and Engineering **243** (2017) 012030 doi:10.1088/1757-899X/243/1/012030

Table 6. Pressure contour for all models

NACA	Gap (mm)	Pressure contour	
		Big flap	Short flap
2408	10	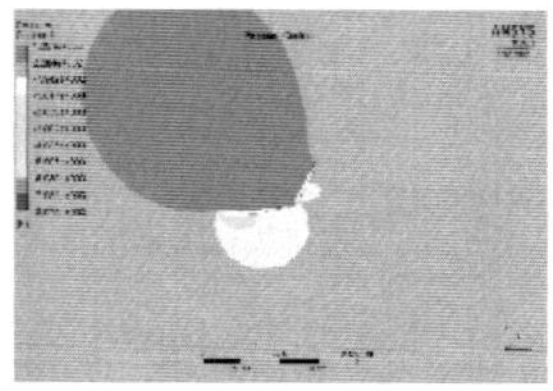	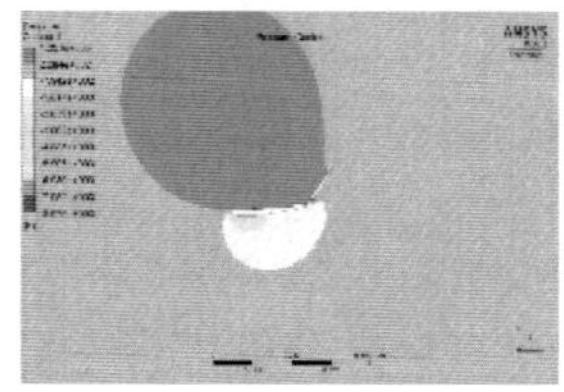
	50	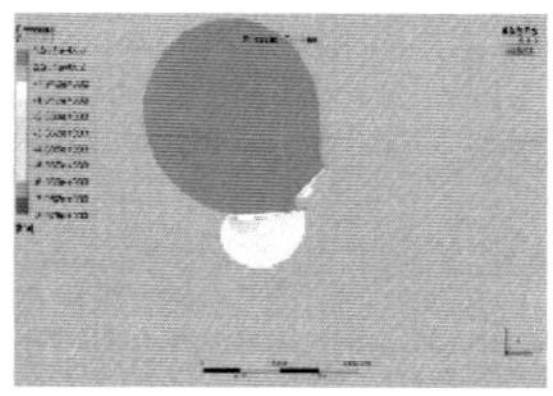	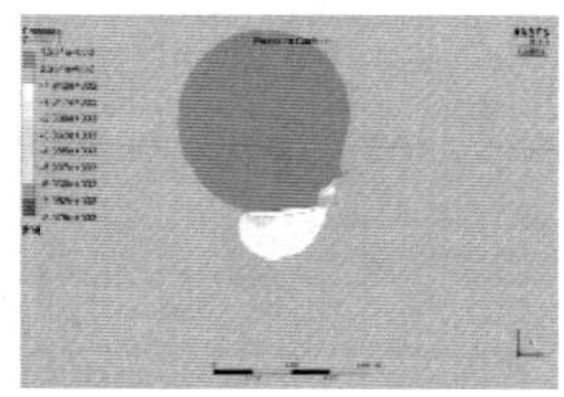
2412	10	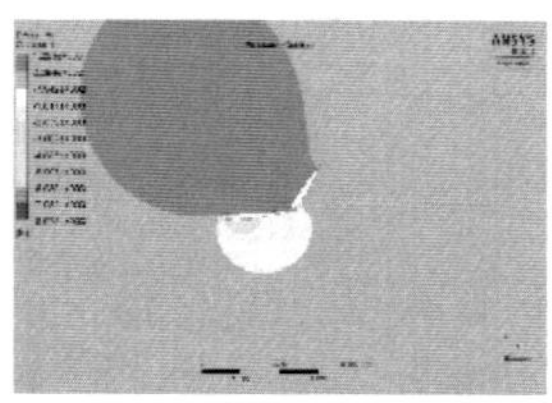	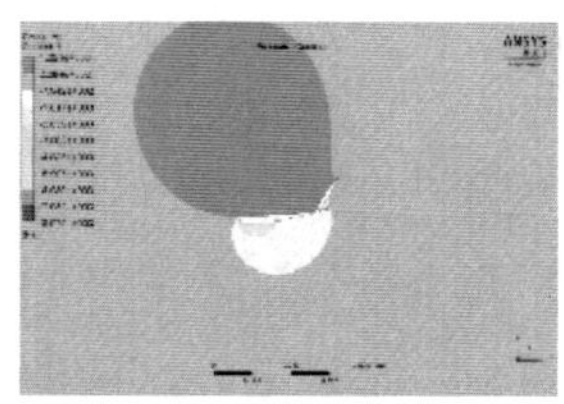
	50	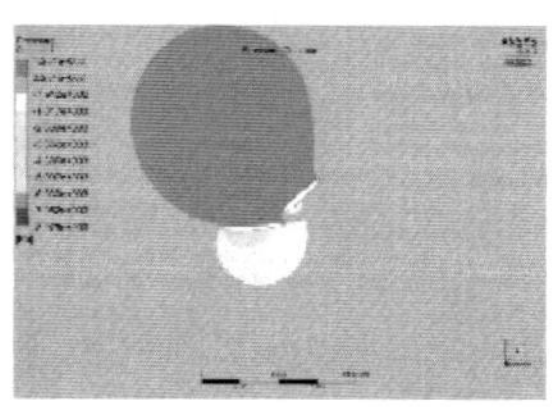	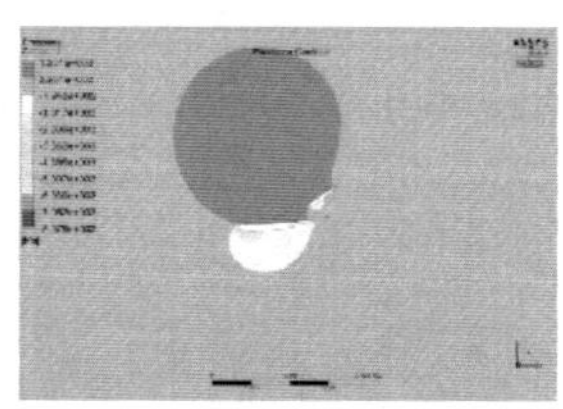
2415	10	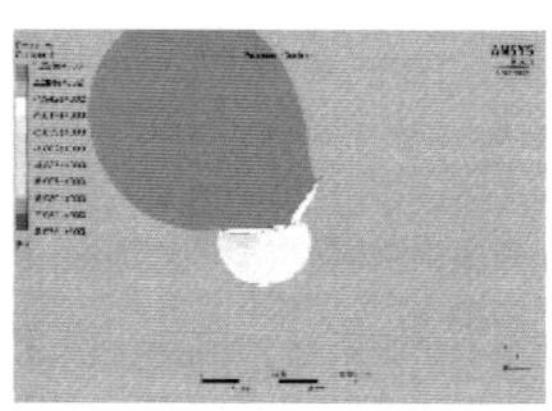	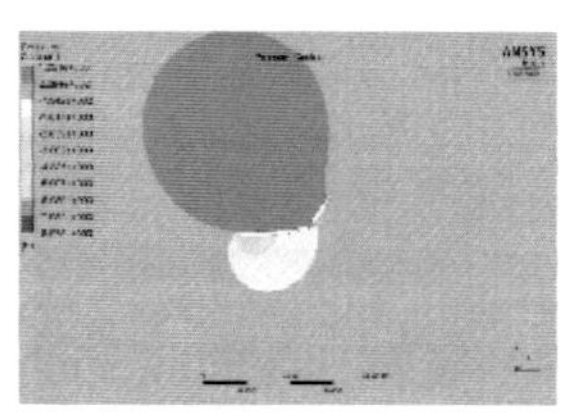
	50	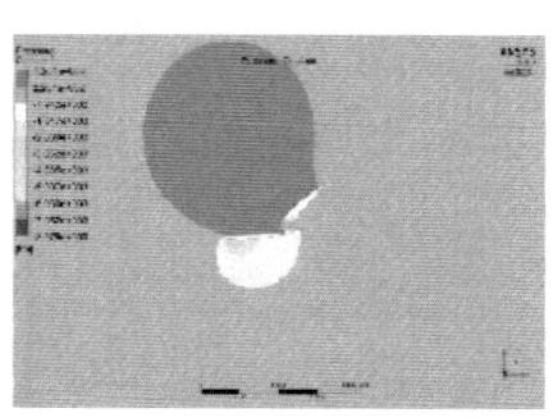	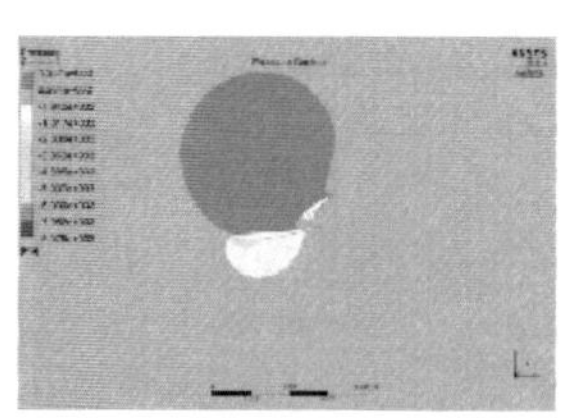

Table 7. Velocity contour for all models

NACA	Gap (mm)	Velocity contour	
		Big flap	Short flap
2408	10	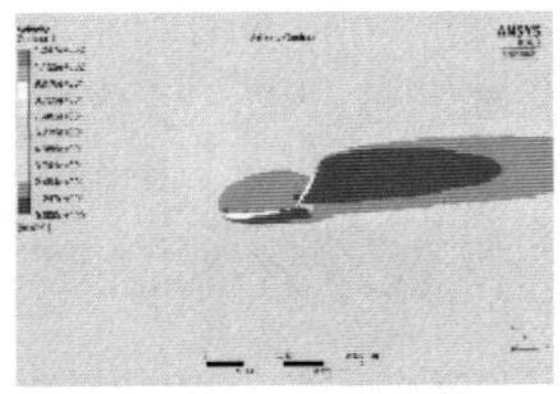	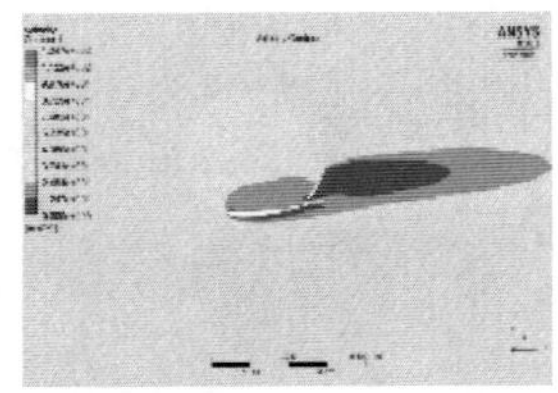
	50	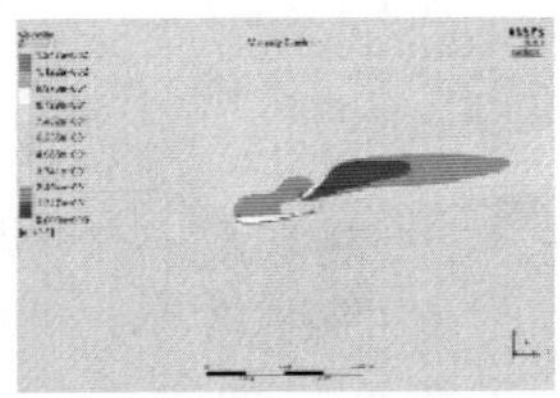	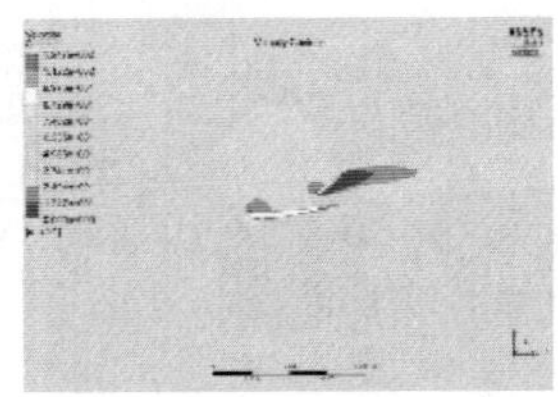
2412	10	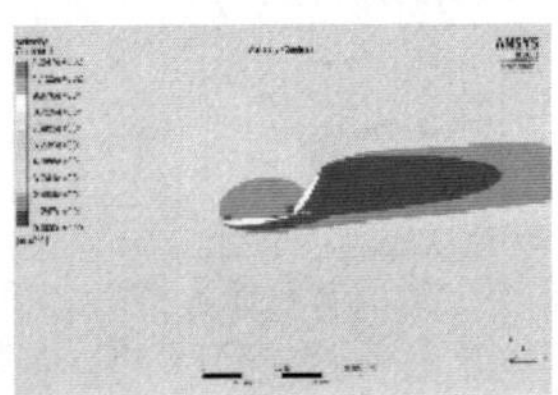	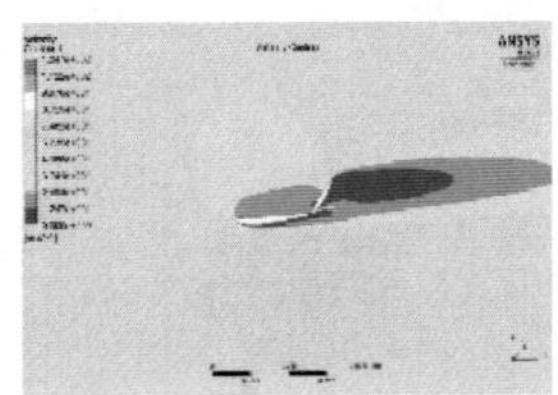
	50	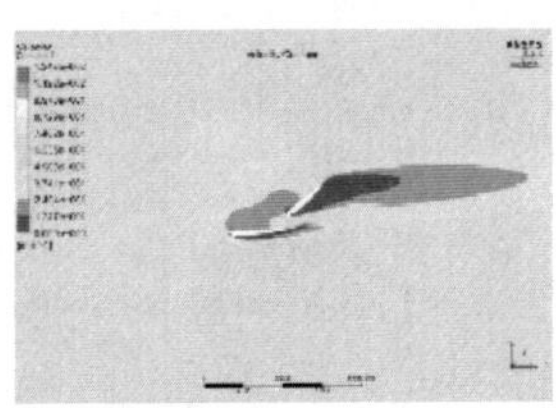	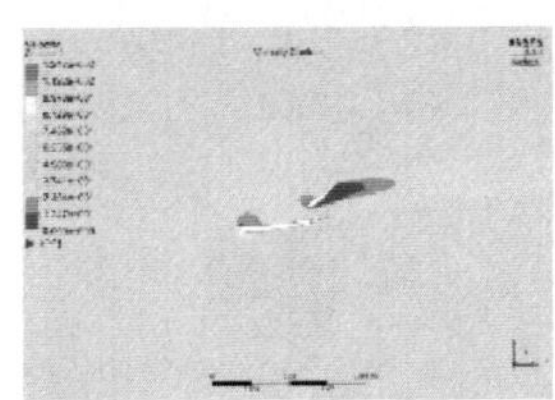
2415	10	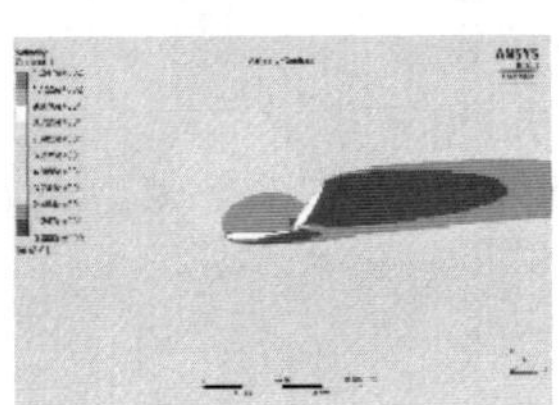	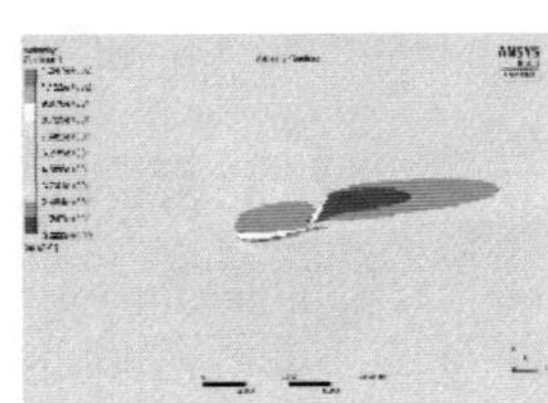
	50	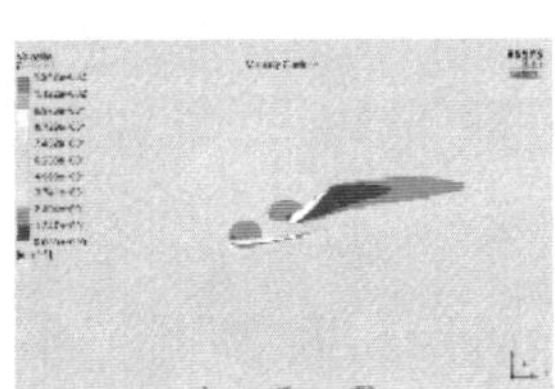	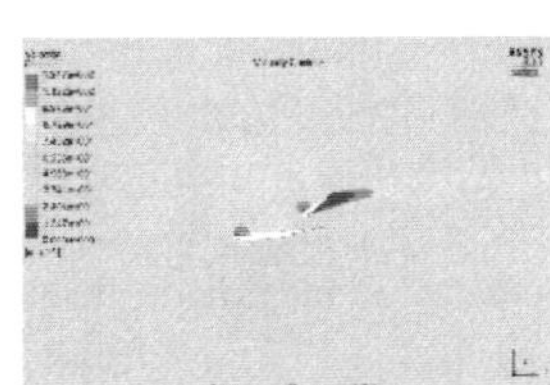

IOP Conf. Series: Materials Science and Engineering **243** (2017) 012030 doi:10.1088/1757-899X/243/1/012030

From the Table 6 and Table 7, it can be seen that the pressure on the upper side of the aerofoil have larger pressure than the lower side of the aerofoil due to the high velocity at the lower side of the aerofoil. The Bernoulli's principle stated that by increasing in the velocity will be resulting in the decreases in pressure. Therefore, a downward force was generated by the wings. It also can be seen that by having less thickness of aerofoil on the flap wing will generate more pressure on the rear wing thus will increase the downforce of the F1 car. When the Drag Reduction System (DRS) is activated, the pressure generated by the wings was reduce therefore resulting in increasing of velocity. This is because, when the flap wing was open with a gap of 50mm, it will allow more air to passed through and decreased the drag compared to the one with a gap of 10mm where a lot more surface exposed to stagnation.

3.2. Pressure against distance

In order to get clearer view of pressure distribution for all of the cases, the graph of pressure distribution versus distance along x-axis were presented in the Table 8 and Table 9 below. Each figure has two shape of pressure because there were two aerofoil involves. The bigger one represent the main plane of the rear wing while the smaller one represent the flap wing. The point for each surfaces is shown in The Figure 2 below.

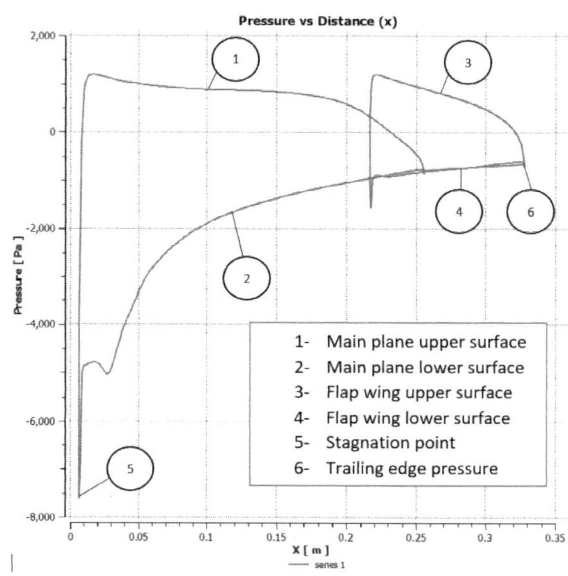

Figure 2. Pressure vs distance graph

IOP Conf. Series: Materials Science and Engineering **243** (2017) 012030 doi:10.1088/1757-899X/243/1/012030

Table 8. Velocity contour for models with 10mm gap

NACA	Gap (mm)	Pressure vs Distance graph	
		Big flap	Short flap
2408	10	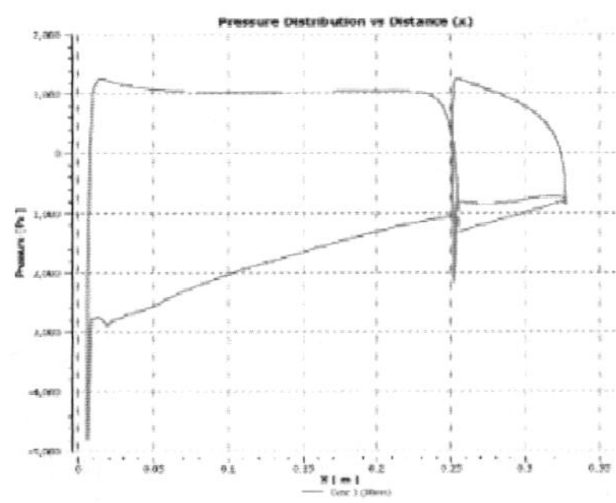	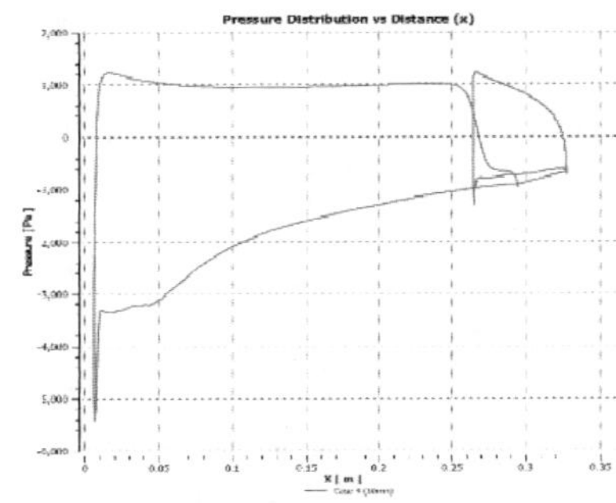
2412	10	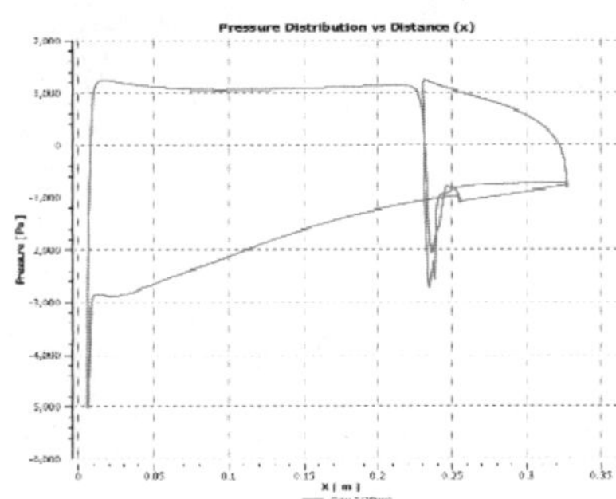	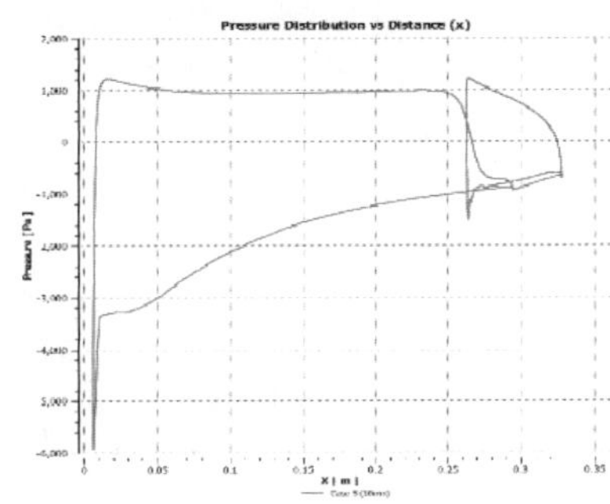
2415	10	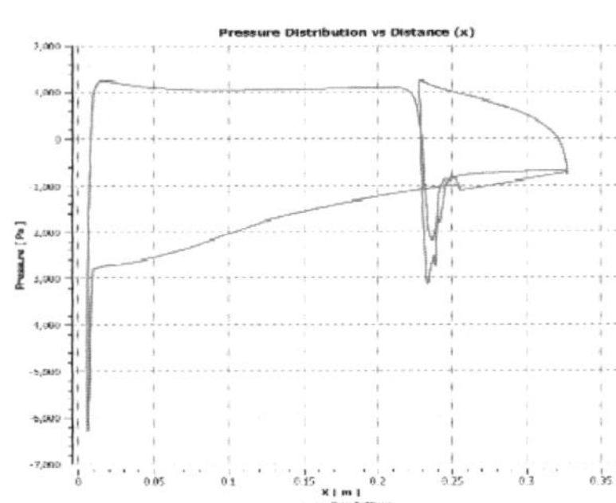	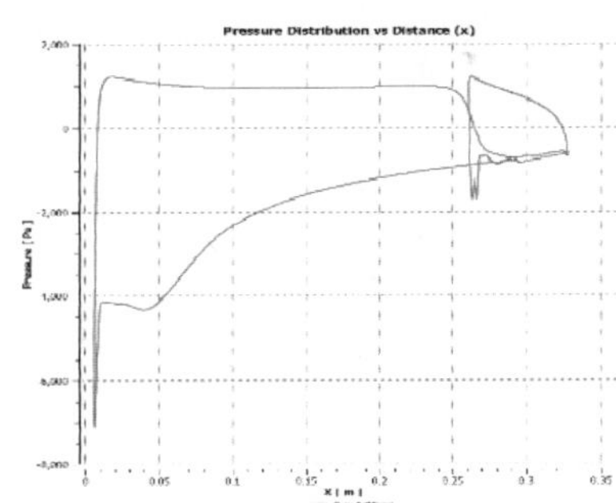

IOP Conf. Series: Materials Science and Engineering **243** (2017) 012030 doi:10.1088/1757-899X/243/1/012030

Table 9. Velocity contour for models with 50mm gap

NACA	Gap (mm)	Pressure vs Distance graph	
		Big flap	Short flap
2408	50	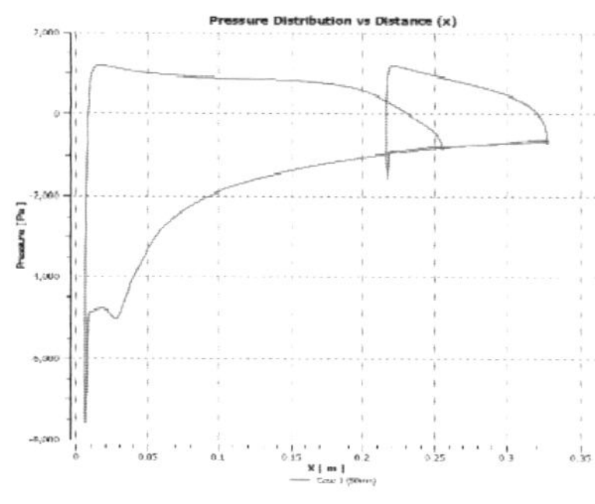	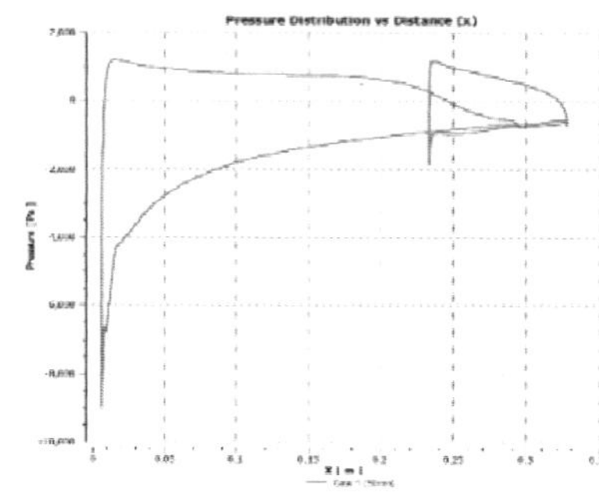
2412	50	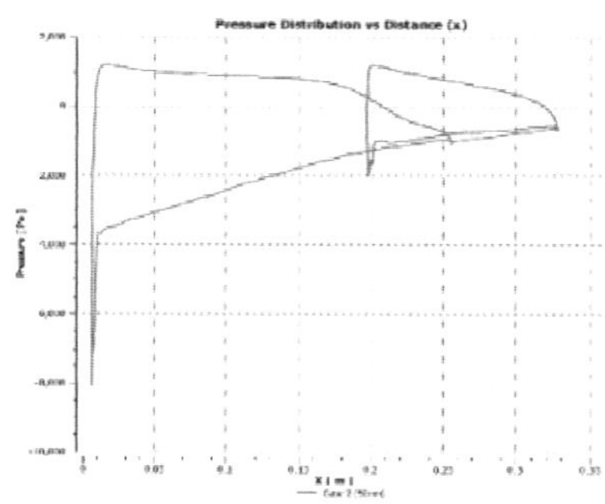	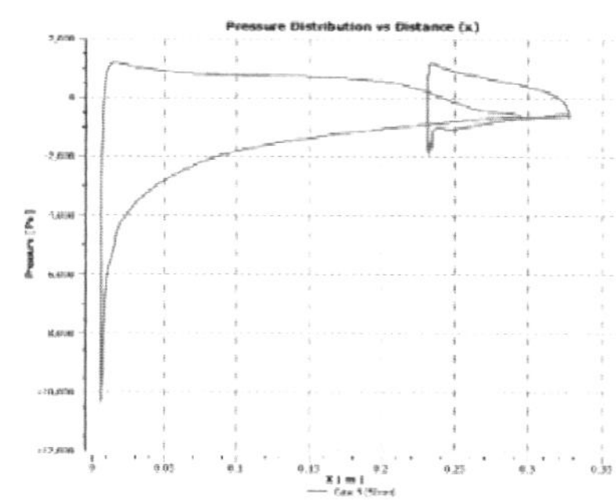
2415	50	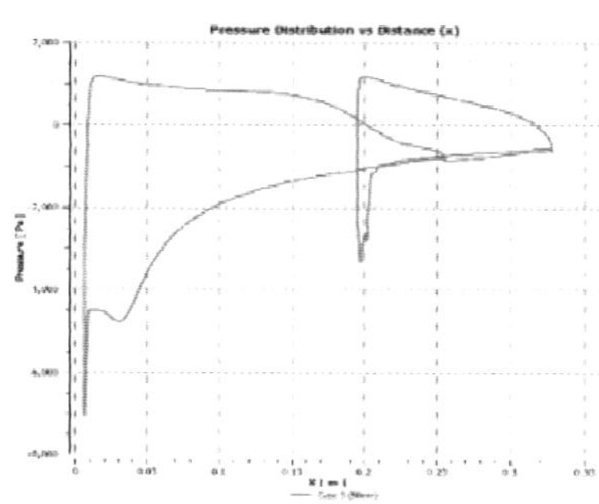	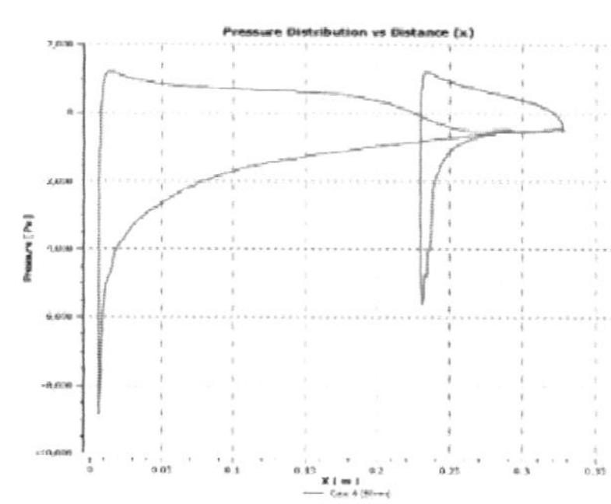

From the Table 8 and Table 9, we can see that the pressure on the lower surface is lower than the upper surface thus causing the higher pressure to press the lower pressure resulting in downward force. As the distance is increasing, the pressure is decreasing as it reached the trailing edge. The point 5 and point 6 is the stagnation point and the trailing edge pressure where it is occurred for both on the main plane and on the flap wing. The stagnation for the flap wing is higher when it is at 10mm because more surface area exposed to the wind.

3.3. Drag and lift forces

Drag and down force is an important factor that need to be taken into consideration when designing rear wing of F1 car. The results for the drag coefficient and force, lift coefficient and force, and the lift over drag ratio of the overall rear wing is presented in the Table 10 below.

Table 10. Forces and coefficient of all cases

Type of flap wing	NACA	Wing's gap (mm)	Drag force (N)	Down force (N)	Drag coefficient, Cd	Lift coefficient, Cl	L/D
Big	2408	10	246.73195	924.27946	0.2098	0.7859	3.746
		50	141.03082	865.79765	0.1199	0.7291	6.139
	2412	10	245.69891	872.072	0.2089	0.7415	3.549
		50	159.00314	849.73987	0.1352	0.7225	5.344
	2415	10	230.13479	857.43091	0.1957	0.7361	3.726
		50	129.17991	839.99689	0.10813	0.7142	6.503
Short	2408	10	176.06802	839.8714	0.1497	0.7141	4.77
		50	84.633378	799.84627	0.07119	0.6801	9.451
	2412	10	168.87273	838.98505	0.1436	0.7133	4.968
		50	67.08332	791.56129	0.057	0.673	11.8
	2415	10	145.51319	819.11694	0.1237	0.6964	5.629
		50	51.32626	745.62411	0.0436	0.634	14.816

From the Table 10, the NACA 2415 with short flap wing has the highest L/D for both gap of 10mm and 50mm. This shows that the NACA 2415 is the best model when compared to the other models as it generates more downforce in addition to less drag force. The yellow colour indicates the highest value for wings with gap of 10mm while the green colour indicates the highest value for 50mm wing's gap.

4. Conclusion

From the simulation using k-kl-omega turbulence model, it can be seen that the drag force, downward force, lift coefficient, and drag coefficient have been obtained. The pressure and velocity for all cases are presented in the form of contours and graphs. From Table 6 and 7, it can be seen that the thickness of the aerofoil for the flap wing influences the down force and the drag force of the rear wing module (combination of main wing and flap wing). Increases of thickness will decreases the downforce and drag force produced by the rear wing. From all the obtain results, it can be concluded that the short flap wing is favourable when facing a circuit that have more straight track. Although the short flap wing produced lower downforce when compared to the big flap wing, but when the DRS is activated, it can reduce the drag force significantly therefore will increases the speed of the F1 car.

References

[1] Katz J 2006 Aerodynamics of race cars *Annu. Rev. Fluid Mech.* **38** 27-63 https://doi.org/10.1146/annurev.fluid.38.050304.092016

[2] Formula 1® - The Official F1® Website 2017 Drag Reduction System. Retrieved from: https://www.formula1.com/en/championship/inside-f1/rules-regs/Drag_Reduction_System.html

[3] Versteeg H K and Malalasekera W 1995 *An introduction to computationnal Fluid Dynamics, The finite volume control, éd* (New York: Longman Scientific and Technical, John Wiley & Sons Inc)

[4] Chandra S 2011 CFD Analysis of PACE Formula-1 Car *Comput. Aided Des. Appl.* PACE **1** 1-14 https://doi.org/10.3722/cadaps.2011.PACE.1-14

[5] Abbott I H and Von Doenhoff A E 1959 *Theory of wing sections: Including a summary of airfoil data* (Mineola, New York: Dover publication)

Development of small scale cluster computer for numerical analysis

N H N Zulkifli[1], **A Sapit**[1], **A N Mohammed**[1]
[1] Flow Analysis, Simulation and Turbulence Research Group, Faculty of Mechanical and Manufacturing Engineering, Universiti Tun Hussein Onn Malaysia, Johor, Malaysia

E-mail: azwans@uthm.edu.my

Abstract. In this study, two units of personal computer were successfully networked together to form a small scale cluster. Each of the processor involved are multicore processor which has four cores in it, thus made this cluster to have eight processors. Here, the cluster incorporate Ubuntu 14.04 LINUX environment with MPI implementation (MPICH2). Two main tests were conducted in order to test the cluster, which is communication test and performance test. The communication test was done to make sure that the computers are able to pass the required information without any problem and were done by using simple MPI Hello Program where the program written in C language. Additional, performance test was also done to prove that this cluster calculation performance is much better than single CPU computer. In this performance test, four tests were done by running the same code by using single node, 2 processors, 4 processors, and 8 processors. The result shows that with additional processors, the time required to solve the problem decrease. Time required for the calculation shorten to half when we double the processors. To conclude, we successfully develop a small scale cluster computer using common hardware which capable of higher computing power when compare to single CPU processor, and this can be beneficial for research that require high computing power especially numerical analysis such as finite element analysis, computational fluid dynamics, and computational physics analysis.

1. Introduction

Cluster computer can be understand as a set or a number of computers that connected to work together and can be form as a single system. As stated in previous research, cluster computer has better performance than single unit computer from time consuming, data storage, data processing, and many more [1]. Therefore, when these computers act as one system, it can be used to solve a problem much faster compared to single unit computer. A study was conducted and the aim of this study is to develop a cluster computer in small scale which involved two units of processors that multicore [2]. It contains four cores each, thus make this whole cluster of two computers have eight processor. These together will form a single cluster unit that will have greater processing power. The cluster computer will provide larger storage capacity, high speed of processing and wider availability of resources. Therefore, this cluster computer can calculate and generate result in a shorter time when compared to single unit computer.

Since numerical analysis require much time especially on difficult and large area problems, time is an important resource. All of this work will surely take longer time to be complete without having access to high computing power. Dedicated custom high performance computer is expensive. Such the idea of using currently access able computing power (available desktop) and by networking and clustering it together, we can develop a system of clustered computer that can combine their computing power, without any expensive cost [1]. Therefore, two main objectives of this study are to develop the cluster computer system and to evaluate the performance of parallel computing system.

The developement of this system will use the ready existing technology with the open source software LINUX installer and the parallelization by MPI program [3]. The writing of the coding will be done by using FORTRAN software. The performance of the system will be test by solving a Laplace Equation in a Square Domain by Jacobi Iteration problem as the benchmarking.

2. Methodology

In order to build a parallel computer, a few steps are needed. Here, the procedure or the steps are divided into five parts which start from Ubuntu installation until the testing part at the end. In this project, both computers are in the same brand, same system and processor type. One of these computers has to become the master and the other one is slave.

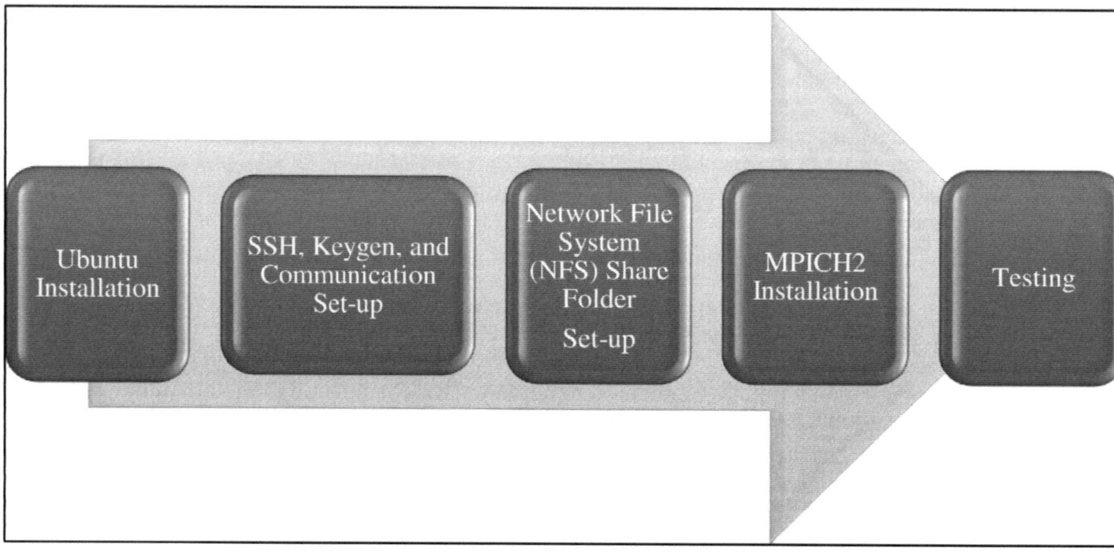

Figure 1. Summary of procedure to set-up parallel computer.

All the components were set up and combine to form a complete set of computers with two monitor, 2 multi-core processors and LAN connection using network switch. This setup is currently located at the CEIES Building (Block C19), University Tun Hussein Onn Malaysia (UTHM). In this research, the software's used are Ubuntu, MPICH2, and FORTRAN. Ubuntu is the operation system (environment), while MPICH2 is a freely available, portable implementation of MPI, a standard for message-passing for distributed-memory applications used in parallel computing. Also, FORTRAN is a well-known programming language.

The MPICH2 will help the Ubuntu or the system to distribute the works or loads to the processors involve. All the distributions are by itself. In order to build a parallel computer, a few steps are needed. Here, the procedure or the steps are divided into five parts which start from Ubuntu installation and will have testing part at the end. All the procedures were followed when the hardware set-up was ready at the working station. In this project, both computers are in the same brand, same system and processor type. Since these computers were from the vendor, both computers are named as UTHM-HP-ProDesk-400-G2-MT from the beginning. One of these computers has become the master and the other one is slave.

The key to parallel processing especially in this case which is the Message-Passing Interface or commonly called as MPI is a message-passing library interface specification. It is stated here that the MPI is specification; which mean it is not an implementation and also not a language. In message-passing, the data is moved from the address space of one process to another process through cooperative operations on each process. Since MPI is specification, this specification is for a library interface and all of its functions are expressed as functions, subroutines, or methods. C, C++, and FORTRAN are the appropriate language binding and are part of MPI standard [4].

MPI is used across the entire spectrum of parallel machines that from small Linux Cluster to the largest parallel machines in the world such as IBM Blue Gene/L and Cray XT3. It has existed for a long time and also has many main advantages such as portability and ease of use. It can provides the vendors with a clearly defined base set of routines that can implement efficiently, or in some cases for which they can provide hardware support, thereby enhancing scalability. Previously, it is stated that MPI is not implementation; so there are numbers of MPI implementations includes MPICH,MPICH2, MVA-PICH, MVAPICH2, LAM, and Open MPI [5].

3. Initial setup and testing

Before the SSH, keygen, and communication setup, the computers were already networked and can communicate with each other validated through "ping" command. The SSH, Keygen, and communication allowed the master computer to log into the slave without any password. The setup start from the update and then downloading the OpenSSH-server.

After the SSH setting, the procedure proceed with NFS Share Folder Setup and MPICH2 setup. Both of this setup had done successfully without any problem. The setting of NFS can be check by using command "ls –a"; which then all files from master that had been mirror and shared to slave are display. In addition, the FORTRAN's compiler also being downloaded and install after MPICH2 setup. When the MPICH2 setup is done, the cluster computer system is ready completely. The system communication is test by using simple Hello World program which utilize MPI capability.

Figure 2. Result on terminal after running MPI Hello World program

The system is then tested to compare the performance of the cluster computer compare to single computer. The benchmarking was done by solving Laplace Iteration of Temperature (Steady Heat) Case that applied MPI subroutine in order to make use of the cluster capability. The code was the run by varying the numbers of processor available for calculation, (1, 2, 4 and 8). The first test used single processor to run the coding. Then, the number of processor used were doubled each time, until all eight processors are call and used by the MPI. From the time taken to complete the calculation, the performance of the system then analyse..

4. Results and discussion

Table 1 shows the results of time needed to solve the code. When the number of processor used increased, the total local nodes were decrease. The number of total nodes decreased exactly half when the number of processor is doubled. This result shows that the increasing number of processor (doubled) in the system when running the code make the work done by one processor decrease significantly.

IOP Conf. Series: Materials Science and Engineering **243** (2017) 012031 doi:10.1088/1757-899X/243/1/012031

Table 1. Result for the performance

NO OF PROCESS OR (npes)	TOTAL TIME TAKEN (seconds)	TOTAL LOCAL NODES	TOTAL NX_LOCAL (x axis)	TOTAL NY_LOCAL (y axis)
1	378.2996	262144	512	512
2	189.4584	131072	512	256
4	102.0725	65536	512	128
8	93.9772	32768	512	64

Also, the total time taken simultaneously decreased when the number of processor increased. This proved that the cluster computer shorten the time needed for calculation when the processor used were increased, and in this study, doubled. It can be state that with each doubling of computer processor used in the calculation process, time needed reduced about 45%-50%. This is true until the reduction in time is becoming less significant, shown in the case of 8 processors.

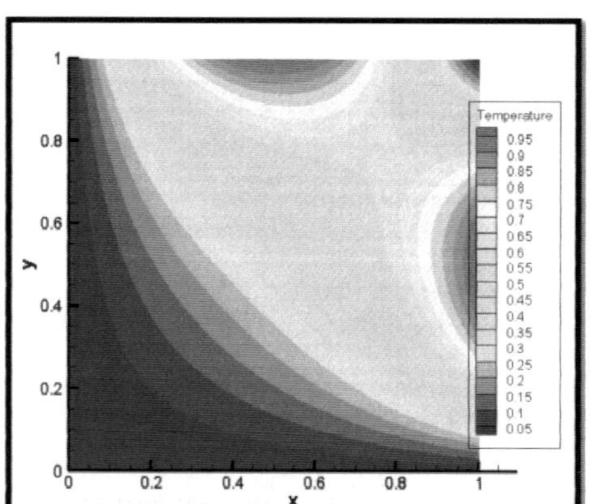

(a) Contour graph from result from 1 processor

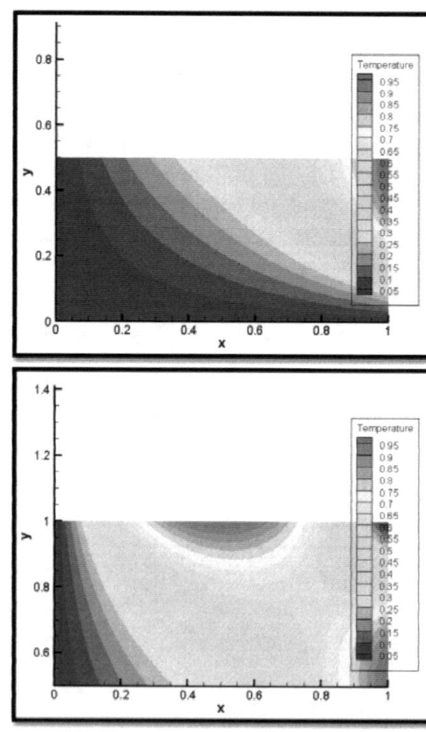

(b) Contour graph from result from 2 processors

Figure 3. Contour graph made using Techplot from the output which indicate the domain and result of the calculation for each processors

IOP Conf. Series: Materials Science and Engineering **243** (2017) 012031 doi:10.1088/1757-899X/243/1/012031

From previous research and study, researcher believes that the time taken for cluster computer when doubled the processor numbers, do not perfectly reduced to half is due to overhead from the network traffic.[3] In addition, the 0-5% of the time loss maybe used for the processors, cores, and nodes communicate for distribution of information by MPICH2. Even with 0-5% of overhead from the network traffic or communication, cluster computer still proved to have higher performance and higher calculation capability when compared to single computer. This result is strengthened by the result from four and eight processors that showed the decreasing in time taken.

The contour of single processor showed that the processor run all of the calculation itself hence the result is a contour graph. However, when 2 processors are used, the contours are then divided into 2 sections, 1 section for every each processor. This is achieved due to the MPICH2 which utilize the processor according to the number of processor that user stated in command.

5. Conclusion

Based on this study, cluster computing proved to be beneficial in by providing extra computing power by harnessing collective computing power of individual computer. This provide the means of low cost high performance computing without the need of specialized custom made expensive super computer. Result also shown that works are distributed among the processors that being utilized according to the command. This is achievable by the method of MPI implementation that will make the system utilize the processors and distribute the work.

From the results, it is also shown that doubling the number of processors will reduce the calculation time to about half. Even with 0-5% of overhead from the network traffic and other factors, cluster computer still proved to have higher performance and higher calculation capability when compared to single computer. As the conclusion, the objectives of this study were successfully achieved and accomplished.

References

[1] Pfalzgraf A M and Driscoll J A 2014 A low-cost computer cluster for high-performance computing education *IEEE International Conf. on Electro/Information Technology (Milwaukee, WI, USA, 5-7 Jun 2014)* (Miwaukee: IEEE) pp 362-366 doi: 10.1109/EIT.2014.6871791

[2] Chai L, Gao Q and Panda D K 2007 Understanding the impact of multi-core architecture in cluster computing: a case study with intel dual-core system *Seventh IEEE International Symposium on Cluster Computing and the Grid (CCGrid '07)* (Rio De Janeiro: IEEE) pp 471-478. doi: 10.1109/CCGRID.2007.119

[3] Bierbaum N 2002 MPI and embedded TCP/IP Gigabit Ethernet cluster computing *Proc. LCN 2002, 27th Annual IEEE Conference on Local Computer Networks* (Tampa, FL, USA: IEEE) pp 733-734. doi: 10.1109/LCN.2002.1181853

[4] Pjesivac-Grbovic J, Angskun T, Bosilca G, Fagg G E, Gabriel E and Dongarra J J 2005 Performance analysis of MPI collective operations *19th IEEE International Parallel and Distributed Processing Symposium* (Denver, CO, USA: IEEE) p 8 doi: 10.1109/IPDPS.2005.335

[5] Gabriel E, Fagg G E, Bosilca G, Angskun T, Dongarra J J, Squyres J M, Sahay V, Kambadur P, Barrett B, Lumsdaine A and Castain R H 2004 *Open MPI: goals, concept, and design of a next generation MPI implementation* In Recent advances in parallel virtual machine and message passing interface vol 3241 ed D Kranzlmüller, P Kacsuk and J Dongarra (Berlin, Heidelberg: Springer) doi: 10.1007/978-3-540-30218-6_19

A study on a swirl type multi-hole fuel injector for a PFI system to improve air-fuel mixture formation

A J Alimin[1], Y KuangHe[2], A Sapit[3] and H Selamat[4]
[1]Head, Automotive Powertrain Research Group (ARG), Faculty of Mechanical & Manufacturing Engineering, UniversitiTun Hussein Onn Malaysia, 86400 Parit Raja, BatuPahat, Johor, Malaysia
[2] Bachelor Degree Project Student, Faculty of Mechanical & Manufacturing Engineering, UniversitiTun Hussein Onn Malaysia, 86400 Parit Raja, BatuPahat, Johor, Malaysia
[3] Principal Researcher,Flow Analysis, Simulation, and Turbulence Research Group (FAST), Faculty of Mechanical & Manufacturing Engineering, UniversitiTun Hussein Onn Malaysia, 86400 Parit Raja, BatuPahat, Johor, Malaysia
[4] Director, Centre for Artificial Intelligence and Robotics (CAiRO), UniversitiTeknologi Malaysia, 81310 Skudai, Johor, Malaysia

E-mail: ajais@uthm.edu.my

Abstract. This paper described the development and simulation works thatwere conducted on an injector that has a combined swirl spray and multi-hole injection functions. It was expected that those two functions would provide beneficial effects to the fuel injector, in order to provide better fuel and air mixing, especially for Port Fuel Injection (PFI) application. The combination of both functions was aimed for achieving high quality spray formation, which might be useful in cases when the combustion time is limited. Swirl spray pattern is known for fine droplets production with relatively smaller SMD and bigger coverage area in the combustion chamber. On the other hand, a multi-hole injector has the advantages of simplicity in manufacturing and operation. Preliminary results on the simulation works that were done separately, demonstrated the ability of the injector's design to produce swirl spray as well as injecting near-symmetrical sprays from the four holes at the injector's end. Nevertheless, these findings must be improved further in the future, especially on simulating the benefits of having a combined swirl and multi-hole functions inside a single injector. Furthermore, detailed assessment on manufacturability and cost effectiveness of this injector must also be addressed.

1. Introduction

According to BP 2016 Statistical Review of World Energy [2] energy consumption for emerging economies accounted for up to 97% of the increase in global energy consumption. Notable emerging economies countries [3] are forecasted to have increase access to transportation: either for commercials or personal usage [1]. As light duty vehicles such as motorcycles and compact cars are among the primary types of vehicles for personal transport, works on improving the fuel economy and reducing hazardous emissions for these types of engines and vehicles are always relevant and important.

A good air-fuel mixture is imperative for achieving efficient combustion for better fuel economy and emissions control, especially for a port fuel injection (PFI) system where the fuel delivery occurs a distance away from the intake port. The quality of the mixture is highly influenced by the injection characteristics as well as the quality of mixing with the entrained air. For a PFI system, spray coverage and penetration are known to be the issues associated with conventional fuel injector. In this paper, the findings from a preliminary CFD work on the operation of a fuel injector with a swirl effect nozzle, which is designed to be used with small engines (capacity below 250cc),are presented, concerning the injector's fuel delivery characteristics: spray and penetration.

2. Literature review

The conventional understanding of spray formation are described by Sazhin (2014) [6] to follow these stages: 1) development of a jet;2) conversion of a jet into liquid sheets and ligaments;3) disintegration of ligaments into relatively large droplets (primary break-up) and 4) break-up of large droplets into smaller ones (secondary break-up). The successful transformation achieved by the injected fuel, upon leaving the injector nozzle, in accordance to the abovementioned stages will ensure the achievement of an efficient combustion.

Prior to that,Mitroglou*et al.*(2006)[4] also emphasised the importance of also assessing the quality of spray structure, tip penetration and fuel droplets distribution, velocities and diameters as a function of nozzle design, as well as the value of the fuel delivery pressure. In their work, on the spray characteristic of a multi holes injector under different spray pressures, they found that the spray angle remains constant and is almost independent of injection and chamber pressures. In addition, they deduced that injector with a multi-hole and central hole will produce uneven flow distribution and consequently created over-penetration and unstable spray pattern.

For port fuel injection engines, mixture formation process starts with the injection of liquid fuel into the intake ports. The fuel droplets interact with the air inside the intake manifolds, as well as intake port walls and intake valves, creating fuel vapour and wall films. Therefore, injector design and intake port geometry are also influential to the fuel injection spray quality [5]. Their findings showed that, as the fuel pressure increased, both jet velocity and spray cone angle increased and as a result it makes the diameter of droplets much smaller and a better atomization was achieved.A better understanding of spray characteristics of port injector is also very important in matching spray parameters with different engine designs and targeted operations.

Recent researches, studies and experiments on the mixture formation and spray pattern are mostly done by using numerical analysis and computational fluid analysis as these methods are more detailed and provide more complete data for analysis. By carrying out the analysis using the correct operating boundaries and practical geometries, injection spray and mixture formation quality can be assessed and further recommended for actual physical experiment could be recommended. For the current work described in this paper, the proposed fuel injector will function using a central injection hole surrounded with swirl channels.

3. Methodology

The study was initiated with the design of the new injector nozzle, which has a central hole with swirl channels using CAD Software, *Solidworks*. The basic injector design was based on combination of a single path flow with swirl paths, which will converge into four outlets at the end of the injector.Then, using FLUENT software, the fluid flow simulations inside the injector were carried out. Based on this preliminary CFD works, the initial design will be improved and upon achieving an optimum performance through simulations, a prototype will be fabricated for future physical experiments.

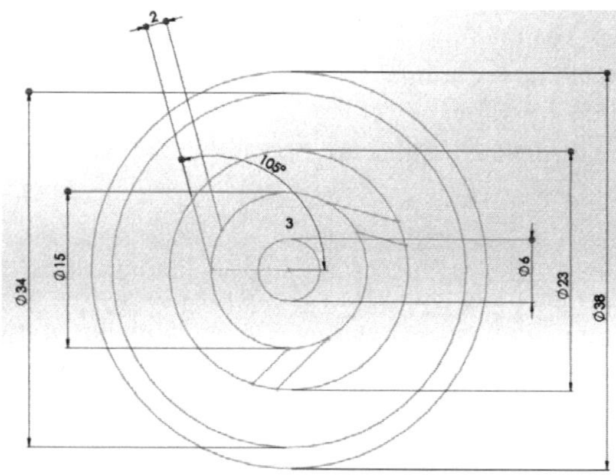

Figure 1. The injector: plan view's dimensions

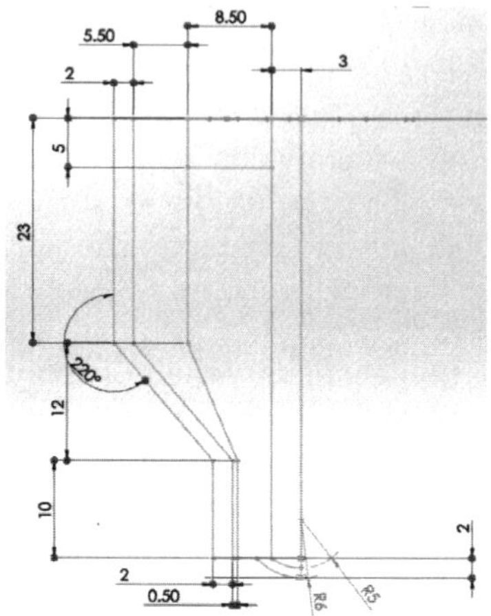

Figure 2. The injector: side view's dimensions

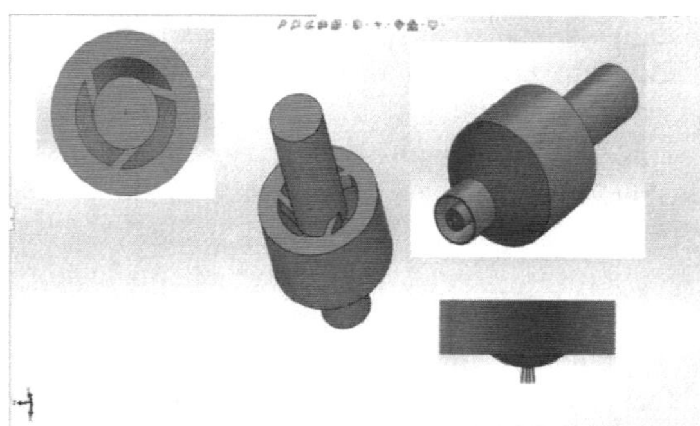

Figure 3. The multiple views of the reversed body of the injector representing the fluid inside

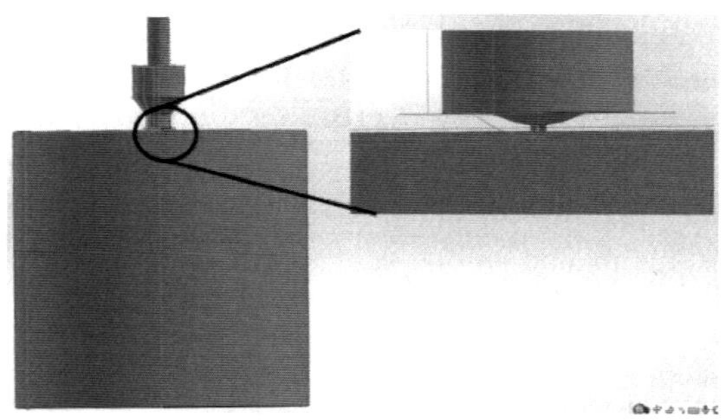

Figure 4. Incorporation of cylinder at the bottom of the injector

Figure 1 and figure 2 show the basic design of the swirl fuel injector and figure 3 depicted the reversed body of the injector representing the fluid inside it.Figure 4 shows the constructed cylinderin order to study the flow of the fuel after leaving the outlets, where the spray pattern can be observed. In this preliminary simulation, due to the difficulty in meshing, the simulations were divided into swirl-cylinder simulations and hole-cylinder simulations.The input and output boundary conditionswere pressure-based, set at 3 MPa and 0 MPa respectively; the operating temperature was set at room temperature, and k-epsilon was the turbulent model used. Although they were made separately, it will form as the basis for future simulation works on the combined spray system. In both simulations, air and n-octane (as liquid and vapour forms) were used as the working fluids. For the swirl path, the initial angle was 105°and was then improved to 90°, to eliminate the formations of stagnation points inside the swirl paths.

4. Results and discussion

Preliminary simulation results are shown in figure 5 and figure 6, for the swirling behaviours inside the injector leading to the outlets. Whereas in figure 7 and 8, the simulation results after the fuel exiting the four outlets are shown, from bottom and side views.

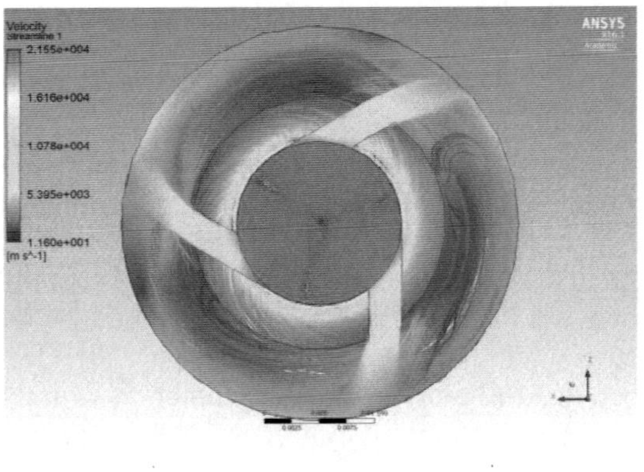

Figure 5. Plan view of the swirling flows with 90° angle.

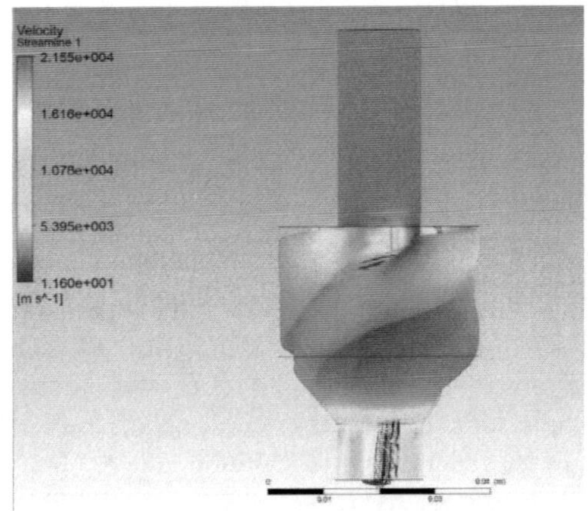

Figure 6. Side view of the swirling flows with n-octane as liquid.

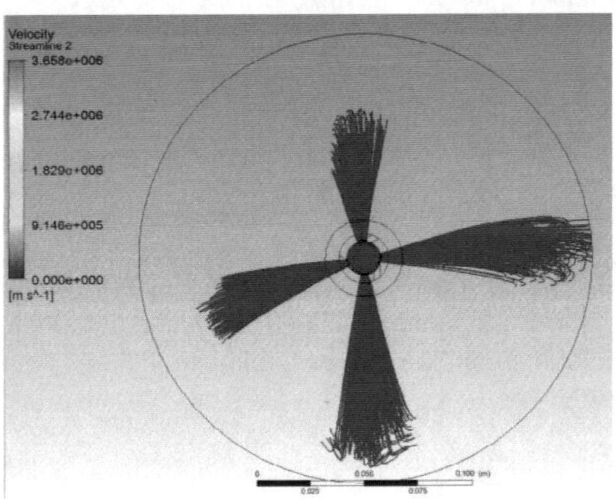

Figure 7. Simulation of the injected fuel from the four outlets – view facing the outlets.

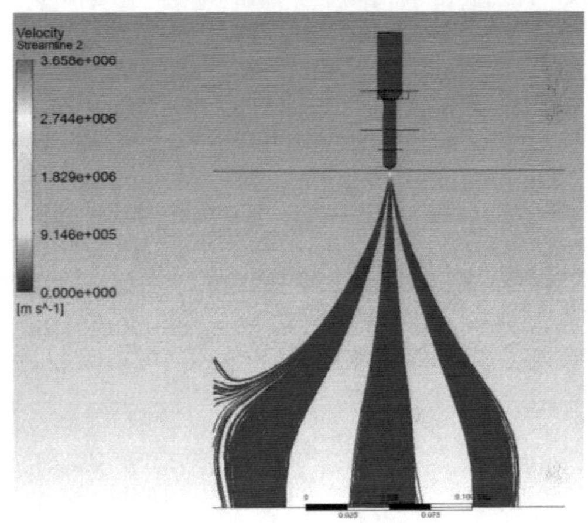

Figure 8. Side view of the injected fuel from the injector.

High velocity flows were observed inside the swirl paths and the selected angle were able to eliminate stagnation points in between the swirl paths. The flows encircled the single injection path and moving downwards towards the four injector's outlets. Nevertheless, the simulations were unable to simulate the combined flows from the swirl paths and the single injection path. High velocity streamlines were observed, as in figure 6, which could be due to the pressure difference between the inlet and outlet conditions, and they were too high in comparison to actual physical events. This observation will be subjected to further refinement in the future. In figure 7 and 8, the fuel flows exiting the four outlets were observed as expected. Even though the flows from the outlets are distinct and almost symmetrical, in accordance to the location of the outlets, the effects from having swirling paths inside the injector are not clearly demonstrated from these simulations.

The results obtained must be benchmarked in the future against simulations from the conventional multi-hole injector and also from swirl-only type injector. Those shortcomingswill need to be investigated for future works for this type of fuel injector. Moreover, the positive effects of combining the swirl paths with the single injection must also be examined closely, in order to ascertain the influential factors for better fuel mixing inside the combustion chamber. Another aspect that must be considered will be from the manufacturing perspective and the cost effectiveness of having this type of fuel injector as opposed to the conventional fuel injectors.

5. Summary

In this current work, a new configuration of fuel injector, which has a combined swirl spray and multi-hole functions had been successfully designed andinvestigated via simulations using CFD. Although the preliminary simulation works were carried out separately for the swirl paths flows and the single injection path flow, the results were useful to serve as the benchmarks for future works on this type of fuel injector. Swirling behaviour was observed and near-symmetrical flows were obtained at the outlets of the injector. It is recommended that for future works on this type of injector, the CFD simulations must attempt to combine and elucidate the benefits of having swirl spray function with the multi-hole injection. Furthermore, detailed assessment on manufacturability and cost effectiveness of this injector must also be addressed.

Acknowledgement

The authors would like to express their gratitude to researchers from Automotive Powertrain Research Group (ARG) and Flow Analysis, Simulation, and Turbulence Research Group (FAST), for the endless valuable support in conducting this work. The financial support from APT TOUCH SDN BHD is also acknowledged.

References

[1] Asif M and Muneer T 2007 *Renewable and Sustainable Energy Reviews*, **11**
[2] Dudley B 2016 *BP Statistical Review*
[3] Hoskisson R E, Eden L, Chung M L and Wright M 2000 *The Academy of Management Journal*, **43** 3
[4] Mitroglou N, Nouri J M, Gavaises M and Acroumanis, C 2006 *International Journal of Engine Research* **7** 3
[5] Movahednejad E ,Ommi F and Nekofar K 2013 *EPJ Web of Conferences*, **45**
[6] Sazhin S 2014 *Droplets and Sprays* (London: Springer)

IOP Conf. Series: Materials Science and Engineering **243** (2017) 012033 doi:10.1088/1757-899X/243/1/012033

Flow behaviour in normal and Meniere's disease of endolymphatic fluid inside the inner ear

Muhammad Sufyan Amir Paisal[1, a], Muhamad Azmi Wahab[1], Ishkrizat Taib[1, b], Norasikin Mat Isa[1], Yahaya Ramli[1], Suzairin Md Seri[1], Nofrizalidris Darlis[2], Kahar Osman[3], Ahmad Zahran Md Khudzari[3], Normayati Nordin[1]

[1]Flow Analysis, Simulation and Turbulence Research Group, Faculty of Mechanical and Manufacturing Engineering, Universiti Tun Hussein Onn Malaysia, Parit Raja, 86400 Batu Pahat, Johor, Malaysia

[2]Faculty of Engineering TechnologyUniversiti Tun Hussein Onn Malaysia, Parit Raja, 86400 Batu Pahat, Johor, Malaysia

[3]Faculty of Mechanical Engineering, Universiti Teknologi Malaysia, 81310 Skudai Johor, Malaysia

Email: muhammadsufyanamir@gmail.com [a], iszat@uthm.edu.my[b]

Abstract. Meniere's disease is a rare disorder that affects the inner ear which might be more severe if not treated. This is due to fluctuating pressure of the fluid in the endolymphatic sac and dysfunction of cochlea which causing the stretching of vestibular membrane. However, the pattern of the flow recirculation in endolymphatic region is still not fully understood. Thus, this study aims to investigate the correlation between the increasing volume of endolymphatic fluid and flow characteristics such as velocity, pressure and wall shear stress. Three dimensional model of simplified endolymphatic region is modeled using computer aided design (CAD) software and simulated using computational fluid dynamic (CFD) software. There are three different models are investigated; normal (N) model, Meniere's disease model with less severity (M1) and Meniere's disease model with high severity (M2). From the observed, the pressure drop between inlet and outlet of inner ear becomes decreases as the outlet pressure along with endolymphatic volume increases. However, constant flow rate imposed at the inlet of endolymphatic showing the lowest velocity. Flow recirculation near to endolymphatic region is occurred as the volume in endolympathic increases. Overall, high velocity is monitored near to cochlear duct, ductus reuniens and endolymphatic duct. Hence, these areas show high distributions of wall shear stress (WSS) that indicating a high probability of endolymphatic wall membrane dilation. Thus, more severe conditions of Meniere's disease, more complex of flow characteristic is occurred. This phenomenon presenting high probability of rupture is predicted at the certain area in the anatomy of vestibular system.

1. Introduction

Meniere's disease is an incurable vestibular disorder in the inner ear characterized by a triad of vertigo, tinnitus or ringing in the ear and sensorineural hearing loss by the Committee of Hearing and Equilibrium of American Academy of Otolaryngology [1]. Meniere's disease usually affects only one ear and attacks of dizziness may come suddenly or after a short period of muffled of hearing or tinnitus.

Meniere's disease is also a form of episodic endolymphatic hydrops that produces recurring set of symptoms as a result of abnormally large amount of a fluid called endolymph collecting in the inner ear. This endolymphatic hydrops is instability of pressure difference between endolymph and perilymph caused by the dysfunction of cochlear [2]. People who effected with the disease have extremely severe dizziness which also known as vertigo lead to loss of their balance and suddenly fall.

The build-up of excess fluid in the vestibular system of the inner ear, called endolymph is also one of the causes for the symptoms of Meniere's disease. As the excess endolymphatic fluid keeps increasing, the volume inside the membranous labyrinth increases. The membranous labyrinth is also called as endolymphatic wall membrane. The membranous labyrinth covers the endolymphatic fluid from the cochlear duct to ductus reuniens, utricle, endolymphatic duct, saccule, ampullae and semicircular canals as seen in figure 1 with black shaded region [3]. A high dilation of the membranous labyrinth is tending to break which allows the mixing of perilymph and endolymph [4]. This mixing is actually one of endolymphatic hydrop factor that cause injury to the hair cells producing characteristic symptoms such as vertigo, tinnitus and hearing loss [5].

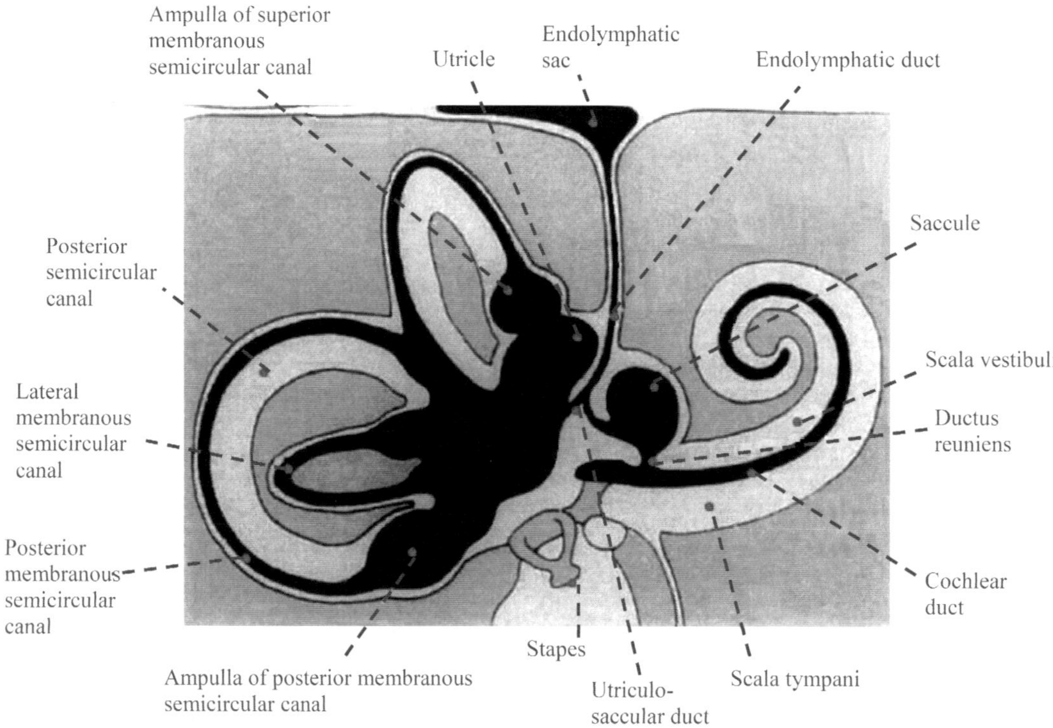

Figure 1. Schematic drawing of inner ear which endolymphatic fluid is black region, temporal bone is grey region and perilymph is white region [3].

Furthermore, the dilation of the endolymphatic wall membrane could be detected through the method of computational fluid dynamic (CFD) which is numerical analysis on flow characteristics. The main parameter of flow characteristic to determine critical area of high dilation probability is wall shear stress which is due to high velocity of the viscous endolymphatic fluid flow [6, 7]. The most commonly used commercial CFD software nowadays is ANSYS (Canonsburg, PA, USA) that can predict the flow characteristics of fluid flow. In 2012, Boselli et al did a study on vestibular system focussing on utricle and ampulla in determining the area with high probability of dilation. From the study, the geometrical model ellipticity of the cross-section plays a fundamental role in the shear stress distribution [8].

The existence of the CFD analysis could facilitate the medical practitioner in order to detect the critical area for treatment. Currently, the Meniere's disease still don't have a cure but there are some recommendation for the treatments that given by the doctor. This recommendation only helps the patient to control the disease. Doctor always prefers medication to control the disease because this way can relieve dizziness and shorten the attack. There is other treatment such as salt restriction and diuretics, cognitive therapy, injections, alternative medicine and most use is surgery. The surgery extremely effective and control vertigo in about 95% of operated patients. There is two way of surgery exist that is through ear canal and one more is through the mastoid bone behind the ear. The operation through the mastoid will removes dysfunctional balance organ and allows the tiny nerves from the balance organ been cut.

From scientist's research, an estimation ratio of six from ten people get better by their own and also control their vertigo with drug, diet and devices, and only some people get relief with Meniere's disease only by going surgery. Many researches are developing an ear device and studying about the relationship between endolymph volume and inner ear function in process of reduce or eliminate dizziness. Thus, the present study is established to investigate the correlation between the increasing volume of endolymphatic fluid and flow characteristics such as velocity, pressure and wall shear stress inside the three dimensional inner ear simplified model by using numerical analysis of CFD.

2. Methodology

2.1. The simplified model of vestibular system

Figure 2 shows the simplified model of vestibular system for all three cases which are normal (N), meniere's disease 1(M1) and meniere's disease 2 (M2). The models are drawn by using computer aided design software which is SOLIDWORKS (Dassault Systèmes Solidworks Corporation, Waltham, MA). M2 model is representing more severe disease than the M1 model. For each of the inner ear model, only seven components are focused in the endolymphatic region of the vestibular systems as shown in figure 3 and table 1. The dimensions of the model are based on the previous research that has been done by Buckingham et al in 2001 and Krombach et al in 2005 [3, 9]. However, a specific model on arrangement of the vestibular anatomy has not been studied anywhere. Thus, the arrangement of the vestibular anatomy for this study is only an approximation based on literature study to facilitate the parametric study [10, 11].

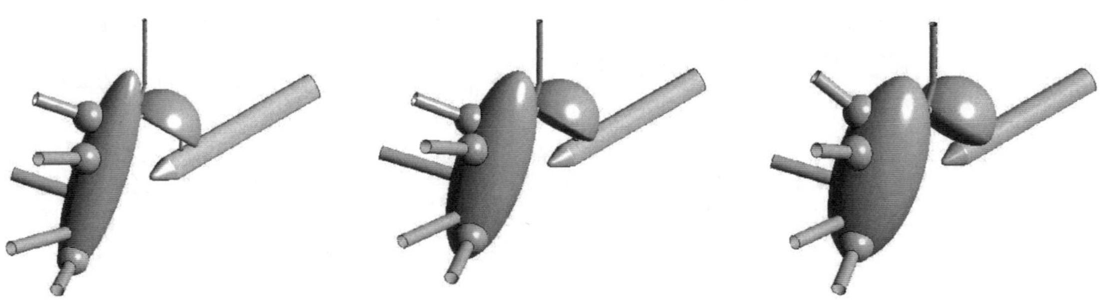

(a) (b) (c)

Figure 2. The different model of simplified vestibular system, (a) normal (N), (b) Meniere's disease (M1) and (c) Meniere's disease (M2).

IOP Conf. Series: Materials Science and Engineering **243** (2017) 012033 doi:10.1088/1757-899X/243/1/012033

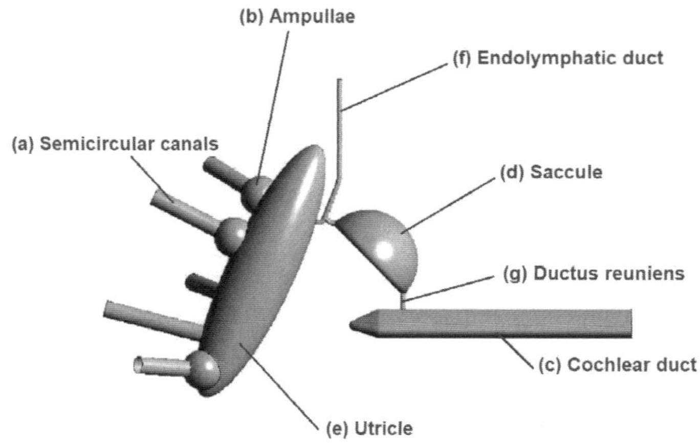

Figure 3. The simplified of vestibular system components.

Table 1. Dimension of the simplified computational model [3, 9].

Anatomy	Shape	Model drawing	Dimensions (mm) Model N	Model M1	Model M2
(a)	cylinder	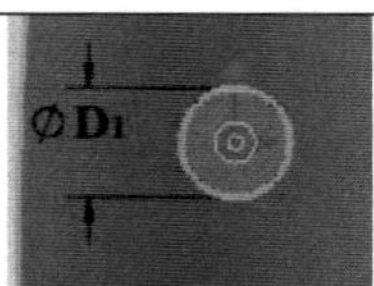	D=0.34	D=0.37	D=0.40
(b)	sphere	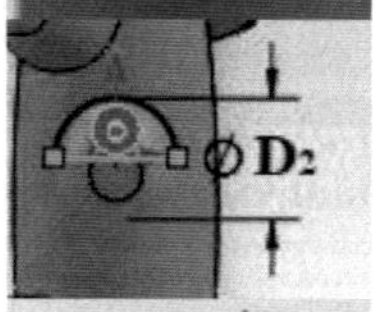	D_2=0.80	D_2=0.90	D_2=1.00
(c)	cylindrical triangle	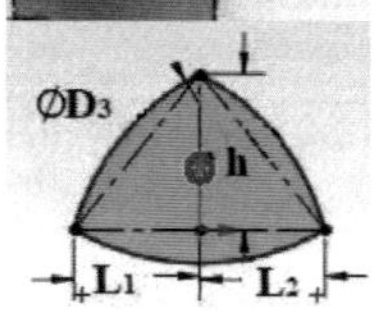	D_3=1.60 h=0.50 L_1=0.40 L_2=0.40	D_3=1.20 h=0.50 L_1=0.40 L_2=0.40	D_3=0.90 h=0.50 L_1=0.40 L_2=0.40
(d)	hemisphere	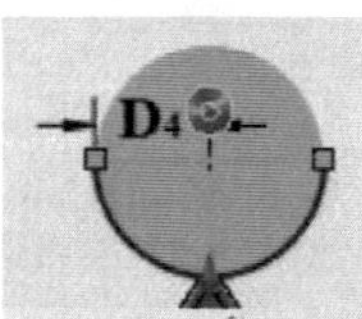	D_4=1.00	D_4=1.10	D_4=1.20
(e)	oval	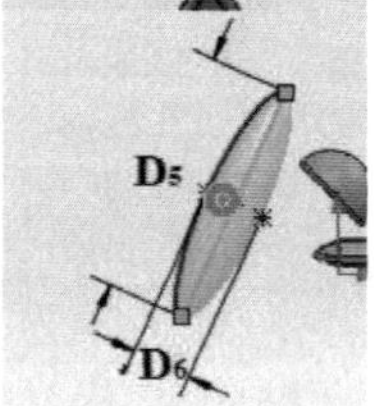	D_5=5.50 D_6=1.35	D_5=5.50 D_6=1.93	D_5=5.50 D_6=2.50
(f)(g)	cylinder	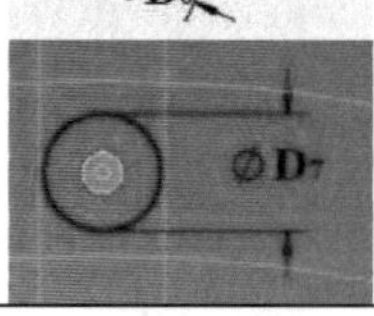	D_7=0.10	D_7=0.15	D_7=0.20

2.2. Meshing of vestibular model

The geometry is meshed up to sufficient number of nodes make sure it can be run in the ANSYS solver which called as ANSYS FLUENT 16.1. Usually, the different numbers of node in the meshed up model affects the result computed by the ANSYS solver. Thus, a Grid Independent Test (GIT) as shown in figure 4 is done to get accurate result without consuming much time by the ANSYS solver. A normalized error is analyzed from the error of current mesh by referencing to the previous mesh. The GIT setting that has 842672 nodes is chosen for the lowest average normalized error of 0.311 percent compared to other settings that have 0.314 percent by the 483961 nodes and 0.529 percent by the 1294742 nodes. The chosen GIT setting of the generated mesh can be seen in Figure 5.

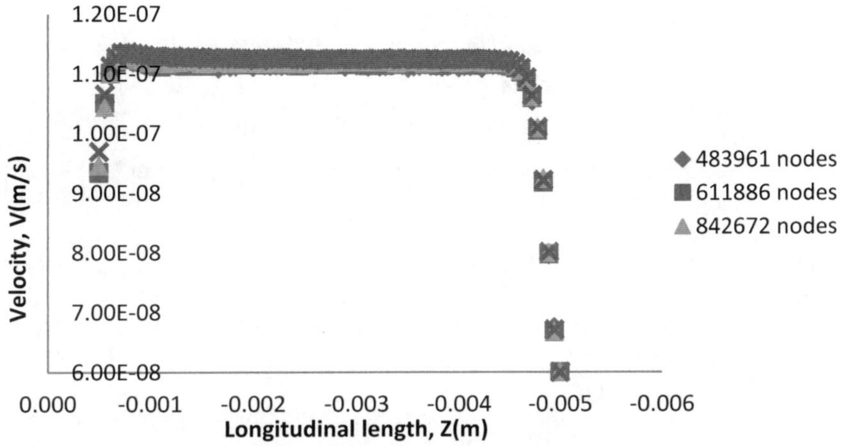

Figure 4. GIT of meshed up model.

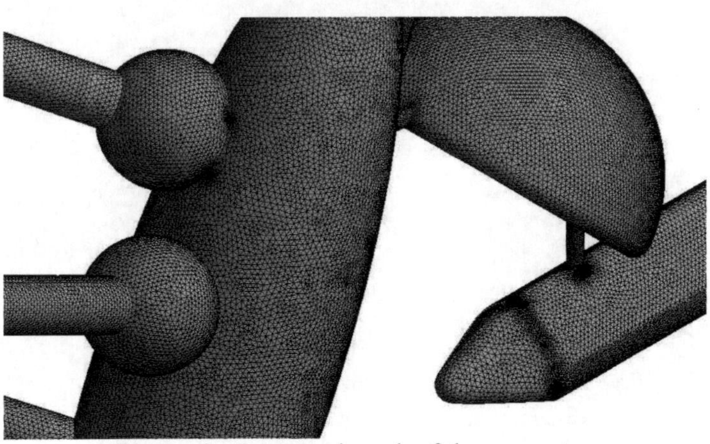

Figure 5. Generated mesh of the geometry.

2.3. Parameter assumptions and boundary conditions

Endolymphatic fluid is assumed as incompressible and Newtonian fluid which mimicking the mechanical properties of water that has an average density of approximately 1000 kg/m^3 and kinematic viscosity of 10^{-6} m^2/s [12, 13]. The endolymphatic membrane wall is assumed to be rigid and no slip condition. The given inflow from the cochlear duct is 0.00112 mm^3/min [14]. The mean outlet pressure of N model condition is 2.8 cm H_2O while the Meniere's disease condition of M1 model is 3.2 cm H_2O and M2 model is 3.6 cm H_2O [15]. Steady flow is considered into the analysis of endolymphatic effect on different volume of endolymphatic region. Figure 6 shows the stated boundary conditions of the computational model. The viscous model used is Shear Stress Transport (SST) in order to capture the visualization of flow recirculation.

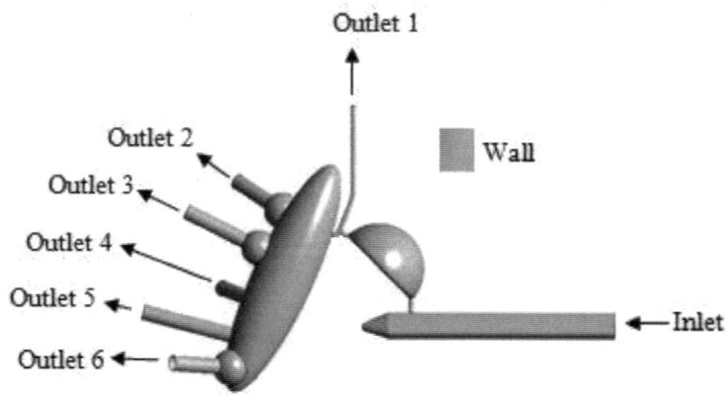

Figure 6. Boundary condition of simplified model.

2.4. Velocity profile validation

Present study is validated through the endolymphatic velocity profile inside semicircular canal of previous research by Boselli et al and Grieser et al in 2012 as shown in figure 7 [8, 16]. From figure 7, the error differences are taken at L_0 equals to 0.125, 0.375, 0.625 and 0.875. Thus, the mean percentage error of present study as compared to the previous study by Boselli et al is 8.12 percent while Grieser et al is 8.36 percent.

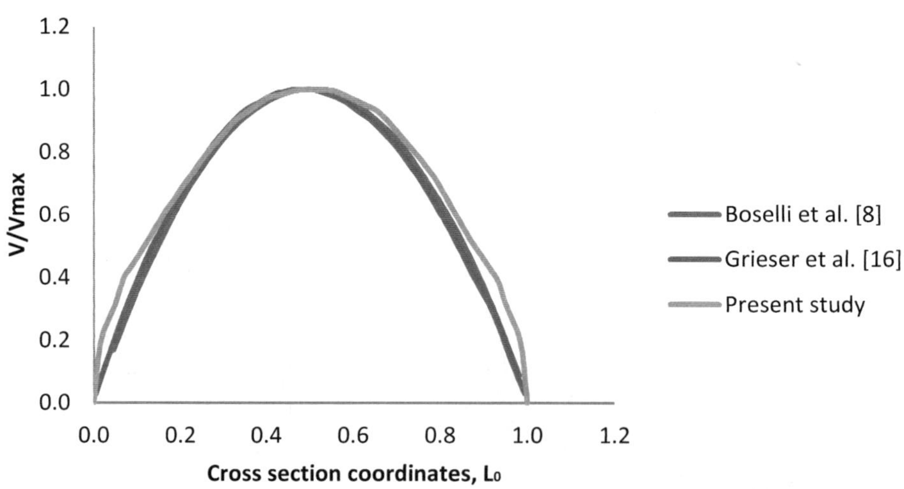

Figure 7. Validation of velocity profile in semicircular canal for both numerical and experimental study.

IOP Conf. Series: Materials Science and Engineering **243** (2017) 012033 doi:10.1088/1757-899X/243/1/012033

3. Result and Discussion

3.1. Pressure drop in vestibular models

M1 model is designed to be 75% larger than N model, while M2 model is 169% larger than N model. The designed model is set along with outlet pressure according to the disease severity and the result at the endolymphatic wall membrane could be seen in figure 8. Although the pressure looks qualitatively constant, the increased volume of endolymphatic fluid gives low pressure drop. If the pressure drop reaching zero as shown in figure 9, the high pressure from inlet will remains constant till reaching the outlets. Thus, the wall membrane becomes easier to be dilated causing more severe Meniere's disease.

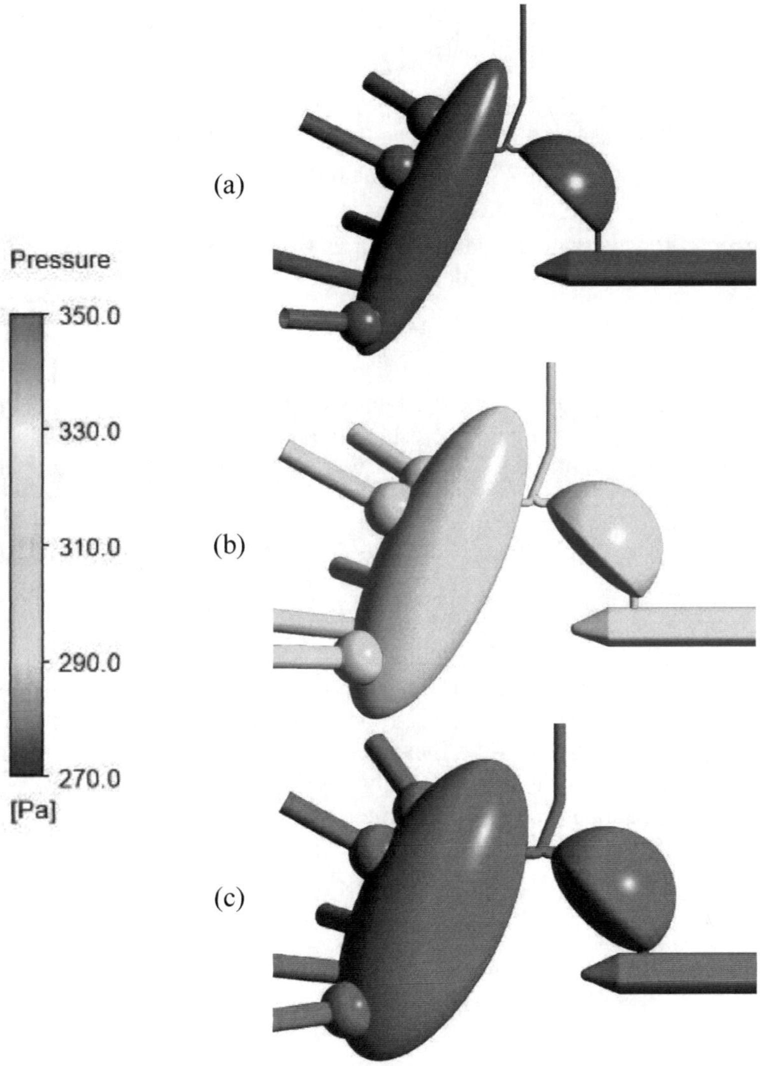

Figure 8. Pressure of endolymphatic wall membrane of (a) N model, (b) M1 model and (c) M2 model.

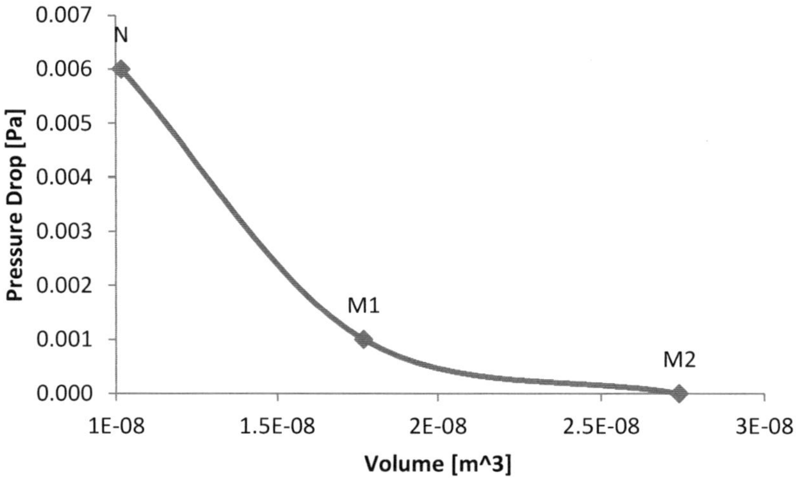

Figure 9. Pressure drop against volume of vestibular systems of N model, M1 model and M2 model.

3.2. Velocity of endolymphatic fluid

Velocity from the inlet is affected by the cochlear duct sectional area and the constant endolymphatic flow rate. If the area of the sectional inlet increases, the velocity of endolymphatic fluid decreases. Figure 10 shows the velocity contour has the lowest velocity in cochlear duct compared to M1 and N model. However, the highest velocity distributions for all models are seen in cochlear duct, ductus reuniens and endolymphatic duct compared to other inner ear anatomies. As the velocity distribution in certain part increases, the wall shear stress (WSS) parameter also will be affected and goes higher too.

The range of detected velocity for N model is from 4.91845e-12 m/s to 4.5651e-6 m/s. The velocity range for M1 model is from 7.09624e-12 m/s to 2.0342e-6 m/s, while M2 model is from 3.3349e-13 m/s to 1.1231e-6 m/s. Thus, the mean velocity for N model is 4.3570e-8 m/s, M1 model is 1.9692e-8 m/s and M2 model is 1.4167e-8 m/s, meaning M2 model have the lowest mean velocity. This is due to many flow recirculation that occurred in the increased endolymphatic volume as shown in figure 10 that give very low velocity distribution. The flow recirculation occurred is one of the highest probabilities to the sudden attack of vertigo in the inner ear.

IOP Conf. Series: Materials Science and Engineering **243** (2017) 012033 doi:10.1088/1757-899X/243/1/012033

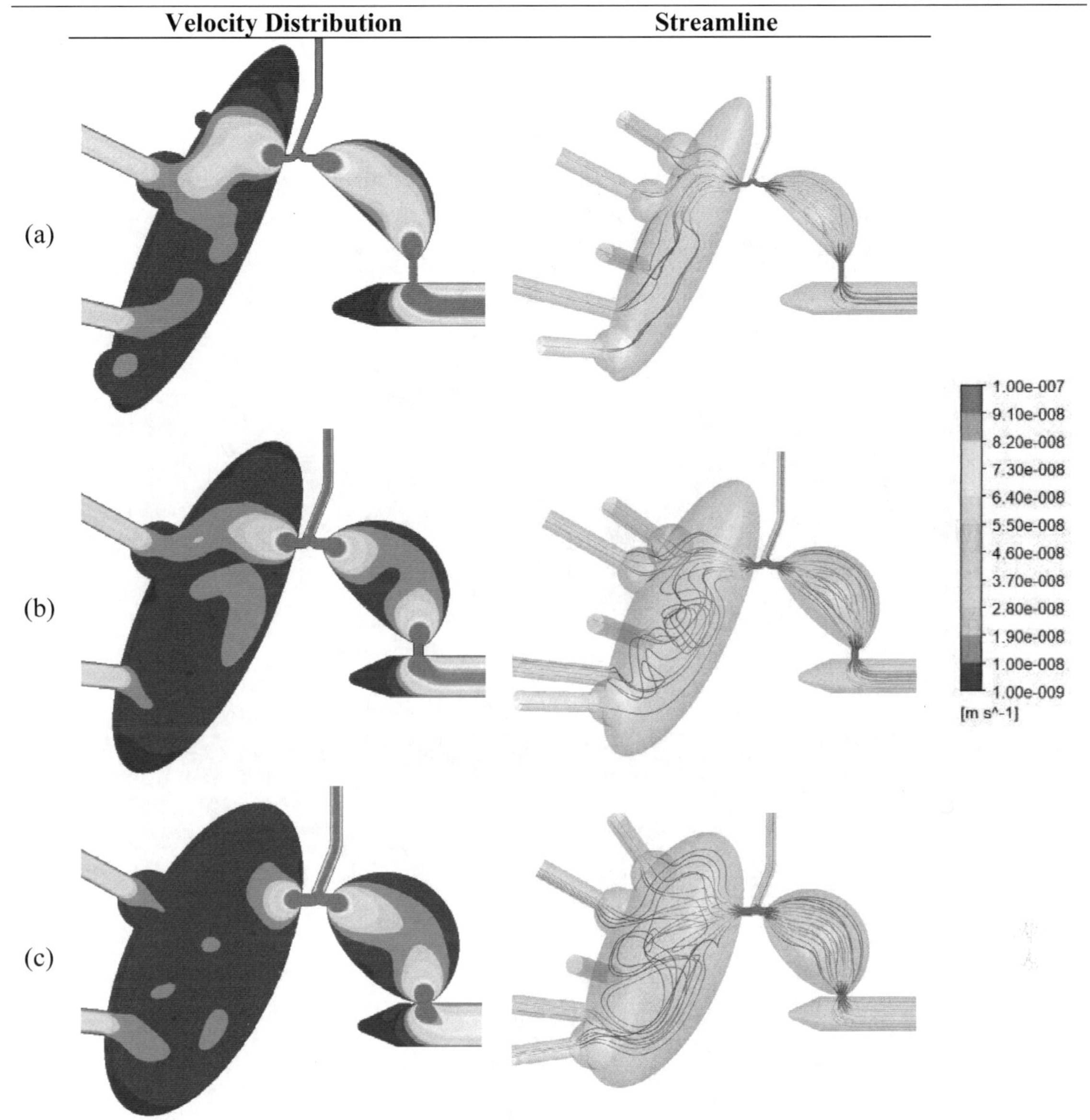

Figure 10. Velocity distributions and streamline of vestibular systems for (a) N model, (b) M1 model and (c) M2 model.

3.3. Wall shear stress (WSS) at endolymphatic wall membrane

WSS is affected by the velocity of the viscous endolymphatic fluid at the wall membrane. High WSS value on soft tissue including the endolymphatic wall membrane tends to rupture. Therefore, endolymph fluid is mixed with perilymph fluid leading to an abnormal increase of fluid within the endolymphatic chamber. This abnormality of the Meniere's disease can be occurred such as vertigo, tinnitus and hearing loss.

In Figure 11, cochlear duct, ductus reuniens, endolymphatic duct and semicircular canals probably tend to rupture due to high WSS distributions. WSS distributions of model N at utricle and saccule is higher compared to M1 and M2 model. N model has mean WSS value of 5.0427e-6 Pa, while M1 model has mean WSS value of 9.8473e-7 Pa and M2 model has mean WSS value of 5.4604e-7 Pa.

This is due to the decreasing of velocity and volume is increase at endolymphatic as the WSS distributions are observed from N to M2 model.

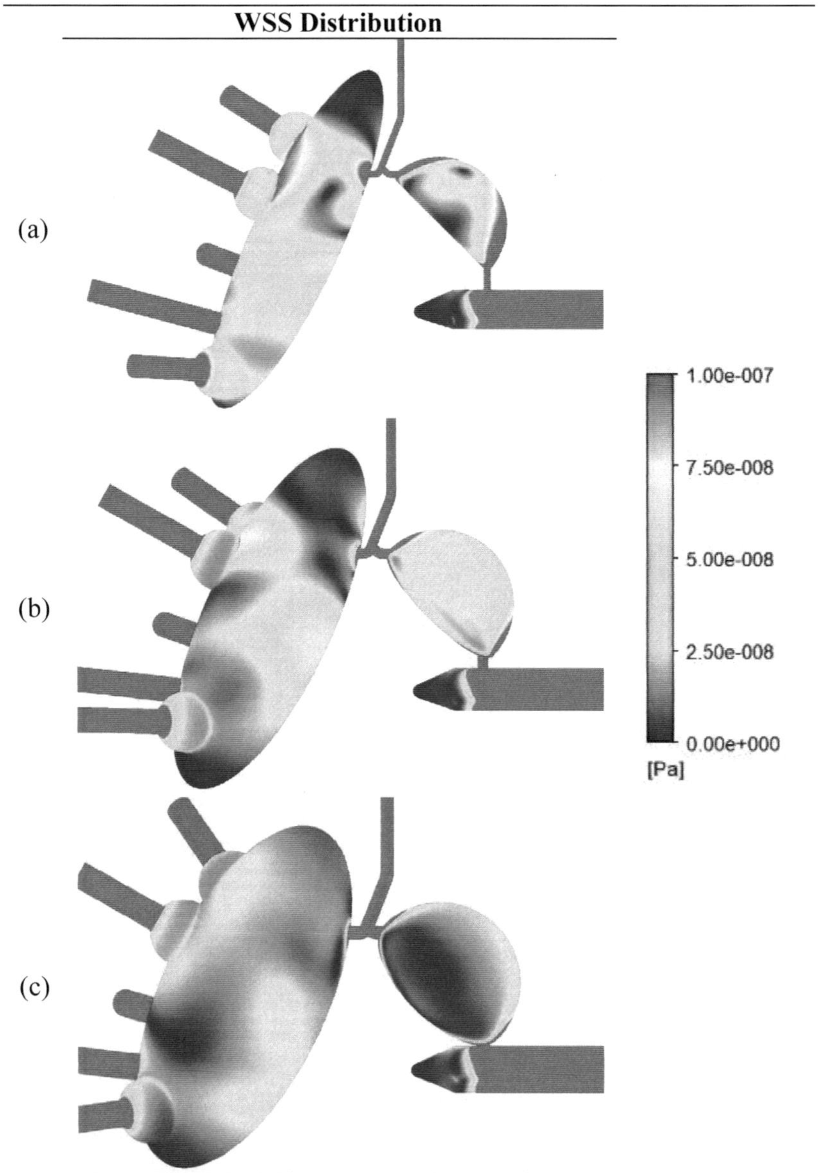

Figure 11. WSS distributions on endolymphatic wall membrane for (a) N model, (b) M1 model and (c) M2 model.

4. Conclusion

From this study, pressure of the outlet is affected by the volume of the endolymphatic fluid. The pressure drop between inlet and outlet of normal condition is high compared to the severe Meniere's disease condition. Since severe Meniere's disease condition has very small negligible pressure drop, the endolymphatic fluid volume fail to control the outlet pressure.

Besides that, the components of vestibular models where the highest velocity distributions occur, is in the cochlear duct, ductus reuniens, endolymphatic duct and semicircular canals. Thus, these components clearly have high WSS distributions compared to other components which also having high probability of wall membrane dilation and rupture. However, if the WSS distributions observed closely from N model to M2 model at the wall membrane of saccule, utricle and ampullae, the distributions become lower due to flow recirculation. The endolymphatic fluid flow recirculation actually is an instable flow because of the increasing volume of the fluid. Based on the analysis that has been made, these could lead to the high probability of vertigo unpredictable attack, tinnitus and hearing loss.

Acknowledgement
The support of the University Tun Hussein Onn Malaysia (UTHM), under the GPPS grant project, led by Dr. Ishkrizat Taib and under grant number **U723** is gratefully acknowledged. This paper was partly sponsored by Flow Analysis, Simulation and Turbulence (FAST) research group of UTHM.

References
[1] Surgery N 1995 *Otolaryngol Head Neck Surg.* **113** 181–185
[2] Klockhoff I and Lindblom 2016 *Acta Oto-Laryngol.* **63**(2-3) 347–365
[3] Buckingham R A and Valvassori G E 2016 *Ann Otol Rhinol Laryngo* **110** 113–117
[4] Horst W L and Rauch S 1961 *Arch. Otolaryngol.* **73**(3) 262–267
[5] Paparella M M, Goycoolea M V and Meyerhoff W L 1979 *Laryngoscope* **89** 43–58
[6] Shuknecht H F, Benitez J T and Beekhuis J 1962 *Trans Am Otol Soc.* **50** 310–328
[7] Lu D and Kassab G S 2011 *J. R. Soc. Interface* **8** 1379–1385
[8] Boselli F, Obrist D and Kleiser L 2013 *Biomech Model Mechanobiol* **12** 335–348
[9] Krombach G A, Martino E D, Schmitz-rode T, Gunther R W and Wildberger 2005 *Eur Radiol* **15** 1505–1513
[10] Richard C, Laroche N, Malaval L and Dumollard J M 2010 *Auris Nasus Larynx* **37** 155–161
[11] Pyykko I, Zou J, Poe D, Naganawa T and Shinji N 2010 *Otolaryngol Clin N Am* **43** 1059–1080
[12] Obrist D 2011 *Fluid Mechanics of the Inner Ear* (Zurich: ETH Zurich)
[13] Rabbitt R D and Damiano E R 1992 *J. Fluid Mech.* **238** 337–369
[14] Harris J P 1999 *Menier's Disease* (Amsterdam: Kugler Publications)
[15] Böhmer A and Dillier N 1990 *Ann. Otol. Rhinol. Laryngol.* **99**(6) 470–476
[16] Grieser B, Obrist D and Kleiser L 2012 *Validation of assumptions on the endolymph motion inside the semicircular canals of the inner ear* (Zurich: ETH Zurich)

IOP Conf. Series: Materials Science and Engineering **243** (2017) 012034 doi:10.1088/1757-899X/243/1/012034

CFD simulation of fatty acid methyl ester production in bubble column reactor

N S Mohd Salleh[1] **and N F Nasir**[1*]

[1] Centre for Energy and Environment Studies, Universiti Tun Hussein Onn Malaysia, 86400, Parit Raja, Batu Pahat, Johor, Malaysia.

E-mail: fitriahnasir@gmail.com

Abstract. Non-catalytic transesterification is one of the method that was used to produce the fatty acid methyl ester (FAME) by blowing superheated methanol bubbles continuously into the vegetable oil without using any catalyst. This research aimed to simulate the production of FAME from palm oil in a bubble column reactor. Computational Fluid Dynamic (CFD) simulation was used to predict the distribution of fatty acid methyl ester and other product in the reactor. The fluid flow and component of concentration along the reaction time was investigated and the effects of reaction temperature (523 K and 563 K) on the non-catalytic transesterification process has been examined. The study was carried out using ANSYS CFX 17.1. The finding from the study shows that increasing the temperature leads to higher amount of fatty acid methyl ester can be produced in shorter time. On the other hand, concentration of the component such as triglyceride (TG), glycerol (GL) and fatty acid methyl ester (FAME) can be known when reaching the optimum condition.

1.0 Introduction

Biodiesel is an alternative diesel fuel that is currently being used as blended fuel for transportation. The most common method to produce biodiesel are through a process known as transesterification of vegetable oils or animal fats and an alcohol in presence of catalyst or without catalyst [1]. Replacing fossil fuels with biodiesel is mostly due to the environmental policies to reduce carbon dioxide emission which are known as the causes of global warming and greenhouse effect [2].

Bubble column reactor is a cylindrical vessel which equipped with a gas distributor at the bottom [3]. The gas bubbles will diffuse into the surrounding liquid and react with the dissolved reactants. The bubble column reactor is developed to produce fatty acid methyl ester (FAME) by blowing bubbles of superheated methanol vapor continuously into vegetable oil without using any catalyst [4]. There are two mode of operation that valid for bubble column which are semi batch mode and continuous mode. For continuous operation the gas and the suspension flow concurrently upward into the column and the suspension that leaves the column will be recycled to feed tank. For semi batch mode of operation, the suspension is in stationary condition which means that it is zero liquid throughputs and the gas is bubbling upward into the column [5]. Characteristics of fluid dynamics in the bubble column reactors has significant effects on the operation and performance of bubble columns. Besides that, the flow regimes in bubble columns is classified and maintained according to the superficial gas velocity employed in the column. Type of flow regimes in bubble column are homogeneous (bubbly flow) regime, heterogeneous (churn-turbulent) regime and slug flow regime [5]. Figure 1 shows the flow diagram of a reactor using non-catalytic transesterification [4].

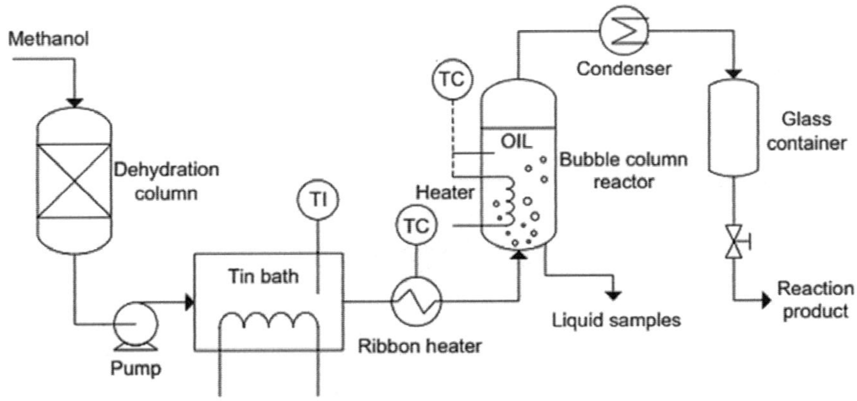

Figure 1. Schematic flow diagram of a reactor using non-catalytic transesterification experiment [4].

Computational fluid dynamic (CFD) model of a reactor developed in ANSYS software can be used to understand the physical and chemical process for the study. Development of CFD model allows different reactor designs (diameter and length) and operating conditions (flow rate) to be easily tested [6]. The ANSYS CFD software give valuable insight into a product's performance where it is delivered very well validated results to a model performance. Besides that, it is also capable of showing geometry for the computational domain and the various boundary conditions incorporated with the actual experiment setup. A study by Nikou and Ehsani (2008) shows the Computational Fluid Dynamics (CFD) model to describe the liquid to liquid flow phenomena observe in a reaction medium. The study has predicted the velocity distribution, pressure, concentration and temperature profiles in CFD model [7]. Therefore, this paper aim to predict the component concentrations in the bubble column reactor for biodiesel production and investigate the fluid flow inside the bubble column reactor during the reaction happens.

2.0 Methodology

Computational Fluid Dynamic (CFD) simulation by using CFX code which is consists in ANSYS 17.1 Workbench was used to investigate the fluid flow inside the bubble column reactor. For modeling the bubble column geometry, SolidWorks 2016 software was used. The diameter for bubble column reactor are referred from study conducted by previous researcher [4]. Material for bubble column reactor was selected to be stainless steel (316SS). The isometric view of bubble column reactor is shown in Figure 2.

The pressure in the bubble column was fixed 0.1MPa with temperature varied at 523 K and 563 K. For transient state solution, appropriate time step up was included to investigate the components concentration and fluid flow during the reaction time. On top of that, the meshing process was used to maintain a good quality of the cell. Then, the meshing model are runs in the solver process at reaction time for 1000 s. The boundary condition was used to determine the fluid flow directly toward the domain. The boundary condition are inlet, outlet and wall. The data setup for inlet was shown in Table 1.

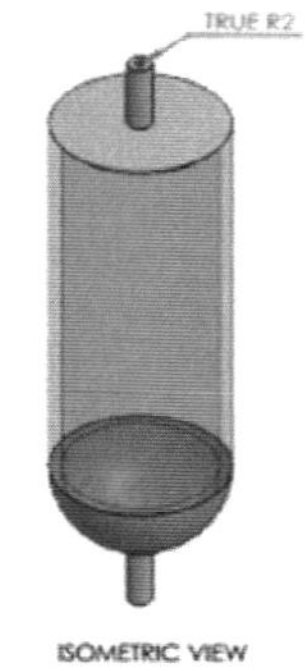

Figure 2. The bubble column reactor

Table 1. Data setup for the inlet condition.

Parameter	Value
Velocity	2 m/s
Component concentration	Triglyceride = 0.59 x 10^3 mol/m^3 Methyl Ester = 12.3 x 10^3 mol/m^3

For this simulation, the outlet pressure was set to standard operating pressure at 101325 Pa. Then, along the wall was set to no slip boundary condition. For the material properties, the domain type are fluid domain which is contains methanol and palm oil (triglyceride, diglyceride and monoglyceride) in fluid database. Table 2 shows the properties for each fluid domains that are used in the simulation.

Table 2. Data for material properties.

Fluid Domain	Material properties
Methanol	Density = 791.8 kg/m^3 Viscosity = 0.59 mPa.s
Palm oil	Density = 870.2 kg/m^3 Viscosity = 16.93 mPa.s

3.0 Results

3.1 Fluid flow inside the bubble column reactor

The bubble column reactor was modelled for fluid flow at high gas velocities. The cross flow velocity within the bubble column were remained stable and did not change by time. In this condition, the bubble column reactor was analyzed at a steady state for different reaction temperature under atmospheric pressure. Figure 4 and 5 show the resulted velocity contour at the temperature of 523 K and 563 K for the bubble column reaction in XY Plane.

By increasing the temperature, it has affected the flow characteristic in the bubble column reactor. Red color in the Figure 3 shows the methanol in gas phase was started by blowing the bubbles of superheated methanol vapor at static pressure (0.1 MPa). The highest velocity was located at the inlet and become lowest when across the bubble column reactor. Then, blue color at Figure 3 represent the palm oil in the bubble column reactor at zero velocity. The effect of temperature was observed when the velocity magnitude increase to 563 K. It is found that by increasing the temperature, the velocity magnitude will increase.

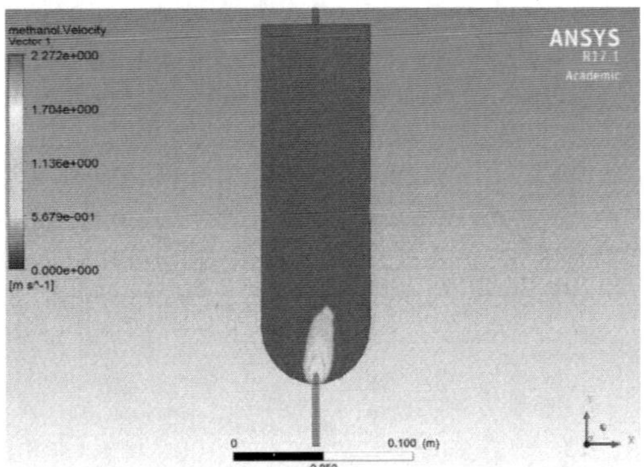

Figure 3. Contour of velocity in the bubble column reactor at steady state at 523 K

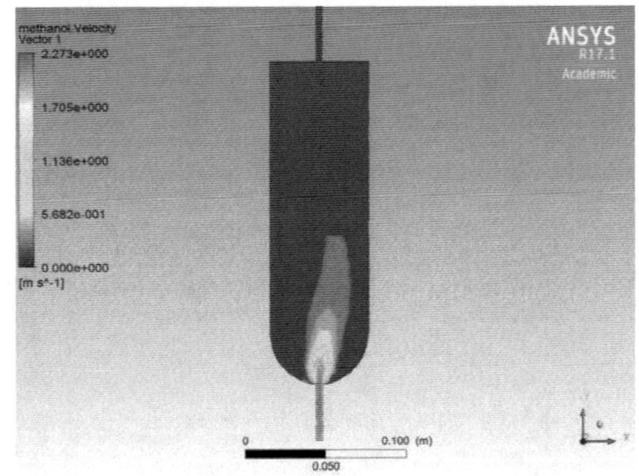

Figure 4. Contour of velocity in the bubble column reactor at steady state at 563 K

3.2 Contour for Component of concentration

The study deals with the component of the concentration from palm oil by using non-catalytic transesterification. Besides that, the result of the reaction in the component of the concentration should happen through of the bubble column reactor. Figure 5 and 6 shows the non-catalytic transesterification reaction for palm oil during the reaction.

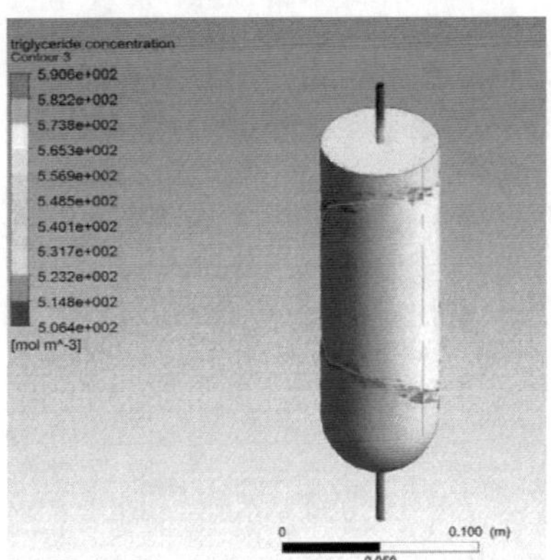

Figure 5. Concentration contour for triglyceride at 523 K

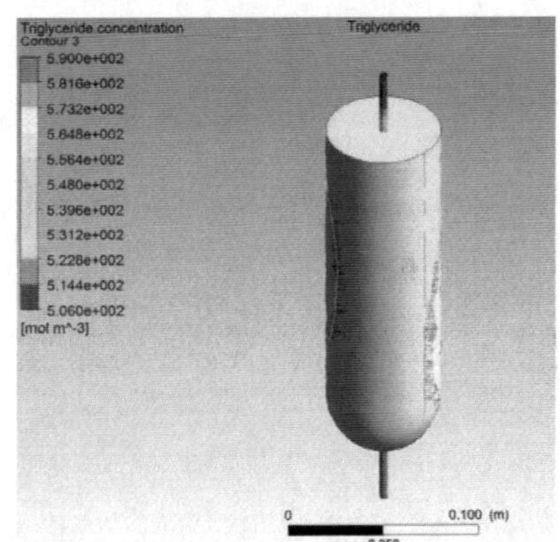

Figure 6. Concentration contour of triglyceride at 563 K

Based on Figure 5 and 6 the changes in color were found because of the movement of materials depends on their viscosity, and there was more residence time for TG than other materials [8]. The difference in temperature increased the rate of chemical reaction [4]. Moreover, the raising in temperature can also increase the concentration of reactant in reactant gas. Methyl ester concentration from the simulation in bubble column reactor are shown in Figure 7 and 8.

IOP Conf. Series: Materials Science and Engineering **243** (2017) 012034 doi:10.1088/1757-899X/243/1/012034

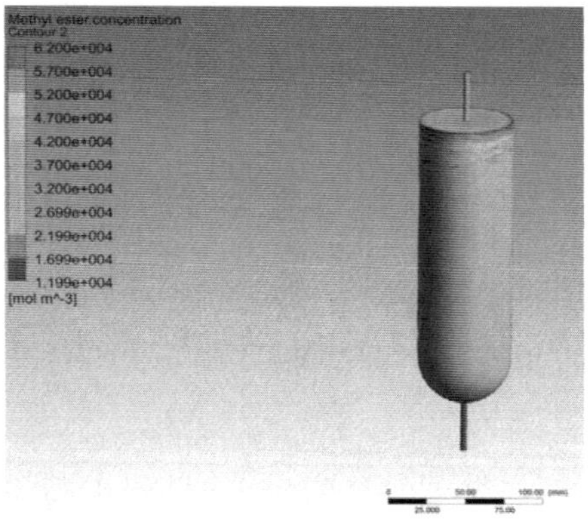

Figure 7. Concentration contour of methyl ester at 523 K

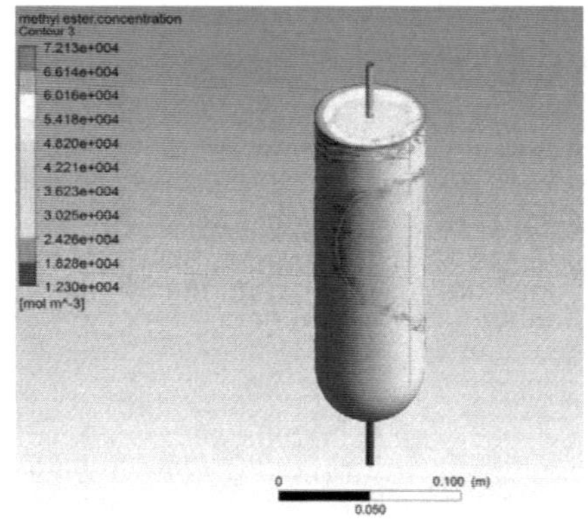

Figure 8. Concentration contour of methyl ester at 563 K

The production of methyl ester was started earlier because of the first and second reaction in transesterification. Concentration of methyl ester are higher than triglyceride and glycerol. It is because methyl ester (ME) was produced by the three reactions in transesterification [4]. The blue contour at the bottom of reactor shows the initial reaction to produce ME.

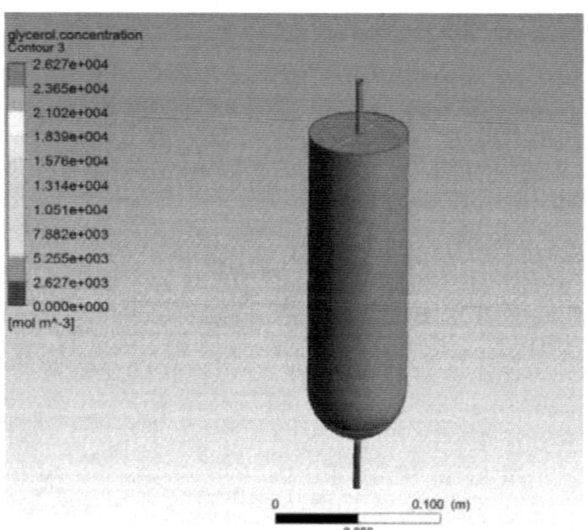

Figure 9. Concentration contour of glycerol at 523 K

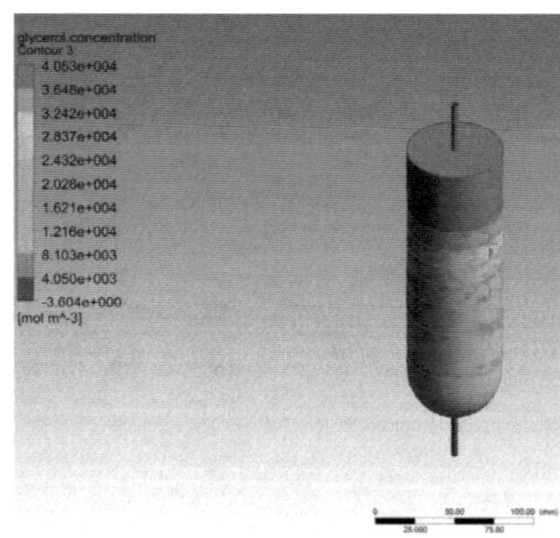

Figure 10. Concentration contour of glycerol at 563 K

Figure 10 shows the contour for concentration of glycerol at 563 K has changed color within the production process. The differences between patterns is depending on the increasing rate of their concentration which is much faster in methyl ester compared to glycerol because glycerol was produced only by the last reaction. On top of that, it was calculated by the interphase mass transfer process and the mixing which is caused by the bubbles. Otherwise, the effect of kinetic were depends on the reaction temperature so the two different temperature was selected to get two different reaction rates. All in all, when the temperature rise, it will make the reactant of particles move more quickly and more energy provided.

3.3 Graph for Component of concentration

The simulation result was compared with the finding of research conducted by Jeolianingsih et. al [4]. The trend indicates the reducing of the triglyceride concentration in experimental and simulation by Computational Fluid Dynamics. Based on Figure 11, the experimental graph has slightly decreased after 800s compared to the CFD graph that remained unchanged at 800s.

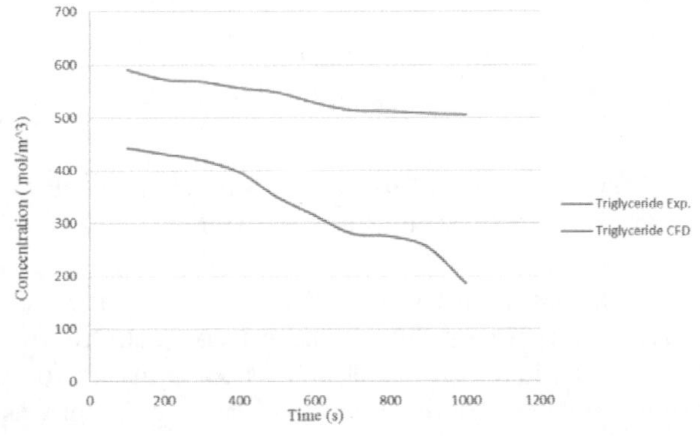

Figure 11. Concentration of triglyceride in bubble column reactor at 563 K

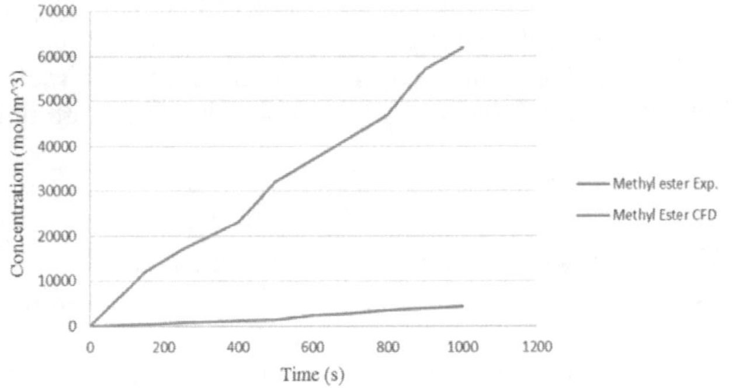

Figure 12. Concentration of methyl ester in bubble column reactor at 563 K

Figure 12 shows the concentration of methyl ester by simulation and experiment. In the initial reaction, the production of methyl ester was rapid by simulation and experimental. Based on Joelianingsih et. al [4], the increase in methyl ester will lead to increase in glycerol concentration at it was liberated from triglyceride molecules. Thus, the conversion of the reaction of methyl ester increased with reaction time and reaction temperature. The simulation result shows that the methyl ester is produced at every 100 s but for the experiment, it started to changes at the 400s. This is due to the velocity magnitude between the experiment and simulation. The velocity will lead to the increasing in the concentration as well as will effect the component concentrations.

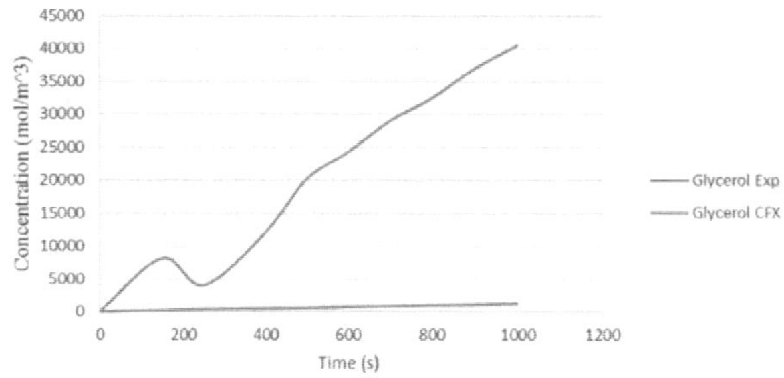

Figure 13. Concentration of glycerol in bubble column reactor at 563 K

Figure 13 shows that the concentration for glycerol was rapidly increased in the simulation and maintained the same level at the experiment. The value for concentration in the experiment shows a slow changes at 500 s. Based on Jeolianingsih et al. [4], vapor pressure for glycerol was higher than triglyceride at the same reaction temperature thus the volatility of glycerol was higher than triglyceride [4].

4.0 Conclusion

The non-catalytic transesterification process shows that the production of biodiesel can be accomplished by using methanol at high pressure and temperature in BCR. By analyzing the component such as triglyceride (TG), diglyceride (DG) and fatty acid methyl ester (FAME) in BCR, the concentration can be shown along the process of producing biodiesel. Further study can be conducted to include the effects of molar ratio of oil to alcohol in simulation of biodiesel production in BCR.

References

[1] Abbaszaadeh, A., Ghobadian, B., Omidkhah, M. R., & Najafi, G. (2012). Current biodiesel production technologies: A comparative review. In *Energy Conversion and Management* (Vol. 63, pp. 138–148). http://doi.org/10.1016/j.enconman.2012.02.027

[2] Abdullah, A. Z., Salamatinia, B., Mootabadi, H., & Bhatia, S. (2009). Current status and policies on biodiesel industry in Malaysia as the world's leading producer of palm oil. *Energy Policy*, *37*(12), 5440–5448. http://doi.org/10.1016/j.enpol.2009.08.012

[3] Kantarci, N., Borak, F., & Ulgen, K. O. (2005b). Bubble column reactors. Process Biochemistry. http://doi.org/10.1016/j.procbio.2004.10.004

[4] Joelianingsih, Maeda, H., Hagiwara, S., Nabetani, H., Sagara, Y., Soerawidjaya, T. H., Abdullah, K. (2008). Biodiesel fuels from palm oil via the non-catalytic transesterification in a bubble column reactor at atmospheric pressure: A kinetic study. Renewable Energy, 33(7), 1629–1636. http://doi.org/10.1016/j.renene.2007.08.011

[5] Kantarci, N., Borak, F., & Ulgen, K. O. (2005a). Bubble column reactors. Process Biochemistry, 40(7), 2263–2283. http://doi.org/10.1016/j.procbio.2004.10.004

[6] Hyndman, C. L., Larachf, F., & Guy, C. (1997). Understanding gas-phase hydrodynamics in bubble columns : a convective model based on kinetic theory, 52(1).3683

[7] Nikou, M. R. K., & Ehsani, M. R. (2008). Turbulence models application on CFD simulation of hydrodynamics, heat and mass transfer in a structured packing. International Communications in Heat and Mass Transfer, 35(9), 1211–1219. http://doi.org/DOI 10.1016/j.icheatmasstransfer.2008.05.017

[8] Sajjadi, B., Abdul Raman, A. A., Baroutian, S., Ibrahim, S., & Raja Ehsan Shah, R. S. S. (2014). 3D Simulation of fatty acid methyl ester production in a packed membrane reactor. *Fuel Processing Technology*, *118*, 7–19. http://doi.org/10.1016/j.fuproc.2013.07.015

IOP Conf. Series: Materials Science and Engineering **243** (2017) 012035 doi:10.1088/1757-899X/243/1/012035

Study of the fluid flow pattern in a bubble column reactor for biodiesel production

A A Suhaimi[1] and N F Nasir[1*]

[1] Centre for Energy and Environment Studies, Universiti Tun Hussein Onn Malaysia, 86400, Parit Raja, Batu Pahat, Johor, Malaysia.

*E-mail: fitriahnasir@gmail.com

Abstract. The applications of bubble column reactor (BCR) are very important as multiphase reactors in process industry. In biodiesel production, one of the reactors that are used is BCR. The advantages of BCR are low operating cost and maintenance due to the compactness and no moving parts. It is important to understand the nature of hydrodynamics and operational parameters to characterize their operation, including gas superficial velocity and bubble rise velocity to make the design and scale-up process. Computational fluid dynamics (CFD) can be used to evaluate the performance of BCR at lower cost compared to the experimental setup for biodiesel production. In this work, a commercial CFD software, FLUENT 14.0 was used for modeling of gas-liquid flow in a BCR for biodiesel production. Multiphase simulations were performed using an Eulerian-Eulerian two-fluid model. In this study, the simulation was conducted by using three different temperature, which are 523 K, 543 K and 563 K. The CFD result predicts the turbulent kinetic energy, gas hold-up and the liquid velocity were fairly good, although the results seem to suggest that further improvement on the interface exchange models and possibly further refine the two-fluid modeling approaches are necessary especially for the liquid velocity and turbulent kinetic energy.

1. Introduction

Biodiesel is a fuel derived from vegetable oil or animal fat. It is a direct petro-diesel replacement that can significantly reduce gas emissions and provide energy independence [1]. Biodiesel is produced under chemical reactions which are transesterification and esterification. Transesterification is a process where triglycerides (TG) are reacted with alcohol, generally methanol (MeOH) to form fatty acid methyl esters (FAME). There are three steps reactions with the intermediate formation of triglycerides (DG) and monoglycerides (MG) resulting in the production of 3 moles of methyl esters (ME) and 1 mole of glycerol (GL) [2].

There are several types of reactors to produce biodiesel. One of the reactor that is widely used in the biodiesel industry is a bubble column reactor (BCR). They are excellent reactors for biodiesel production which require a large interfacial area for gas-liquid mass transfer and efficient mixing for reacting species. Producing biodiesel by using superheated methanol vapor in the BCR take place under atmospheric pressure and requires high temperature (250°C - 300°C), therefore reactor was made of stainless-steel and covered by insulation [3].

IOP Conf. Series: Materials Science and Engineering **243** (2017) 012035 doi:10.1088/1757-899X/243/1/012035

2. Material and Methods

2.1. Geometry Modeling

In this study, all the CFD simulation was conducted by using ANSYS FLUENT 14.0. For the geometric modeling, the design is constructed by using SOLIDWORKS 2014. The model is then imported to ANSYS Workbench for further analysis. The references of geometry were referred based on previous studies and analysis [2-3]. This interactive process is the first stage in pre-processing before generating the mesh. The dimensions and physical conditions of the BCR are given in Table 1.

Table 1. Geometry and physical conditions for BCR

Parameter	Definition
Size of column	L x d = 192 mm x 55 mm
Size of inlet pipe	L x d = 45 mm x 4.8 mm
Size of outlet pipe	L x d = 25 mm x 10.5 mm

Figure 1 shows the geometry of the BCR model.

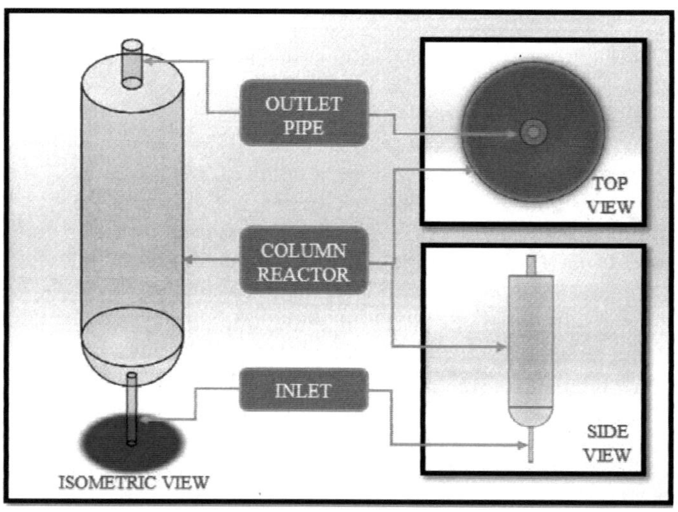

Figure 1. Drawing of BCR

2.2. Meshing

Meshing is one of the most important parts to determine the efficiency and effectiveness of an analysis. The function of meshing is to perform the Finite Element Analysis. In order to obtain more accurate results, the mesh must be refined. The easiest mesh refinement method is by reducing the element size throughout the modeling domains. After trials and errors, the suitable grid size and mesh type for the BCR simulation were determined which can obtain accurate results and shorten the calculating time. The mesh configurations are shown in Figure 2 and the number of nodes and elements are shown in Table 2.

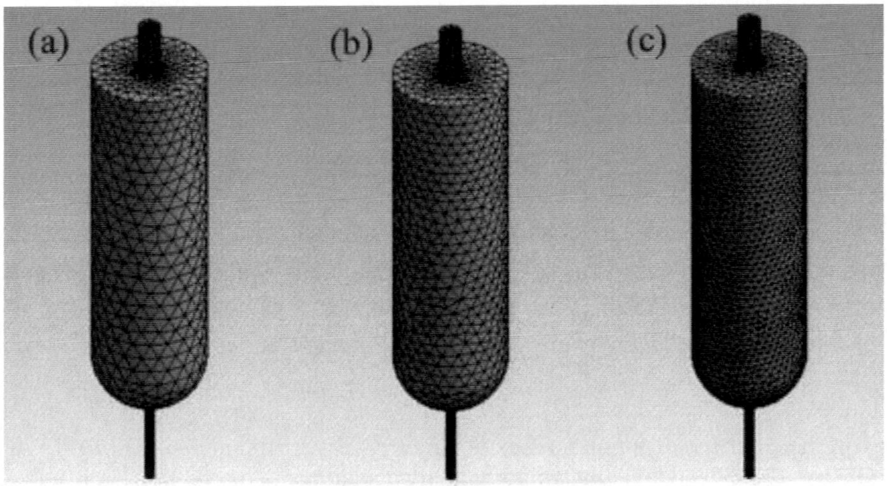

Figure 2. Type of mesh applied: (a) Coarse; (b) Medium; (c) Fine

Table 2. Total nodes and elements

	Coarse Mesh	Medium Mesh	Fine Mesh
Nodes	7352	8162	12820
Elements	34865	38691	62022

Finer mesh can capture the flow details efficiently since it has the highest number of nodes and elements. That is the reason why the finer meshes can provide a better result compared to coarse mesh. In addition, the calculation of the problems also can be converged by using a finer mesh.

2.3. Boundary Condition

The boundaries of the BCR are either an inlet, outlet, or wall. The inlet is dispersed with a uniform gas velocity and a volume fraction of 1. The velocity at the inlet is 2 m/s. Besides, the outlets have a pressure equivalent to the atmospheric pressure, which is 0.1 MPa. The turbulent kinetic energy and dissipation rate are set to 0 at the inlets and outlets because the effects of turbulence at the boundaries are difficult to predict. The walls are specified with a no-slip condition for the gas and liquid phases. The Euler – Euler model treats both phases as continuous phase, means that the gas phase does not move along the wall, therefore, no-slip boundary condition is applied to the gas phase along the wall. Table 3 summarize the boundary condition for BCR simulation.

Table 3. List of boundary condition for simulation

Boundary condition	Parameter
Domain / interior	FAME
Wall	Reactor (no-slip)
Velocity inlet	2 m/s
Inlet pressure gauge	100 000 Pa

3. Computational Model

The Eulerian multiphase model allows for multiple interacting phases by assuming that each phase behaves as interpenetrating continua [4]. The continuity equation is formulated for each phase without exchange between phases after Reynolds averaging term

$$\frac{\partial}{\partial t}(\alpha_q \rho_q) + \nabla.(\alpha_q \rho_q \vec{u}_q) = 0 \quad (1)$$

where u_q and α_q is the velocity and volume fraction of phase q^{th} respectively.

By comparing to the mass conservation, the momentum conservation is described by the Navier-Stokes equation extended by the phase volume fraction. Such an equation for phase 'q' is as follows:

$$\frac{\partial}{\partial t}(\alpha_q \rho_q \vec{u}_q) + \nabla.(\alpha_q \rho_q \vec{u}_q) = -\alpha_q \nabla p + \nabla.\bar{\bar{\tau}}_q + \alpha_q \rho_q \vec{g} + \sum_{p=1}^{n} k_{pq}(\vec{u}_p - \vec{u}_q) + \vec{F}_{Lorentz} \quad (2)$$

The terms on the right-hand side describe all forces acting on the phase 'q' of a fluid element in the control volume. The effect of virtual mass and lift force in this work has been neglected. From equation (2), τ_q is the stress tensor of phase q^{th} and calculated as follows:

$$\bar{\bar{\tau}}_q = \alpha_q \mu_q (\nabla \vec{u}_q + \nabla \vec{u}_q^T) + \alpha_q \left(\lambda_q - \frac{2}{3}\mu_q\right) \nabla.\vec{u}_q \bar{\bar{I}} \quad (3)$$

where μ_q and λ_q are shear and bulk viscosity of phase q respectively.

The exchange coefficient for gas-liquid flow has been considered as following equation:

$$k_{pq} = \frac{\alpha_q \alpha_p \rho_p f}{\tau_p} \quad (4)$$

The momentum exchange due to drag forces is:

$$F_{drag} = \frac{3}{4}(\alpha_q \rho_p)\frac{C_D}{d_p}(u_p - u_q)|u_p - u_q| \quad (5)$$

$$f = \frac{C_D Re}{24} \quad (6)$$

$$C_D = \begin{cases} \dfrac{24(1 + 0.15Re^{0.687})}{Re} & Re \le 1000 \\ 0.44 & Re > 1000 \end{cases} \quad (7)$$

$$Re = \frac{\rho_q |\vec{u}_p - \vec{u}_q| d_p}{\mu_q} \quad (8)$$

where C_D is the drag coefficient, to compute this coefficient the Schiller-Naumann relation has been used. To apply the effect of magnetic field on the fluid flow, the Lorentz force source term has been added to the momentum equation [5].

4. Result and Discussion

In this section, some of the characteristics of bubble dynamics were explored. It is now necessary to examine the regime of validity of those analyses. The simulation was conducted under 3 different temperatures; 523 K, 543 K and 563 K. Then the flow regime and velocity profile can be obtained from each temperature. Turbulence modeling is an important factor when modelling two-phase flows. The standard *k-ε* model was also tested for turbulent dispersion.

IOP Conf. Series: Materials Science and Engineering **243** (2017) 012035 doi:10.1088/1757-899X/243/1/012035

4.1. Flow Pattern

The flow pattern inside the reactor can be obtained from each temperature. By using XY Plane as the viewing angle, the production of bubbles and flow pattern inside the reactor can be observed. Figure 3 shows the contours of vapor volume fraction which describes the distribution of methanol bubble in the oil as result of CFD simulations. Red color appears when the volume fraction of vapor is 1 and the blue color shows that vapor volume fraction is 0. It is assumed that the methanol bubble was in vapor form while the oil was in liquid form.

At 2 seconds flow, the bubbles have reached at the top of the reactor and begins to circulate in the reactor. Based on the three conditions from Figure 3, the circulation behavior of the liquid is the same as the model made by Joshi and Sharma which is a Donut model [6]. The bubbles dispersed from the sparger will always move upwards in the form of swarm in a priority direction, and then creating a large liquid vortex structure. As the time increase, the priority direction suddenly changes over into another which could be any direction inside the reactor.

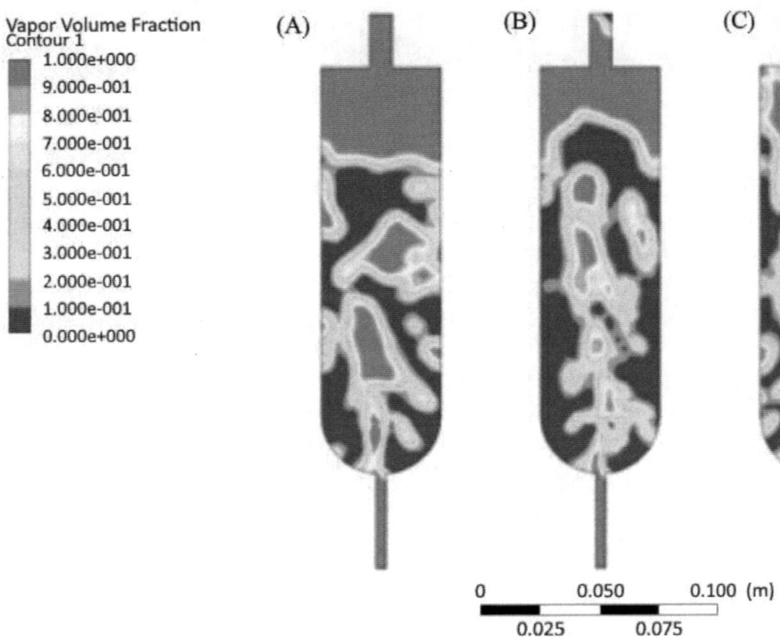

Figure 3. Flow pattern at (A) 523 K, (B) 543 K and (C) 563 K

4.2. Velocity Vector

To understand the flow characteristic in the reactor, velocity vectors can provide an excellent visualization of the flow and can also show the direction of the flow. In order to simulate the hydrodynamic behavior of the system in turbulent regime, the calculations were started with the laminar model with the liquid at rest [7]. When the liquid velocity in most parts of the reactor was greater than zero, the k and ε fields were initialized and the turbulence model was started. Figure 4 shows the velocity vector at the wall and the center of the reactor at 523 K.

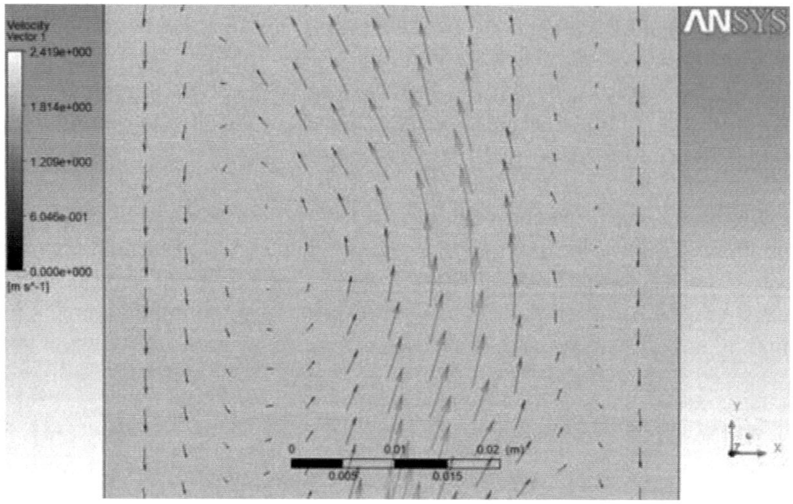

Figure 4. Velocity vector at the center and wall region of the column.

From Figure 4, the direction of the vector in the center region is in reverse direction compared to the wall region. This is because most part of the center region in the BCR was filled with the upward liquid flow and it returned to the bottom along the column wall.

4.3. Gas Holdup

The effect of gas velocity inlet to gas holdup and surface contact area was analyzed. Gas holdup is the basic parameter indicating the hydrodynamic characteristics of BCR [8]. It affects directly the geometric sizes, and the gas-liquid interfacial, thus affect the mass-transfer rates of BCR. Therefore, it is one of the important parameters for the BCR design. As a result, the contact surface area and gas holdup increase with increasing gas velocity. The effect of temperature on gas holdup is shown in Figure 5.

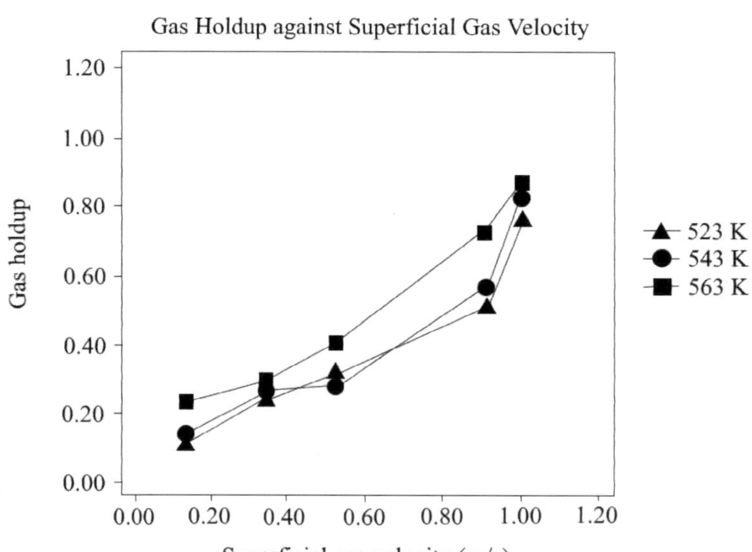

Figure 5. Gas holdup vs superficial gas velocity

4.4. Velocity Analysis

The velocity analysis based on the graph has been plotted according to the vertical plane and horizontal plane as shown in Figure 6. By plotting the results from both planes, the observation between all conditions can be compared with each other. From the analysis, the best and suitable temperature can be selected. In addition, the graph only focused on the velocity result because the velocity can determine the flow pattern in the BCR either it is highly turbulence or less turbulence regime during the transesterification process. The direction of the velocity flow is upward towards the top of the BCR. Figure 5 shows the plotted velocity against distance at 523 K, 543 K and 563 K.

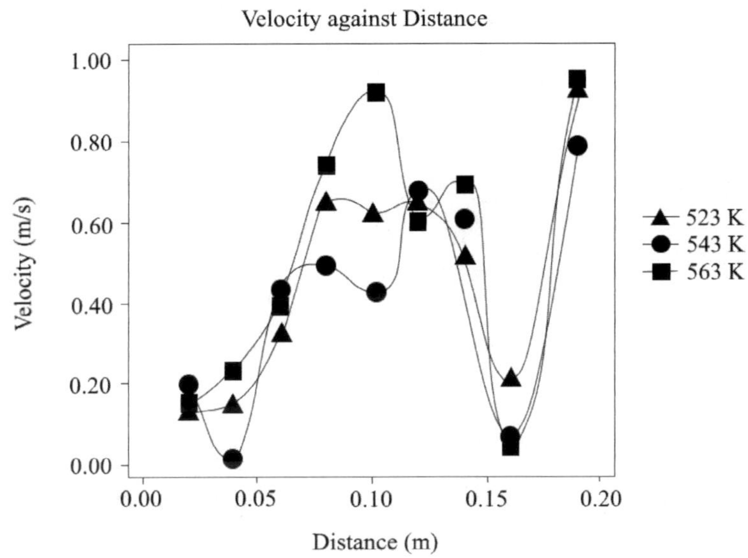

Figure 6. Graph of velocity against distance

From Figure 6 above, it shows that the velocity of the fluids in mostly increase until it reach 0.14 m reactor level, then the velocity reduces ranging below 0.2 m/s. This might due to the bubble has reached at the entrance of the outlet pipe. After that, the flow velocity begins to increase again until they reach the top of the outlet of the reactor.

4.5. Validation

Validation may refer to the comparison between the experimental data and the simulation result based on the study in Ref. [2] and [7]. From the observation, the velocity was resulted in different value of simulation from the experiment. In these problems, the error might cause by a high flow circulation by the fluids in the reactor, thus making the standard k-ε turbulence model cannot deliver the exact predictions for the velocity results. In addition, the geometry drawing of the reactor might be less accurate from the exact dimensions. Therefore, it will affect the result obtained from the simulation. Meanwhile, the errors might come from the experiment itself. During the measurements, some errors might be occurred that affect the experimental result. Figure 7 shows the comparison between experimental results with the CFD simulation in the term of gas holdup.

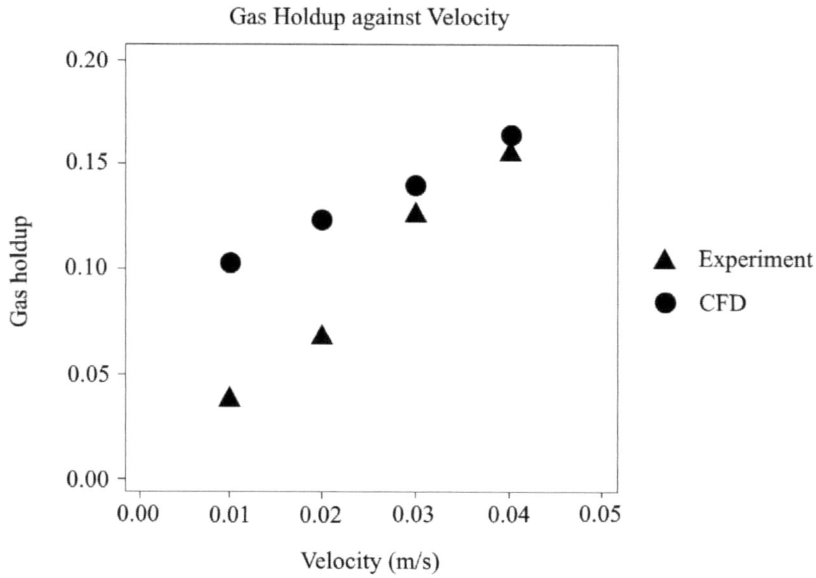

Figure 7. Comparison between Experiment and CFD results

5. Conclusion

In this study, the fluid flow inside the BCR has been simulated and analyzed by using CFD software. There are two phases which are triglyceride (liquid) and MeOH (vapor), therefore using Volume of Fluid (VOF) model is suitable in this simulation. Moreover, the standard k-ε model is used to predict the turbulence flow inside the BCR during the mixing process. The results obtained with the 3D version of the k-ε turbulence model are good and acceptable agreement with experimental results. The inclusion of the standard k-ε turbulence model is useful to describe the instantaneous large-scale vertical flow structure correctly.

The effects of inlet gas velocities to the gas holdup were accounted. This parameter is very important to describe the interaction between two-phase fluid flow in the BCR. Many correlations for predicting gas hold up can be found in literatures. Experimental data shows that the total gas hold up is increased with increasing superficial gas velocity. From the simulation result, it is proven that the increasing of inlet gas velocities cause increasing of the gas holdup. The large bubbles have higher rise velocity and decrease the gas hold up.

Acknowledgment

This study was supported by University Tun Hussien Onn Malaysia, Johor, Malaysia, under the Research Acculturation Grant Scheme (RAGS) Vot R071.

References

[1] Sheehan, J.J., Biofuels and the conundrum of sustainability. Current Opinion in Biotechnology, 2009. 20(3): p. 318-324.
[2] Joelianingsih, et al., A continuous-flow bubble column reactor for biodiesel production by non-catalytic transesterification. Fuel, 2012. 96: p. 595-599.
[3] Wulandani, D., et al., Determination on CFD Modeling for Bubble Column Reactor to Improve Biodiesel Fuel Production. SustaiN'2010, 2010: p. 41-44.
[4] Studley, Allison F. Numerical Modeling of Air-Water Flows in Bubble Columns and Airlift Reactors. Diss. Virginia Tech, 2010.

[5] Sivasankaran, S. and C.-J. Ho, Effect of temperature dependent properties on MHD convection of water near its density maximum in a square cavity. International Journal of Thermal Sciences, 2008. 47(9): p. 1184-1194.
[6] Joshi JB, Sharma MM. A Circulation cell model for bubble columns. Trans Inst Chem Eng 1979;57:244–51.
[7] Irani, M., & Khodagholi, M. A, Investigation of bubble column hydrodynamics using CFD simulation (2D and 3D) and experimental validation. Petroleum & Coal, 2011. 53(2), 146-158.
[8] Yuanxin,W., Chen One,B., Al-Dahhan,M.H.: Predictions of Radial Gas Hold up Profiles in Bubble Column Reactors, Chem. Eng. Sci., 56 (2001) 1207-1210.

IOP Conf. Series: Materials Science and Engineering **243** (2017) 012036 doi:10.1088/1757-899X/243/1/012036

CFD simulation of flow through an orifice plate

M M Tukiman[1], M N M Ghazali[1], A Sadikin[1], N F Nasir[1], N Nordin[1], A Sapit[1], M A Razali[1]

[1]Department of Energy and Thermofluid Engineering, Faculty of Mechanical and Manufacturing Engineering, Universiti Tun Hussein Onn, Malaysia

E-mail: azmah@uthm.edu.my

Abstract. In this present paper, the commercial Computational Fluid Dynamics (CFD) is used to predict the flow features in the orifice flow meter. Outcomes of the CFD simulations in terms of profiles of velocity and pressure are discussed in detail. It is observed that the flow is jet-like flow in the core region and the presence of recirculation, reattachment and shear layer regions flow features downstream the orifice. The location of vena-contracta was also estimated from CFD simulations. These results are consistent with other published data.

1. Introduction

The orifice plate is very inexpensive for it is just a flat plate and a thin orifice plate. It is also very easy to install in the pipeline. It works on simple principle of using effects of velocity and pressure variation caused by reduction of the available area for flow. Orifices are widely used in flow metering because of their ruggedness, simple mechanical construction, and other known advantages [1]. The study on effects of orifice to pipe diameter ratio are extensive. The effects of ratio on the mass transfer rate behind the orifice numerically studied by [2] and [3]. The characteristic length and velocity scales in different regions of orifice flows using PIV was studied by [4] and found out that the position of the vena contracta shows a slight orifice to pipe diameter ratio dependence.

CFD tools are also widely used in modelling and analyzing orifice meters [5-8]. The flow through conventional single-hole orifice meter and integrated CFD simulation with measurements to analyze oscillating air flow through an orifice in a circular pipe was studied using used Open FOAM-1.6 by [1]. Some studies have focused on obtaining the associated discharge coefficients [9] and pressure drops [10]. Although the orifice plate is cheap, robust and easy to implement, but one disadvantage of this meter is it maximizes form friction. And it is very serious that by using this kind of meter a large percentage of pressure drop is unrecovered. The fluid velocity is increased at the opening of the orifice plate and not much energy is lost but as it flows through and starts slowing down, much of the excess energy is lost. With the above distinct advantages of a flow meter of high industrial importance, it is necessary to understand the flow pattern of orifice meter to further improve its performance in terms of flow measurement with better accuracy. Therefore, in this paper, CFD simulations are carried out to predict the flow pattern and vena contracta in the orifice meter. The outcomes of the CFD simulations in terms of velocity profile and pressure profile. CFD would candidate as a valuable engineering tool for the design and use of these devices in all those situations, quite common in the oil and gas fields, where experiments may be too complex or time consuming.

IOP Conf. Series: Materials Science and Engineering **243** (2017) 012036 doi:10.1088/1757-899X/243/1/012036

2. Governing equations and modelling assumptions

The governing equations (continuity and momentum) with the appropriate Reynolds stress closure need to be solved. In the present work, standard *k-epsilon* turbulence model by [11] has been used. The *k-epsilon* turbulence model is simple to use, cheap, most widely validated and has excellent performance for many industrial flows [12]. Since the problem is assumed to be steady and incompressible flow, time dependent parameters are dropped from the equations. Below shown the governing equation for the flow in orifice plate;

Conservation of mass;

$$\nabla.\left(\rho\bar{V}\right)=0 \tag{1}$$

y-momentum;

$$\nabla.\left(\rho v\bar{V}\right)=-\frac{\partial p}{\partial y}+\frac{\partial\tau_{yy}}{\partial y}+\frac{\partial\tau_{zy}}{\partial z} \tag{2}$$

z-momentum;

$$\nabla.\left(\rho w\bar{V}\right)=-\frac{\partial p}{\partial z}+\frac{\partial\tau_{yz}}{\partial y}+\frac{\partial\tau_{zz}}{\partial z} \tag{3}$$

3. Geometry, grid and computational domain

A set of CFD simulations is performed to investigate the flow through an orifice meter with a diameter throat of 5 mm and the length is 2 mm. The length of the upstream flow is 246 mm and total length of the pipe is 494 mm. The pipe has 12.30 mm inlet diameter. Adequate length at the upstream and downstream of the orifice plate is to avoid the effect of patched boundary conditions on the flow through the orifice. This also to provide fully developed turbulent flow in the upstream of the orifice plate. The model of the orifice is shown in figure 1.

. The meshing gave a total of 595 000, nodes and had 740,000 elements that consisted of unstructured hexahedral. It was ensured that the viscous sublayer ($y^+ < 5$) and several nodes in the buffer and turbulent zones for fine meshes in the orifice region. The working fluid is water. Static pressure is assigned to the outlet pipe in order to obtain relative pressure drop between inlet and outlet. The inlet velocity profile is assumed to be uniform and set to subsonic flow of 0.38 m/s. The pipe is

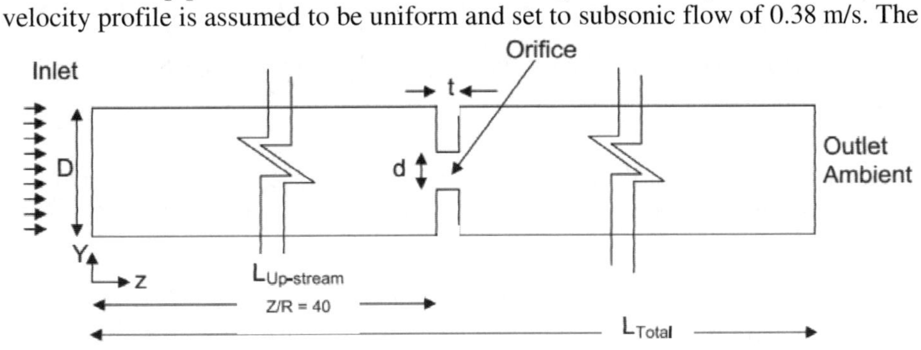

Figure 1. The orifice geometry (orifice is at z/R = 40 from inlet).

IOP Conf. Series: Materials Science and Engineering **243** (2017) 012036 doi:10.1088/1757-899X/243/1/012036

modelled as solid wall with no slip condition. The *k-epsilon* model was chosen for the turbulence model and the numerical accuracy was set to first order. All the discretized equations were solved in a segregated manner with the SIMPLE (Semi Implicit Pressure Linked Equation) algorithm. The simulation was run until the residual of the pressure and velocities were less than 0.00001.

4. Results of calculations

Velocity contour plot and velocity vectors are given in figure 2. As we can see form the figure, as the water flows from a narrowing orifice plate, the flow forms a free flowing jet in the downstream fluid. The velocity is at its maximum. The separation of boundary layer is seen at the downstream side of orifice plate. The turbulent and wake region, together with recirculation zones can be seen just downstream of orifice meter. The flow pattern for orifice flow is well captured in the numerical simulations. These observations are excellent agreement with [1].

(a)

(b)

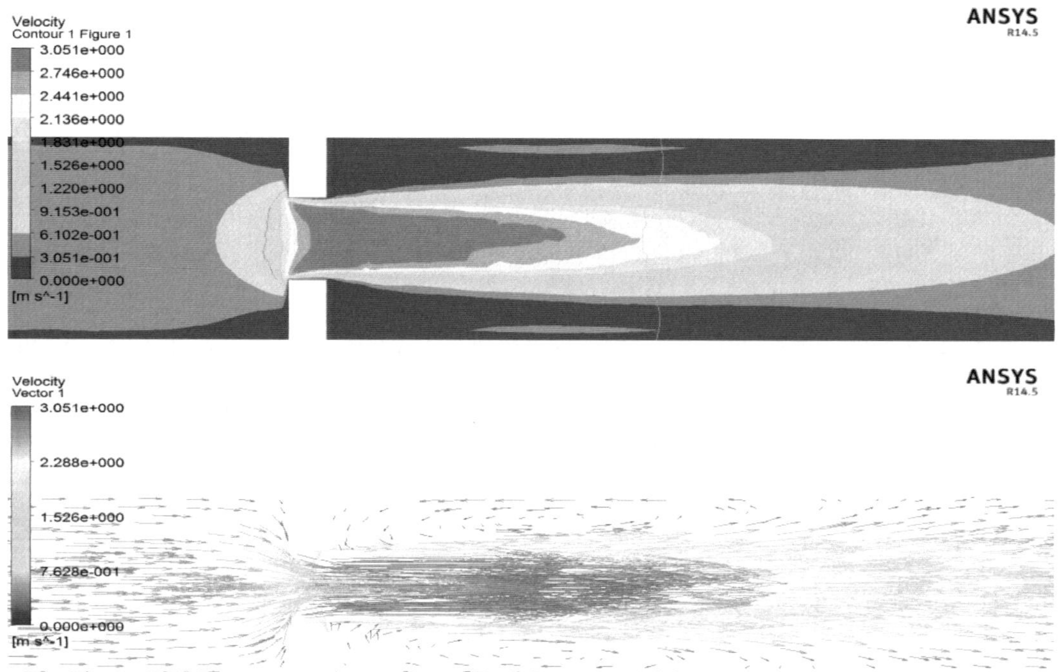

Figure 2. (a) Velocity contours and (b) velocity vectors

Figure 3 shows the velocity profile upstream of orifice at z = 0.15 m from the entrance. The strong velocity gradient near the wall because of the no-slip condition is observed. This region is often very thin, and it is then called a boundary layer. The boundary layers grow in thickness as more and more of the fluid is affected by viscous friction originating from the gradients set up by the no-slip condition. The layers meet in the middle and merge, and the flow reaches an asymptotic state called fully-developed flow. This figure also indicates that the fully turbulent flow is developed at the upstream of the orifice meter. This shows that the length is adequate for pipe upstream for fully developed flow. Nonetheless, the length for pipe downstream of the orifice meter is also adequate for the recovery of pressure.

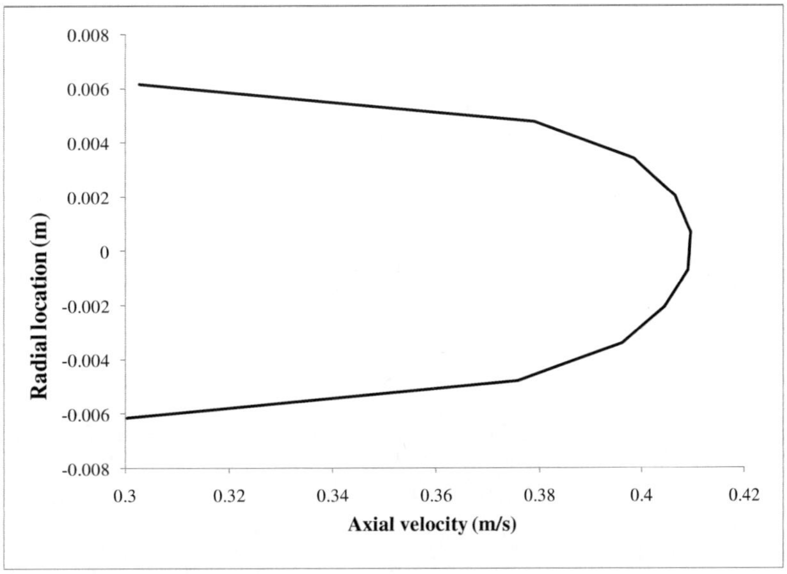

Figure 3. Radial profile velocity upstream of orifice at axial distance of $z = 0.15$ m from entrance

Centerline axial velocity profiles and centerline pressure profile are given in figure 4 (a) and (b) respectively. Centerline axial velocity increases as the flow approaches the throat of orifice meter. The further it travels downstream of orifice meter and reaches at its maxima at z = 0.234 m. This point of maximum velocity is called vena-contracta. Beyond this point the velocity decreases. As for pressure, as we can see from figure 4 (b), energy has to be conserved at all the points in the domain, accordingly the pressure decreases when the flow approaches orifice meter, reaching minimum at vena-contracta and starts recovering as the flow moves further downstream. This is supported by the work of [1].

(a)

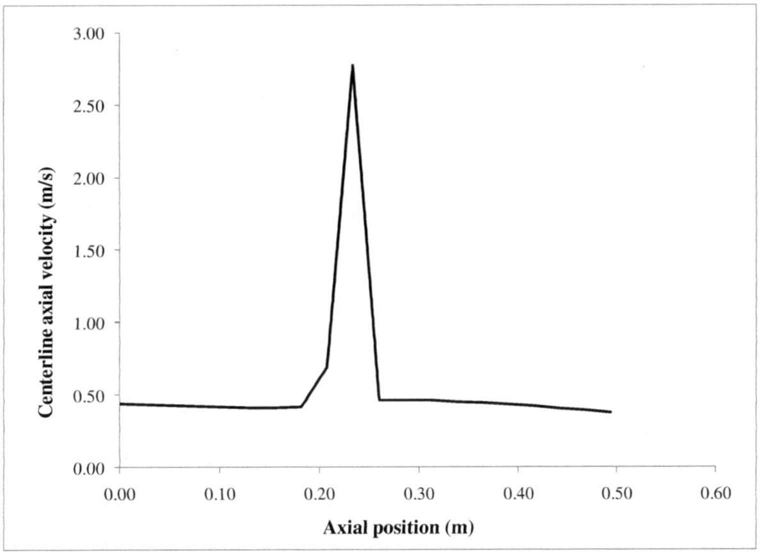

(b)

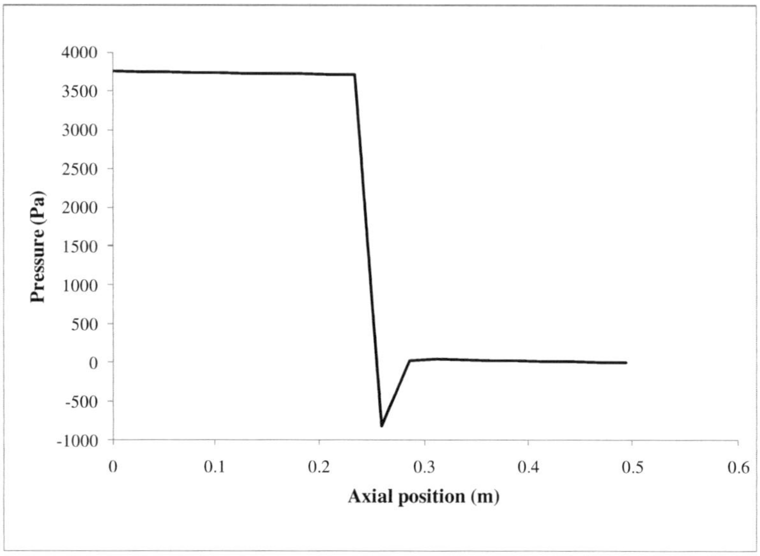

Figure 4. (a) Centerline axial velocity and (b) centerline pressure profile

5. Conclusion

The computational fluid dynamic simulations for orifice meter in a pipeline was carried out. The results agree well with the published data in terms of flow pattern, velocity profiles and pressure profile. The location of vena-contracta was estimated from CFD simulations. It is believed that flow physics, such as the location of the vena contracta and characteristic length and velocity scales in orifice flow, are very interesting for researchers and engineers in the study of flow metering, particularly the effects of the ratio on these flow physics. It is also concluded that the CFD technique can be used as an alternative and cost effective tool towards replacement of experiments required for estimating discharge coefficient, empirically. Further work to use orifice meter on different liquids and gases at different beta ratios are to be carried out in future.

Acknowledgement

The author would like to thank both Universiti Tun Hussien Onn Malaysia (UTHM) and Ministry of Higher Education of Malaysia for the financial support through Research Accultration Grant Scheme (Vot R071).

References

[1] Shah M S, Joshi J B, Kalsi A S, Prasad C S R and Shukla D S 2012 Analysis of flow through an orifice meter: CFD simulation Chem. Eng. Sci. 71 300-309

[2] El-Gammal M, Ahmed W H and Ching C Y 2012 Investigation of wall mass transfer characteristics downstream of an orifice Nucl. Eng. Des. 242 353-360

[3] Ahmed W H, Bello M M, El Nakla M and Al Sarkhi 2012 Flow and mass transfer downstream of an orifice under flow accelerated corrosion conditions Nucl. Eng. Des. 252 52-67

[4] Shan F, Liu Z, Liu W and Tsuji Y 2016 Effects of the orifice to pipe diameter ratio on orifice flows Chemical Engineering Science 152 497-506

[5] Reis L, Carvalho J, Nascimento M, Rodrigues L, Dias F and Sobrinho P 2014 Numerical modeling of flow through an industrial burner orifice Appl. Therm. Eng., 67 201-213

[6] Shaaban S 2014 Optimization of orifice meter's energy consumption Chem. Eng. Res. Des. 92 1005-1015

[7] Singh V and Tharakan T J 2015 Numerical simulations for multi-hole orifice flow meter Flow Meas. Instrum. 45 375-383

[8] Kumar P and Bing M W M 2011 A CFD study low pressure wet gas metering using slotted orifice meters, Flow Meas. Instrum. 22 (1) 33-42

[9] Reader-Harris M 2015 Orifice Discharge Coefficient, Orifice Plates and Venturi Tubes, Springer 127-186

[10] Cioncolini, Scenini F and Duff J 2015 Micro-orifice single-phase liquid flow: pressure drop measurements and prediction Exp. Therm. Fluid Sci. 65 33-40

[11] Launder and Spalding D 1974 The numerical computation of turbulent flows Comput. Methods Appl. Mech. Eng. 3 269-289

[12] Versteeg H K and Malalasekera W 1995 An introduction to computational fluid dynamics: The finite volume method (England:Pearson Education Limited)

IOP Conf. Series: Materials Science and Engineering **243** (2017) 012037 doi:10.1088/1757-899X/243/1/012037

Investigation of wind behaviour around high-rise buildings

Norasikin Mat Isa[1,a], Nurul Fitriah Nasir[1,b], Azmahani Sadikin[1,c] and Jamil Ariff Hairul Bahara[1]

[1]Energy and Technologies Research Group (EnRG), Department of Thermofluid Engineering, Faculty of Mechanical and Manufacturing Engineering, Universiti Tun Hussein Onn Malaysia (UTHM)

E-mail: sikin@uthm.edu.my[a], fitriah@uthm.edu.my[b], azmah@uthm.edu.my[c]

Abstract. A study on the investigation of wind behaviour around the high-rise buildings is done through an experiment using a wind tunnel and computational fluid dynamics. High-rise buildings refer to buildings or structures that have more than 12 floors. Wind is invisible to the naked eye; thus, it is hard to see and analyse its flow around and over buildings without the use of proper methods, such as the use of wind tunnel and computational fluid dynamics software.The study was conducted on buildings located in Presint 4, Putrajaya, Malaysia which is the Ministry of Rural and Regional Development, Ministry of Information Communications and Culture, Ministry of Urban Wellbeing, Housing and Local Government and the Ministry of Women, Family, and Community by making scaled models of the buildings. The parameters in which this study is conducted on are, four different wind velocities used based on the seasonal monsoons, and wind direction. ANSYS Fluent workbench software is used to compute the simulations in order to achieve the objectives of this study. The data from the computational fluid dynamics are validated with the experiment done through the wind tunnel. From the results obtained through the use of the computation fluid dynamics, this study can identify the characteristics of wind around buildings, including boundary layer of the buildings, separation flow, wake region and etc. Then analyses is conducted on the occurance resulting from the wind that passes the buildings based on the velocity difference between before and after the wind passes the buildings.

1. Introduction

Buildings are built higher and higher nowadays to support the human's needs and desire, and it is also because of limited space in cities that are crowded and filled with other buildings. Many of the buildings in crowded cities cannot be built anywhere else but up. When building high rise buildings, there are a few design standards that must be taken into consideration such as handling wind issues. As the building increases in height, it subjected to stronger winds that can cause the building to sway. It is known that building structures can be demolished, collapsed or even blown off when it is put against high winds such as typhoons and hurricanes [1].

Thus to prevent such an event from occurring, engineers have to conduct tests on models of the building before hand in order to see whether or not the building can withstand strong winds that flow around the building. Wind or air flow is moving fluid that cannot be seen by the naked eye, and this makes it a bit difficult to study its behaviour [2]. However, through the use of wind tunnel, such study is able to be conducted in order to study the effect of wind around objects, where streamlines are created after the wind flow passes the objects. Smoke is used to make the flow visible to the naked eye. Through numerical analysis using a software called Computational Fluid Dynamics of CFD, engineers are able to predict and simulate the behaviour of winds around buildings using a computer.

1.1. Literature Review

In modern day, buildings are considered to be the life-support machine, as it is required to provide the necessities for the survival of the human race, such as shelter, security and safety. Not only are buildings needed for humans to be safe from the natural elements that can harm us, such as earthquakes, storms and snow, it is also needed to protect humans from fire. When providing a building of character and aesthetic appeal, it is important that the building is easily, economically and well constructed, with the requirements that it can be easily renovated, maintained, as well as has a sustainable form of construction that can be adapted to changing trends and legislation [3]. Wind, also known as the invisible motion of air with respect to the earth's surface, is caused by the variable solar heating of the earth's atmosphere and is initiated by the difference of pressure between points of equal elevations. The effects on the wind due to friction along the ground can be neglected at a sufficiently great height and the relative horizontal motion of air to the surface is said to be in an unaccelerated flow because of the balance of pressure gradient, the Coriolis and the centrifugal force. However, when near to the ground, the earth's surface exerts a horizontal drag force to the moving air in order to retard the flow. This effect however, decreases, as the height of the flow increases above ground, and as mentioned earlier is negligible above the height δ known as the boundary layer of the atmosphere. It is above this height that the frictionless wind balance takes place with the wind flow with gradient wind velocity along the isobars. Thus, it is of direct interest of the civil engineering structures designer that the wind regime is within the boundary layer of atmosphere [4].

According to Varkute & Maurya, 2013, "CFD has become an influential tool in the building industry. It is used to justify the selection of design option. Not only as a justification or confirmation tool. It has emerged as a shaping and moulding tool also on the drawing board where design is refined" [5].

2. Methodolody

2.1 Scale Models

Using the dimensions of the buildings intended for the study provided by the Malaysian Public Works Department located in Putrajaya, a scaled model of 1:1300 is drawn by using SolidWorks. The drawings of the scale model of the buildings are then sent to be 3D printed to produce a solid scaled model of the buildings for the wind tunnel testing. Scale models of the drawing are then simplified to be imported into the ANSYS FLUENT Workbench for the computational fluid dynamics simulation.

2.2 Wind Speed and Direction

According to Malaysia's Metrological Department, there are four seasons in Malaysia that can be distinguished by the wind direction obtained from the recorded wind flow patterns. These seasons are known as southeast monsoon, northeast monsoon, and two shorter periods of inter-monsoon seasons. The southeast monsoon usually occurs in late June till the end of September while Northeast monsoon occurs in early November till March. The inter-monsoon occurs in between the other two monsoons which are from April to May and September to October (Malaysian Meteorological Department, 2013). The Metrological Department has provided data regarding the maximum daily wind speed near Putrajaya area for the year 2014 until April of 2015. From the data, the wind speed and direction is separated in four groups according to the type of monsoon season. The highest velocity of wind for southeast monsoon is 9.81 m/s with the direction of the wind blowing from the west, the northeast monsoon is 16.9 m/s with the wind blowing from the east and the inter-monsoon of April to May is

10.8 m/s with the wind blowing from the west, and lastly the inter-monsoon of September to October is 13.3 m/s with the wind blowing from the south.

2.3 Wind Tunnel Testing

Wind tunnel testing is conducted on the printed 3d models of the buildings. The building is oriented by a turntable which moves the buildings according to the direction of wind. A Pitot tube is placed at the centre of the buildings. 32 points were marked on the Pitot tube to ease in the taking of velocity data from each of the marked positions. The tests are conducted based on the four seasonal monsoons that occur in Malaysia.

2.4 Computational Fluid Dynamics

The Computational Fluid Dynamics software used for the numerical analysis method is the Workbench15.0. In CFD, the process conducting the numerical analysis is divided into three stages, which are the pre-processing, solvers and post-processing. The type of analysis conducted using this software is Fluid Flow (Fluent). CFD has been proven to be beneficial and useful in various industries as it provides a way to analyse data of a system while saving cost and time from doing the actual experiment. The simplified models are imported into the Workbench 15.0 and the simulation is conducted on the buildings based on the seasonal monsoons that occur in Malaysia.

2.5 Validation and Grid Independent Study

A test is conducted on buildings based on the Southwest Monsoon in order to validate the CFD result with the results obtained from the wind tunnel testing. Furthermore, verification also been done in order to determine which k-ε turbulence models to be used. The grid independent study is also conducted to choose the type of mesh to be used for the simulation. From the validation, it is found that the error between the simulation and the wind tunnel testing is about 17%. The k-ε turbulence model used is the Realizable k-ε turbulence model and the mesh type were determined to be Medium with 70000 elements and 14139 nodes.

3. Result and Discussion

Four wind velocities are used as the initial velocity based on the seasonal monsoons and the buildings are oriented in accordance with the direction the wind. The results obtained from the computational fluid dynamics are separated into three different layers or planes, which are 0.00154 m, 0.066 m and 0.127 m as shown in Figure-1.

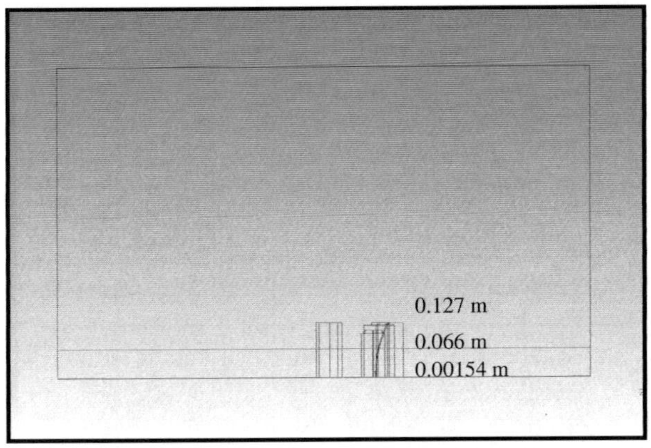

Figure 1: Example of the three planes made for the velocity contour

These reference planes are made to observe the wind behaviour at 2 m from the ground for pedestrian comfort (scale wise), the behaviour in the middle of the buildings and the behaviour that occurs 3 m above the building. From these planes, the velocity contour and the velocity vector can be seen, which represents the wind behaviour around the buildings. The effect of the wind passing the buildings can be seen based on the seasonal monsoons.

The velocity of the wind decreases as it hits the surface of the building due to boundary layer that occurs at the surface. As the wind is stagnated at the surface, it begins to separate, causing the wind to flow around and over the building. This causes wake regions to develop behind the buildings. Recirculation also occurs behind the buildings and irregular vortices mixed together, causing a turbulence behind the buildings. The wind velocity also increases as the separated wind around building mixes together. The wind at the top of the building also increases in flow over the building. An example of the occurring behavior can be seen in Figure-2.

Figure 2: An example of side view of the wind flow behaviour that occurs in one of the Monsoons.

The wind flow velocity contour as seen in Figure-3 shows that, at 0.00154m, the wind velocity is not more than 5 m/s. This means that the wind velocity is suitable for pedestrians, as it is known that only wind velocity of over 20 km/h or 5.56 m/s can bring discomfort to pedestrians [6].

IOP Conf. Series: Materials Science and Engineering **243** (2017) 012037 doi:10.1088/1757-899X/243/1/012037

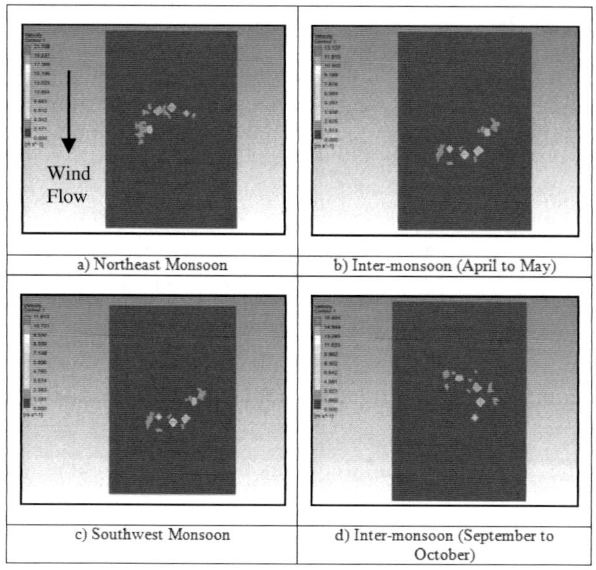

Figure 3: The top view of the velocity Contour of the monsoons at 0.00154 m

The results obtained for all the planes show the different behaviour of the wind as the height of the buildings increase, which can be seen in Figure-4 at the height of 0.06m, and Figure-5 at 0.127 m. Figure-6 shows one of the side views of the velocity contour obtained from one of the monsoons.

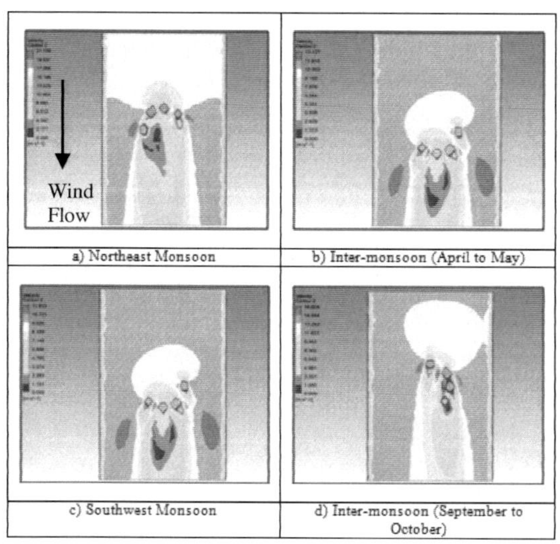

Figure 4: The top view of the velocity contour of the monsoons at 0.06 m

At 0.06 m, the wind flow velocity can be seen changing from its intial velocity as the colour indicates. For the Northeast Monsoon, the initial velocity is 16.9 m/s. However, when the wind hits the surface of the building, the velocity changes to approximately 21 m/s in areas indicated by the red colour. This can be seen with all the seasonal monsoons, such as Inter-monsoon (April to May) with an increase from 10.8 m/s to approximately 13 m/s, Southwest Monsoon from 9.8 m/s to approximately 11 m/s, and Inter-monsoon (Sept to Oct) with an increase from 13.3 m/s to approximately 16 m/s. This increase in velocity is indicated by the red colour as depicted in Figure-6.

The wake region generated behind the building can also be observed, which is indicated by the blue colour, that proves that the velocity is very low – close to 0 m/s – in that area.

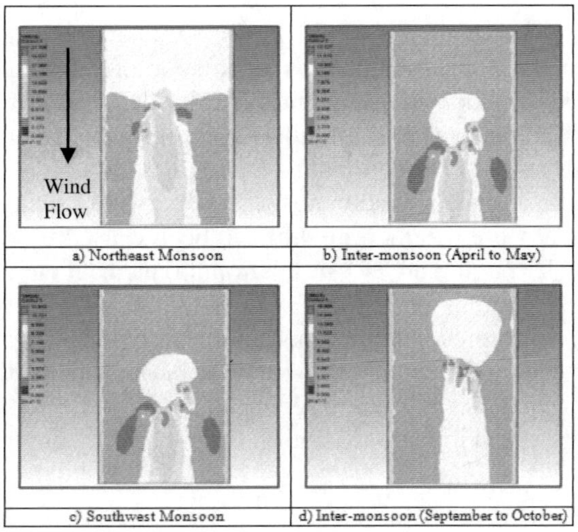

Figure 5: The top view of the velocity contour of the monsoons at 0.127 m

At 0.127m, the wind behaviour that takes place just 3m above the height of the tallest buildings can be seen. The only difference between the wind flow characteristics at this level when compared to that of at 0.06 m is the size of the wake region which decreases and becomes more slender, almost converging back into the uniform flow.

Figure 6: An example of the side view of the velocity contour obtained from one of the monsoons.

In Figure-6, the wind can be seen moving in a uniform flow as it heads towards the building as indicated by the orange colour. As it nears the building, the velocity decreases and becomes stagnated at the surfaces of the building which is caused by the surface boundary layer of the buildings. This forces the flow to separate, which results in the forming of the wake region behind the buildings. The velocity of the wind flow increases as it hit the top edge of the building. This might have been due to the mixing of the separation flow with the uniform flow that occurs above the buildings.

4. Conclusion

From this study, the overall conclusion that can be made is that the flow characteristic and behaviour of wind around high-rise buildings can be seen through the use of the ANSYS Fluent software. There are three focus points that are taken from the y-axis, which are at 0.00154 m, 0.066 m, and 0.127 m. These points represent the pedestrian height of 2 m, half the height of the total buildings, and 3m above the buildings. From these points, the flow behaviour of the wind can be seen through the velocity vector and velocity contour of the wind that hits the buildings based on the seasonal monsoons. There are a lot of advantages when conducting simulation through this software in terms of time and cost. Based on the results obtained from the experiment and the simulation done, there are some differences between them, and that both ways can be used to analyse wind behaviour around buildings. The differences that were obtained might have been caused by the simplified models of the buildings used in the CFD. Through the CFD simulation, boundary layer, wake region, separation of flow, and recirculation of flow can be seen. The layout of the building also plays an important role in reducing the negative effect of winds, such as gust and whirlwind. Thus, the distance between the buildings must be taken into consideration when building high-rise buildings. The results that were obtained also show that the wind that hits the building is below the discomfort velocity. This proves that a lot of aspects have to be taken into consideration when building these high-rise buildings.

5. Acknowledgement

The work was partly supported by University Tun Hussien Onn Malaysia, Johor, Malaysia, under the RAGS Grant Vot R071.

References

[1] Holmes, J. D., 2007. *Wind Loading of Structures.* 2 ed. s.l.:Taylor and Francis Group.
Malaysian Meteorological Department, 2013. *Official Website Malaysian Meteorological Department.* [Online] Available at: http://www.met.gov.my/index [Accessed 17 2 2015].
[2] Sato, M. & Takaya, K., 2012. A fundamental study of the flow past a circular. *SIMULIA Community Conference,* pp. 1-5.
[3] Osbourn, D. & Greeno, R., 2007. *Mitchell's Introduction to Building 4th Edition.* 4 ed. s.l.:Pearson Education Limited.
[4] Simiu, E. & Scanlan, R. H., 1996. *Wind Effects on Structures.* s.l.:A Wiley-Interscience Publication
[5] Varkute, N. S. & Maurya, R., 2013. CFD Simulation in Township Planning- A Case Study. *International Journal of Computational Engineering Research,* 3(3).
[6] Planning and Building Department, Development and Design Division, City of Mississauga, June 2014. Urban Design Terms of Reference. *Pedestrian Comfort and Safety Studies.*

Recent advances in fluidized bed drying

N S Haron[1], J H Zakaria[1] and M F Mohideen Batcha[1]
[1]Centre for Energy and Industrial Environment Studies,
Faculty of Mechanical Engineering and Manufacturing,
Universiti Tun Hussein Onn Malaysia,
86400 Parit Raja, Batu Pahat, Johor, Malaysia

Email: mfaizal@uthm.edu.my

Abstract. Fluidized bed drying are very well known to yield high heat and mass transfer and hence adopted to many industrial drying processes particularly agricultural products. In this paper, recent advances in fluidized bed drying were reviewed and focus is given to the drying related to the usage of Computational Fluid Dynamics (CFD). It can be seen that usage of modern computational tools such as CFD helps to optimize the fluidized bed dryer design and operation for lower energy consumption and thus better thermal efficiency. Among agricultural products that were reviewed in this paper were oil palm frond, wheat grains, olive pomace, coconut, pepper corn and millet.

1. Introduction

Fluidized bed dryer (FBD) has been used widely for drying various products due to its many advantages, including high drying rate due to an excellent gas-solid contact, high thermal efficiency, relatively low cost of operation etc [1]. In fluidized bed drying, hot air is forced through a distributor into bed at a sufficiently high velocity to overcome the gravitational forces on the products. When the air velocity is greater than the gravitational force and the bed resistance, the products will suspend. This condition is called fluidization which is a physical process that transforms solid particles into a fluidized state through suspension in a liquid or gas [2, 3]. The particles are fluidized in bed when the drag force created by the gas flow through the bed is equal to the weight of the particles [4]. When fluidization occurs, the solid particles have many properties of a liquid. One noticeable property is the fluidized particles seek to level and assume the shape of the containing vessel. Large, heavy objects sink when added to the bed, and light particles float [5]. Before drying process was conducted in fluidized bed system, it is importance to understand the system characteristics and bed behaviour during fluidization. Among that aspects identified are regimes of operation, minimum fluidization velocity (U_{mf}) particle entrainment velocity, bed voidage (ε), bed pressure drop (ΔP_b), superficial velocity (V_s) and minimum fluidization velocity (U_{mf}). Pressure fluctuations in the fluidized bed column to characterize the bed behaviour were measured at a certain location in the axial direction in the bed above the distributor [6-11].

2. Drying of agro products in FBD

As reported by many researchers, fluidized bed technology has been successfully used in drying of wet solid particles as shown in Table 1. Table 2 summarizes the FBD performances studies.

Table 1. Drying of agro product in fluidized bed dryer

Particles	Aim	Findings
Oil palm frond [12]	Investigation of drying kinetics in agitated fluidized bed dryer.	- Agitation in fluidized bed allowed fluidization of the fibrous material and hence homogeneous mixing in the bed was achieved. - Drying rate increased with the increase in drying temperature and superficial air velocity, however, decreased with an increase in bed load. - Drying rate was greatly affected by drying temperature, followed by air velocity, bed load and lastly agitation speed. - Proposed drying rate model provided the best fit model among all of the models.
Wheat grains [13]	Present drying kinetics of wheat grain and the effects of the swirling flow field on the drying performance.	- Moisture removal rate, the specific moisture removal rate and the dryer efficiency effected by air mass flow rate, drying temperature and flow field characteristics - Drying rate in swirling flow fields are higher than in non-swirling flow fields. - Swirling motion increase drying rate hence, decrease drying time.
Olive pomace [14]	Investigate the effect of drying air temperature and bed height on the drying kinetics of olive pomace in a fluidized bed dryer, and to fit the experimental data to known mathematical models.	- Drying process of olive pomace occurred in falling rate period. - Effective diffusivity values were calculated between 0.68 and 2.15 x 10^{-7} m^2/s, increasing with increase of temperature or bed height of the sample. - Higher activation energy values resulted by higher bed height of the sample.
Coconut [15]	Effects study on drying kinetics with different operating parameters such as inlet air velocity and drying or air temperature and quality of coconut colour and surface oil content.	- Drying rate become higher since higher drying temperature or higher inlet air velocity was used. - Less drying temperature remove a lot of surface oil content, while higher inlet air velocity led to higher quantity of surface oil content. - Colour of coconut was affected by air temperature.
Pepper corn [16]	Comparison on drying kinetics between typical distributor plate and helical distributor plate in fluidized bed.	- Different velocities leads to no significant effect on drying rate for typical distributor plate but provides a substantial influence for helical distributor plate. - Swirling drying bed reduce drying time than the typical fluidized-bed
Millet [17]	Study on drying kinetics of millet in a batch fluidized bed.	- Drying rate and effective diffusivity were found to increase with increase in drying temperature and air velocity while decrease with increase in solids holdup. - Effective diffusivity of millet is found to be lower than that of ragi (food grain) and corn.

Table 2. Study on fluidized bed dryer performance

Particles	Aim	Results/Outcome
Spherical PVC particles [6]	Study on hydrodynamic characteristics in the swirling regime of operation.	- Three or four regime of operation observed depend on bed weight. - Pressure drop in swirling regime increase with increase in air flow rate. - Low distributor pressure drop recorded in swirling fluidized bed.
Jowar seed, ragi seed, wheat [18]	Understanding the hydrodynamics behaviour of swirled fluidized bed.	- Minimum swirl velocity increase with higher bed height, bigger column diameter, smaller fluid inlet size and less fluid inlets. - Minimum swirl velocity also affected by particle properties. - No bubble formation, gas channeling, solids segregation and slugging of bed.
PVC material [19]	Heat transfer study in swirling fluidized bed dryer.	- There are a few regime of operation; incipient mode, partial swirling, stable swirling and two-layer fluidization. - Heat transfer coefficient, h, dissipates with higher bed height. - Higher heat transfer coefficient resulted with smaller particles than bigger particles. - Cylinder shaped center-body having higher heat transfer coefficient than cone shaped center-body.
Granular material [20]	Study the variation in diffusion coefficient with different sizes of solid holdup.	- Higher bed temperature indirectly increases the rate of diffusion of moisture, thus increasing the drying rate. - Higher effective diffusivity coefficient resulted from smaller bed loading. - Fick's diffusion equation only applied to individual particle with assumption that all the materials are exposed to uniform drying process conditions.
Paddy [21]	Investigates the drying characteristics of a fluidized bed dryer by inclining the bed, subsequent using spiral inside the drying chamber.	- Energy consumptions for inclined bed were lower compared to vertical bed dryer. - Blower energy consumption and heat input decreased with inclination of bed and the use of spiral. - The results indicate the highest drying efficiency and minimum energy consumption for the inclined bubbling fluidized bed dryer with spirals.
Wheat [22]	Study on drying kinetics of millet in a batch fluidized bed.	- Energy efficiency and exergy efficiency was found to be decrease as the bed hold up decreased. - No significant difference in drying time for different initial moisture ratio. - Energy and exergy efficiency increase with decreasing air velocity or increasing drying temperature.

IOP Conf. Series: Materials Science and Engineering **243** (2017) 012038 doi:10.1088/1757-899X/243/1/012038

3. Assisted fluidized bed dryer

Reports by [23, 24] presented potato slices and macaroni beads dried using microwave assisted FBD. FBD were placed in microwave oven by cutting circular holes with diameter of 10 mm at top and base of the microwave oven. The microwave+FBD then covered with a removable rectangular container which made of perforated stainless-steel sheets to ensure that no microwaves escaped into the surroundings. The other parts of the system were the same as those in the FBD [23]. The microwave oven assisted FBD significantly reduced drying time of the particles [24].

Paddy was dried in multi-stage dryer consist of FBD, far-infrared (FIR) irradiation and ambient air ventilation. The experimental results showed that FIR assisted FBD reduced moisture content of paddy about 2% d.b. difference rather than FBD drying process [25]. A spouted bed dryer was assisted by solar collector [26]. This type of combination consists of solar collector which covered with glass sheet, blower and spouting column to dry peas. Moisture removal in this dryer was increased compare to the sun drying method. Another hybrid FBD were reviewed by [27] whereby the FBD dryer was immersed in a heater, heat pump assisted FBD dryer and ultrasonic transducer assisted FBD dryer.

4. Computational fluid dynamics (CFD) study in FBD

CFD was used as numerical tool to get better understanding of airflow distribution through stainless steel woven wire mesh distributor as reported by [28]. In the report, CFD validation and experimental results were in good agreement in velocity and pressure drop characteristics. Other than that, a uniform airflow contacting upon the distributor, modelling of the distributor was better performed using the 2-D porous jump boundary condition. However, when dealing with unequal airflow conditions contacting upon the distributor, the distributor should be modelled using the full 3-D porous media model. CFD modelling also reveals that non-homogenous airflow through the distributor cause by tangential air inlet.

CFD simulation carried out to compare with experimental hydrodynamics study of coarse and fine particles. Bed height and pressure drop characteristics of both coarse and fine particle were compared to the CFD simulation show less than 7% and 14% deviations respectively. For CFD simulation, a few assumptions such as isothermal non-reactive, no lift force, no mass transfer between gas and solid phase, constant pressure gradient and constant density of each phase were considered [29].

A multifluid Eulerian model and kinetic theory of granular flow for solid particles was applied to simulate the gas–solid flow in Fluent 6.0. Momentum exchange coefficients were calculated using drag functions from literature while, solid-phase kinetic energy fluctuation was characterized by varying the restitution coefficient values from 0.9 to 0.99. 2D computational modelling well predicted with experimental bed hydrodynamics such as pressure drops at different superficial velocities and minimum fluidization velocity, U_{mf} [30].

Hydrodynamic and thermal behaviour are studied to verify particles temperature distribution in pseudo-2D fluidized bed. Experimental results were validated with computational fluid dynamics and discrete element model (CFD-DEM) concerning heat transfer in the fluidized bed with heat production in the particle phase [31].

Eulerian–Granular multiphase model were used in CFD to predict gas-solid flow of fluidization in circulating fluidized bed. Prediction on main features of the complex gas–solids flow, coexistence of up-flow in the lower regions and downward flow in the upper regions at the wall for high gas velocity and solid mass flux were found. [32].

5. Conclusion

In conclusion, vast study of agro product drying in FBD has been proposed in the literature. These FBDs are unique in design and usually product specific. Among studied parameters were hydrodynamics, energy and exergy consumption apart from proposal of various mathematical models to predict drying characteristics. The design and operation of FBD were also studied by researchers using CFD for optimization of the system.

Acknowledgement
The authors would like to extend theirs deepest gratitude to Centre for Graduate Studies, Universiti Tun Hussein Onn Malaysia (UTHM) for sponsoring this research. Also our sincere thank to the Centre for Energy and Industrial Environment Studies in the completion of this study.

References

[1] Mujumdar A S and Devahastin S 2003 Applications for fluidized bed drying In ed W. C. Yang, *Handbook of fluidization and fluid particle systems* (New York: Marcel Dekker) p 469-84
[2] Geldart D 1973 Types of gas fluidization *Powder Technology* **7(5)** 285-92
[3] Kunii D and Levenspiel O 1991 *Fluidization Engineering* 2nd *Ed.* (Butterworth-Heinemann London) p 1-92
[4] Perry R H, Green D W and Maloney J O 1997 *Perry's Chemical Engineers Handbook* (New York: McGraw-Hill)
[5] Doymaz İ and İsmail O 2010 Drying and rehydration behaviors of green bell peppers *Food Sci. Biotechnology* **19**(6) 1449-55
[6] Sreenivasan B and Raghavan V R 2002 Hydrodynamics of a swirling fluidised bed *Chemical Engineering and Processing: Process Intensification* **41**(2) 99-106
[7] Lim K S, Zhu J X and Grace J R 1995 Hydrodynamics of gas-solid fluidization *International journal of multiphase flow* **21** 141-93
[8] Felipe C A S and Rocha S C S 2007 Prediction of minimum fluidization velocity of gas-solid fluidized beds by pressure fluctuation measurements-analysis of the standard deviation methodology *Powder Technology* **174**(3) 104-13
[9] Chyang C S and Lin Y C 2002 A study in the swirling fluidizing pattern *Journal of Chemical Engineering of Japan* **35**(6) 503-12
[10] Mohideen M F, Md Seri S and Raghavan V R 2012 Fluidization of Geldart Type-D Particles in a Swirling Fluidized Bed *Applied Mechanics and Materials* **110-116** 3720-27
[11] Zakaria J H, Zaid M H H M, Batcha M F M and Asmuin N 2015 Feasibility study on pliant media drying using fluidized bed *IOP Publishing* **88**(1) p 012071
[12] Puspasari I, Talib M Z M, Daud W R W and Tasirin S M 2012 Drying kinetics of oil palm frond particles in an agitated fluidized bed dryer *Drying Technology* **30**(6) 619-30
[13] Özbey M and Söylemez M S 2005 Effect of swirling flow on fluidized bed drying of wheat grains *Energy Conversion and Management* **46**(9) 1495-512
[14] Meziane S 2011 Drying kinetics of olive pomace in a fluidized bed dryer *Energy Conversion and Management* **52**(3) 1644–649
[15] Niamnuy C and Devahastin S 2005 Drying kinetics and quality of coconut dried in a fluidized bed dryer *Journal of Food Engineering* **66**(2) 267-71
[16] Boonloi A and Promvonge P 2006 Drying kinetic of pepper corns in a fluidized–bed with helical distributor plate In *The 2nd joint international conference on sustainable Energy and Environment*
[17] Srinivasakannan C and Balasubramanian N 2009 An investigation on drying of millet in fluidized beds *Advanced Powder Technology* **20**(4) 298-302
[18] Kumar S H and Murthy D V R 2010 Minimum superficial fluid velocity in a gas–solid swirled fluidized bed *Chemical Engineering and Processing: Process Intensification* **49**(10) 1095-100
[19] Mohideen M F, Sreenivasan B, Sulaiman S A and Raghavan V R 2012 Heat transfer in a swirling fluidized bed with geldart type-D particles *Korean Journal of Chemical Engineering* **29**(7) 862-67
[20] Srinivasakannan C and Balasubramanian N 2008 An analysis on modeling of fluidized bed drying of granular material *Advanced Powder Technology* **19**(1) pp.73-82
[21] Thant PP, Robi P S and Mahanta P 2015 Performance Enhancement of Inclined Bubbling FBD on Paddy, *International Journal of Engineering and Applied Sciences (IJEAS)* **2**(8) 2394-3661

[22] Syahrul S, Hamdullahpur F and Dincer I 2002 Exergy analysis of fluidized bed drying of moist particles *Exergy, an International Journal* **2**(2) 87-98
[23] Reyes A, Ceron S, Zuniga R and Moyano P 2007 A comparative study of microwave-assisted air drying of potato slices *Biosystems Engineering* **98**(3) 310-18
[24] Goksu E I, Sumnu G and Esin A 2005 Effect of microwave on fluidized bed drying of macaroni beads *Journal of Food Engineering* **66**(4) 463-68
[25] Meeso N, Nathakaranakule A, Madhiyanon T and Soponronnarit S 2004 Influence of FIR irradiation on paddy moisture reduction and milling quality after fluidized bed drying *Journal of food engineering* ***65***(2) 293-301
[26] Sahin S, Sumnu G and Tunaboyu F 2013 Usage of solar-assisted spouted bed drier in drying of pea *Food and Bioproducts Processing* **91**(3) 271-78
[27] Sivakumar R, Saravanan R, Perumal A E and Iniyan S 2016 Fluidized bed drying of some agro products–A review *Renewable and Sustainable Energy Reviews* **61** 280-301
[28] Depypere F, Pieters J G and Dewettinck K 2004 CFD analysis of air distribution in fluidised bed equipment *Powder Technology* **145**(3) 176-89
[29] Sahoo P and Sahoo, A 2014 A Comparative Study on Fluidization Characteristics of Coarse and Fine Particles in a Gas-Solid Fluidized Bed: CFD Analysis *International Journal of Engineering Science and Innovative Technology* **3** 246-52
[30] Taghipour F, Ellis N and Wong C 2005 Experimental and computational study of gas–solid fluidized bed hydrodynamics *Chemical Engineering Science* **60**(24) 6857-867
[31] Li Z, Janssen T C E, Buist K A, Deen N G, van Sint Annaland M and Kuipers J A M 2017 Experimental and simulation study of heat transfer in fluidized beds with heat production *Chemical Engineering Journal* **317** 242-57
[32] Almuttahar A and Taghipour F 2008 Computational fluid dynamics of a circulating fluidized bed under various fluidization conditions *Chemical Engineering Science* **63**(6) 1696-709

IOP Conf. Series: Materials Science and Engineering **243** (2017) 012039 doi:10.1088/1757-899X/243/1/012039

Numerical simulation of tangential inlet configuration for plenum chambers

A N Mohammed[1], J M Sheriff[2], M F Mohideen Batcha[1], A Sapit[1] and M A Razali[1]

[1] Faculty of Mechanical and Manufacturing Engineering, Universiti Tun Hussein Onn Malaysia,

[2] Faculty of Mechanical Engineering, Universiti Teknologi Malaysia

E-mail: akmaln@uthm.edu.my

Abstract. Swirling fluid motion in enclosed chambers was studied using Computational Fluid Dynamics. Using the tangential inlet configuration as the basic design, 3 swirl generator models was created using Computer Aided Design software. The aim was to see whether a modified design from the original configuration could provide a reduction in the backflow effect that is constantly present in swirling flows. Simulations show that swirl generator inlets at different angles from the original tangential position results in a change in velocity profiles across the flow cross section. From the simulations performed, it was found that the swirl generator model with inlets set to 45 degrees produced the least backflow compared to other models.

1. Introduction

Swirl is a three-dimensional phenomenon in fluids flowing in a vessel with a circular cross-section at speed, where some of the particles follow a spiralling trajectory around the longitudinal axis of the vessel as they move along in the flow. Combustion systems that operate based on the principles of swirling fluid motion to facilitate the combustion process have increasingly been in extensive use [1]. Swirl flows are now utilized in modern combustion machinery such as refinery or power station burners, gas turbine combustors, and internal combustion engines.

An interesting phenomenon of swirling flows in enclosed boundaries is the existence of a precessing vortex core along the longitudinal axis of the flow [2]. The core of the swirl flow is visually distinct from the rest of the fluid movement, and can be described as having the shape of a continuous corkscrew spinning in the opposite direction of the general flow. The behaviour of this vortex core is a significant occurrence in high velocity flows, as it affects the aerodynamic and thermal performance of the downstream turbine [3]. In the combustion process, the existence of the vortex core in the combustion mixture flow will cause the flame itself to have the form of a vortex [4].

The current study is principally based on a conference paper entitled 'Computational Analysis of Turbulent Swirling Flows for Gas Turbine Combustor Applications' by Benim et al. (2007) [2]. Consistent with the title of their presentation, the authors' work centres on investigating the existence of highly rotating vortex core in the flow within the combustion chamber of a modern gas turbine combustor and the impact of this phenomenon on the nozzle guide vanes at the combustor/turbine interface. They firstly sought out to validate the predictive capability of current modelling procedures for turbulent swirling flows, especially near the sub/supercritical vortex core transition. They then analysed the effectiveness of these predictions as compared to experimental results done on two

IOP Conf. Series: Materials Science and Engineering **243** (2017) 012039 doi:10.1088/1757-899X/243/1/012039

different laboratory test rigs with appropriately arranged parameters. It has been found that the tangential inlet configuration that is used to produce swirl flow can lead to undesirable backflow.

Thus, for the current study, a newer design is proposed in order to seek possible improvements that can be made compared to the conventional design [5]. For the project, the number of inlets is increased from three to six, whilst the inlet orientation is varied from the original tangential position. The difference of results from three different inlet orientations is then examined and compared.

2. Methodology

For the model case considered, simulations were run for at least three different speeds. Other simulation parameters were kept constant for all simulation runs.

Table 1. Simulation parameter values.

Parameter	**Value**	**Units**
Density	1.225	kg/m^3
Viscosity	1.789×10^{-05}	kg/ms^{-1}
Specific Heat, Cp	1006	J/kg-K
Molecular Weight	28.97	kg/kg-mol
Reynolds Number [a]	2300 ~ 18000	
Strouhal Number	~ 0.86	
Swirl Number [a]	0.07 ~ 1.0	

[a] The range given are based on the minimum and maximum inlet velocities set in the simulation.

All of the turbulent models used in the study were picked based on their capability of resolving three-dimensional transient motion of coherent flow structures without assuming a scalar turbulent viscosity at all scales; they are the URANS-RSM, LES and DES modelling approaches.

The model geometry used for simulation work has inlets that are positioned on the sides of the primary cylinder (main body) as shown in Figures 1, 2, and 3 depicting the three configurations named Model 1, Model 2, and Model 3 respectively. An example of the computer generated image for the model is shown in Figure 4, whilst in Figures 5 and 6, sample mesh illustrations of the simulation model is shown. The length of the inlet tubes for this geometry is 124.88 mm at maximum and 20 mm minimum, whilst the main body has a total length of 500 mm as shown in Figure 7. Each of the inlet diameter is set to 30 mm, with the main body being 305 mm in diameter. This geometry was chosen since tangential inlet is a traditional design commonly found for this type of swirl generators.

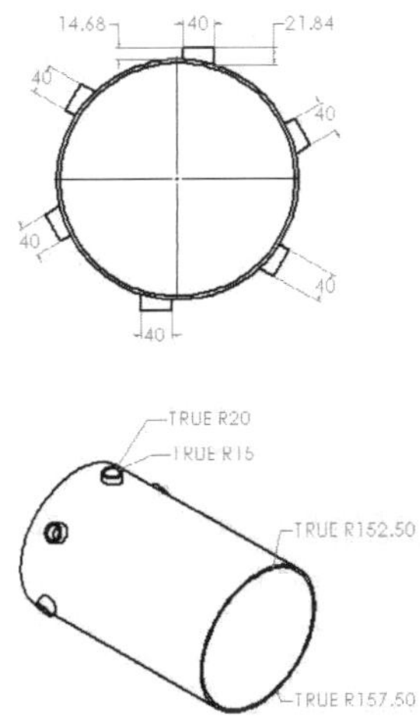

Figure 1. In Model 1, the six inlets are offset by 10° from perpendicular to the plenum.

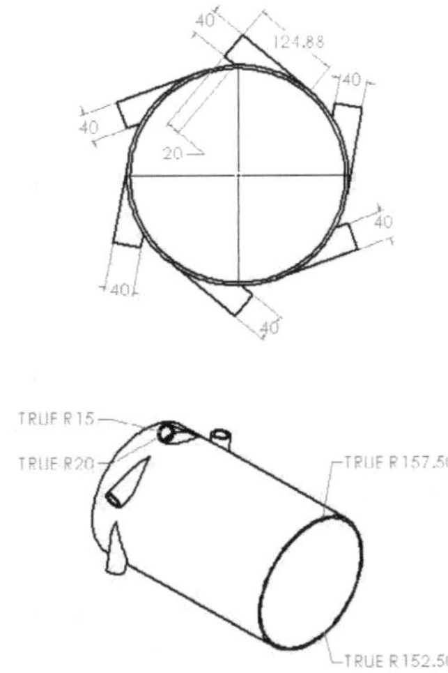

Figure 2. In Model 2, the inlets are tangential to the main plenum chamber.

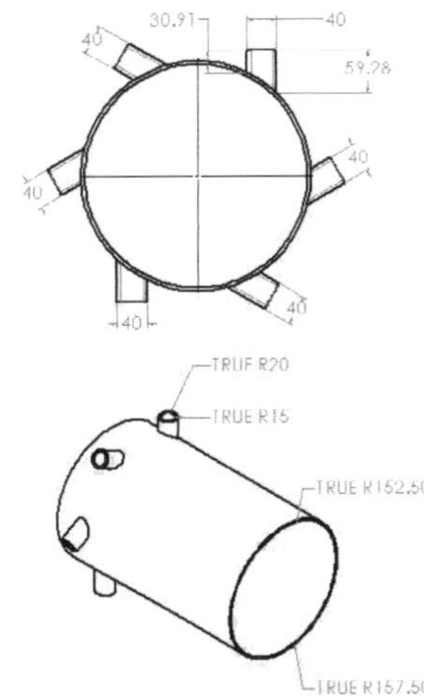

Figure 3. In Model 3, the inlets are slanted by 45° from perpendicular to the plenum.

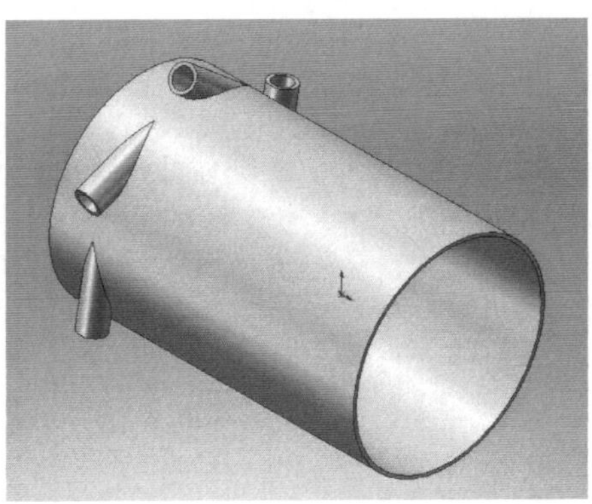

Figure 4. Computer generated model of the swirl generator.

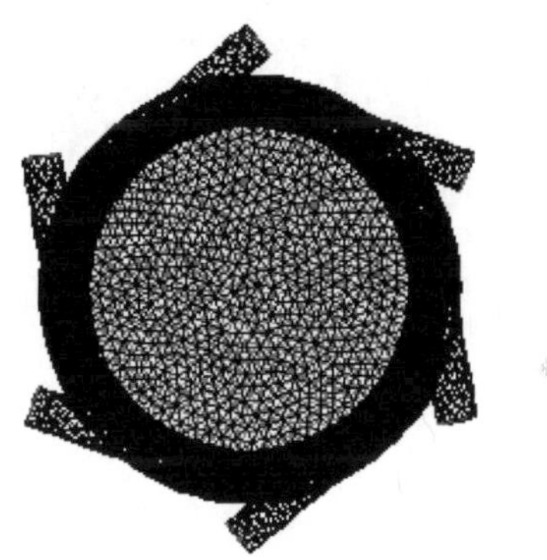

Figure 5. Sample model mesh of the flow domain viewed from the front.

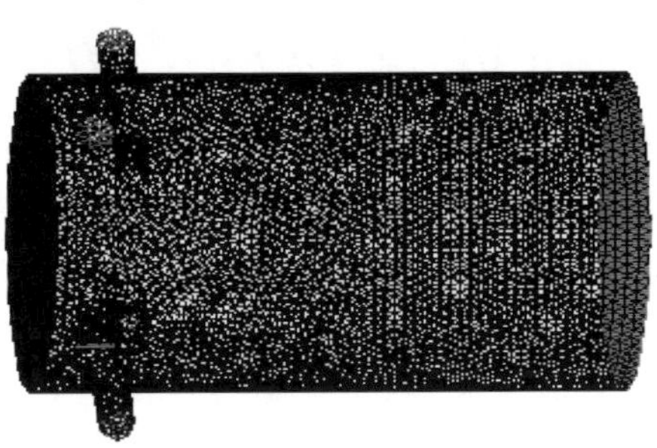

Figure 6. Sample model mesh of the flow domain viewed from the side.

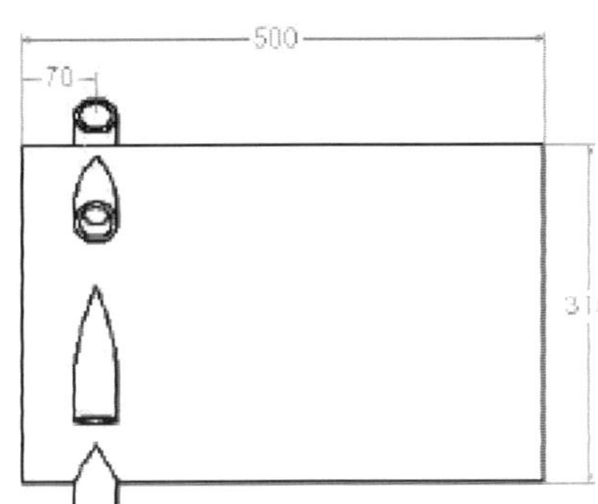

Figure 7. Dimensions of the main plenum and the position of the inlets.

IOP Conf. Series: Materials Science and Engineering **243** (2017) 012039 doi:10.1088/1757-899X/243/1/012039

3. Results and discussion

Figure 8 shows the velocity vector plot for the finished simulation run at an inlet velocity of 10 m/s. Following that is Figure 9, which shows a velocity contour plot at the outlet for the same case.

Figure 8. Velocity vector plot for the case of Model 2 at inlet velocity of 10 m/s.

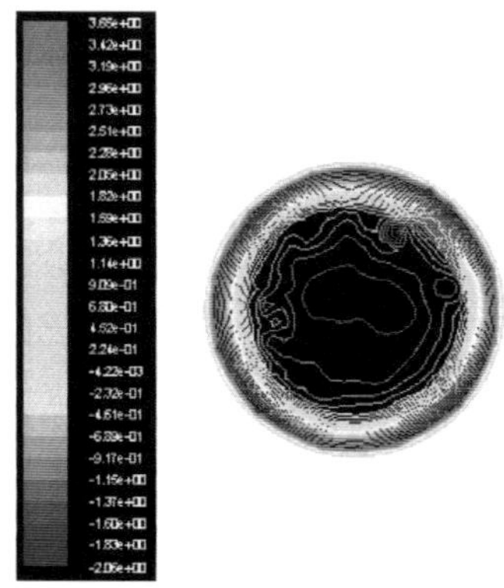

Figure 9. Contour plot for Model 2 at 10 m/s inlet velocity.

Figure 10 shows the velocity profile along the centre axis of the cross section at the exit of Model 1. The high velocity contours occur around the outer regions of the cross-section, which eventually drops off around the centre. This indicates backflow occurs along the longitudinal axis of the primary cylinder, an expected result for this particular geometry model. However, significant differences in profiles are depicted for all three velocities, which point toward the different phases that the flows of the three cases are experiencing near the exit. The combination of these patterns indicates the possibility of a secondary flow inside the chamber. This recirculation zone induces a backflow along the length of the cylinder.

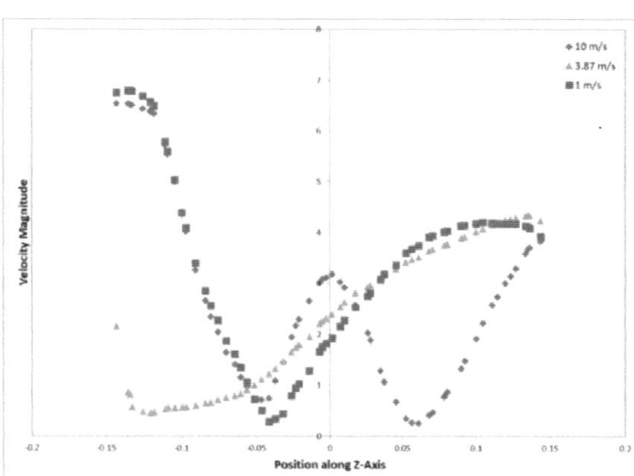

Figure 10. Velocity profile along the centre axis of the cross section at the exit of Model 1.

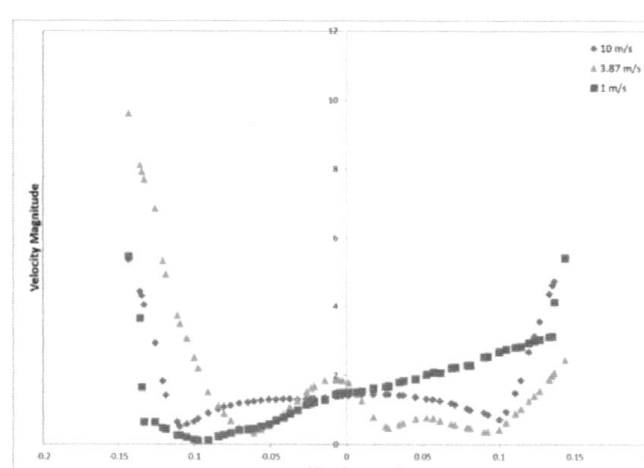

Figure 11. Velocity profile along the centre axis of the cross section at the exit of Model 2.

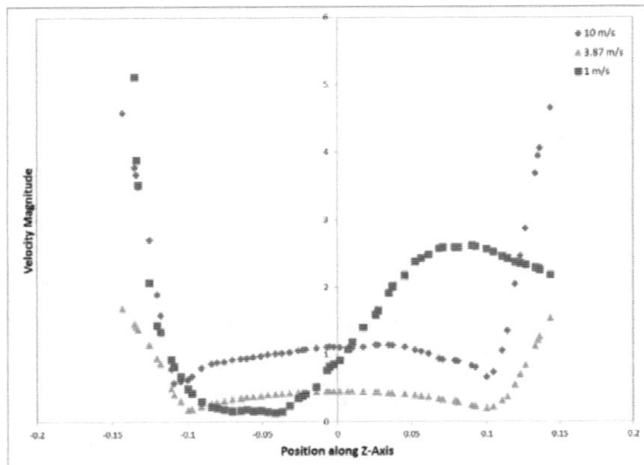

Figure 12. Velocity profile along the centre axis of the cross section at the exit of Model 3.

In Figure 11 representing Model 2, the profiles for inlet speeds of 10 m/s and 3.87 m/s are closer in shape to each other, whilst the profile for the inlet speed of 1 m/s seems skewed, having lower speeds close to the bottom of the cylinder. The 'double-valley' pattern that occurred for only the high-speed inlet flow in the previous geometry now occurs also for the medium inlet speed case, indicating the existence of the central recirculation zone. However, the recirculation zone for the high-speed inlet flow seems to be somewhat wider that the medium speed case, but a bit weaker.

In Figure 12, there exist similarities in the profile pattern between this model geometry and the previous two models. Again, the profile at low speed is somewhat misleading due to its skew, but profiles at medium and high inlet speeds exhibit similar characteristics to what is expected of a swirling flow. Comparing all three models, it has been calculated that Model 2 consistently provides the highest average velocity magnitude around the central area of the cylinder at all speeds considered, which means that the backflow for Model 2 is the lowest of the three configurations tested. On the other hand, Model 1 exhibits significant peaks and valleys whilst Model 3 has average velocity magnitudes that are simply lower that Model 2.

4. Conclusions and recommendations

Swirl flow behaviour in enclosed chambers has been studied using both Computational Fluid Dynamics and Particle Image Velocimetry. Geometry models of swirl generators were created, meshed, and simulated. It has been found from simulations that using the semi tangential inlet geometry instead of the traditional tangential inlet provides a discernible reduction in the backflow. Further reductions of backflow could possibly be achieved using the near perpendicular inlet geometry, but the flow pattern in this model deviates from the conventional swirling flow pattern, and is less predictable.

Acknowledgments

This research was made possible through funding from Universiti Tun Hussein Onn Malaysia (UTHM) IGSP Grant U413 and UTHM STG Grant U126. The authors would also like to acknowledge the contributions made by members of Flow Analysis, Simulation and Turbulence Research Group (FAST), the Centre for Energy and Industrial Environment Studies (CEIES) of UTHM, the Faculty of Mechanical and Manufacturing Engineering at UTHM, and the Faculty of Mechanical Engineering at Universiti Teknologi Malaysia (UTM).

IOP Conf. Series: Materials Science and Engineering **243** (2017) 012039 doi:10.1088/1757-899X/243/1/012039

References

[1] Syred N 2006 *Progress in Energy and Combustion Sci.* **32** (2006) 93-161

[2] Benim A, Escudier M P, Stopford P J and Syed K J 2007 Computational analysis of turbulent swirling flows for gas turbine combustor applications *9th Asian Int. Conf. Fluid Machinery* AICFM9-IL06

[3] Batcha M F M, Hafiz M A, Sulaiman S A and Raghavan V R 2013 *Asian. J. Sci. Research.* **6**(2) 157-66

[4] Hafiz M A, Batcha M F M and Asmuin N 2013 *Material Sci. Eng.* **50** 012021

[5] Mohammed A N 2008 *Swirl Flow in Combustion Chambers* (Universiti Teknologi Malaysia)

[6] Binu T S and Mathew E M 2015 *Int. J. Sci. Eng. Tech. Research* **4**(10) 3601-10

[7] Sabudin S, Wan Salim W S I and Batcha M F M 2016 *ARPN J. Eng. Applied Sci.* **11**(18) 11129-34

[8] Naz M Y and Sulaiman S A 2016 *J. Inst.* **11** 05019

[9] Cloos F J, Zimmermann A L and Pelz P F 2016 A second turbulent regime when a fully developed turbulent flow enters a rotating pipe *Proc. ASME Turbo Expo 2016: Turbomachinery Tech. Conf. and Expo. (Seoul, South Korea, 13–17 June 2016)* (ASME) GT2016-57499

[10] Othman S, Wahab A A and Raghavan V R 2008 Numerical study of the plenum chamber of a swirling fluidized bed *Proc. Int. Conf. Mech. Manuf. Eng. 2008 (Johor Bahru, Malaysia, 21–23 May 2008)* 97–98 –2963–59–2

[11] Herrmann-Priesnitz B, Calderón-Muñoz W R, Valencia A and Soto R 2016 *Appl. Thermal. Eng.* **109** 22-34

[12] Mufeed V E and Goutham R 2016 *Int. J. Sci. Research and Management* **4**(6) 4333-37

IOP Conf. Series: Materials Science and Engineering **243** (2017) 012040 doi:10.1088/1757-899X/243/1/012040

Numerical study of canister filters with alternatives filter cap configurations

A N Mohammed[1], A R Daud[1], K Abdullah[1], S M Seri[1], M A Razali[1], M F Hushim[2] and A Khalid[2]
[1] Faculty of Mechanical and Manufacturing Engineering, Universiti Tun Hussein Onn Malaysia
[2] Faculty of Engineering Technology, Universiti Tun Hussein Onn Malaysia

E-mail: akmaln@uthm.edu.my

Abstract. Air filtration system and filters play an important role in getting good quality air into turbo machinery such as gas turbine. The filtration system and filters improve the quality of air and protect the gas turbine part from contaminants which could bring damage. During separation of contaminants from the air, pressure drop cannot be avoided but it can be minimized to help reduce the intake losses of the engine. This study is focused on the configuration of the filter in order to obtain the minimal pressure drop along the filter. The configuration used is the basic filter geometry provided by Salutary Avenue Manufacturing Sdn Bhd. and two modified canister filter caps which were designed based on the basic filter model. The geometries of the filter were generated by using SOLIDWORKS software, and Computational Fluid Dynamics (CFD) software was used to analyse and simulate the flow through the filter. In this study, the parameters of the inlet velocity are 0.032 m/s, 0.063 m/s, 0.094 m/s and 0.126 m/s. The total pressure drop produced by the basic, the modified filter 1 and the modified filter 2 are 292.3 Pa, 251.11 Pa and 274.7 Pa respectively. The pressure drop reduction for the modified filter 1 is 41.19 Pa and is 14.1% lower compared to the basic filter, whilst the pressure drop reduction for modified filter 2 is 17.6 Pa and is 6.02% lower compared to the basic filter. The pressure drops for the basic filter are slightly different with the Salutary Avenue filter due to limited data and experiment details. Indeed the CFD software is very reliable in running simulations as an alternative to producing prototypes and conducting experiments, thus reducing overall time and cost in this study.

1. Introduction

A top quality gas turbine system is rated by their high availability and reliability, low maintenance cost and low performance degradation [1, 2]. The gas turbine is known as the engine that acts as the heart of the power plant that provides electric current. Gas turbine is a machine that has the ability to convert natural gas to mechanical energy. The energy produced is then supplied into generators that will produce electrical energy. This kind of energy is supplied to buildings and houses via power lines and also used for military purposes [3]. Operation of a gas turbine uses huge amount of air during its operation [4], thus the quality of air that enters the system is crucial as it influences the life cycle and performance of the gas turbine. Air is sucked into the combustion chamber using a compressor. In the combustion chamber, fuel is sprayed into the air thus resulting in an ignition and generates a high-temperature flow.

IOP Conf. Series: Materials Science and Engineering **243** (2017) 012040 doi:10.1088/1757-899X/243/1/012040

Air filtration system plays a crucial role in achieving low maintenance cost and high performance gas turbines [2]. It helps to capture liquids, sub-micron particles and dissolved contaminants such as water-borne salt from entering the engine. Designers reconstructed, optimized and rebuild the pleat geometry thus leading to a filter that has high particle collection efficiency [5]. If a filtering system is not applied before the air enters the engine, it can lead to severe problems such as blade erosion, compressor fouling and hot gas fouling which can reduce the efficiency and become a threat to the turbine [6]. Any particle that has a size larger than 10μm in size can cause an erosion threat towards the internal gas turbine parts. Currently, no treatment can be applied to the damaged part due to erosion; these turbine parts need to be replaced with the new ones thus increasing operational cost.

Fouling is the abrasive removal of material from the flow path by hard particles impinging on flow surfaces. The compressor usually is the most affected component by this kind of defect. Fouling of compressor blades is an important mechanism leading to performance deterioration in gas turbines over time. Particles that cause fouling are typically smaller than 2 to 10m. Compressor fouling is due to the size, amount, and chemical nature of the aerosols in the inlet air flow, dust, insects, and organic matter such as seeds from trees, rust or scale from the inlet ductwork. Deposits from dissolved solids in a water spray inlet cooling system, oil from leaky compressor bearing seals, ingestion of the stack gas or plumes from nearby cooling towers also can lead to fouling. The effect of fouling is that it can increase surface roughness, decreasing the flow passage area, change in air foil shape of the turbine blade and overall performance degradation [6]. Fouling can be removed by off line water washing and slowed down by online water washing. Theoretically, the engine can be kept at a very small degradation level at all times, if it is frequently washed on-line, and the cost of shutting the engine down for off-line washing (typically a half day) is carried. The decision to shut the engine down for off-line washing is a balance between lost productions due to the lower power vs. the lost production for shutting the engine down for a certain amount of time.

2. Methodology

In this study, Computational Fluid Dynamics (CFD) is used to analyse the filter performance. The geometry of the filter is provided by Salutary Avenue Manufacturing Sdn. Bhd. The geometry of the filter is modelled using SolidWorks and imported into the ANSYS WORKBENCH. The simulation and analysis process is being run by using the fluid flow (CFX).

The parameters used for this study are also obtained from Salutary Avenue (SA) which is the air flow rate, surface area of the filter and pressure drop of the Salutary Avenue filter from the experiment. The filter classes that are used are F9 from European Standard 779:2002. Table 1 shows the data that are used in this simulation. The velocity is used as the inlet parameter of the simulations.

Table 1. Parameter Used for Basic and Modified Filter.

Flow Rate (m^3/s)	Pressure Drop (Pa)	Area (m^3)	Inlet Velocity (m/s)
0.236	25.0	7.5	0.032
0.472	47.2	7.5	0.063
0.708	77.1	7.5	0.094
0.944	121.8	7.5	0.126

The three processes involved in CFX are pre-processing, solver, and post processing. The geometry of the filter is developed in the SolidWorks and then imported into the CFX design modeller. The surface of the filter is then named as inlet, enclosure, outlet, symmetry, porous and wall. Figure 1 shows the filter model imported into the design modeller. After assigning the surface names, meshing

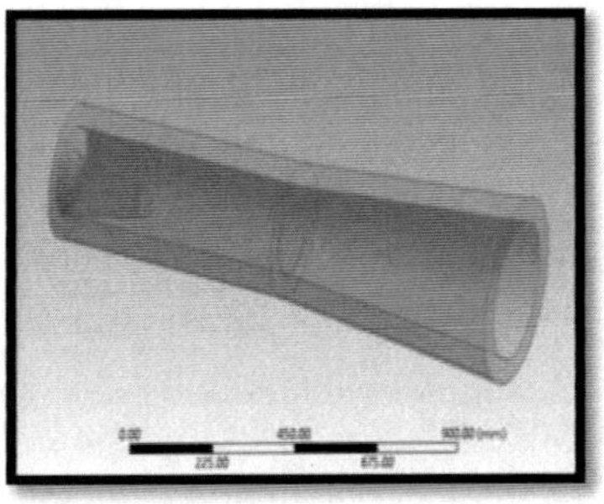

Figure 1. Filter Model in Design Modeller.

Figure 2. Filter Model in Meshing.

takes place. In this part, the minimum and maximum are set up for the geometry which is 8mm and 11 mm. Inflation is set at the outlet of the filter with maximum 5 layers. Figure 2 shows the meshed filter model.

The next step in pre-processing is the setup where all the parameters of the filter are declared. There are two main domains set in setup which is the air and filter domain. Figure 3 shows the air and filter domain. Next, the boundary condition is assigned. The inlet and outlet boundary condition of the filter setting is shown in figure 4 and 5.

The inlet and outlet setting for the CFX setup have 2 different configurations. The first case applies the inlet velocities at 0.032 m/s, 0.064 m/s, 0.094 m/s and 0.126 m/s and using outlet pressure at 30 kPa. In the second case, the inlet velocities are set at 0 m/s and outlet pressure at -30 kPa.

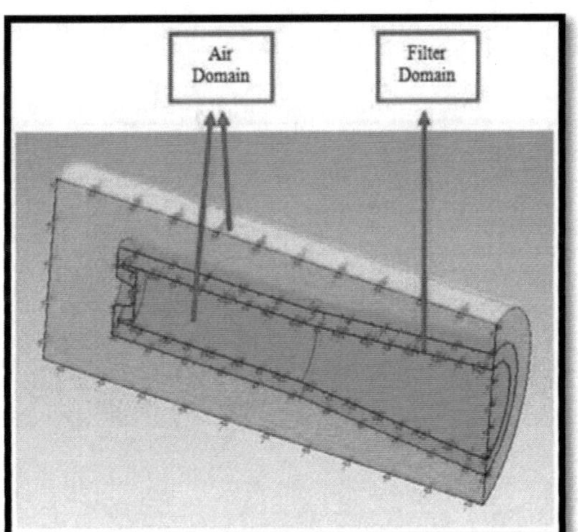

Figure 3. Air and Filter Domain.

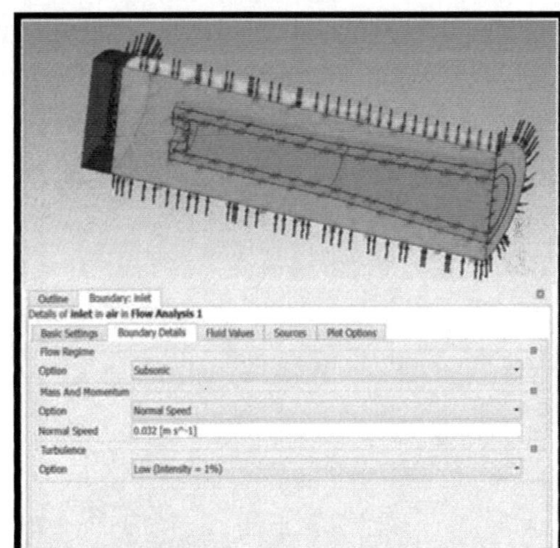

Figure 4. Inlet Boundary Setting.

Lastly, after setup processes completed, solver process takes place. In solver, the number of iteration is set at 1000 iteration and 0.0001 RMS value. This shows that the iteration will stop if the RMS value is converged even if the number if iteration did not reach 1000 iteration. Figure 6 shows the solver and output control setting.

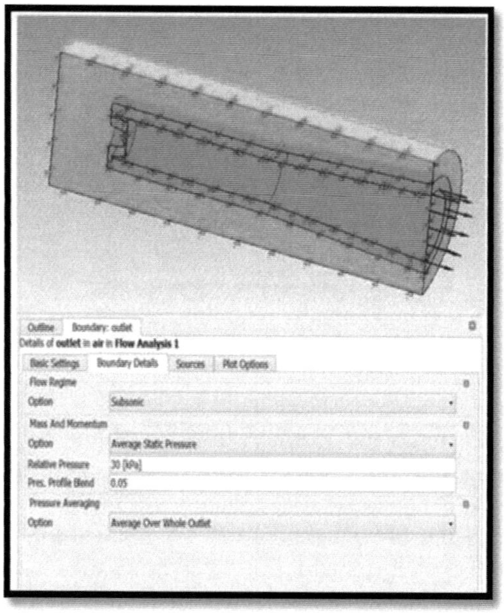

Figure 5. Outlet Boundary Setting.

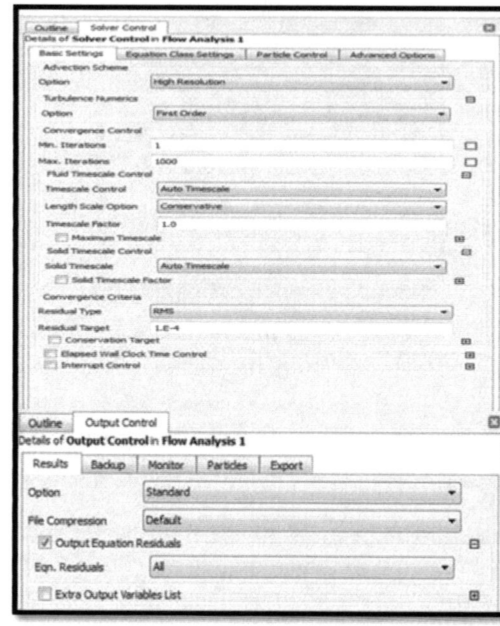

Figure 6. Solver and Output Control Setting.

Simulations are then ready to run and post-processing process takes place. This is where the result of the simulation is generated. From this section, graphs and contours such as pressure, velocity vector and velocity contour will be generated. Line location is also applied on the centre of the filter by placing the coordinates in order to determine the pressure along the filter as shown in figure 7.

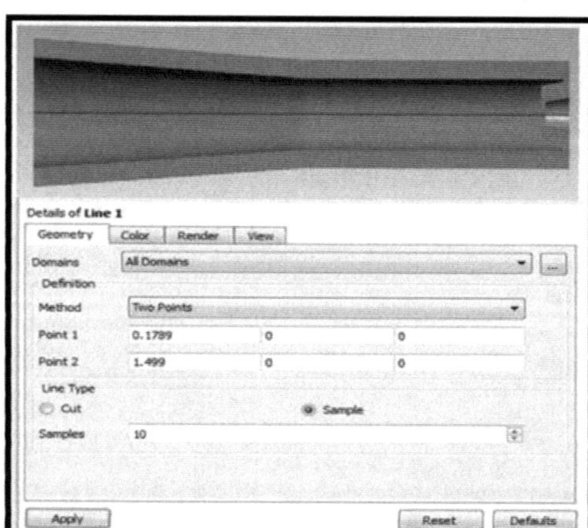

Figure 7. Line Location Setting.

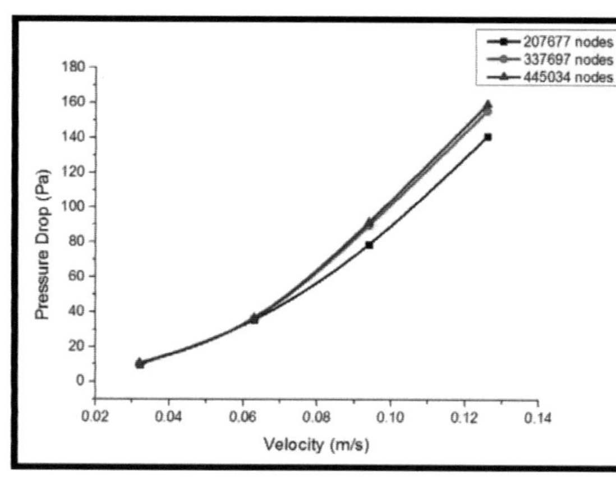

Figure 8. Pressure Drop for Each Sizing.

Grid independent test is then conducted. The goal is to determine which sizing is to be used. In this test, basic filter is used as the benchmark model. The result obtained from different meshing sizes are then compared to the selected meshing size which is minimum at 8 mm and maximum at 11 mm. this size is selected because of the result are near to the smaller sizing. The selection is based on the number of nodes generated during meshing process and the time consumption by the software to generate the mesh and solves the simulation and the result is compared between the different sizing in the figure 8. Table 2 shows the number of nodes and elements at different sizing.

Table 2. Nodes and element at different sizing.

Grid Size (Min)	Grid Size (Max)	No of Nodes	No of Element
8	13	207677	1118435
8	11	337697	1845530
7	10	445034	2451446

3. Results and discussion

In this section, the velocity, pressure vector and contour extracted from simulation is presented. The filtration assembly from Salutary Avenue (SA) is compared with the simulation of basic, modified 1 and modified 2 filters and discussed. Pressure drop is determined by calculating the pressure different at the inlet and outlet of the filter. The pressure drop is determined by using equation (1).

$$\Delta P = P_{inlet} - P_{outlet} \quad (1)$$

Here, ΔP = Pressure Drop, P_{inlet} = Pressure Inlet and P_{outlet} = Pressure Outlet.

Figure 9. Basic Filter Case 1 Streamline.

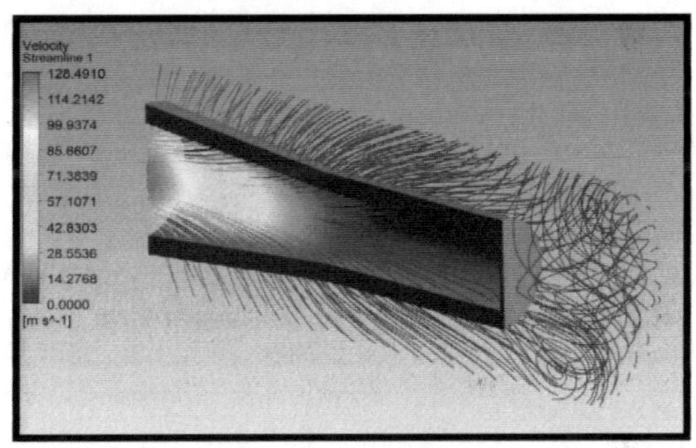

Figure 10. Basic Filter Case 2 Streamline.

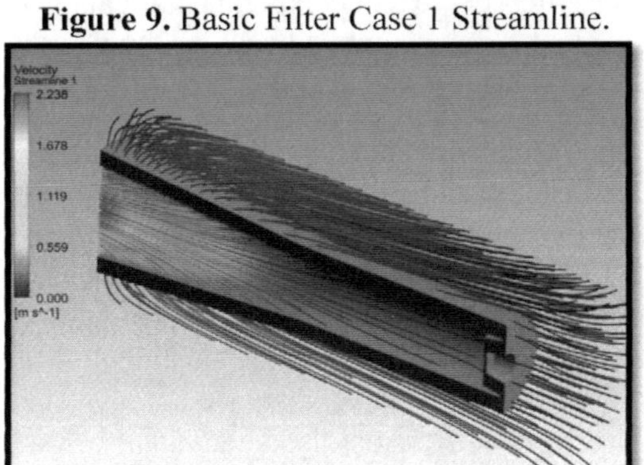

Figure 11. Modified Filter 1 Case 1.

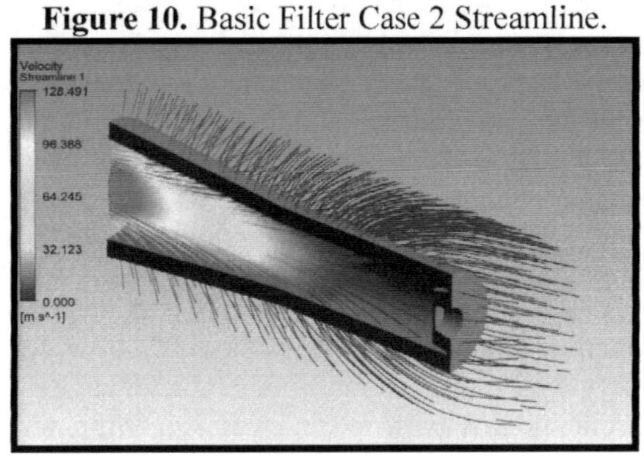

Figure 12. Modified Filter 1 Case 2.

Figure 13. Modified Filter 2 Case 1.

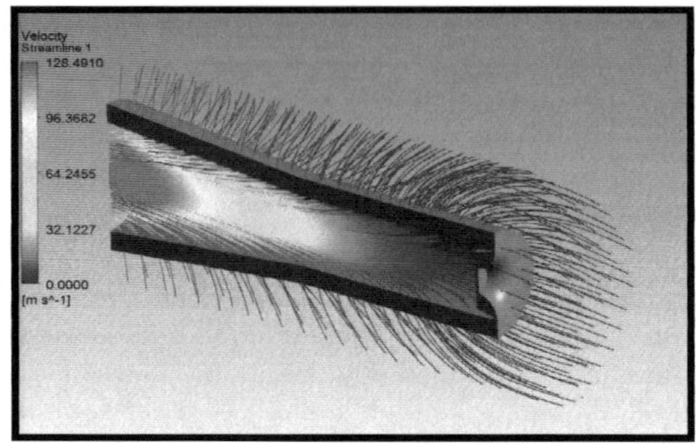

Figure 14. Modified Filter 2 Case 2.

For the case 1, figures 9, 10 and 11 shows the velocity streamline for basic, modified filter 1 and 2 at 0.032 m/s. The streamline figure is show how the air is flows around and inside the filter. From the figure air is flows through filter near the outlet section due to suction pressure. For the modified filters, it can be seen that air is flow in the front side of filter compare to the basic filter.

For the case 2, which are using 0 m/s as inlet velocity and suction pressure at -30 Kpa. Figure 15, 16 and 17 shows the velocity streamline for basic, modified filter 1 and 2. The streamline figure show how the air is flows around and inside the filter. From the figure, the basic filter shows that air cannot flows through the filter through the inlet. This is because, in that particular location, it has a solid structure and does not allow air to flow through it. For the modified filters, it can be seen that air is flow in the front side of filter compare to the basic filter.

The comparison between Salutary Avenue (SA) filter and simulation of the basic filter is conducted. For the Salutary Avenue filter, the results is based from the experiment that conducted by the industry. Figure 18 shows the graph of pressure drop versus velocity for both of the filter.

It can be seen that the pressure different for both filter is increased as the inlet velocity increasing. The pressure drop for both filters is not same with each other. The total pressure drop for basic filter is much higher compare to the Salutary Avenue filter. The pressure drop for basic filter at the lower velocity which is 0.032 and 0.063 m/s is lower than Salutary Avenue and for the last two velocities the pressure drop for basic filter become higher than Salutary Avenue filter.

There is different between the pressures drop of the both filter due to some experimental data from Salutary Avenue that is not fully revealed which is confidential for the company such as the pressure on the outlet, data measurement of the pressure, enclosure details and air that is used for the experiment. Thus, the assumption is been made which lead to the differences of the results. Even the pressure drop is different between those two filters the data is still between the ranges with the Salutary Avenue filter.

From the figure 19, it is clearly can be seen that as the inlet velocity increase the pressure drop also will be increase for the filters. Modified filter 1 recorded better pressure drop compare to the basic and modified 2 filter. According to the theory, the lower pressure drop is much better for filtration systems. The modified filter 1aand 2 produce the lower pressure drop due to the geometry factor where the cap is been modified and allow air to flow through the filter from the front side.

The improvement of the pressure drop for the modified filter 1 are from 11.47% to 14.82%, compared to the basic filter at the inlet velocity from 0.032 to 0.126 m/s. The total pressure drop produce by basic filter is 292.3 Pa and 251.11 Pa for modified filter 1. The pressure drop reduction is 41.19 Pa and 14.1% from basic filter. The total pressure drop for the modified filter 2 model is 274.7 Pa which is higher than pressure drop in modified filter 1 but lower than basic filter. The pressure drop reduction compared to the basic filter is 17.6 Pa and 6.02% lower.

Figure 20 show the pressure drop for four of the filters which are Salutary Avenue (SA), basic, modified 1 and modified 2 filter. According to the total pressure drop the basic filter produce the higher total pressure drop which is 292.3 Pa then follow by modified 2, modified 1 and Salutary Avenue (SA) which 274.7 Pa, 251.11 Pa and 270.9 Pa respectively. From the figure, at 0.032 m/s to 0.063 m/s, the lowest pressure drop is from the modified 1 filter compare to other filters while at 0.094 m/s to 0.126 m/s, the lowest pressure drop is from the salutary avenue filter.

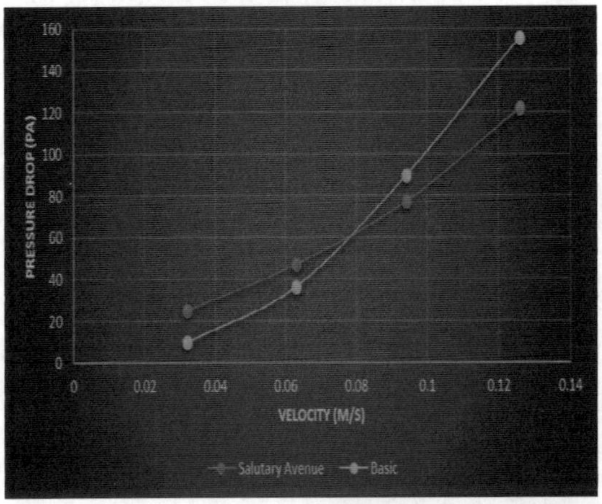

Figure 15. Filter Model in Design Modeller.

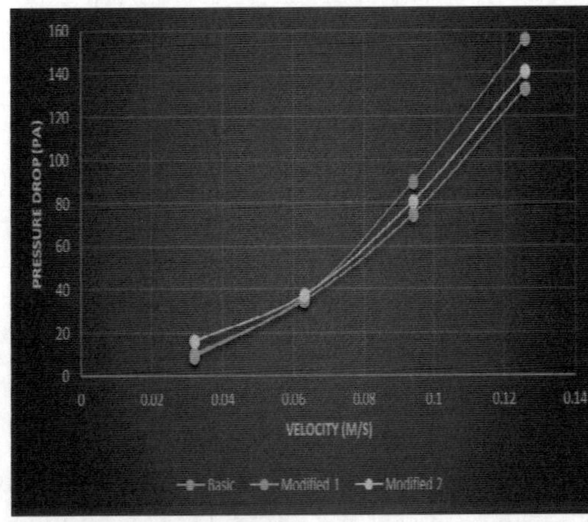

Figure 16. Filter Model in Meshing.

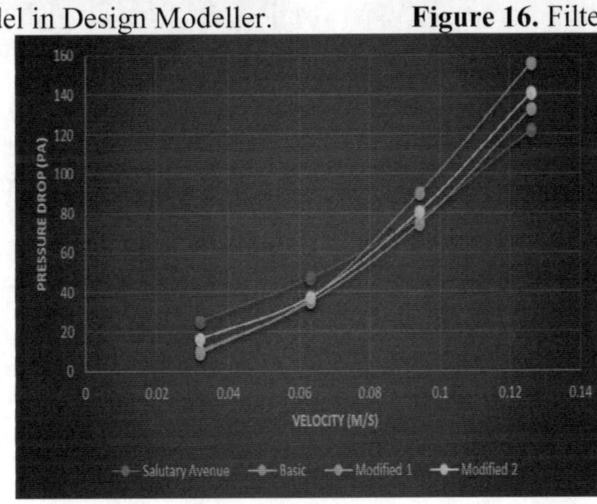

Figure 17. Filter Model in Design Modeller.

For the case 2, the pressure drops for modified 2 filter are lower compare to the basic and modified 1 filter. This is related to the shape of the inlet at modified 2 filter that is equipped with a bell-mouth shaped inlet which have the ability to help divert the surrounding air into the filter more efficiently. Pressure drop on modified 1 filter is lower than the basic filter because of the filter cap that helps to increase the area of filtration for the modified filter 1. Table 3 shows the pressure drop for Basic, modified 1 and Modified 2 Filter.

Table 3. Pressure drop for Basic, Modified 1 and Modified 2 Filter (Case 2).

Inlet Velocity (m/s)	**Pressure Drop (Pa)**		
	Basic Filter	**Modified 1**	**Modified 2**
0.0	34985.80	31265.89	30725 .77

4. Conclusion

In conclusion, in case 1, modified filter 1 produce the minimal pressure drop compare to basic and modified filter 2. There were four values of inlet velocity used to obtain the pressure drop for each configuration which is 0.032 m/s, 0.063 m/s, 0.094 m/s and 0.126 m/s. The improvement of the pressure drop for the modified filter 1 are from 11.47% to 14.82% compared to the basic filter. On the other hand, improvement of pressure drop for modified filter 2 compared to basic filter were from 10.46% to 9.75% at the inlet velocity from 0.032 to 0.126 m/s, even though the modified filter 2 pressure drop were higher compared to the basic filter at slower inlet velocity. The total pressure drop produce by basic, modified filter 1 and 2 are 292.30 Pa, 251.11 and 274.70 Pa. The pressure drop reduction for the modified filter 1 is 41.19 Pa and 14.1% compared to basic filter and the pressure drop reduction for modified filter 2 is 17.6 Pa and 6.02% compared to the basic filter. The pressure drops for the basic filter are slightly different with the Salutary Avenue filter due to limited data and experiment details from the industry.

Based on the case 2, we can see that the lowest pressure drop recorded by the modified 2, modified 1 and the basic filter at 30725 .77, 31265.89 and 34985.80 Pa. The velocity streamline also shows that there are higher rate of air entering the filter for the modified filters while the basic filter configuration did not allow air to flow through the front inlet of the filter.

The geometry of the filter is vital to minimize the pressure drop along the filters. The modified filter 1 and 2 are proved to be better at minimizing the pressure drop compared to the basic filter as they are designed to have more porous area that enable air to flow through them. The larger filtration area tends to helps in decreasing the velocity restriction thus producing lower pressure drop.

Acknowledgments

This research was made possible through funding from Universiti Tun Hussein Onn Malaysia (UTHM) IGSP Grant U413 and UTHM STG Grant U126. The authors would also like to acknowledge the contributions made by members of Flow Analysis, Simulation and Turbulence Research Group (FAST), the Centre for Energy and Industrial Environment Studies (CEIES) of UTHM, the Faculty of Mechanical and Manufacturing Engineering, and the Faculty of Engineering Technology at UTHM.

References

[1] Abam F I, Effiom S O and Ohunakin O S 2016 CFD evaluation of pressure drop across 3-D filter housing for industrial gas turbine plants. *Frontiers in Energy* **10**(2) 192–202

[2] Orhon D, Hiner S D, Kurz R and Benson J 2015 Gas turbine air filtration systems for offshore applications *Turbosymposium* (Houston: Tutorial)

[3] Jansohn P 2013 *Overview of gas turbine types and applications* (Philadelphia: Woodhead Publishing Limited).

[4] Effiom S O, Abam F I, and Ohunakin O S 2015 Performance modelling of industrial gas turbines with inlet air filtration system. *Case Stud. Therm. Eng* **5** 160–7.

[5] Fotovati S, Tafreshi H V and Pourdeyhimi B 2012 A macroscale model for simulating pressure drop and collection efficiency of pleated filters over time. *Sep. Purif. Technol.* **98** 344–55

[6] Santini M, Marchetti G and Giuntini F 2008 Gas turbine high efficiency filtration systems *GE Oil & Gas*

IOP Conf. Series: Materials Science and Engineering **243** (2017) 012041 doi:10.1088/1757-899X/243/1/012041

Computational fluid dynamics simulation of pressure and velocity distribution inside Meniere's diseased vestibular system

N F H Shamsuddin[1,a], N M Isa[1,b], I Taib[2,c] and A N Mohammed[2,d]
[1]Energy and Technologies Research Group (EnRG), Department of Thermofluid Engineering, Faculty of Mechanical and Manufacturing Engineering, UniversitiTun Hussein Onn Malaysia (UTHM)
[2]Flow Analysis, Simulation and Turbulence Research Group (FAST), Universiti Tun Hussien Onn Malaysia (UTHM)

E-mail:fatinhannahshamsuddin@gmail.com[a], nora_matisa@hotmail.co.uk[b], iszat@uthm.edu.my[c], akmaln@uthm.edu.my[d]

Abstract. Meniere's disease or known as endolymphatic hydrops is an incurable vestibular disorder of the inner ear. This is due to the excessive fluid build-up in the endolymphatic sac which causing the vestibular endolymphatic membrane to start stretching. Although this mechanism has been widely accepted as the likely mechanism of Meniere's syndrome, the reason for its occurrence remains unclear. Thus, the aims of this study to investigate the critical parameters of fluid flow in membranous labyrinth that is influencing instability of vestibular system. In addition, to visualise the flow behaviour between a normal membranous labyrinth and dilated membranous labyrinth in Meniere's disease in predicting instability of vestibular system. Three dimensional geometry of endolymphatic sac is obtained from Magnetic Resonance Images (MRI) and reconstructed using commercial software. As basis of comparison the two different model of endolymphatic sac is considered in this study which are normal membranous labyrinth for model I and dilated membranous labyrinth for model II. Computational fluid dynamics (CFD) method is used to analyse the behaviour of pressure and velocity flow in the endolymphatic sac. The comparison was made in terms of pressure distribution and velocity profile. The results show that the pressure for dilated membranous labyrinth is greater than normal membranous labyrinth. Due to abnormally pressure in the vestibular system, it leads to the increasing value of the velocity at dilated membranous labyrinth while at the normal membranous labyrinth the velocity values decreasing. As a conclusion by changing the parameters which is pressure and velocity can significantly affect to the instability of vestibular system for Meniere's disease.

1. Introduction

Prosper Meniere first described the symptom complex known as Meniere's disease in 1861. In 1938, Hallpike and Caims first proposed an underlying mechanism of periodic endolymph fluid accumulation or known as endolymphatic hydrops as a cause of the hearing loss and vertigo in the disorder. Although this mechanism has been widely accepted as the likely mechanism of Meniere's syndrome, the reason for its occurrence remains unclear [1].

Meniere's disease or endolymphatic hydrops is a disorder of the inner ear that causes severe dizziness or known as vertigo, ringing in the ears or tinnitus, hearing loss, and a feeling of fullness or congestion in the ear. Meniere's disease usually affects only one ear. Attacks of dizziness may come on suddenly or after a short period of tinnitus or muffled hearing. Based on Sajjadi & Paparella (2008), some people will have single attacks of dizziness separated by long periods of time while others may experience many attacks closer together over a number of days. Some people with Meniere's disease have vertigo. so extreme that they lose their balance and fall. These episodes are called "drop attacks" [2].

There are four symptoms that causes by elevated pressure in the endolymph which are (a) fluctuating sensorineural hearing loss, (b) occasional episodic vertigo, (c) tinnitus, and (d) aural fullness. Meniere's disease is alternatively called as endolymphatic hydrops due to the swelling seen in histologic sections of the endolymphatic space [3].

Meniere's disease is a common inner ear disease characterized by vertigo, hearing loss, and tinnitus. According to Yokota et al (2015), 80-90% of cases with Meniere's disease can be controlled by medication and/or modification of their life style. However, 10-40% of cases unilateral Meniere's disease gradually becomes bilateral year by year. Bilateral disease is associated with the development of profound bilateral sensor neural hearing loss, resulting in mental illness more often than unilateral one. Bilaterally and mental disorder are the main causes to make treatments for Meniere's disease much more difficult. For this reasons, Meniere's disease was designated as a specific disease by Japan's Ministry of Health, Labour, and Welfare in 1957 [4].

1.1. Membranous labyrinth

Like most of the hollow organs, the membranous labyrinth is lined with epithelium which is a sheet of specialized cells that covers internal and external body surfaces. It is filled with a fluid called endolymph, which has a markedly different ionic content from perilymph. As shown in the figure 1, the endolymph and perilymph are not mix because the membranous labyrinth is a closed system.

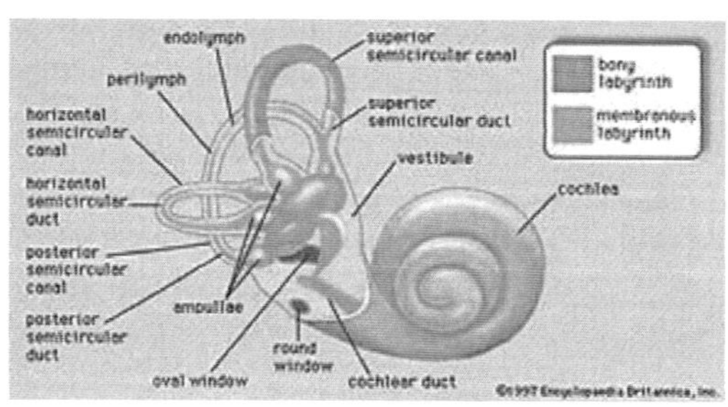

Figure 1: The labyrinths in the inner ear [5]

The two membranous sacs of the vestibule, utricle and saccule, are known as the otolith organs. Each sac has on its inner surface a single patch of sensory cells called a macula, which is about 2 millimetres (0.08 inch) in diameter and which monitors the position of the head relative to the vertical. In the utricle the macula projects from the anterior wall of that tubular sac and lies primarily in the horizontal plane.

In the saccule the macula is in the vertical plane and directly overlies the bone of the inner wall of the vestibule. In shape it is elongated and resembles the letter J. Each macula consists of neuroepithelium, a layer that is made up of supporting cells and sensory cells, as well as a basement membrane, nerve fibres and nerve endings, and underlying connective tissue. The sensory cells are called hair cells because of the hairlike cilia-stiff, nonmotilestereocilia and flexible, motile kinociliathat project from their apical ends. The nerve fibres are from the superior, or vestibular, division of the vestibulocochlear nerve [6].

2. Methodology

2.1. Computational fluid dynamics

Computational Fluid Dynamics (CFD) is based on the solution of the fundamental conservation laws of mass, momentum and energy in a set of partial differential equations (PDEs). The set of PDEs, coupled to initial and boundary conditions, is solved by numerical methods and discretization techniques of the domain such as the finite volume method and finite different equation method [7]. The governingequation for the endolymph flow is theNavier-Stokes equations. As shown below, equation 1 represented for momentum equation while equation 2 is continuity equation for incompressible.

$$-\nabla \mathrm{p}(\mathrm{x}) + \mu\nabla^2 u(x) = f(x,t) \tag{1}$$

$$\nabla \cdot u(x) = 0 \tag{2}$$

With no slip boundary conditions at the wall; t is the time, $x = (x_1,x_2,x_3)$ are the coordinates, and pand $u = (u_1,u_2,u_3)$ are the pressure and the velocity of the fluid, respectively.

2.2. Geometry of the inner ear

A magnetic resonance imaging (MRI) scan is an imaging test that uses powerful magnets and radio waves to create pictures of the body. This study used MRI of human inner ear where is the image is importedfrom commercial software so called 3D Slicer software (http://www.slicer.org) [8].

The 3D Slicer software shows in three different planes where it mimicsthe real model. To get the image of human inner ear this study use crop volume properties from the software and crop from the MRI scan of human head. To generate the model it follows the selected region and the model is generated by the software. The 3D modeling was used in this study as geometry as shown in the figure 2.

Figure 2: Geometry of the inner ear

2.3. Modelling and meshing

The modeling of magnetic resonance image (MRI) is reconstructed using commercial software. The modelingis converted to .iges format to make it compatible to the ANSYS software (Canonsburg, PA, USA) which is used for the simulation. This software manages to analyse multiple, automated parametric.

The mesh of model is started by setup the size. The more advance size function is set on the proximity and curvature while the reference canter is set as fine. To get more accurate resultanalysis, the total of node is determined and the model is meshed smoothly. Two main commandswhich are patch conforming method and inflationare finalizedto make the model meshed completely and suitable for the analysis. The process of meshing is very important stage where the total value of node, element, and the shape of each element connected can give different result. From the shape element it also shows the pattern flow in the model. The mesh operation are performed with the all the data requirement such as the spacing for each element and the surface at wall. Figure3 show the final meshing of the model geometry.

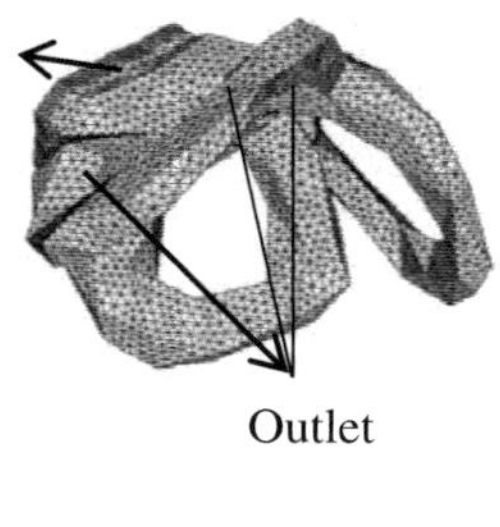

Figure 3: The meshing of the inner ear

2.4. Boundary condition setup

Before running the solver, all boundaries must be initialized and specialized by specified parameters as illustrated in Table 1. Four different pressure were imposed in this study are 200 Pa, 313.6Pa, 400 Pa and 500Pa. Normal pressure of 313.6 Pa was simulated as a basis of comparison. All of the pressure values are absolute pressure.

Table 1: Physical and geometrical parameters [9, 10]

Fluid Properties	Boundary Condition Value
Rate Production (Flow Rate), Q	0.36 l/min
Density, ρ	998 kg/m^3
Viscosity, μ	0.001003 kg/ms^2
Hydrostatic pressure for healthy inner ear, p	205.8 Pa
Hydrostatic pressure for Meniere's Disease, p	200 Pa 313.6 Pa (Reference Value) 400 Pa 500 Pa

3. Result and discussion

3.1. Flow characteristics

The analysis was divided into two models which is model I for normal membranous labyrinth in Meniere's disease while model II represent dilated membranous labyrinth in Meniere's disease ear. The models are classified as a closed system. Flow characteristics are done to investigate the critical parameters of fluid flow which is the pressure distribution and velocity profile in membranous labyrinth that is influencing instability of vestibular system. The visualization flow is shown by the contour results images from the post-processing.

3.1.1 Normal membranous labyrinth. In normal condition the endolymphatic hydrostatic pressure at the inlet is 205.8Pa. Figure 4illustrates the pressure contour obtained for model I. As shown in figure 4, the pressure distribution of the inner ear is constant shown by green colour contour which the value obtained is 4.508×10^{-15}Pa. The changing of colour pressure occurred mostly in the outlet part.

Figure 5 shows the velocity contour results for model I. The endolymph enters through the inlet and exit to three outlets. From the figure, it shows that high velocity occurs at one of the outlet which the value obtained is 5.266 m/s. At the other parts the endolymphs flow with very low velocity movement.

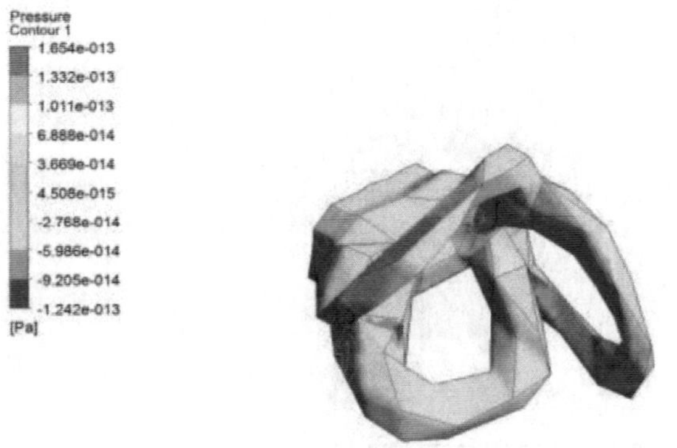

Figure 4: Pressure distribution of endolymph flow in model I

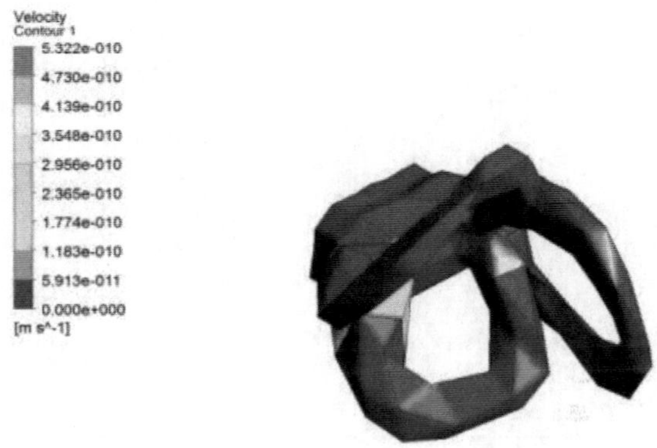

Figure 5: Velocity profile of endolymph flow in model I

3.1.2 Dilated Membranous Labyrinth. Next, comparison between pressure distribution and velocity profile in dilated membranous labyrinth are shown below.For dilated membranous labyrinth in Meniere's disease condition the endolymphatic hydrostatic pressure at the inlet is 313.6Pa.

Figure6 shows the pressure distributions in the inner ear with different pressure value. From the observed, the pressure distribution of inner ear is not much different for each pressure value. The differentpressure values are expected to influence the pressure distribution in the inner ear. However, the higher pressure is expected to increase the possibility of abnormality in the inner ear which leads to dulled hearing. When look at the graph contour at the outlet area, it can see that at the pressure of 200 Pa it shows redder contour then the other pressure does. In addition, the colour contour are changing vary with the distance due the movement of the endolymph in the inner ear.

Figure 7 shows the velocity profile in the inner ear based on different pressure value. Similar to pressure distributions, the velocity profile of the inner ear shows similardistributions for each pressure value. The distribution of the inner ear is observed to be dominated near the outlet part as illustrated in figure 7. The velocity profile at the wall is mostly constant shown by the blue contour and the higher velocity obtained at the outlet.

IOP Conf. Series: Materials Science and Engineering **243** (2017) 012041 doi:10.1088/1757-899X/243/1/012041

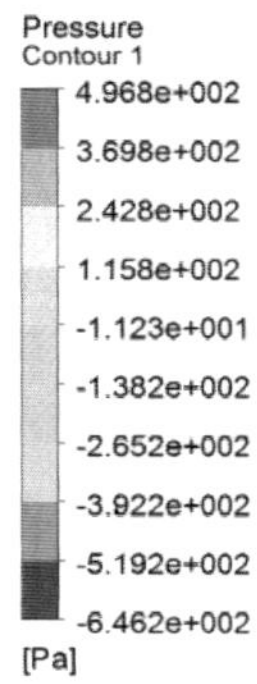

Pressure (Pascal)	Pressure Distribution
200	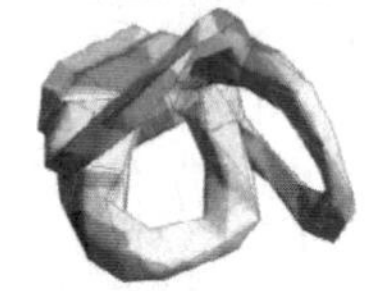
313.6	
400	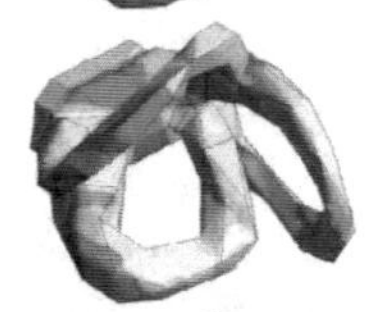
500	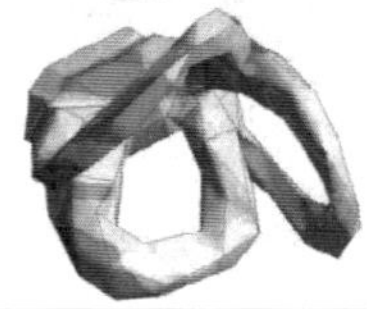

Figure 6: Pressure distributions of inner ear as illustrate 200 Pascal, 313.6 Pascal, 400 Pascal and 500 Pascal.

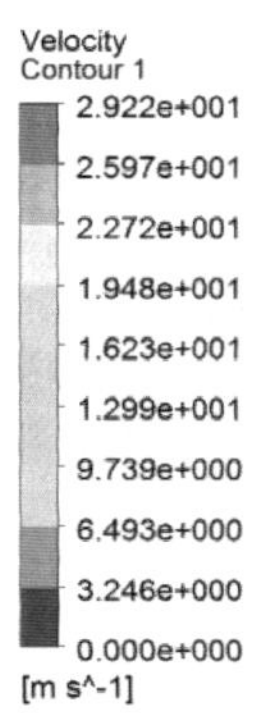

Pressure (Pascal)	Velocity Profile
200	
313.6	
400	
500	

Figure 7: Velocity profile of inner ear as illustrate 200 Pascal, 313.6 Pascal, 400 Pascal and 500 Pascal.

3.2. Pressure distribution

Pressure comparison is done to compare the flow behaviour between a normal membranous labyrinth and dilated membranous labyrinth in Meniere's disease in predicting instability of vestibular system. Based on the figure 8, it shows the pressure comparison between normal membranous labyrinth and dilated membranous labyrinth against distance from inlet to outlet part. The figure shows a pressure fluctuation pattern for both models. For the dilated membranous labyrinth graph, it shows when the distance is increased, the pressure values is increase then decrease and vice versa. The pressure values are not the same at all points in the inner ear. Considering Pascal's Law, a change in pressure at any point in an enclosed fluid at rest is transmitted undiminished to all points in the fluid but it holds for static fluids essentially.

Clearly illustrated in the figure, for a normal membranous labyrinth, the pressure values obtained do not have so much difference between each other's. In normal membranous labyrinth, there is no endolymph build-up in the labyrinth that interferes with normal balance and hearing signals which results to a constant pressure. Based on the figure, the pressure for dilated membranous labyrinth is greater than normal membranous labyrinth. The highest pressure value achieved for both model is at distance 0.109 meter where the pressure value for model one is 0.568 Pascal and the value for model two is 5.036 Pascal. Meanwhile, the lowest pressure obtained for model one is 0.107 Pascal at distance 0.136 meter and for model two the lowest pressure is 1.091 Pascal at distance 0.096 meter.

Therefore, we can say that for a person having a Meniere's disease there will be increasing pressure inside the inner ear compared to normal person. The increasing pressure inside the inner ear is a reason why the vestibular endolymphatic membrane to start stretching and causes swelling of the labyrinth. This abnormality is the factors that lead to instability of vestibular system.

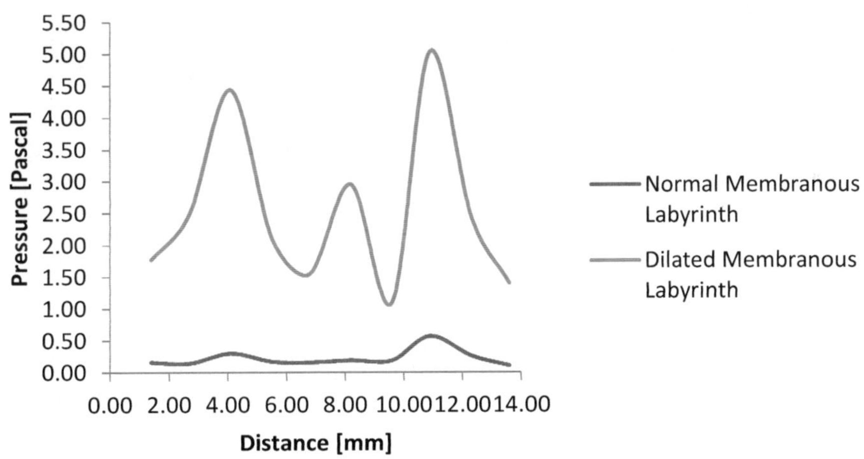

Figure 8: Pressure distribution between model I and model II

3.3. Velocity Profile

Velocity comparison is done to compare the velocity flow between a normal membranous labyrinth and dilated membranous labyrinth in Meniere's disease in predicting instability of vestibular system.

Figure 9 show the velocity comparison between model I and model II. The pattern of graph between dilated membranous labyrinth and normal membranous labyrinth is likely similar at distance between 0 to 0.08 meter which is the value is decreasing. However, at distance 0.082 meter the normal membranous labyrinth continued to decrease while for dilated membranous labyrinth the value increase rapidly and it decrease back at distance 0.136 meter. Due to abnormally increased pressure in the vestibular system it leads to the increasing value of the velocity at dilated membranous labyrinth.

The lowest velocity value for normal membranous labyrinth is 0.004 m/s while for dilated membranous labyrinth the value is 0.012 m/s.

From the figure 9 it shows that the velocity for dilated membranous labyrinth is lower than normal membranous labyrinth. Moreover from the pressure comparison as discussed earlier the dilated membranous labyrinth having higher pressure compare to normal membranous labyrinth. We can conclude that, the results obtained are valid as it applied the concept of Bernoulli's Principle. The higher the pressure of the fluid the lower the velocity it exerts. Fluid pressure is caused by the random motion of the fluid molecules. When the fluid speeds up, some of the energy from that random motion is used to move faster in the fluid's direction of motion. This results in a lower pressure.

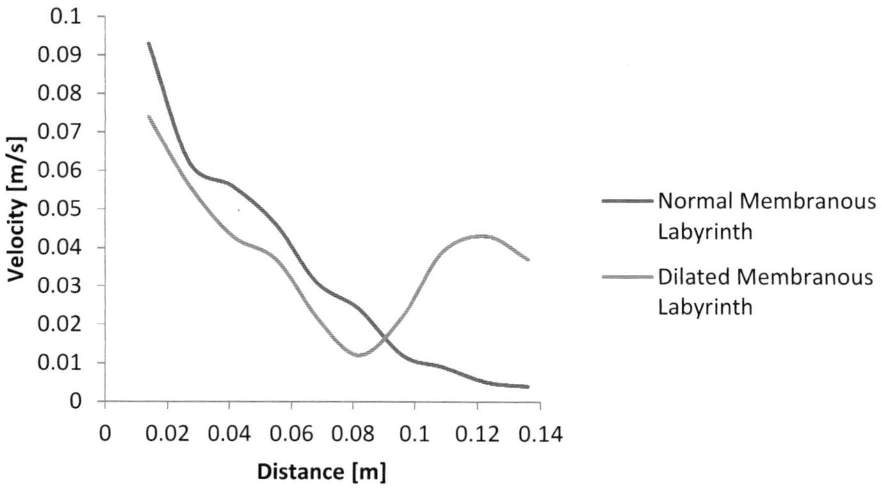

Figure 9: Velocity profile between model I and model II

4. Conclusion

The pressure and velocity of fluid flow can significantly affect to the instability of vestibular system for Meniere's disease. Since there is limited previous study about pressure and velocity parameters, this study refers based on the concept of Bernoulli's principle and Pascal's Law. Based on the results obtained, it shows that by changing the parameters can affect the fluid flow inside the vestibular system. The surface and space of vestibular labyrinth is also a factor that can affect the endolymph flow. For dilated membranous labyrinth it has abnormal increasing pressure compare to normal membranous labyrinth. As the pressure in the inner ear increase the velocity of the fluid flow is decrease.

Acknowledgement

The author would like to thank Universiti Tun Hussien Onn Malaysia (UTHM) for the financial support through IGSP vot U413.

References

[1] Fife T D, and Ippolito J V 2014 Ménière's Disease. *Encyclopedia of the Neurological Sciences* **2** 1057–1059 http://doi.org/10.1016/B978-0-12-385157-4.00139-1
[2] Sajjadi H and Paparella M M 2008 Meniere's disease *Lancet* **372**(9636) 406–14 http://doi.org/10.1016/S0140-6736(08)61161-7
[3] Seyed Tootoonchi S J, Ghiasi S, Shadara P, Samani S M and Fouladi D F 2015 Hearing function after betahistine therapy in patients with Ménière's disease *Brazilian Journal of Otorhinolaryngology* (xx) http://doi.org/10.1016/j.bjorl.2015.08.021

[4] Yokota Y, Kitahara T, Sakagami M, Ito T, Kimura T, Okayasu T, Yamashita A and Yamanaka T 2015 Surgical results and psychological status in patients with intractable Ménière's disease *Auris Nasus Larynx* http://doi.org/10.1016/j.anl.2015.10.007

[5] Encyclopaedia Britannice, Inc 1997 The labyrinth in the inner ear.

[6] Inamoto R, Miyashita T, Matsubara A, Hoshikawa H and Mori N 2017 The difference in endolymphatic hydrostatic pressure elevation induced by isoproterenol between the ampulla and the cochlea *Auris Nasus Larynx* **44**(3) 282–287 http://doi.org/10.1016/j.anl.2016.07.018

[7] Teepakorn C, Grenier D, Fiaty K and Charcosset C 2016 Characterization of hydrodynamics in membrane chromatography devices using magnetic resonance imaging and computational fluid dynamics. *Chemical Engineering Research and Design* **113** 61–73 http://doi.org/10.1016/j.cherd.2016.06.027

[8] Fedorov A, Beichel R, Kalpathy-Cramer J, Finet J, Fillion-Robin J-C, Pujol S, Bauer C, Jennings D, Fennessy F, Sonka M, Buatti J, Aylward S R, Miller JV, Pieper S and Kikinis R 2012 3D Slicer as an Image Computing Platform for the Quantitative Imaging Network *Magnetic Resonance Imaging* **30**(9) 1323-41 PMID: 22770690

[9] Obrist D 2008 Fluidmechanics of semicircular canals - Revisited. *Zeitschrift Fur Angewandte Mathematik Und Physik* **59**(3) 475–497 http://doi.org/10.1007/s00033-007-6037-7

[10] Böhmer A 1993 Hydrostatic pressure in the inner ear fluid compartments and its effects on inner ear function. *Acta Oto-Laryngologica* **507** 3–24 http://doi.org/10.3109/00016489309130250

IOP Conf. Series: Materials Science and Engineering **243** (2017) 012042 doi:10.1088/1757-899X/243/1/012042

Numerical study of flow past a solid sphere at high Reynolds number

C H Yen[1], U J Hui[1], Y Y We[1], A Sadikin[1], NNordin[1], I Taib[1], K Abdullah[1], A N Mohammed[1], A Sapit[1], M A Razali[1]

[1] Department of Energy and Thermofluid Engineering
Faculty of Mechanical and Manufacturing Engineering, UniversitiTun Hussein Onn Malaysia, Johor

E-mail: azmah@uthm.edu.my

Abstract. The present study gives a detail description of separation flow and its effect under high Reynolds number. The unsteady three dimensional flow simulation around sphere using numerical simulation computational fluid dynamics for high Reynolds number between 300 000< Re < 600 000 is discussed. The separation angle and drag coefficient are also presented. The results show that the increasing Reynolds number affecting the formation of vortex shedding, separation point and drag coefficient.The agreement was good, confirming the reliability of the predicted data from computational fluid dynamic in flow analysis around sphere at high Reynolds number.

1. Introduction

The flow separation around simple and complex bluff body is one of the mostimportant and challenging problems in fluid mechanics. The separated flow around a body is difficult to predict and results in many undesirable phenomena such as drag increase, lift loss and fluctuations in the pressure field. As a drag estimates to alarge amount, the running costs of transport of fluids, drag reduction is a crucialdesign issue in engineering. In industry, these custom applications are use ininternal combustion engine, chemical processing and aero industry.The nature of the flow around a sphere changes dramatically as the Reynoldsnumber of the flow increases. Special attention is given to unsteady phenomena inaerodynamic and hydrodynamic. The flows around the sphere or the stability of thewake is the possible flow separation or boundary layer separation that occur behind the sphere will affect significantly the mass, momentum and energy transfer and to determine the values of the transport coefficient.

The visualization studies by [1] have shown that the fluid flow is attached to the sphere and there is no visible recirculation behind itat ($0 \leq Re \leq 20$). This is called attached flow region.Both theoretical arguments [2] and experimental velocity measurement demonstrate that there is a velocity defect region behind the sphere, which is the characteristic of a weak and stable wake. In this flow regime, the flow is attached on the surface of the sphere and separation does not occur. The presence and characteristics of the wake are noticeable in the asymmetry of the vortices field and to a lesser extent, the asymmetry of the streamline contours. In a steady-state wake region ($20 \leq Re \leq 210$), the wake become, wider and longer and its point of attachment on the sphere moves forward, as the value of Reynolds number increases. The separation angle for the wake, which starts at 180° at Re = 20 decreases monotonically with increasing Re to a value of approximately 120° at Re = 130. Above the transition flow (210 ≤ Re ≤ 270), where at point near Re=210, [3] noted that the flow remained attached and stable but was no longer axisymmetric. The natural of the flow in this regime consists of two streamwises vortices tail of equal strength and opposite sign. Although the flow no longer possesses axial symmetry the flow still exhibits planar symmetry in the plane containing the two vortices tails. As the Reynolds number increases within this range a transition from the steady planar symmetric wake to a time-dependent planar symmetric wake occurs**.** The flow becomes unsteady flow (270≤Re≤1000) as the Reynolds number increases. The unsteadiness first appears as waviness in the double threaded wake [4]. The shedding of vortices started to become irregular near Re= 420 and was completely random at Re = 480 [4]. In aerodynamics, flow separation tends to increase the air drag on the front and rearsurfaces of a moving object with a pressure difference between the front and rear thatwill cause some drag results. Knowledge and understanding of the principles andconcepts of fluid mechanics is important to analyze any fluid system is working. Thereason is also much research on aerodynamic and hydrodynamic design that can slowthis flow separation. Conversely, as the Reynolds Number increases, flow separationgradually develops, and this corresponds to a change from a regime of flowdominated by viscous effects, with viscous forces and pressure forces about equallyimportant, to a regime of flow dominated by flow-separation effects, with pressureforces far larger than viscous forces. Therefore, the work of [5] is extended to investigate the flow phenomena occurring around thesphere at high Reynolds Number which is 300 000 to 600 000 past through the sphere. The angle of flow separation and drag coefficient is also reported.These results can be use to capitalize on the trend in thedesign of all engineering applications of fluid mechanic.

2. Computational Fluid Dynamics (CFD)

The flow is simulated over a 38 mm rigid sphere in the flow direction in a 500 x 180 x 200 mm to ensure fully developed flow is achieved. The tube bundle was created in DesignModeler. Two dimensional model for the sphere were produced in CFX-PRE. The model is show in figure 1.

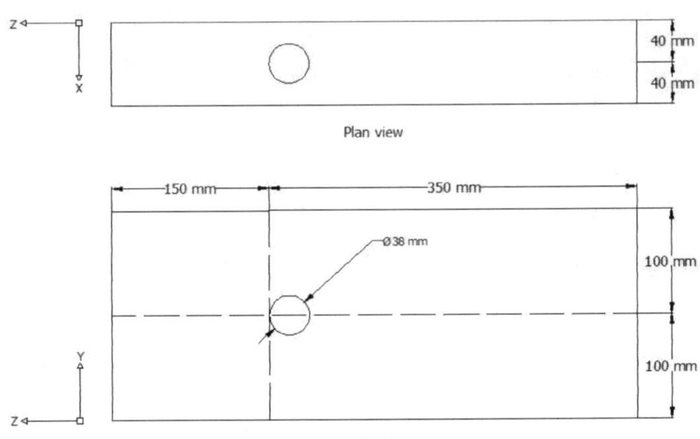

Figure 1. The model

The flow is simulated over a rigid sphere in the flow direction in a 500 x 180 x 200 mm to ensure fully developed flow is achieved. The tube bundle was created in DesignModeler. Two dimensional model for the sphere were produced in CFX-PRE. The computer model and the boundary conditions are shown in figure2. Two mesh configurations of 72 000 and 158,000 cells were conducted for the grid independency test. The pressure coefficient around the sphere was analyzed. The results show there is no significant difference between the two mesh configurations as all lines of both configurations are almost overlapped. These indicate that using finer mesh does not improve the model prediction. Thus, meshing with lower number of mesh cells does not sacrifice the solution accuracy. Since the Central Processing Unit (CPU) time increases exponentially with the number of grids, the lower mesh cells, 72 000 were chosen. Less mesh cells reduce CPU time during CFD simulation which permits a significant number of cases to be run.

The model was constructed with a grid 0.5 mm in length. The meshing gave a total of 15,000 nodes and had 72,000 elements that consisted of prisms. The sphere was set to solid surfaces with no slip. The working fluid is air. The opening boundary condition at the outlet was set to atmospheric pressure and the inlet boundary was set to a normal velocity minimum of 115 m/s – 230 m/sto give Reynolds number more in a range of 300 000 to 600 000 which is turbulence flow. An inflation layer of 1.0 mm thickness and containing 20 layers with an expansion factor of 1.2 was inserted between the tube walls and the bulk fluid to capture the effects near the wall. The transient simulation was set and SIMPLE was chosen for the pressure correction method. The turbulence model is using realizable k-epsilon because itprovides superior performance for flows involving separation and the flow wasassumed to be fully turbulent. The simulation was run until the residual of the pressure and velocities were less than 0.00001.

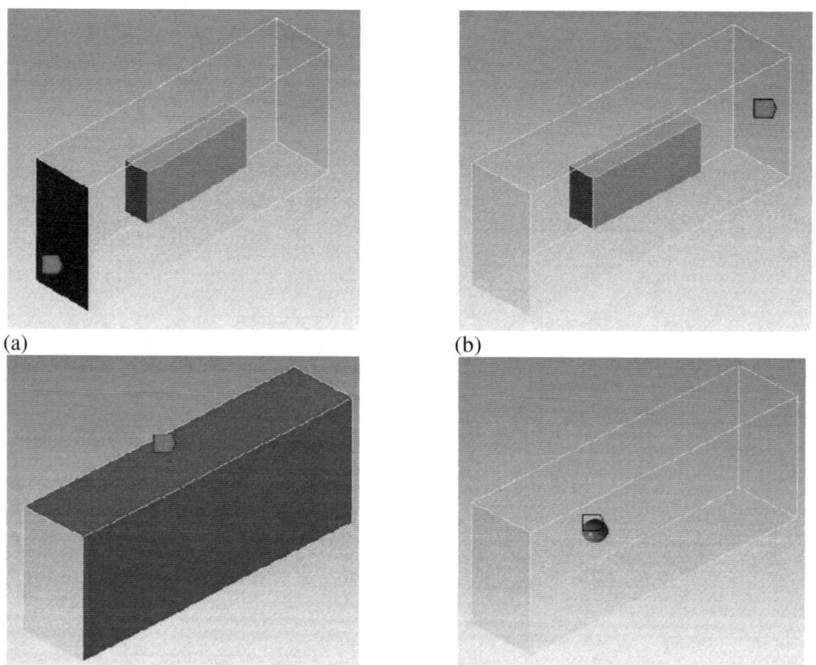

(a) (b) (c) (d)

Figure 2. The boundary conditions (a) velocity inlet (b) pressure outlet (c) wall (d) sphere wall

3. Results and discussion

3.1.Flow regimes around solid sphere

Figure3and figure 4 shows the velocity streamline and velocity contour around the sphere for Reynoldsnumber ranged from 300 000 to 600 000 respectively. The velocity streamline for Reynolds number 300000 until 600000 only had slightly different. The increase in Reynoldsnumber caused circulation flow formed in the wake region.

IOP Conf. Series: Materials Science and Engineering **243** (2017) 012042 doi:10.1088/1757-899X/243/1/012042

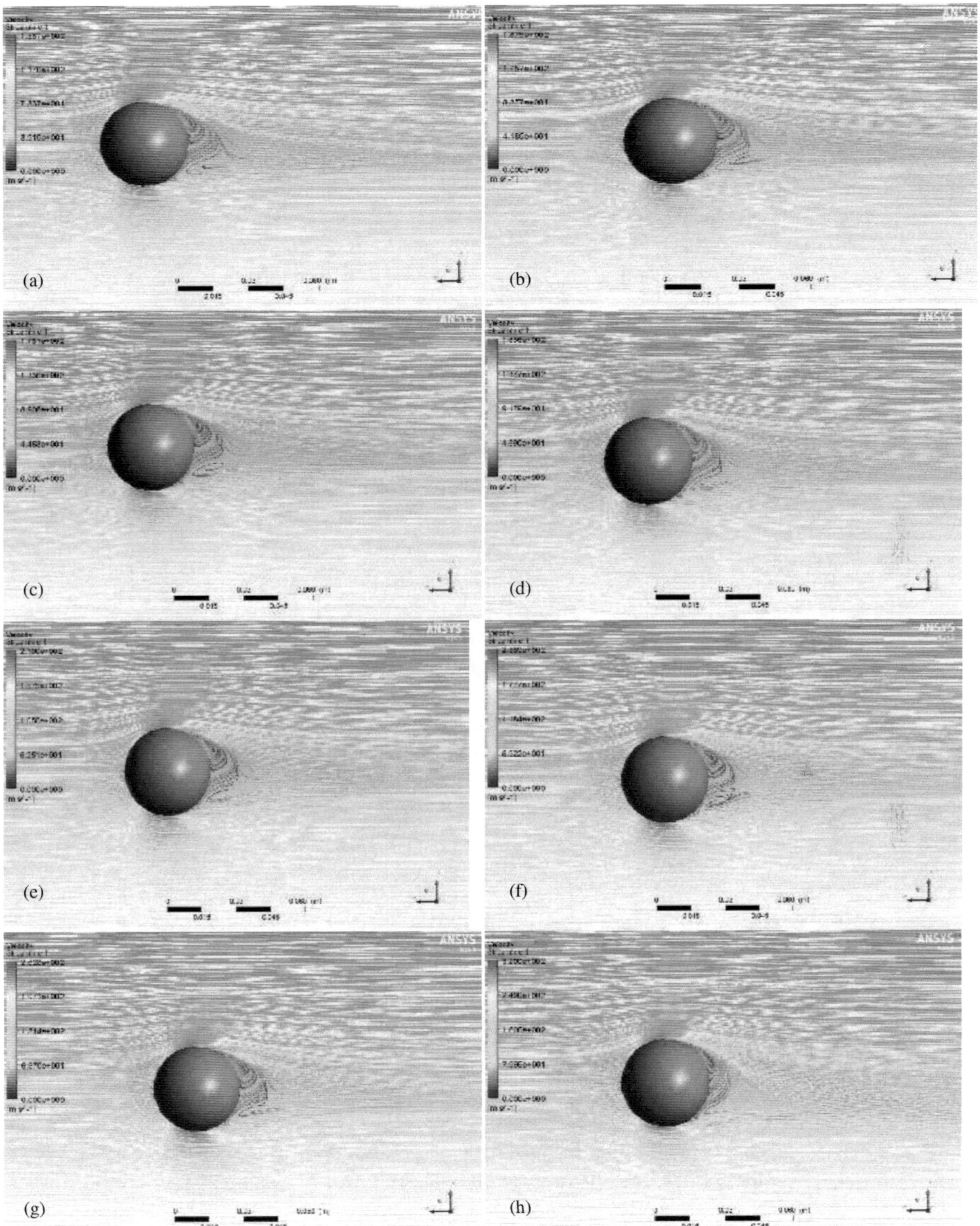

Figure 3. The velocity streamline around the spheres (a) Re = 300 000 (b) Re = 320 000 (c) Re = 340 000 (d) Re = 350 000 (e) Re = 400 000 (f) Re = 450 000 (g) Re = 500 000 and (h) Re = 600 000

IOP Conf. Series: Materials Science and Engineering **243** (2017) 012042 doi:10.1088/1757-899X/243/1/012042

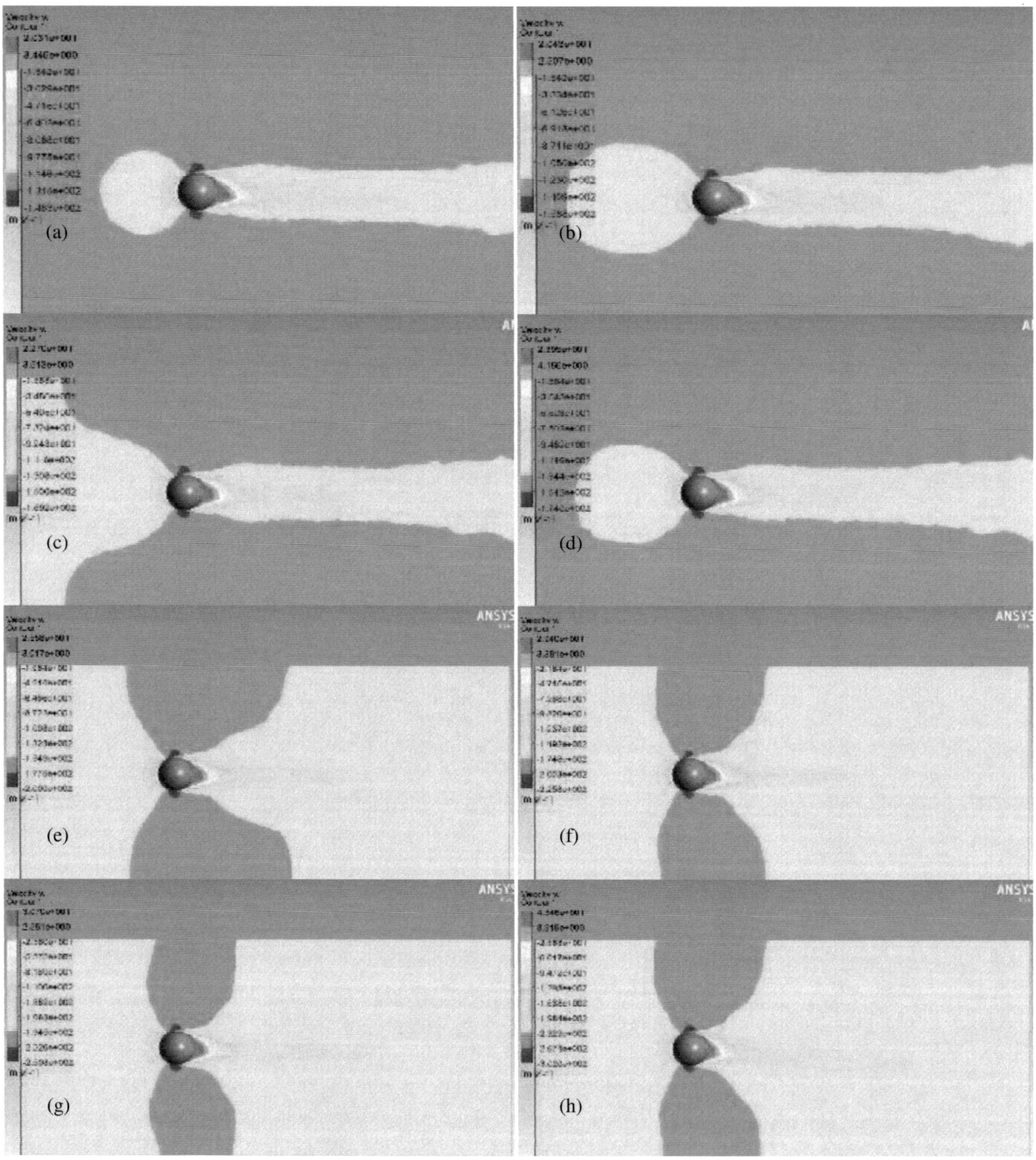

Figure 4. The velocity contour around the spheres (a) Re = 300 000 (b) Re = 320 000 (c) Re = 340 000 (d) Re = 350 000 (e) Re = 400 000 (f) Re = 450 000 (g) Re = 500 000 and (h) Re = 600 000

The near-wake re-circulating region shrinks and the wavy motion of the wake vanishes as the Reynolds number increase.The vanishing of progressive wavy motion of the wake behind the sphere is obvious at Reynolds number 400000. There were reverse flow which form circulation flow at the wake region.

3.2.Pressure around solid sphere and separation angle
The fluid particles on the midplane strike the cylinder at the stagnation point, bringing the fluid to a complete stop and thus raising the pressure at that point. It is observed that it is maximum at this point for all the Reynolds number (300 000 – 600 000). The flow separation occurs when a point near the surface of the sphere began to diverge or fall to the minimum flow. The separation point occurs when the shear stress at the wall is zero. It showed that as the Reynolds number is increases, the flow separation occurs earlier. It is due to the fact that the flow has difficulty to attach longer to the sphere as the velocity increases. This is because the inertia effects more dominant than the viscous effect as the Reynolds number increases, so that the boundary layer separates from its wall more quickly. This finding is found to be in close agreement with the past researchers [6,7,8].

3.3. Drag coefficient
The component of the resultant pressure and shear forces that acts in the flow direction is called the drag force. The drag coefficient was deduced from the drag force from the CFD simulation. The drag coefficient is shown in figure5. From the graph it was concluded the drag coefficient decrease as the Reynolds number increase due to the flow in the boundary layer becoming turbulent, which moves the separation point further on the rear of the body, reducing the size of the wake and thus the magnitude of the pressure drag.

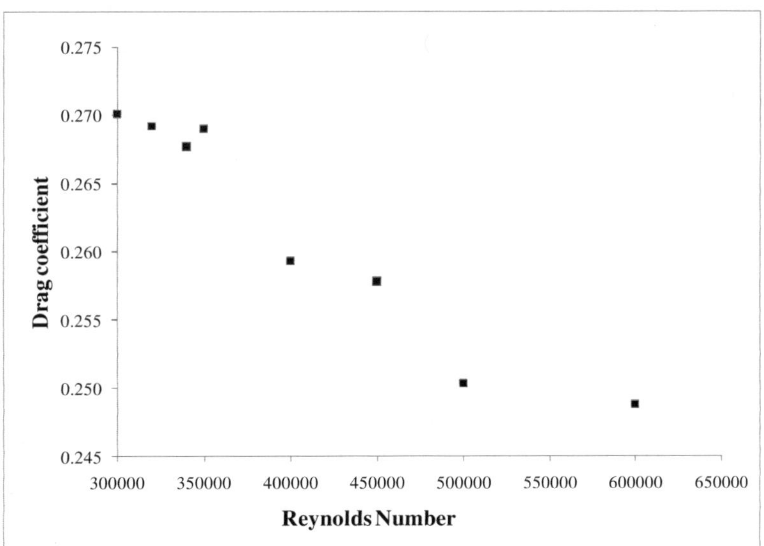

Figure 5. Drag coefficient versus Reynolds number

4. Conclusion

The flow behaviour around sphere is changing as the Reynolds number is increases. The flow is found to be in vortex shedding at all cases. The flow separates early as the Reynolds number increases. This finding is found to be in close agreement with the past researchers [6,7,8]. These predictions are ought to help engineers to improve the aerodynamic and hydrodynamic design application.

References

[1] Taneda S 1956Experimental investigation of the wake behind a sphere at low reynold number*Journal of the Physics Society of Japan***11**(10) 1104-1108
[2] Hinch E J 1993The approach to stady state in oseen flow*Journal Fluid Mechanics***256** 601-605
[3] Margavey R H, Bishop R I 1961 Transition ranges for three-dimention wakes *Can.J.Phys.*, **39** 1418-1422
[4] Jones D A, Clarke D B 2008 Simulation of flow past a sphere using the fluent code*Maritime Platform Division DSTO*
[5] Sadikin A, Mohd Yunus N A , Abdullah M K, Mohammed A N 2014 Numerical study of flow past a solid sphere at moderate reynolds number *Applied Mechanics and Materials***660** 674-678
[6] Michaelides E E 2006 Particles, bubbles and drop, theirmotion, heat and mass transfer*World Scientific Publishing Co. Pte. Ltd*107-116
[7] Lee S 2000A numerical study of the unsteady wake behind a sphere in a uniform flow at moderate reynolds number*Journal Computers and Fluid* **29** 639-667.
[8] Seeley L E, Hummel R L, Smith J W1975 Experimental velocity profiles in laminar flow around sphere at intermediate reynolds numbers*Journal Fluid Mech.***68** 591-608

IOP Conf. Series: Materials Science and Engineering **243** (2017) 012043 doi:10.1088/1757-899X/243/1/012043

New approach to reducing water consumption in commercial kitchen hood

N Asmuin [1] and M R Pairan [2]
[1-2]Department Energy & Thermofluid Engineering, Faculty of Mechanical and Manufacturing Engineering, Universiti Tun Hussein Onn Malaysia (UTHM), Malaysia

E-mail: [1] norzela@uthm.edu.my, [2] uncle_sid@yahoo.com.my

Abstract. Water mist sprays are used in wide range of application. However it is depend to the spray characteristic to suit the particular application. The modern commercial kitchen hood ventilation system was adopted with the water mist nozzle technology as an additional tool to increase the filtration efficiency. However, low level of filtration effectiveness and high water consumption were the major problems among the Commercial Kitchen Ventilation expert. Therefore, this study aims to develop a new mist spray technology to replacing the conventional KSJB nozzle (KSJB is a nozzle's name). At the same time, an appropriate recommended location to install the nozzle in kitchen hood system was suggested. An extensive simulation works were carried out to observe the spray characteristics, ANSYS (FLUENT) was used for simulation wise. In the case of nozzle studies, nozzles were tested at 1 bar pressure of water and air. In comparison with conventional nozzles configuration, this new approach suggested nozzle configuration was reduce up to 50% of water consumption, which by adopted 3 numbers of nozzles instead of 6 numbers of nozzles in the commercial kitchen hood system. Therefore, this nozzle will be used in industry for their benefits of water consumption, filtration efficiency and reduced the safety limitations.

1. Introduction

The term "water mist" is refers to very fine particle of water sprays in which 99% of the volume of the spray is in drops with diameters less than 1000 microns and that remains suspended in air for an extended period of time [1]. The first finding of the application on water mist is during 1950's and 1960'S about water mist fire protection system [2]. This water mist system is rather cheap and effective system compared to available system such as conventional sprinklers and halon gaseous agent [3]. The principle of this system is by applied the high pressure to the water to generating very fine droplet of water and delivering them to the fire zone in fact due to its high specific heat and heat of vaporation coupled with the increases surface area allowing faster heat absorption [4]. Technically the mist system raised concerns due to the high pressures required to produce a fine spray, the potential for blocking of the small orifice nozzles and doubts the long-term ability to maintain the equipment [5].

Nowadays, Deflector atomizer was widely used as a tool to degreasing and washing application [6]. Commercial kitchen hood system were adopted this deflector nozzle concept as an additional tools in filtration system, this type of nozzle known as KSJB nozzle. Manufacturer implemented a water mist spray nozzle as an additional tool which used nontoxicity fluids (water) to be used in the kitchen area.

IOP Conf. Series: Materials Science and Engineering **243** (2017) 012043 doi:10.1088/1757-899X/243/1/012043

Since the nozzle is utilised in a premise, thus the nozzles which produces a fine mist spray in low pressure condition is a major factor in consideration. Halton Company has selected a nozzle which complies the concept of deflector atomizer. This nozzle functioned as the first stage filtration system to trap the plumes before it passes through the second stage filtration system.

As a conventional nozzle adheres to the concept of deflector wall, it's usually sprayed pointing downwards, causing the restricted in limitation of nozzle configuration. For conventional commercial kitchen hood ventilation system, 6 units of nozzle are required to ensure that entire surface of filter is fully covered. This configuration limitation will lead to water consumption issues. Therefore, it is required to discover an appropriate nozzle design and it's characteristic (water droplets size, water droplets velocity, spray angle and spray penetration) for new development of new mist spray. Thus this research applied low pressure of water and air to produce water mist by using external mixing technique to suitable used in industry sector.

2. Methodology

In this study, ANSYS software was used to simulate the nozzle in order to observing the spray pattern generated by the KSJB nozzle and new development nozzle. Simulation was conducted by using ANSYS fluent, the new development nozzle model and KSJB nozzle were constructed using SOLIDWORKS. Two types of fluids were used in this simulation and experiment which are water and air. Water and air temperature are set to 293K and 300K, the flow rate of the water for pressure 1bar was set as 0.00128 kg/s meanwhile for air pressure 1 bar the flow rate were set as 0.00087kg/s.

Figure 1 show the details of the KSJB nozzle and new development nozzle dimension. The computational domain of KSJB nozzle and new development water mist model were illustrated in Figure 2.

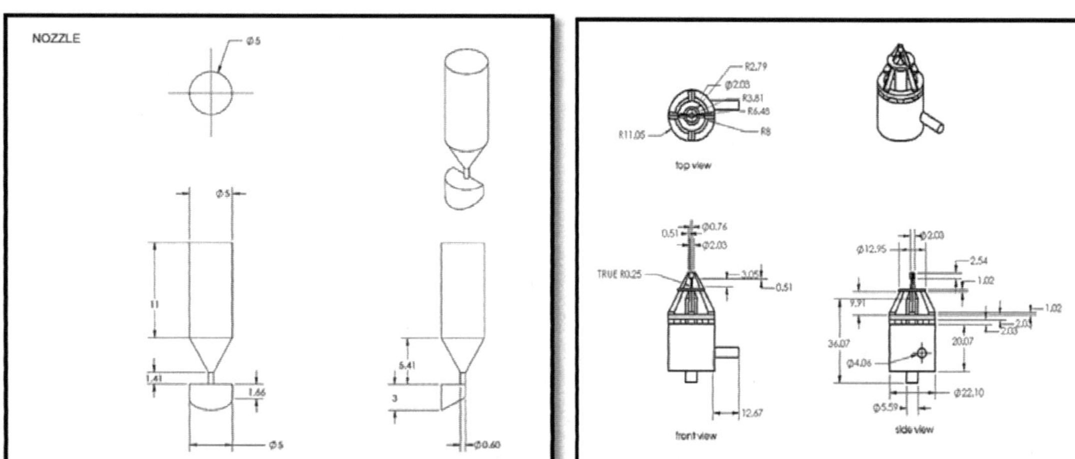

Figure 1. Computational domain new development nozzle

IOP Conf. Series: Materials Science and Engineering **243** (2017) 012043 doi:10.1088/1757-899X/243/1/012043

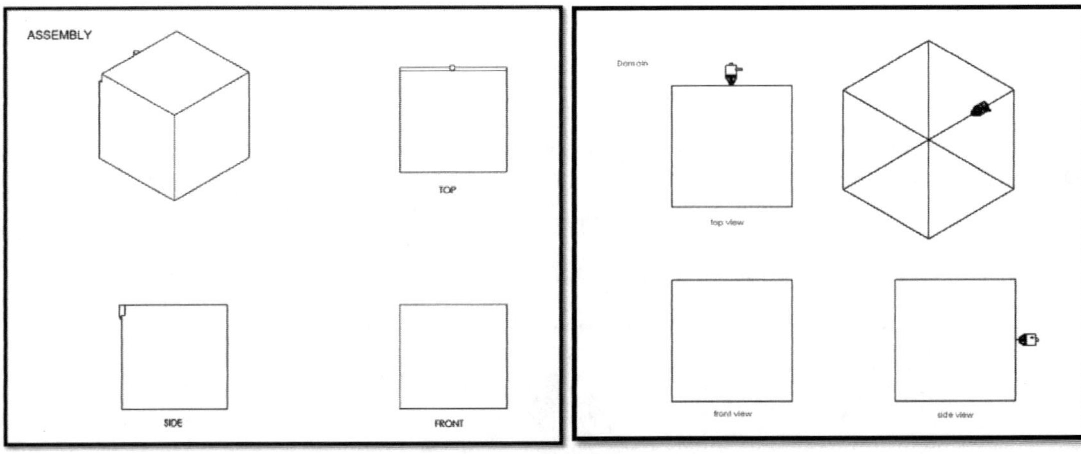

Figure 2. Computational domain new development nozzle

To run this simulation few boundary condition have to apply to the model. For the KSJB nozzle has only one inlet which is water inlet meanwhile for new development nozzle, it consist of two inlets which are water inlet and air inlet. Table 1 and Table 2 indicate the boundary condition details for KSJB nozzle and new development nozzle.

Table 1. Boundary conditions and numerical setup for KSJB nozzle

Item	Physical Properties		
General Condition	Steady State Incompressible Non radiation Multiphase		
Turbulence Model	k-epsilon realizable		
Boundary Condition	Water inlet	Water flow rate	0.00128 kg/s
	Air inlet	Air flow rate	0
	outlet	Pressure (Kpa)	0

Table 2. Boundary conditions and numerical setup for water mist nozzle

Item	Physical Properties		
General Condition	Steady State Incompressible Non radiation Multiphase		
Turbulence Model	k-epsilon realizable		
Boundary Condition	Water inlet	Water flow rate	0.00128 kg/s
	Air inlet	Air flow rate	0.00087kg/s.
	outlet	Pressure (Kpa)	0

IOP Conf. Series: Materials Science and Engineering **243** (2017) 012043 doi:10.1088/1757-899X/243/1/012043

3. Result

Figure 3 show the velocity contour of KSJB nozzle, based on the velocity contour, the KSJB nozzle produce flat spray pattern. The highest velocity of water was at the tip of the nozzle and obviously when the water flowing to the surrounding, the velocity of water becomes slow and widely spreading. This phenomenon happened because of the relation between the impingement between liquid with the wall. It was observed that after water with high velocity flowing out through the small nozzle orifice; it strikes the nozzle wall, thus its result the change of velocity and direction of water. The spray penetration angle of this type of spray is 125 degree with short penetration distance and this type of spray has limited configuration which is it has to spray pointing to the downwards direction.

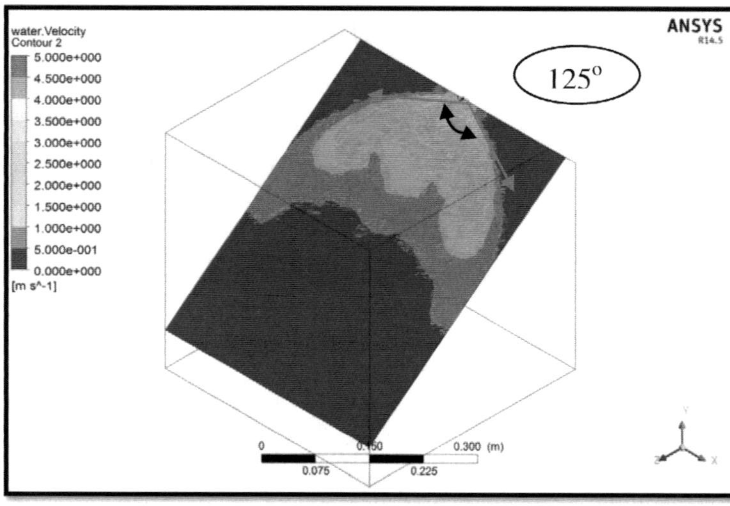

Figure 3. velocity contour of KSJB nozzle

In a real condition, there were 6 numbers of KSJB nozzle is required in order to fully cover the entire filter surface. The KSJB nozzle were placed at the top of the KSA filter and spray pointing to the downwards direction as shown in in Figure 4. The problem of this configuration is the water consumption issue. Thus to overcome this issue, new development nozzle were develop in order to replacing the conventional nozzle to reducing the water consumption.

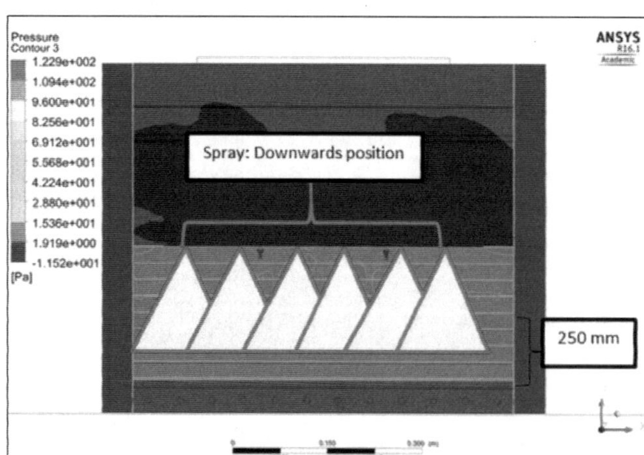

Figure 4. Conventional position of KSJB nozzle in commercial kitchen hood system

IOP Conf. Series: Materials Science and Engineering **243** (2017) 012043 doi:10.1088/1757-899X/243/1/012043

Figure 5 shows the water velocity contour generated by the new development mist spray; this new development nozzle has the advantage and disadvantage compared to the KSJB nozzle which follows the deflector atomiser concept. Since this new development nozzle utilise the external mixing technique, it produce jetting effect which helps to increase the air velocity which will push the water droplet move a way further from the orifice, this phenomenon resulted the effect to the spray penetration length. Even though it has smaller spray penetration angle which is 32 degree, it has a advantages on spray penetration length, this type of spray able to penetrate more than 750mm, by this penetration advantage, it was suggested to install only 3 number of new development nozzle horizontally to fully cover the entire filter surface which has the 1000mm length and 330mm height dimension. The recommendation location was illustrated in Figure 6.

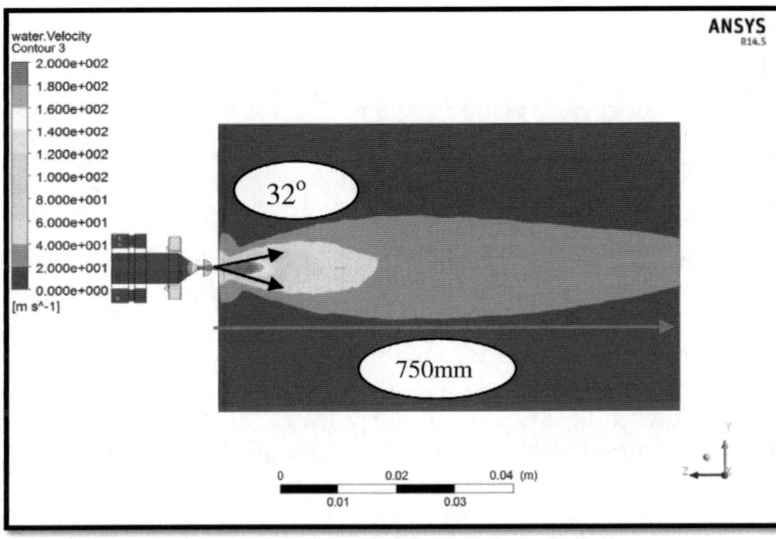

Figure 5. velocity contour for new development nozzle

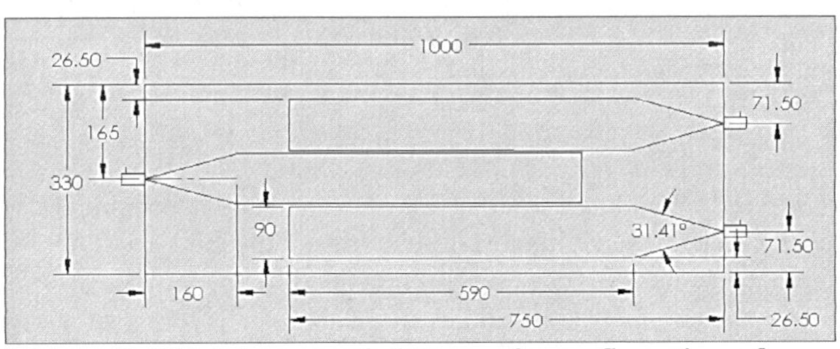

Figure 6. The recommendation nozzle configuration of new development nozzle

Other advantage of this new development spray is the spray performance, in term of water droplets size, under optimum condition which is at 1 bar pressure water and air, this type of nozzle able to produce very fine water droplets size which is below than 2 micron as shown in Figure 7. This become the additional advantage of this nozzle which expected to increase the filtration efficiency of the CKV system since the combination of the KSJB nozzle and KSA filter fails to trap particles size less than 2.5 micron.

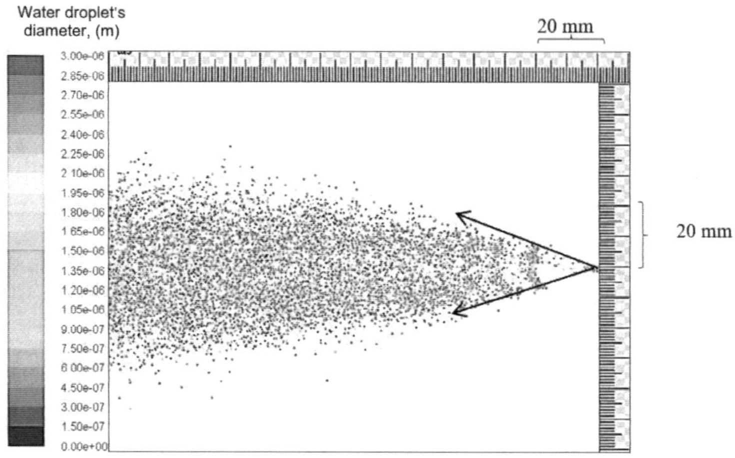

Figure 7. Water droplets diameter generated by new development nozzle

4. Conclusions

As a conclusion, since the nozzle has the ability to penetrate more than 750mm, the suggested locations to install the nozzle were at the side of the KSA filter which the nozzle will be spraying horizontally. There were 3 pieces of nozzle were suggested to install in the system in order to the spray pattern will cover the entire surface. By installing the new nozzle following this configuration water consumption reduced to 50% compared to the previous nozzle. The advantages of placing the nozzle at the side of the kitchen hood was to form a streamline which was generated by the water droplet are not parallel with the streamline generated by the plumes. This method was expected to increase the collision between the water droplet and the contaminant.

5. Future Scope

This study is for the future scope to implement the new development mist spray to the commercial kitchen hood system as a tool to traps the small particles. It is very important to know the physical characteristic of the new development nozzle.

Acknowledgements

The authors are grateful to University Tun Hussein Onn Malaysia (UTHM) and Kementerian PendidikanTinggi Malaysia for funding the grant FRGS vote 1540 and grant KTP vote 1277 for this research project. Special thanks to Halton Manufacturing Sdn Bhd for their contribution in this research and their continued support is very much appreciated by our beneficiaries.

References

[1] Mawhinney J R and Solomon R 1997 *Water Mist Fire Suppression Systems, Fire Protection Handbook*, 18th ed. National Fire Protection Association, Quincy, MA
[2] Milke J A 1996 Comparison of the Performance of Water Mist System Designs For Library Stack Areas University of Maryland, College Park, Baywood Publishing Co. Inc.
[3] Benjamin Piers Hume 2003 Water Mist Suppression in Conjunction with Displacement Ventilation, School of Engineering University of Canterbury Christchurch, New Zealand
[4] Jones A and Nolan P F 1955 Discussion on the Use Of Fine Water Sprays or Mist for Fire Suppression, *Journal Loss Prev. Process ind.* **8** Buterworth-Hainemann Ltd 17-21
[5] Mawhinney J R and Eng P *Characteristics of Water Mist for Fire Suppression in Enclosures National Research Council* (Canada: Institute for Research in Construction, National Fire Laboratory)
[6] Bill J Jones 2006 Selecting Nozzles for Hand-held Applicators, Ae Village

IOP Conf. Series: Materials Science and Engineering **243** (2017) 012044 doi:10.1088/1757-899X/243/1/012044

A numerical analysis of flat fan aerial crop spray

Siti Natasha Malik Fesal[1], **Mas Fawzi**[2] **and Zamri Omar**[3]
[1] Mechanical Engineering Department, Politeknik Ungku Omar, Malaysia
[2] Faculty Mechanical and Manufacturing Engineering, Universiti Tun Hussein Onn Malaysia

E-mail: stmf_tasha@yahoo.com

Abstract. Spray drift mitigation, in the agriculture aerial spraying literature, and spray quality in the application of plant protection products, still continues as two critical components in evaluating shareholder value. A study on off-target drift and ground deposit onto a 250 m strip were simulated through series of Computational Fluid Dynamic (CFD) simulations. The drift patterns for evaporating droplets were released from a constant aircraft velocity at 30 m/s (60 mph) carrying 20 m swath width spray boom with 12 fan-type nozzles at released height from the ground ranging from 3.7 m to 4.7 m. Droplet trajectories are calculated from the given airspeed with a Lagrangian model for particle dispersion excluding any wind effect perturbation. The proposed CFD's model predictions agreed well with cited literatures for a wide range of atmospheric stability values. The results revealed that there is considerable increased in spray drift and droplets trajectories with the increased in spray released height. It suggested that a combination of low aircraft spray released height with low airspeed is essential to improve spray quality and maximizing uniform deposition on the target area are significant in minimizing spray drift risks.

1. Introduction

In the last three decades, the growing concern of aerial spray deposits and proxy of pesticides residue is continuing to pose major threat to human being and environmental [1], adaptation to its negative effects has moved to the forefront of the stakeholder. Aerial spraying works are poorly understood by public and persistently criticized globally [1]. Aerial spraying works involving pesticide released into atmosphere at certain height above the crop or forest to the ground. Thus it requires atomizers that are able to meet with stringent environmental law and also ensuring pesticide stays in the intended area [2, 3, 4]. Spray drift is undesirable, but still the important role of pesticides or plant protection products in modern agriculture is undeniable in terms of reducing pest attacks and increasing high yield [5, 6].

Uneven spray distribution and drifted to adjacent areas has been pronounced as serious problems regarding aerial spray delivery [7]. Uniformity spray coverage area are desired in aerial crop spraying, thus improving aerial pesticide application to reach a satisfactory level of spray deposition in most cropping system is essential. Typically, aerial spray drift can be classified into two categories namely as downwind deposition and airborne or vapor drift [8]. The potential aerial spray drift can passively be improved predetermined by notable controllable factor such as boom length, nozzle type, spray angle, pressure, air speed, remain constant based on real application scenario [9], however the uncontrolled factor limitation atmospheric conditions such as wind speed, relative humidity and temperature are yet still remained [4,10]. Notable spray characteristics of aerial crop spraying is dependent on spray equipment such as atomization methods, sprayer nozzle types, spray adjuvants, meteorological conditions, crop characteristics and buffer zones. Amongst these variables, larger

nozzle tips, low spray pressures and low spray boom height apparently significant in reducing spray drift [11]. Nozzle size and design selection is the most critical parameter in aerial spraying as it influenced the spray droplets spectrums and droplet velocity characteristics [12, 13].
Studies on the aerial crop spray quality has been very limited especially the droplet dispersal process; an area of research within the spray characteristics of aerial pesticide delivery [4, 14]. On the other hand, there are plentiful reports on the outdoor analysis available on aerial drift affecting factor; for example, the works originates from spray drift task force (SDTF) known as agricultural dispersion models (AGDISP) in United States [5, 9].
For assessing the aerial pesticide application, various methods can be used at the outdoor and indoor environments such as (i) full-scale measurements, (ii) wind-tunnel measurements or spray chambers, and (iii) numerical simulation with Computational Fluid Dynamics (CFD) [10, 15]. Each of the sampling techniques has advantages and disadvantages especially in terms of numbers, time constraint and cost. Recently, computer modelling such as CFD provides whole-flow field data, fast and reproducible, repeatable and controllable conditions, reliable data, rapid, economical and accurate means however at the expense of incompatibility [16]. CFD is increasingly used to study a wide range of atmospheric and environmental processes [17].
The objective of this study is to investigate on the effect of low velocity, variation by lowering the boom spray released to the ground from flat fan 90-degree atomizers, the particle mass concentration released, nozzles orientation and the droplet trajectories. Regions of interest that depicts the most important unsteady flow would be a common series of divisions from horizontally projected from 0 to 250 m distance away from the spray sources [10]. Thus, a 3D CFD, which is a useful approach to predict the probable unsteady flow characteristics is recommended in understanding the spray deposition based on the effect of spray technology and spray liquid properties as reported by Baetens et.al [15]. According to Ryan et al. [4], the most suitable technique to represent the actual event is by using 3D volume of fluid multiphase model, which consists of mainly two phases: (i) air and (ii) water.

2. Methodology

In the present work, an unsteady 3D CFD using Fluent version 15.0 from ANSYS® modelling was designed to simulate two-sub model; an airflow model [3] and a particle tracking model which are mutually coupled [4]. A Lagrangian approach representatives of approximate of 40000 droplets trajectories was utilized for all simulations.

2.1. Aerial Aircraft Configurations and Schematic Diagram

The present simulation aims to highlight the configuration parameters that represent the actual event of spray depositions of nozzle in controlled conditions without environmental factors perturbation. The commercial GT500® agriculture aircraft specification as shown in Figure 1a) was chosen as reference due to its favorability in agriculture aircraft pesticides delivery in most countries. It mainly used to cover widespread infestations, or inaccessible areas in steep or remote terrain. Engineering boom spraying technology, the state of art need to be tailored well fit the target specific infestations and reduces the risk of pesticides drifted to non-target or over spray surrounding areas. In this simulation, the aircraft speed was purposely conducted for 30 m/s (60 mph),comparative lower from previous studies which employed at higher airspeed between 54 m/s (120 mph) to 72 m/s (140 mph) for agriculture aircraft spraying [8, 9]. The reason was to configure the effect of low airspeed in agriculture aerial works spraying.
In this study, twelve set nozzles coordinates were allocated, mounted on an aircraft boom spray unit releasing injections at various height (altitude) ranging from 3.7 m to 4.7 m above ground of 30 m/s speed imitates as aircraft spraying pesticide over plantation area. The nozzles schematic diagram mounted on aircraft boom as shown in Figure 1b) complete with the physical configurations dimension of GT 500® agriculture aircraft for the simulation as listed in Table 1 below.

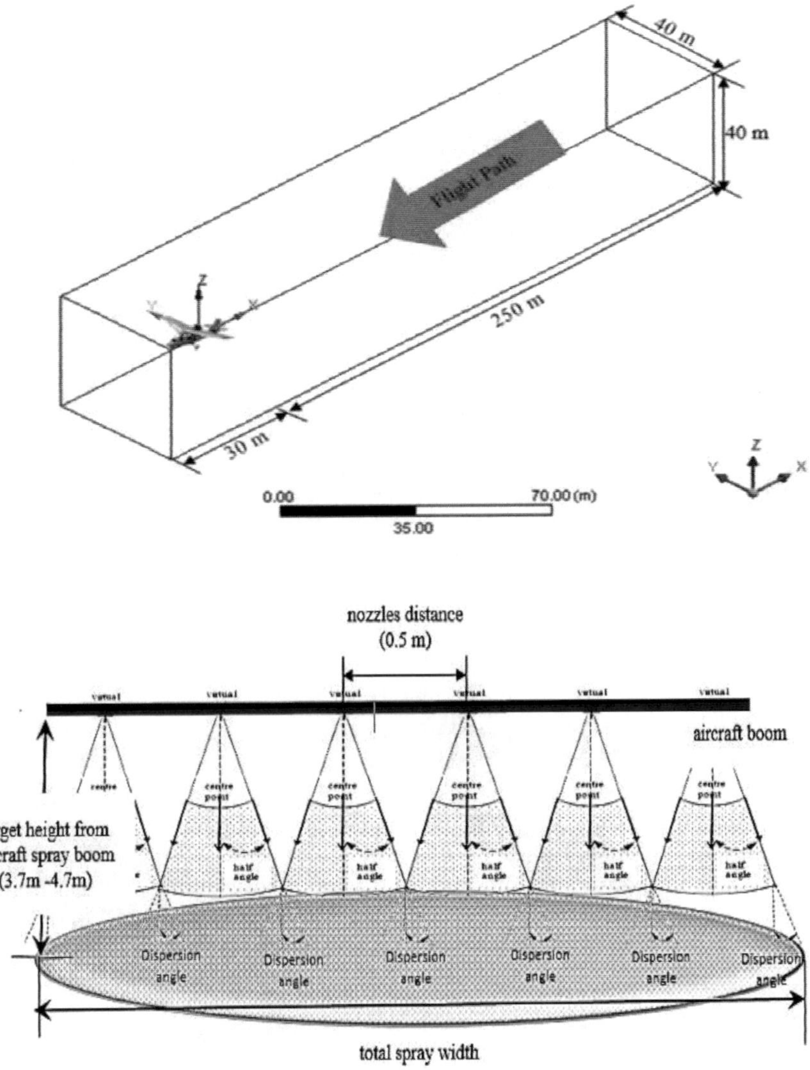

Figure 1. a) Schematic diagram of GT500 aircraft flight path **b)** Schematic diagram of 6 nozzles mounted on aircraft boom.

Table 1. GT 500® Aircraft general configurations

General GT500® Aircraft	Configurations (dimension in meter)
Length	9.94
Height	6.23
Wingspan	9.14
Boom Length	8.50

2.2. Computational domain setup and grid

Based on Figure 2, a 3D rectangular computational domain was developed with the dimensions of 280 x 40 x 40 (m) representing a sample of a flight path, with an assumption the centerline of the aircraft body is located 30m from the x axis of the total flight path length. The working domain was set to ensure that the domain size was not disturbed by the length of wing span and the turbulent airflow within flight path. The set up computational domain adapting the existing agriculture aircraft GT 500 in market is shown in Figure 2. Geometry and grid generated was executed and discretized using an appropriate mesh sizes consist of 549331 elements and 1100932 nodes as seen in Figure 3. The quality report indicated that meshed result had a good condition according to the skewness parameter, with an average values located at 0.23 as shown in Table 2.

Table 2. Mesh quality criteria parameters: skewness and element quality.

Mesh Metric	Skewness	Element quality
Minimum	5.806×10E-4	1.1881×10E-2
Maximum	0.9400	0.9996
Average	0.2318	0.4600
Standard Deviation	0.1289	0.3653

2.3. Boundary Conditions

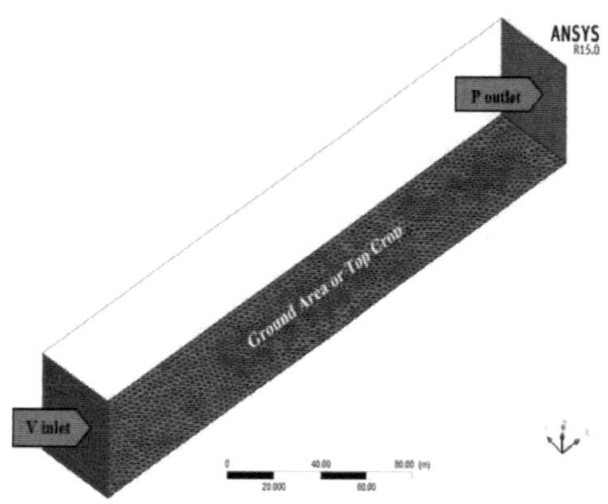

Figure 2. 3D Computational domain with boundary conditions assigned.

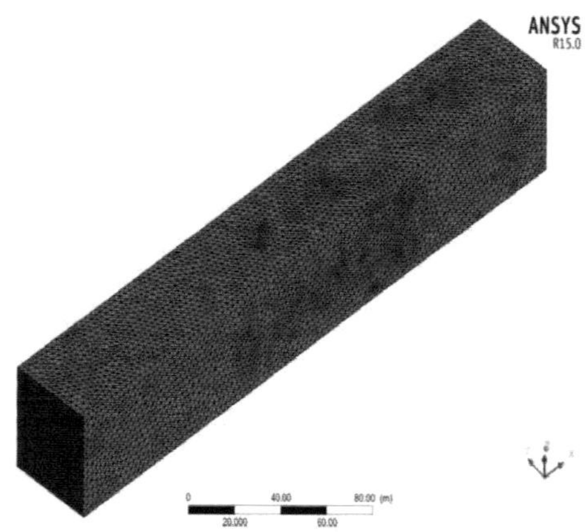

Figure 3. Computational grid with meshing of 549331 elements and 1100932 nodes

With reference to Figure 2, in all simulations, a uniform air velocity profile was set based on the constant 'velocity inlet' and the opposite faces at atmospheric pressure condition was located at the outlet. At the inlet velocity, the air was forced to move inside the domain at the given velocity which is 30 m/s and enabling freely motion of the air, as well as particles. The bottom surface, considered as 'Floor Wall' was set as 'stationary, no slip' condition as substitute of the top of the plantation areas, trapping all the particles that collide with them, while the remaining three outer faces, the top and surround faces modelled as 'symmetry' properties. In addition, the operation pressure was defined as standard atmospheric pressure which is 101325 Pascal. The operation density was 1.225 kg/m^3, the air density in standard conditions. The boundary condition assigned for each faces of the computed domain and the spray injection point and the exit of the flat fan's nozzle are appointed at the origin point (0, 0, 0), The total spray injection points of 1200 droplets of particle tracks of the replicates 12 nozzles at an air velocity of 30 m/s were created by inserted coordinate with assumption the axial position of the aircraft located at the central of the computational domain.

2.4. Numerical parameters in Fluent and Spray Atomizer

The reference spray parameters used in this simulation was in the same conditions with previous study done by Fritz et.al [9]. The flat fan configurations were obtained from Lechler Nozzles Catalogue [19], that were operated at 276 kPa (40 psi) and 0-degree orientation. The spray particles were set as water-liquid droplets, injected into the computational domain from a 12 nozzle of 0.009 m of width, respectively. When the particles injections exit the nozzle tip, it undergoes a flash evaporation. The main spray parameters used to configure the solver are summarized in Table 3, such as the spray angle, maximum velocity of the spray plume, shape of the plume, as well as, the dimension of the nozzle. The standard particle size data were utilized associates with the Flat Fan Atomizer parameters. Data were fitted to the cumulative function of standard particle size data distribution. The corresponding parameters CFD solver as shown in Table 3 and Figure 4.

Table 3. Solver spray of flat fan atomizer parameter

Variable	Value
X,Y,Z-Center (m)	0, 0.5, 3.7
X, Y,Z-Virtual Origin (m)	-0.00045,0.5,3.7
X,Y,Z-Fan Normal Vector	0,0,-1
Flow Rate (kg/s)	0.00915
Spray Half Angle (degree)	45
Orifice Width (m)	0.0009
Flat Fan Sheet Constant	3
Atomizer Dispersion Angle	6

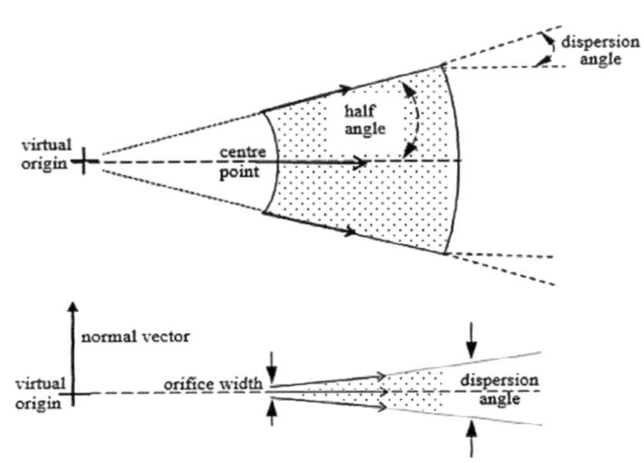

Figure 4. Flat Fan Atomizers Viewed from Above and from the Side

The solution of the differential equations for mass and momentum was done in a sequential manner, using the COUPLED algorithm [12]. The standard discretization scheme was used for the pressure and the second order upwind scheme for the momentum, turbulent kinetic energy and turbulent dissipation rate equations. The time step was 0.1 s, with a maximum limit of 200 iterations for total of 20 s simulation time including gravity effects with 9.81 m/s^2 in the vertical z axis negative direction. DPM model configuration the in the solver are listed in Table 4. The unsteady particle tracking was used until convergence was reached in the simulation by using a default criterion value of 1.0e-3 for the continuity (pressure), x, y and z velocity, and for k and ε turbulence parameters. The interaction with the continuous phase was also included to approximate the simulation to the real event, since the velocity difference between the two phases is high.

Table 4. Parameters for DPM setup

Parameter	Value
Interaction with Continuous Phase	On
Unsteady Particle Tracking	On
Inject Particles at	Particle Time Step
Particle Time Step Size (s)	0.1
Drag Law	Spherical
Two-way coupling turbulence	On

The drag law used (spherical) is the simplest and most used law, once it is the one that best fits the presented model amongst the four different options available in the solver. This drag law considers the particle as a sphere, which is an acceptable simplification for the water particles that exit the flat fan nozzle [4]. The total number of particle parcels streams injected for each the simulation was approximately 40000.

3. Results and Discussion

3.1. Spray Deposition.
The relations of spray released height on spray quality and spray drift deposition was further analyzed from the plotted CFD contours in horizontal travel direction as function of particle mass concentration in various elevated height as shown in Figure 5. Results are consistent for unsteady phase contour at each spraying, corresponding to the standard flat fan 90 degree atomizers. It was observed that for each cases the height of emission or spray released height plays an important role whereby an increase of the boom height increased the level of deposits at different downwind distance or the swath width [16]. For released height of 4.7 meter the measured distance travelled from the sprayer was 50 m compared to 38 m for 3.7 m of released height. The results were in good agreement with previous study reported by Lebeau et.al [5]. These results agreed the same explanation reported by Dorr et.al [2], the droplets velocity was higher at close range near the nozzle released point and reduced as the particles travelled away from the spray sourced. The droplet particles were highly dense at the exit of the nozzle tip, decayed into turbulent atmosphere reaching a constant velocity of 30 m/s an equilibrium state in less than 20 seconds, decreased its velocity and diminished by air resistance which breaks up the droplets.

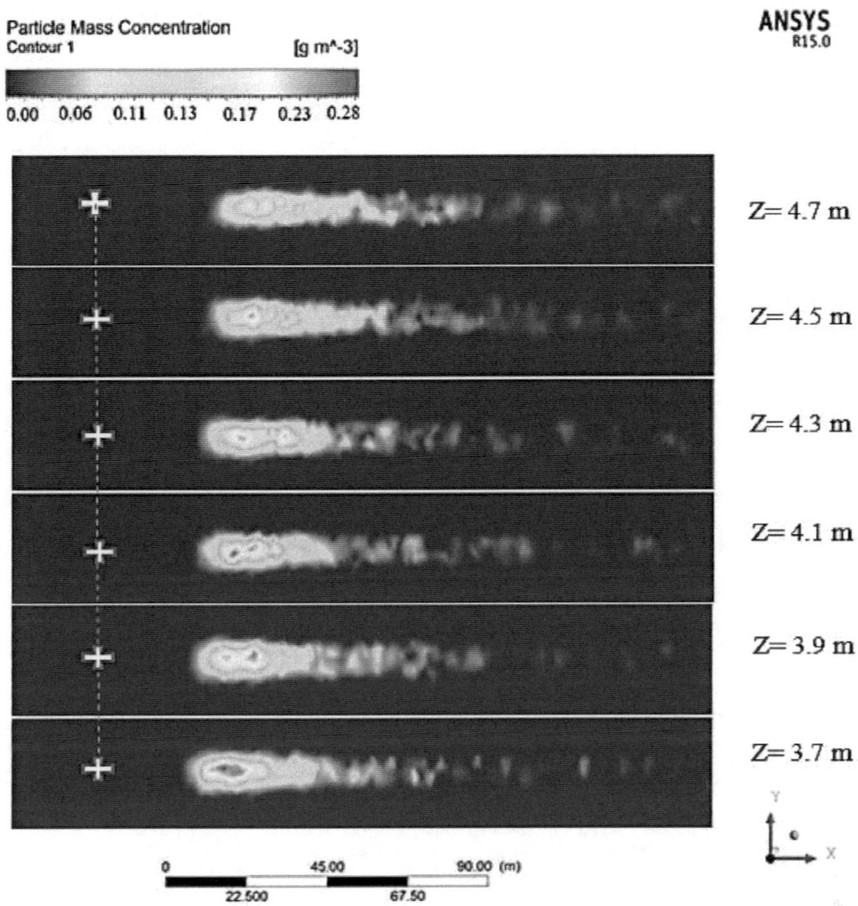

Figure 5. Spray deposition as function of particle mass concentration in height variation. Cross reference indicates nozzle location (spray source).

The sprayed deposition pattern is critical in aerial spraying works, illustrated the effects of aerial spraying works with optimum distance ranging not less than 2.4 m above the crop or ground compare to forestry application which is 6 m or higher. The turbulent dispersion, subjected to change and continue falling in effect of gravity magnitude which coincides at the end of the injection of the spray plume as proposed by Zhang et.al [3]. Generally, it was found that the maximum sprayed deposition area was achieved at the highest released height of 4.7 m compared for a released height of 3.7 m. Thus, a good combination of spray released height with low airspeed is significantly affected the aerial spraying deposition pattern. The results were further analyzed in Figure 7.

3.2. Spray Coverage.

The trajectory of scatterred sprayed droplets of each cases was plotted along a horizontal travel direction to evaluate the evaporated droplets for different spray released height as seen in Figure 6. For each cases,there was a significant reduction trend in the particle mass concentration level as the spray released height elevated. Generally, the mass concentration in each time step reduces with the application rates.This is one attributed factor that relates between mass concentration with droplets diameter formation as the amount of droplets moisture was absorbed by the airflow present in the domain [2]. From Figure 6, the high mass concentration were located at 50 m to 100 m at the central of the plume and reduced as the droplets transported downstream positions at 250 m as the droplets dispersed outwards. Results showed that for a released height of 4.7 meter, the particle mass concentration had the lowest concentration which is 39% reduced from 0.1963 g/m^3 compare to 0.2743 g/m^3 for a released height of 3.7 m.These findings showed the same result explained by Ryan et.al [4] that at lowest spray released height, smaller droplets were formed and scattered outwards at quicker rates but less contribution in droplets volume compared to higher released height. Since uniformity spray distribution was the main aim of this simulation, it is suggested that aerial spraying at low released height ensuring that the particles trajectories is sufficient to reach the intended targets before drift occurs due to the initial airspeed as mentioned in the earlier studies by Dorr et.al [2] .

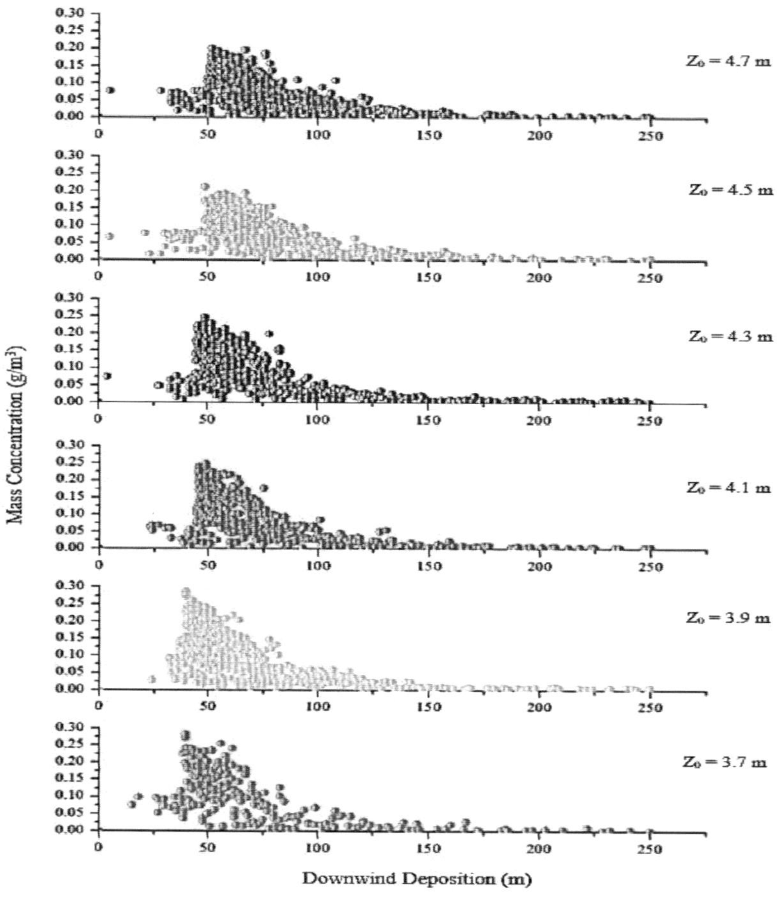

Figure 6. Downward deposition as function of constant velocity at various vertical height.

IOP Conf. Series: Materials Science and Engineering **243** (2017) 012044 doi:10.1088/1757-899X/243/1/012044

3.3. Spray Droplet Characterization.

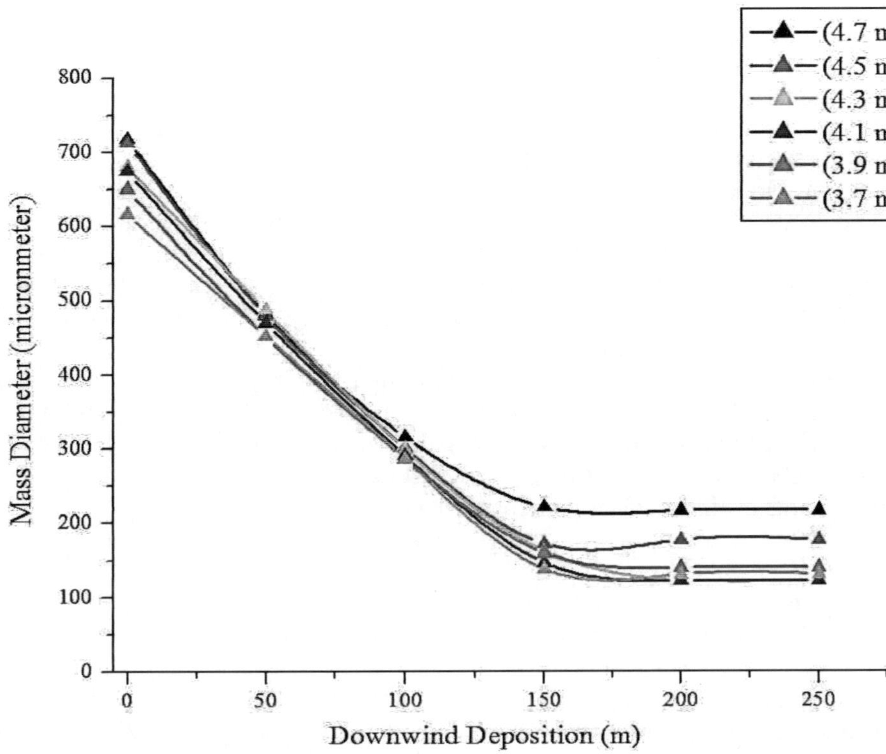

Figure 7. Downward deposition as function of constant velocity at various vertical height.

Figure 7 represents the decreasing trend of particle mean diameter at the horizontal positioned of the plume which is 250 m from the spray source. For each cases , the curves shows a downward slope indicates that the particle mean diameter decreases with the increment of the downwind position. Higher spray released height at 4.7 m produce large droplets which is 216 microns compare to 130 microns for a released height of 3.7 m. The droplets produced range was consistent with classical theory reported by Fritz et.al [9],classifying the agricultural aerial pesticide spraying were targeting droplets spectrum for medium range from 100 to 300 microns (100 – 300μm) with minimal smaller droplets categories which is 50 to 100 microns (< 50 -100 μm). The results indicated the overall droplet formation falls under medium range droplets which is given under ASAE S572 Standard Nozzles Classification by Droplet Spectra [8].The influence of spray released height on the droplets spctrum is significant corresponding due to high transport time and evaporation. Smaller droplets tends to fall much slower compared to large droplets,thus gives good reliability towards the aim to minimize the drift potential. For all simulations, the speed for all falling droplets is governed by the effect of gravitational pull and the airflow interaction within the domain . These findings supports previous theories made by Dorr et.al [2]. These findings also elucidates the importance of reference nozzles especially the orifice size used in aerial spraying works. A good combination of operating pressure will increased in nozzle output as well as droplets spectrum production in terms of uniform distribution as well as minimizing spray drift.

4. Conclusion and Recommendations

The proposed CFD Lagrangian model was able to predict the aerial spray dispersion using flat fan atomizer at various sprays released height of constant 30 m/s aircraft speed was studied. This simulation is a necessary step towards the simulation of crop spray efficacy in aerial spraying delivery in field-realistic conditions regarding operational variables, spray equipment and spray deposition. From the results obtained, several conclusions can be deduced:

a) The results revealed that there is considerable increased in spray drift and droplets trajectories with the increased in spray released height. These findings support the same result obtained by Viret et.al [7]. The highest downwind deposition was found at 4.7 m at the distance of 50 m from the spray sourced, while the lowest was found at 3.7 m above the ground. The overall downwind deposition is in gradually decrease trend in each simulation. Thus, the aerial spray released height should be maintained not less than 2.4 m above crop or ground [9].

b) The result demonstrated that the increase in spray released height will increase the swath width. Two notable contributed factor which is the combining factor of operational variables with the spray equipment's. Thus, emphasizing the reference nozzles is one of the critical factor in aircraft spray equipment's. The nozzles feature such as orifice sizes influenced droplet spectrum, distribution and spray pattern.

c) Spray deposition can be improved by adjusting the spray boom settings by means of influencing spray equipment such as spray types, angle, pressure and volume to the target crop or ground effect. A care selection of nozzle types is essential to ensure that the agriculture aerial spray droplet is maintain in medium droplets ranging from 100 to 300 microns so that it reach uniformly towards the intended crop or coverage area [9].

4.1. Recommendations

These initial step could be viable alternative to use in future research in terms of improving the aerial spray distribution with minimize spray drift potential. Further suggestions can be done for future improvements:

a) The spray simulation is recommended to include with the meteorological conditions inclusive wind speed, temperature and relative humidity to study the real applications towards the spray deposition.

b) Introduce wing aircraft or boom spray in the computational domain to study the effect of vortex towards the spray deposition.

Acknowledgements

This research was conducted in collaboration with Universiti Tun Hussein Onn Malaysia, funded by Ministry of Higher Education Malaysia, under Fundamental Research Grant Scheme (FRGS) No.1548. The authors would like to acknowledge Politeknik Ungku Omar, for their support and assistance given to the authors in carrying out this research study.

References

[1] Guler H, Zhu H, Ozkan HE, Derksen RC, Yu Y and Krause CR 2007 *Spray characteristics and* drift reduction potential with air induction and conventional flat-fan nozzles *Transactions of the ASABE* **50**(3) 745-754

[2] Dorr GJ, Forster WA, Mayo LC, McCue SW, Kempthorne DM, Hanan J, Turner IW, Belward JA, Young J and Zabkiewicz JA 2016 Spray retention on whole plants: modelling, simulations and experiments *Crop Protection, Elsevier* **88** 118-130

[3] Zhang B, Tang Q, Chen LP and Xu M 2016 Numerical simulation of wake vortices of crop spraying aircraft close to the ground *Biosystems Engineering, Elsevier* **145** 52-64

[4] Ryan SD, Gerber AG and Holloway AGL 2016 A time-dependent Eulerian model of droplet diffusion in turbulent flow *Computers and Fluids* **131** 1-15

[5] Lebeau F, Verstraete A, Stainier C, and Destain MF 2011 Computers and Electronics in Agriculture, Elsevier **77**(2) 161-174

[6] Wilson C and Tisdell C 2001 Why farmers continue to use pesticides despite environmental, health and sustainability costs *Ecological Economics, Elsevier* **39**(3) 449-462

[7] Viret O, Siegfried W, Holliger E and Raisigl U 2003 Comparison of spray deposits and efficacy against powdery mildew of aerial and ground-based spraying equipment in viticulture *Crop Protection, Elsevier* **22**(8) 1023-1032
[8] Hoffmann W, Fritz B, Thornburg J, Bagley W, Birchfield N and Ellenberger J 2011 Spray drif reduction evaluations of spray nozzles using a standardized testing protocol *In Pesticide Formulations and Delivery Systems* **30** ASTM International
[9] Fritz BK 2006 Meteorological effects on deposition and drift of aerially applied spray *Transactions of the ASABE* **49**(5) 1295-1301
[10] Donkersley P and Nuyttens D 2011 A meta-analysis of spray drift *sampling Crop Protection Elsevier* **30**(7) 931-936
[11] Nuyttens D, De Schampheleire M, Baetens K and Sonck B 2007 The influence of operator controlled variables on spray drift from field crop sprayers *Transactions of the ASABE* **50**(4) 1129-1140
[12] Sidahmed MM, Taher MD and Brown RB 2005 A virtual nozzle for simulation of spray generation and droplet transport *Biosystems engineering* **92**(3) 295-307
[13] Fox RD, Derksen RC, Zhu H, Brazee RD and Svensson SA 2008 A history of air-blast sprayer development and future prospects *Transactions of the ASAE* **51**(2) 405
[14] Antuniassi UR, Motta AAB, Chechetto RG, Carvalho FK, Velini ED and Carbonari CA 2014 Spray drift from aerial application *Aspects of Applied Biology* **122** 279-284
[15] Baetens K, Nuyttens D, Verboven P, De Schampheleire M, Nicolai B and Ramon H 2007 Predicting drift from field spraying by means of a 3D computational fluid dynamics model *Computers and Electronics in Agriculture. Elsevier* **56**(2) 161-173
[16] Omar Z, Qiang KY, Mohd S and Rosly N 2016 *CFD* Simulation of aerial crop *Spraying In IOP Conference Series: Materials Science and Engineering* **160**, No. 1, p. 012028
[17] Montazeri H, Blocken B and Hensen JL 2015 CFD analysis of the impact of physical parameters on evaporative cooling by a mist spray system *Applied Thermal Engineering. Elsevier* **75** 608-622
[18] Gil Y and Sinfort C 2005 Emission of pesticides to the air during sprayer application *Atmospheric Environment, Elsevier* **39**(28) 5183-5193
[19] www.lechler-agri.de/PDF/agrarkatalog_e.pdf
[20] 148.204.81.206/Ansys/150/ANSYS%20Fluent%20Tutorial%20Guide.pdf

Determination of particle-bound polycyclic aromatic hydrocarbons emitted from co-pelletization combustion of lignite and rubber wood sawdust

R Kan[1], T Kaosol[1,4], P Tekasakul[2,4] and S Tekasakul[3,4]
[1] Environmental Engineering Program, Department of Civil Engineering, Faculty of Engineering, Prince of Songkla University, Hat Yai, Songkhla 90112, Thailand
[2] Department of Mechanical Engineering, Faculty of Engineering, Prince of Songkla University, Hat Yai, Songkhla 90112, Thailand
[3] Department of Chemistry, Faculty of Science, Prince of Songkla University, Hat Yai, Songkhla 90112, Thailand
[4] Air Pollution and Heath Effect Research Center, Prince of Songkla University, Hat Yai, Songkhla 90110, Thailand

E-mail: thaniya.k@psu.ac.th

Abstract. Determination of particle-bound Polycyclic Aromatic Hydrocarbons (PAHs) emitted from co-pelletization combustion of lignite and rubber wood sawdust in a horizontal tube furnace is investigated using High Performance Liquid Chromatography with coupled Diode Array and Fluorescence Detection (HPLC-DAD/FLD). The particle-bound PAHs based on the mass concentration and the toxicity degree are discussed in the different size ranges of the particulate matters from 0.07-11 μm. In the present study, the particle-bound PAHs are likely abundant in the fine particles. More than 70% of toxicity degree of PAHs falls into $PM_{1.1}$ while more than 80% of mass concentration of PAHs falls into $PM_{2.5}$. The addition of lignite amount in the co-pelletization results in the increasing concentration of either 4-6 aromatic ring PAHs or high molecular weight PAHs. The high contribution of 4-6 aromatic ring PAHs or high molecular weight PAHs in the fine particles should be paid much more attention because of high probability of human carcinogenic. Furthermore, the rubber wood sawdust pellets emit high mass concentration of PAHs whereas the lignite pellets emit high toxicity degree of PAHs. By co-pelletized rubber wood sawdust with lignite (50% lignite pellets) has significant effect to reduce the toxicity degree of PAHs by 70%.

1. Introduction

Thailand is the largest global natural rubber production and exportation. Most of 700 cooperatives is located in the Southern Thailand. In order to dry the rubber sheets, the rubber woods are burnt to produce heat and smoke. The rubber woods are also burnt in order to generate the electricity like coal. The rubber wood burning was reported that it emitted very high particulate matters and Polycyclic Aromatic Hydrocarbons (PAHs) [1,2]. Unlike, lignite combustion generated high bottom ash and toxic heavy metals. However, the toxic elements in lignite may relate or increase the toxicity degree of PAHs. The emitted particles and the particle-bound PAHs can be the significant sources of the human health effects and the ambient air.

The PAHs concentration could be found in the raw materials and output residues. Verma et al [3] reported the PAHs in the coal power plant. The coal, raw material, existed 4,542 μg/kg of total PAHs, while the fly ash and bottom ash existed 32.4 and 10.1 μg/kg of total PAHs, respectively. Most of high molecular weight PAHs concentrated on the coal and the fly ash, whereas the low molecular weight PAHs on the bottom ash. The high molecular weight PAHs cause more human health impacts than the low molecular weight. Meanwhile, the rubber wood burning in the workplace was reported that it emitted 10 folds of particle-bound PAHs in comparison with the ambient air [2]. Most of particle-bound PAHs accumulated in the fine particles [1,2,4]. The high concentration of PAHs either in the workplace or ambient air should be concerned especially the workers' health. By blending lignite and rubber wood sawdust, the PAHs emission might be traded. Al-Naiema et al [5] reported that co-firing coal with oat hulls by 50% would lead to the reduction of particulate matters and PAHs by 90% and 40%, respectively but there was insignificantly effect of pollutants above on 3.8% wood chips blends. The particulate matters and PAHs will increase, decrease or remain stable depending on the properties/blending ratios of fuels and combustion conditions.

The PAHs formation has been studied on the various types of fuels, raw chunk blending, excess air, combustion temperature, moisture, and combustion system [1–6]. However, the study of particle-bound PAHs formation emitted for co-pelletization combustion is still remained in the initial information point and rarely identified the risk evaluation. The understanding of PAHs is important since they can accumulate in the particulate matters, which deposit surrounding the workplace or transport in a long distance in the ambient air. The PAHs impact seriously to the human health especially the carcinogenic and repository symptom depending on the size fraction of particles/particulate matters. The fine particles can penetrate deeply into the lung especially in the end of bronchus and alveolus areas rather than the coarse particles. The different size fraction of particles leads to the different health outcome. Therefore, the PAHs components related to the emitted particle size fraction are essentially determined.

In the present study, the particle-bound PAHs emitted from co-pelletization combustion of lignite and rubber wood sawdust in a horizontal tube furnace are investigated using High Performance Liquid Chromatography with coupled Diode Array and Fluorescence Detection (HPLC-DAD/FLD). The particle-bound PAHs are determined in different size ranges of particulate matters from 0.07-11 μm. Moreover, their aromatic rings and toxicity degree are also discussed.

2. Materials and Methods

2.1. Materials
Lignite and rubber wood sawdust were adapted from Mae Moh, Lampang and Rattapoom, Songkhla, Thailand, respectively. Both materials were co-pelletized with 10% starch as binder by screw press. Further detail of raw materials and co-pelletization could be taken from a previous study [7].

2.2. Particle sampling from combustion experiment
The pellets were burnt in a horizontal tube furnace with an inserted quartz column (ID = 55 mm, OD = 60 mm, and L = 850 mm) and heating zone (OD = 60 mm and L = 650 mm) (figure 1). Eleven-selected pellets (3.5-4 L:D ratio) were inserted into the central heating zone of quartz column and burnt at room temperature to 700 °C with 5 °C/min of heating rate and 20 min holding time. Dry clean air approximately 1.6 L/min was purged into the furnace to combust the pellets. In order to reduce the temperature and moisture content, the exhaust was diluted with a dilutor (OD = 35 mm and L = 800 mm) include a mixing tube (L = 70 mm) by the dry clean air. The particulate matters and the flue gas were respectively sampling using an Anderson cascade impactor with an initial filter stage (ANIF) model Dylec NA200 (Tokyo Dylec, Japan) and a gas analyzer model Testo 350XL (Testo, Germany).

The binder-less quartz filters Pallflex 2500QAT-UP (D = 80 mm, Pall Corporation, USA) were used in ANIF. The flow rate of ANIF was controlled at 28.3 L/min by a rotameter at the staked top of ANIF. The filters were treated for 72 h in room temperature and 50% of relative humidity.

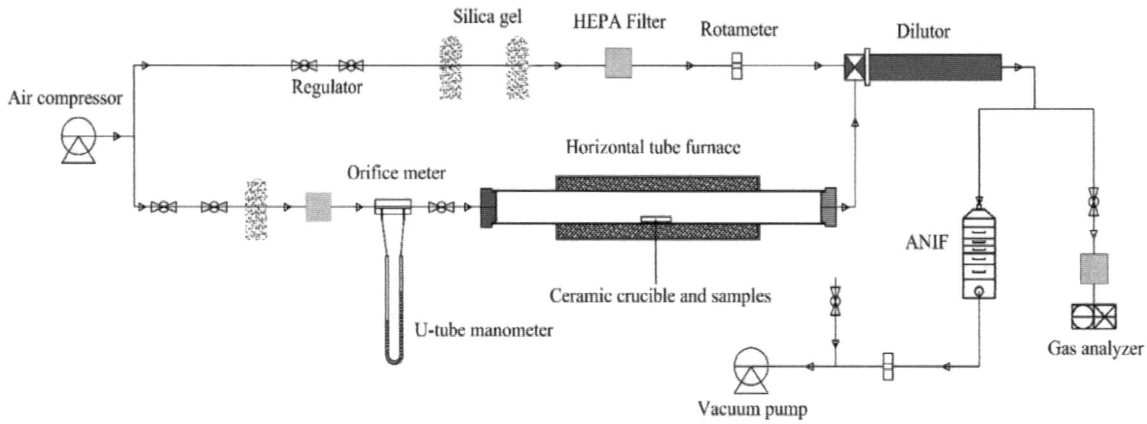

Figure 1. Schematic diagram of combustion system.

2.3. Analysis of particle-bound polycyclic aromatic hydrocarbons

The particulate matters on each stages of ANIF were extracted by ultrasonic extraction technique in order to analyze the PAHs. The extraction was operated with dichloromethane in the cooled ultrasonic cleaner twice for 20 min (99 Hz and <10 °C). The dimethyl sulfoxide (DMSO) was added into the extraction after filtering with the filler papers and 0.45-µm PTFE syringes. After that, the solution was evaporated by a rotary evaporator model Rotavapor-R (Büchi, Switzerland) with cooling system to remove dichloromethane. After evaporating, acetonitrile was added and homogeneously mixed. The PAHs solution was finally filtrated with the 0.45-µm PTFE syringes and stored in the refrigerator at 4±2 °C for the next PAHs analysis.

Table 1. Aromatic rings and toxic equivalence factors of sixteen priority PAHs listed by US EPA.

PAHs	Abbr.	Aromatic rings	TEF
Naphthalene	NAP	2	0.001
Acenaphthylene	ACY	3	0.001
Fluorene	FLU	3	0.001
Acenaphthene	ACE	3	0.001
Phenanthrene	PHE	3	0.001
Anthracene	ANT	3	0.01
Fluoranthene	FLA	4	0.001
Pyrene	PYR	4	0.001
Benzo(a)anthracene	**BaA**	**4**	**0.1**
Chrysene	**CHY**	**4**	**0.001**
Benzo(b)fluoranthene	**BbF**	**5**	**0.1**
Benzo(k)fluoranthene	**BkF**	**5**	**0.01**
Benzo(a)pyrene	**BaP**	**5**	**1**
Dibenz(a,h)anthracene	**DahA**	**5**	**1**
Benzo(g,h,i)perylene	BghiP	6	0.01
Indeno(1,2,3-cd)pyrene	**IcdP**	**6**	**0.1**

Bold words: Probably human carcinogenic PAHs listed by US EPA [8–10]

PAHs are hydrocarbon compounds with two or more aromatic rings released from the combustion of fuels, or found in the emission and fuels. The sixteen particle-bound PAHs as shown in table 1 were analyzed by High Performance Liquid Chromatography with coupled Diode Array and Fluorescence

Detection (HPLC-DAD/FLD) model Hewlett Packard series 1100 (Hewlett Packard, USA) and Inersil ODS-P C-18 reversed-phase column connected with 5-μm Inersil guard (5-μm particle size, 4.6-mm diameter, 250-mm length). The combination of DAD and FLD detection was the powerful technique when NAP, ACE, FLU, and ACY provided the weak signals on the FLD detection, especially ACY. The portion of extracted PAHs was automatically injected into the column approximately 20 μL while the mixture mobile phases of acetonitrile and ultra-pure water were operated with the gradient condition mode of acetonitrile at flow rate 0.5 mL/min (table 2). Timing of PAHs detection was set following the optimum excitation and emission wavelengths of detectors (table 3). In addition, the PAHs peaks were calibrated with the standard PAHs ($R^2 = 0.99$). The blank filters were prepared for the PAHs analysis. The corrected PAHs were applied by subtracting the blank PAHs. The PAHs recovery was 72-138% depending on individual PAHs.

Table 2. Gradient condition of PAHs analysis.

Time (min)	Acetonitrile (%)	Ultra-pure water (%)	Flow rate (mL/min)
0	70	30	
10	70	30	
25	90	10	
27	99	1	0.5
30	100	0	
75	100	0	
76	70	30	
80	70	30	

Table 3. Optimum wavelength of sixteen PAHs with diode array and florescence detection.

Time (min)	Fluorescence		Diode Array	PAHs
	λ Excitation (nm)	λ Emission (nm)	λ Wavelength (nm)	
0	280	340		NAP, ACE, FLU, ACY (except ACY not detected in florescence)
23	250	400		PHE
29	286	433		ANT, FLA
31	331	392	254	PYR
34	264	407		BaA, CHR, BbF, BkF BaP, DahA, BghiP
65	294	482		IcdP
72	280	340		-
80	280	340		-

3. Results and Discussion

The sixteen PAHs such as NAP, ACE, FLU, ACY, PHE, ANT, FLA, PYR, BaA, CHR, BbF, BkF, BaP, DahA, BghiP, and IcdP are examined in the different size ranges of particulate matters from 0.07-11 μm. As seen in figure 2, the pellet combustion emits the low molecular weight PAHs (< 228 g/mol) higher than the high molecular weight PAHs (> 228 g/mol). The 0%, 50% and 100% lignite pellet combustion emits 97%, 91%, and 75% of low molecular weight PAHs, whereas 3%, 9%, and 25% of high molecular weight PAHs, respectively. In addition, the mass fraction of PAHs is mostly dominated by 25-67% of NAP and 23-29% of ACY. It is interested that the more $PM_{2.1}$ increases, the greater high molecular weight PAHs increases. The consequences of high molecular weight PAHs accommodated in the fine particles are the diffusivity of molecular weights, the vapor pressures of PAHs, the absorption between particle surface areas and organic matters, and transformation of gas phase to particle phase [11,12].

The PAHs can accompany either in the gas phase or particulate phase due to its vapor pressure (function of temperature) and its molecular weight. In general, the low molecular weight PAHs are anticipated in gas phase, whereas the high molecular weight in particulate phase with which the size distribution of particles are associated [11]. Most of PAHs is found in the fine particulate matters or in the gas phase anticipated with the small particles [1,2,13]. Moreover, more than 85% of total PAHs is accumulated on the particulate matters according to the literatures [11,14].

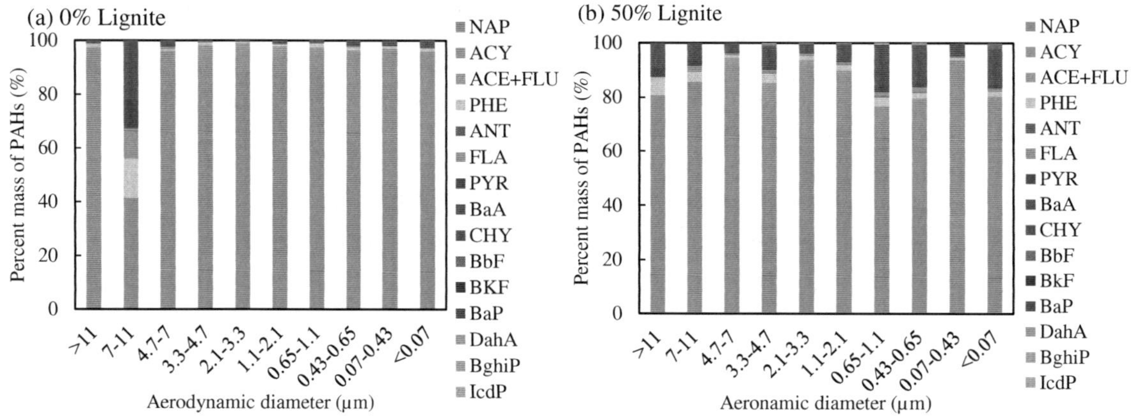

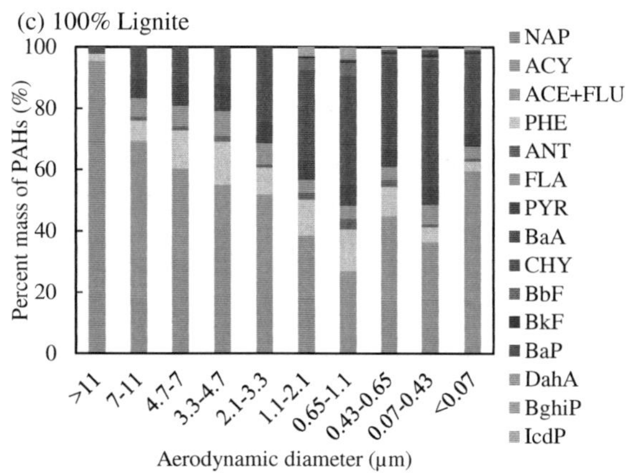

Figure 2. Composition profile of PAHs in different size ranges of particulate matters emitted from (a) 0% lignite pellet, (b) 50% lignite pellet, and 100% lignite pellet combustion.

In general, the non-carcinogenic and carcinogenic PAHs have been recognized to 2-3 aromatic rings and 4-6 aromatic rings of PAHs, respectively. As seen in figure 3, the 0% and 50% lignite pellet combustion are mostly abundant the 2-3 aromatic ring PAHs while the 100% lignite pellet combustion are mostly abundant the 3-4 aromatic ring PAHs. The 5-6 aromatic ring PAHs also increase when the lignite mount increase in the co-pelletiztion. The 4-6 aromatic ring PAHs contributed in the order of 0% (1181 ng/g fuel) > 50% (3349 ng/g fuel) > 100% (8200 ng/g fuel) lignite pellets combustion. The increment of mass concentration of 4-6 aromatic PAHs will more cause human carcinogenic, mutagen, and repository symptom. Meanwhile, Shen [11] reported that there was no significant effect of mass concentration of PAHs between the raw chunk fuel and densified fuel combustion; however, the densified fuel combustion significantly leads to the reduction of particulate matters and also the other pollutants. Hence, the toxicity degree of PAHs emitted from the co-pelletization combustion will be taken into accounts.

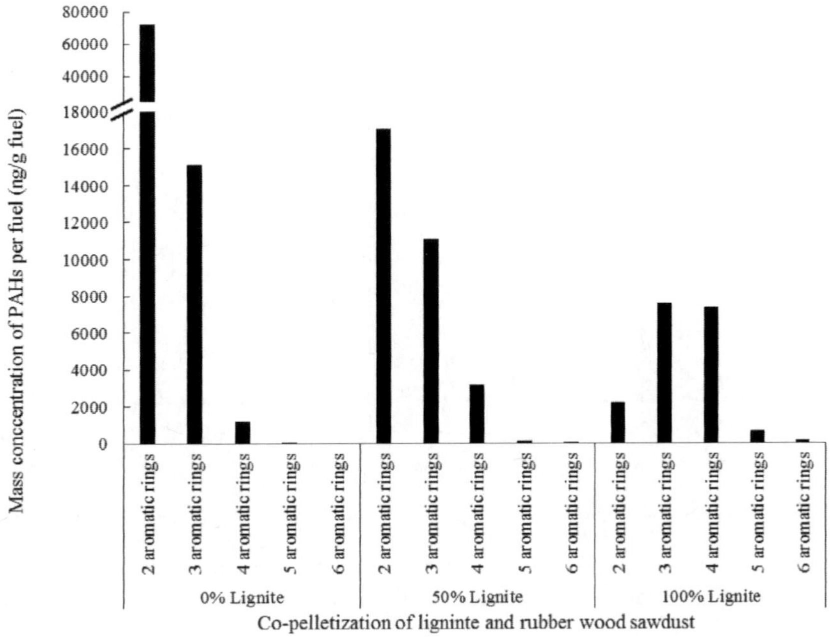

Figure 3. Total mass concentration of different aromatic ring PAHs emitted from co-pelletization combustion of lignite and rubber wood sawdust.

The mass concentration and toxicity degree of PAHs per fuel burning in the different size ranges of particulate matters emitted from 0%, 50% and 100% lignite pellet combustion are shown in figure 4. The particle-bound PAHs would consider to be abundant in $PM_{2.5}$ more than 80%, which presents a good agreement with the literatures [11,15]. The distribution of PAHs of 0% lignite pellet combustion contributes very high amount in each sizes as compared to the 50% and 100% lignite pellet combustion. The most abundant mass concentration of PAHs is in the particle sizes of 1.1-2.1 μm of 0% lignite pellet combustion, whereas in the finer particle sizes of 0.43-0.64 μm of 50% and 100% lignite pellet combustion (figure 4.a). On the other hand, the mass concentration of PAHs cannot judge its toxicity degree, so the toxicity degree of PAHs is calculated in BaP-TEQ basis (Benzo(a)Pyrene Toxic Equivalence Quotient) as following:

$$PAH_{total} = \sum (PAH_i \times TEF_i) \tag{1}$$

where PAH_{total} is the total PAHs (BaP-TEQ), PAH_i is the congener i of PAHs, and TEF_i is the toxic equivalence factor of PAHs congener i.

In contract to the mass concentration of PAHs, the toxicity degree of PAHs is massively abundant in 0.43-0.64 μm of 100% lignite pellet combustion (137.36 BaP-TEQ ng/g fuel) following 50% lignite pellet combustion (53.94 BaP-TEQ ng/g fuel). The 0% lignite pellet combustion is abundant in the particle sizes of 1.1-2.1 μm with less toxicity degree of PAHs (25.27 BaP-TEQ ng/g fuel) (figure 4.b). Furthermore, the total toxicity of PAHs of the 100% lignite pellet combustion is 3.7 folds of the 0% lignite pellet combustion while the total toxicity of PAHs of the 50% lignite pellet combustion is 1.3 and 0.35 folds of the 0% and 100% lignite pellet combustion, respectively. When the high molecular weight PAHs considered 4-6 aromatic ring of PAHs tend to emit in the fine particles, they will influent on the ambient air and human health. The fine particle-bound PAHs can undergo in the long distance with the atmospheric movement in contrast to the large particle-bound PAHs that tend to deposit in a short time around the combustion workplace. Therefore, the particle-bound PAHs should get much more attention on the fine particles.

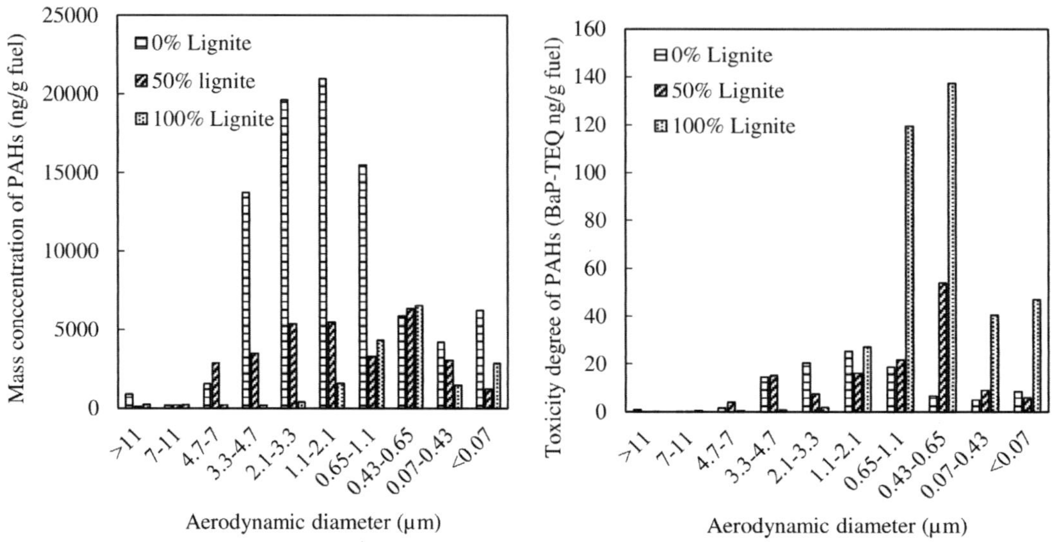

Figure 4. Concentration of PAHs emitted from co-pelletization combustion of lignite and rubber wood sawdust in different size ranges of particulate matters: (a) mass concentration of PAHs and (b) toxicity degree of PAHs.

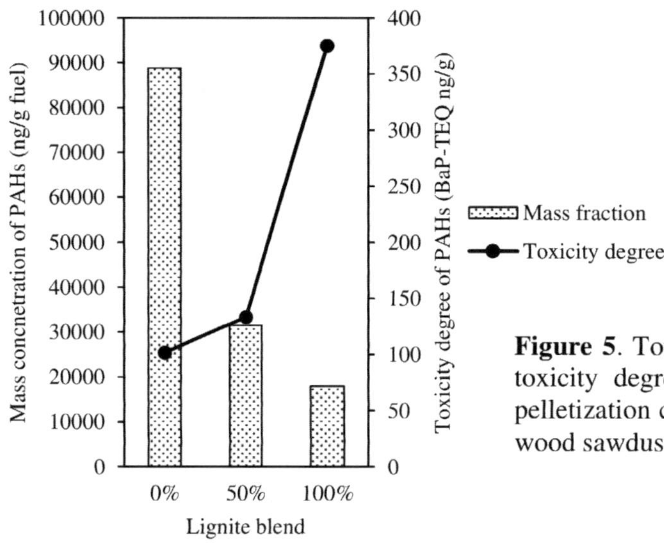

Figure 5. Total mass concentration and total toxicity degree of PAHs emitted from co-pelletization combustion of lignite and rubber wood sawdust.

As shown in figure 5, the mass concentration of PAHs are presented in the order of 0% > 50% > 100% lignite pellet combustion in contracted to the toxicity degree of PAHs. However the total mass concentration of PAHs from the 0% lignite pellet combustion emits higher value than the 100% lignite pellet combustion, the toxicity degree of PAHs from the 100% lignite combustion is more effect to the human health. It is widely documented that the high moisture content, high volatile matter, and unstable medium combustion condition would happen the incomplete combustion and produce more particulate matters and even the mass concentration of PAHs while the toxicity degree of PAHs causes from the complex structures of co-pellets especially the complex aromatic structures of lignite and the mineral association of lignite [6,11,13]. It is worth noting that the lignite combustion alone should be much more concerned. By pelletized rubber wood sawdust with lignite (50% lignite pellets), the toxicity degree of PAHs can be reduced by 70% of total toxicity degree of PAHs.

4. Conclusion

By sampling the particulate matters associated with PAHs in the horizontal tube furnace and further analysing with HPLC-DAD/FLD, the characteristics of particle-bound PAHs emitted from co-pelletization combustion of lignite and rubber wood sawdust are obtained that: (1) the PAHs components are abundant by 25-67% of NAP and 23-29% of ACY; (2) the mass concentration of particle-bound PAHs decrease in addition of lignite, whereas 4-6 aromatic PAHs increase and even the toxicity degree of PAHs; and (3) by co-pelletized rubber wood sawdust with lignite (50% lignite pellets), the toxicity degree of PAHs can be reduced by 70% of total toxicity degree of PAHs, and the large contribution of PAHs appears in the fine particle size of 0.43-0.65 μm while the PAHs are massively abundant approximately 70% in $PM_{1.1}$.

Acknowledgements
This work was supported by the Higher Education Research Promotion and the Thailand's Education Hub for Southern Region of ASEAN Countries Project Office of the Higher Education Commission, and Prince of Songkla University (ENG590662S).

References

[1] Tekasakul P, Furuuchi M, Tekasakul S, Chomanee J and Otani Y 2008 *Aerosol Air Qual. Res.* **8** 265–78
[2] Furuuchi M, Tekasakul P, Murase T, Otani Y, Tekasakul S and Bai Y 2006 *J. Ecotechnology Res.* **12** 135–39
[3] Verma S K, Masto R E, Gautam S, Choudhury D P, Ram L C, Maiti S K and Maity S 2015 *Fuel* **162** 138–47
[4] Hata M, Chomanee J, Thongyen T, Bao L, Tekasakul S, Tekasakul P, Otani Y and Furuuchi M 2014 *J. Environ. Sci.* **26** 1913–20
[5] Al-Naiema I, Estillore A D, Mudunkotuwa I A, Grassian V H and Stone E A 2015 *Fuel* **162** 111–20
[6] Chomanee J, Tekasakul S, Tekasakul P, Furuuchi M and Otani Y 2009 *Aerosol Air Qual. Res.* **9** 404–11
[7] Kan R, Kaosol T and Tekasakul P 2016 *KKU Eng. J.* **43** 259–62 (*Preprint* 10.14456/kkuenj.2016.104)
[8] ATSDR 2009 *Agency Toxic Subst. Dis. Regist.* (accessed April 20, 2016) http://www.atsdr.cdc.gov/HAC/pha/pha.asp?docid=59&pg=2
[9] Bojes H K and Pope P G 2007 *Regul. Toxicol. Pharmacol.* **47** 288–95
[10] ATSDR 2013 Polycyclic aromatic hydrocarbons (PAHs): What health effects are associated with PAH exposure? *Agency Toxic Subst. Dis. Regist.* (accessed April 20, 2016) http://www.atsdr.cdc.gov/csem/csem.asp?csem=13&po=11
[11] Shen G 2014 *Emission Factors of Carbonaceous Particulate Matter and Polycyclic Aromatic Hydrocarbons from Residential Solid Fuel Combustions* (Springer) p 92-195
[12] Venkataraman C and Rao G U M 2001 *Environ. Sci. Technol.* **35** 2100–7
[13] You X 2008 *Waste Manag.* **28** 1543–51
[14] Riva G, Pedretti E F, Toscano G, Duca D and Pizzi A 2011 *Biomass Bioenergy* **35** 4261–7
[15] Shen G, Wei S, Zhang Y, Wang B, Wang R, Shen H, Li W, Huang Y, Chen Y, Chen H and Tao S 2013 *Biomass Bioenergy* **55** 141–7

Parametric study on flow dispersion of water sprinkle

R C Tan[1], M H A Khafar[1], N I S Abdullah[1], R N Chendang[1], I Taib[1], N Asmuin[1], Y Ramli[1], S M Seri[1] and A N Mohammed[1]
[1]Faculty of Mechanical and Manufacturing Engineering, Universiti Tun Hussein Onn Malaysia (UTHM), Parit Raja, Batu Pahat, Johor, Malaysia

E-mail: rongcheng93@hotmail.com

Abstract. Although water sprinkler is used extensively in agriculture, little effort had been made to improve its performance, resulting in many sprinkler head available at market having less optimum design. Thus, this study aims to improve the basic design of water sprinkler head by conducting a parametric study on the effect of model geometry due to different flow characteristics. A common type of water sprinkler is modelled with computer aided design software and various changes such as enlarging nozzle diameter from 4mm to 8mm, changing vane angle from 70 degrees to 45 degrees are made to the original model. The models were simulated with computational fluid dynamics (CFD) software to investigate how the variation in geometry affects the flow of water and the performance of sprinkler head. The performance of water sprinkler is compared to original model in terms of watering distance, area of spray and velocity of water jet in air. The result of this study shows that enlarge the nozzle diameter have a positive effect on the velocity of water jet in air and the area covered by water jet but it drastically decreases the watering distance of sprinkler. Besides that, changing the angle of vane from 70 degrees to 45 degrees decrease the watering distance slightly and it concentrates the water into a fine jet that cover a small area. To reduce the effect, grooves can be added to the vane to increase the divergence of water spray. Reducing the angle of curvature from 10 degrees to 5 degrees improves the watering distance. The angle of curvature can be reduced more to increase the watering distance further.

1. Introduction

Water sprinkler is a device to distribute water to a large area automatically. Irrigation technology involving water sprinkler had been developed since late 1800s and later modified for orchards, commercial green houses and high-profit vegetable crops [1]. Nowadays, it is mostly used in agriculture field to irrigate crops or fire safety field to extinguish fire [2].

Water sprinkler came in many shape and sizes, but it can be classified into the following two major types on the basis of the method of water distribution, which is rotating sprinkler system and drip irrigation system. In rotating sprinkler system, nozzles are placed on riser pipe at uniform intervals. The riser pipe is connected to lateral pipe, and the lateral pipe to a water source. When water flow through rotating sprinkler, part of the flow energy is converted into rotating kinetic energy to change the direction of flow. Examples of rotating sprinkler are impact sprinkler, which rotate the sprinkler with a small impact hammer; and rotating vane sprinkler [3]. The advantage of rotating sprinkler is it the lower initial installation cost [4]. However, rotating sprinkler consume more power and is less water efficient compared to perforated pipe system [5].

The second type of water sprinkler is drip irrigation. This method consists of drilled holes that act as nozzles along pipe length through which water is sprayed. The advantage of perforated water sprinkler is it requires less pumping power since the pipes are laid near to the crops' root. Furthermore, this irrigation system can maximise water efficiency by reducing water loss by vaporisation [5].

There are many ways to improves the performance of water sprinkler system, such as upgrade the flow rate and pressure of water supply [6]. The use of drip irrigation with center pivot and lateral move sprinkler system is another method to save cost in water usage [7]. However, little study had investigated the possibility of using a sprinkler head that has better performance to reduce cost on installation. Thus, many type of water sprinkler commonly found on market did not have the optimum design to improve water usage and its performance in terms of distance and coverage area.

Although there are many innovations made to rotating water sprinkler, such as an adjustment ring and throttling element to adjust the distance and coverage area in one recent patent [8], the basic design of water sprinkler head had remain the same for many years [9]. Its basic component includes a deflector (or vane), a nozzle and a housing (or holder). Therefore, current study aims to improve the performance of water sprinkler by carrying out a parametric study with the basic component of a common type of rotating water sprinkler. This study employed the technique of computational fluid dynamics to simulate the flow characteristics of water sprinklers to compare the effect of various changes made to the basic design of water sprinkler.

To simulate the flow of water jet from nozzle and its dispersion from deflector, a similar model of water sprinkler head [10] is used as the reference. The boundary condition and fluid domain definition is mainly based on the previous model.

2. Methodology

2.1. Geometry of water sprinkler

A common type of water sprinkler available at the market was "825". It has the basic design of a rotating water sprinkler with a vane, a nozzle and a holder. The water sprinkler is shown in figure 1. Its size is about 80mm width, 30mm length and 100mm height.

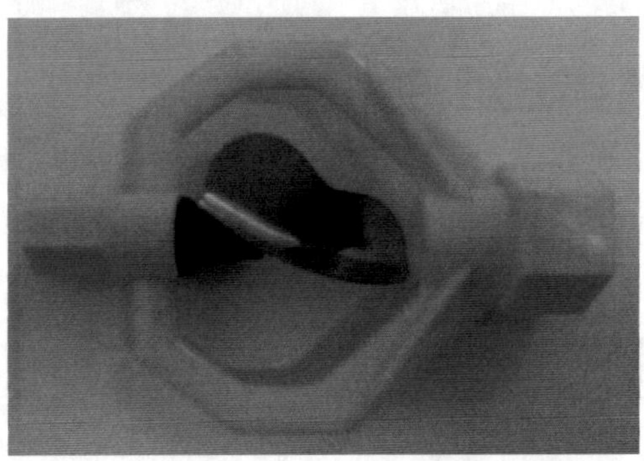

Figure 1. Original model of rotating water sprinkler.

Three dimensional models of water sprinkler (the original model and the modified models) were developed by using computer aided design software (CAD) SOLIDWORKS (Dassault Systemes Ltd). The comparison between original model and modified models are shown in table 1. The models were exported into ANSYS Workbench (ANSYS Ltd) in STEP file format to be simulated with computational fluid dynamics CFX software.

IOP Conf. Series: Materials Science and Engineering **243** (2017) 012046 doi:10.1088/1757-899X/243/1/012046

Table 1. Comparison of original model and modified models

Model name	Modification
a	Original geometry. Control model.
b	Enlarge the nozzle diameter from 4mm to 8mm. 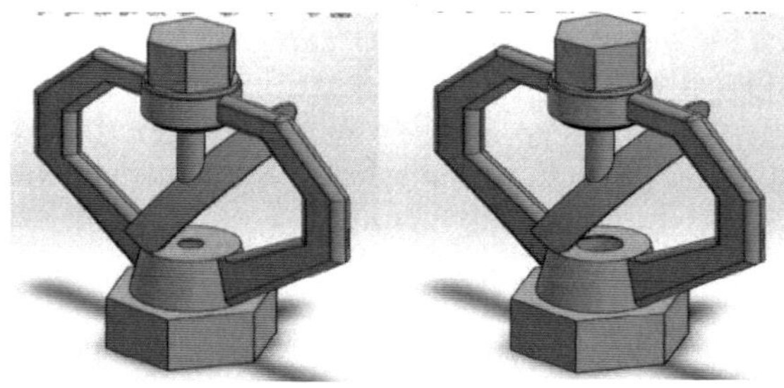
c	Change the angle of vane from 70° to 45°. 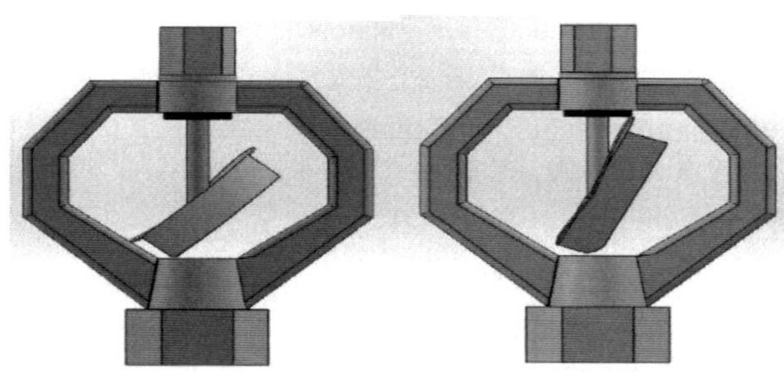
d	Change the vane's angle of curvature from 10° to 5°. 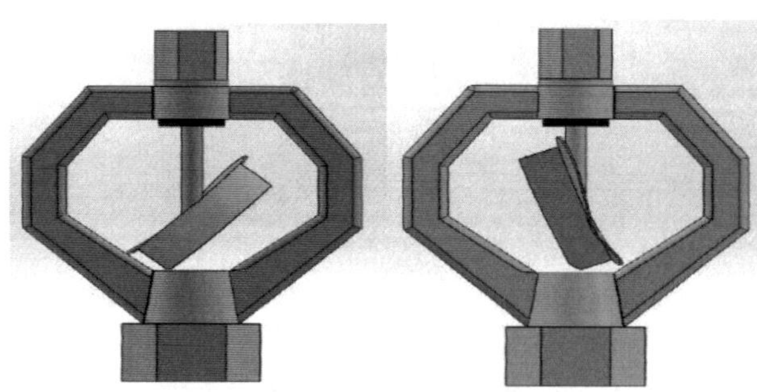

IOP Conf. Series: Materials Science and Engineering **243** (2017) 012046 doi:10.1088/1757-899X/243/1/012046

2.2. Governing equations

The governing equations of the models are continuity equation and momentum equation. The energy equations are not included since the temperature factor is not investigated in this study.

Continuity equation (differential form):

$$\frac{\partial \rho}{\partial t} + \nabla \cdot (\rho \boldsymbol{u}) = 0 \quad (1)$$

Where ρ is the fluid density, t is time, ∇ is the vector differential operator, and $\boldsymbol{u}$ is the velocity of fluid.

Momentum equation:

$$\frac{\partial}{\partial t}(\rho \boldsymbol{u}) + \nabla \cdot (\rho \boldsymbol{u} \otimes \boldsymbol{u}) = -\nabla \cdot \mathrm{p}\mathbf{I} + \nabla \cdot \tau + \rho \boldsymbol{g} \quad (2)$$

Whereρ is the fluid density,$\boldsymbol{u}$ is the flow velocity,∇ is the del operator,p is the pressure,$\boldsymbol{I}$ is the identity matrix,τ is the deviatoric stress tensor, which has order two, and $\boldsymbol{g}$represents body accelerations acting on the continuum, for example gravity, inertial accelerations, electrostatic accelerations, and so on.

2.3. Boundary condition

A square enclosure of 2m each side was chosen as the fluid domain. For fluid model, homogeneous model was switched on. Standard free surface model was selected to simulate the free flow of water spray in air without any solid obstacle.Since water sprinkler is fully turbulent, k-epsilon turbulence model was used to reduce the simulation time required. The reference density for the buoyancy model is set as density of air, $1.2 kg\ m^{-3}$. In fluid pair model, the surface tension coefficient of water-air interphase is set as $0.072 N\ m^{-1}$. The interphase transfer type of free surface waschosen.

The sprinkler head solid surface is defined as standard no slip wall. Since the vane is smooth steel sheet, smooth wall roughness was selected. Hence, the vane can act like a solid surface that redirect the water jet.

A circular area where water should flow in was defined as the inlet. The inlet is a circle with diameter of 26mm.The boundary of the enclosure was defined with opening pressure with zero relative pressure to simulate the open air condition that water jet flow. The direction of flow is normal to the surface.

The fluid volume fraction of air at the boundary of enclosure is one while the volume fraction of water is zero because the fluid at enclosure surface is air. The inlet location is shown in figure 2.

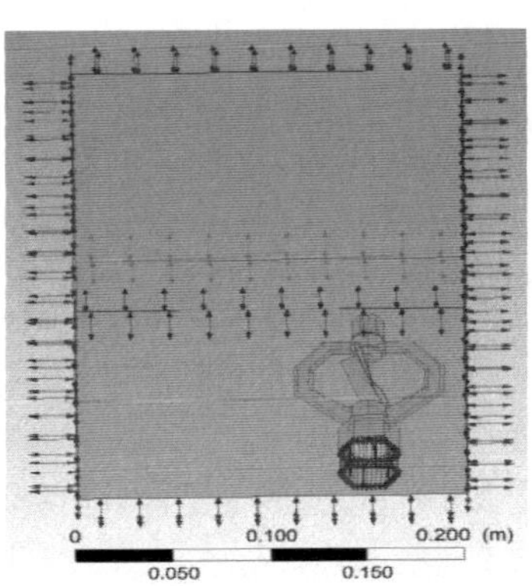

Figure 2. Inlet location of water sprinkler model (highlighted in green).

IOP Conf. Series: Materials Science and Engineering **243** (2017) 012046 doi:10.1088/1757-899X/243/1/012046

2.4. Grid independence test

To eliminate the effect of grid size on the simulation result, grid independence test was carried out. The model is mesh with an initial mesh of medium size. The simulation is run with increasing number of refinement. The result is shown in table 2 and plotted in figure 3. From figure 3, the maximum velocity at the nozzle increase slightly from one refinement onward. Thus, one refinement on the model is sufficient to produce accurate result.

Table 2. Maximum velocity, number of nodes and maximum velocity of test model 1 through 4

Test model	number of nodes	refinement	maximum velocity
1	99117	0	12.9936
2	150046	1	13.102
3	240532	2	13.1014
4	462635	3	13.1035

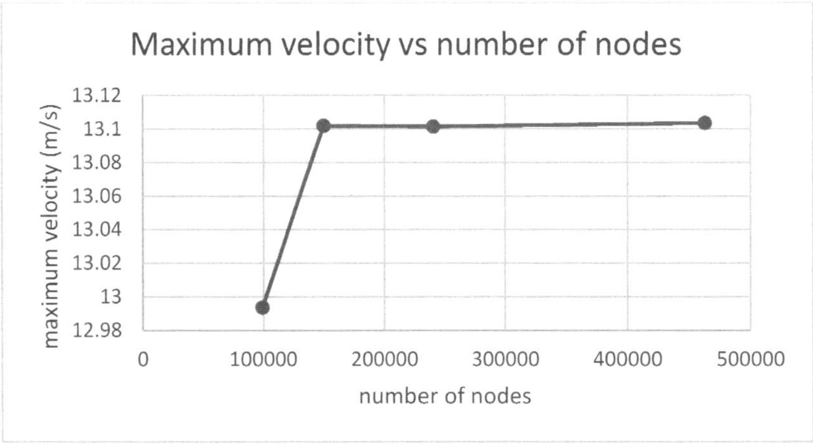

Figure 3. Maximum velocity against number of nodes.

3. Result and discussion

3.1. Experimental data

An experiment is carried out to validate the accuracy of simulated data. The water sprinkler is connected to pipe with same mass flowrate and the valve is opened. Distance of water sprayed is measured.

Table 3. Experiment data (continued)

Experiment	Distance
1	2.09
2	2.17
3	2.15
4	2.03
5	2.06
Average	2.11

The average value of water jet distance in experiment is 2.11m. Compared to the simulated result of model (a), which is 1.983m, the experimental result is higher than expected.

$$error(\%) = \frac{2.11 - 1.983}{2.11} \times 100\% = 6.02\%$$

However, the error is small (about 5%). Thus, the model can be considered to be accurate for the purpose of simulation experiment.

3.2. Velocity distribution

High velocity in water spray for agriculture use is not desirable. The water spray may damage plant leaves and erode soil, reducing crop yield. The velocity of water jet in air from simulation is shown in figure 4 as streamline,contour in figure 5, and tabulated in table 4.

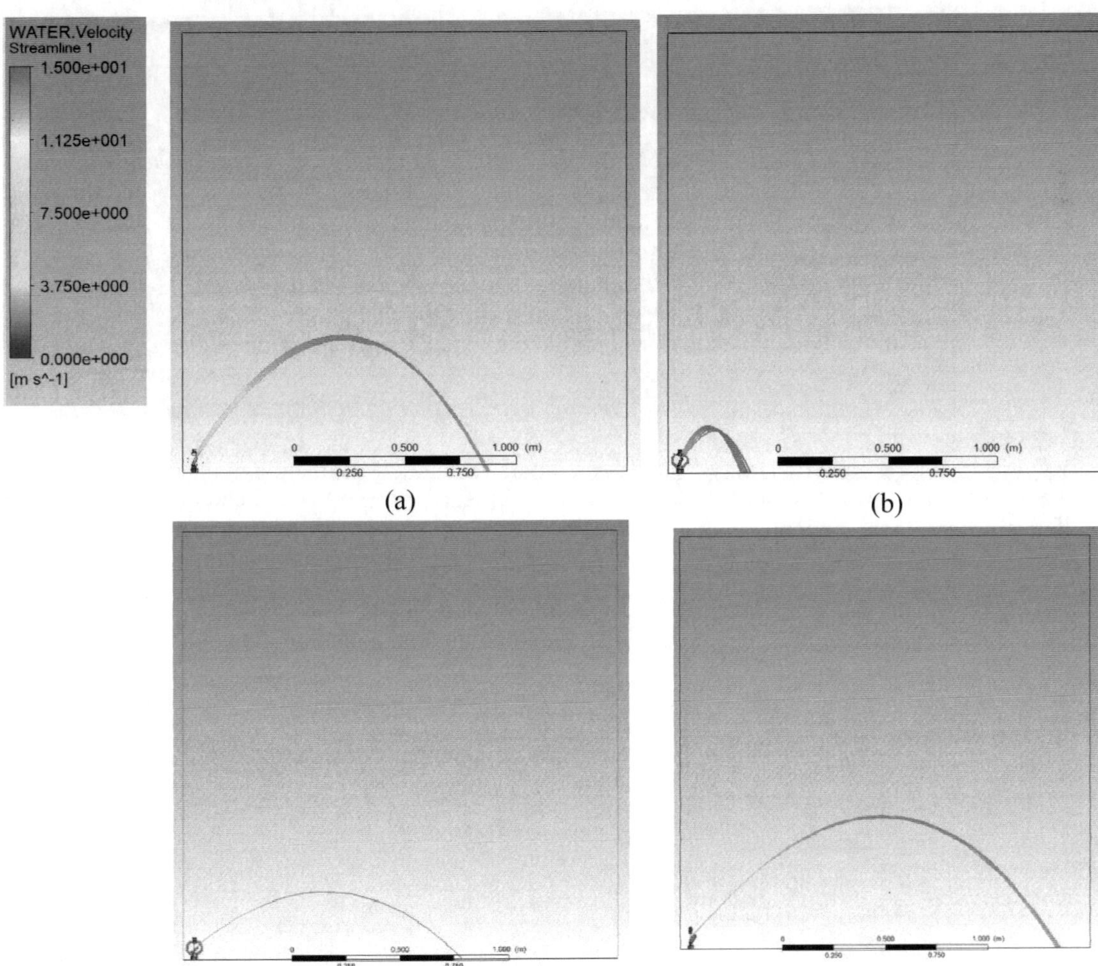

(a) (b)

(c) (d)

Figure 4. Streamline of water flow from sprinkler head with velocity of water along the streamline. (a) Original model, (b) Enlarged nozzle, (c) 45-degree vane angle, (d) 5-degree vane curvature.

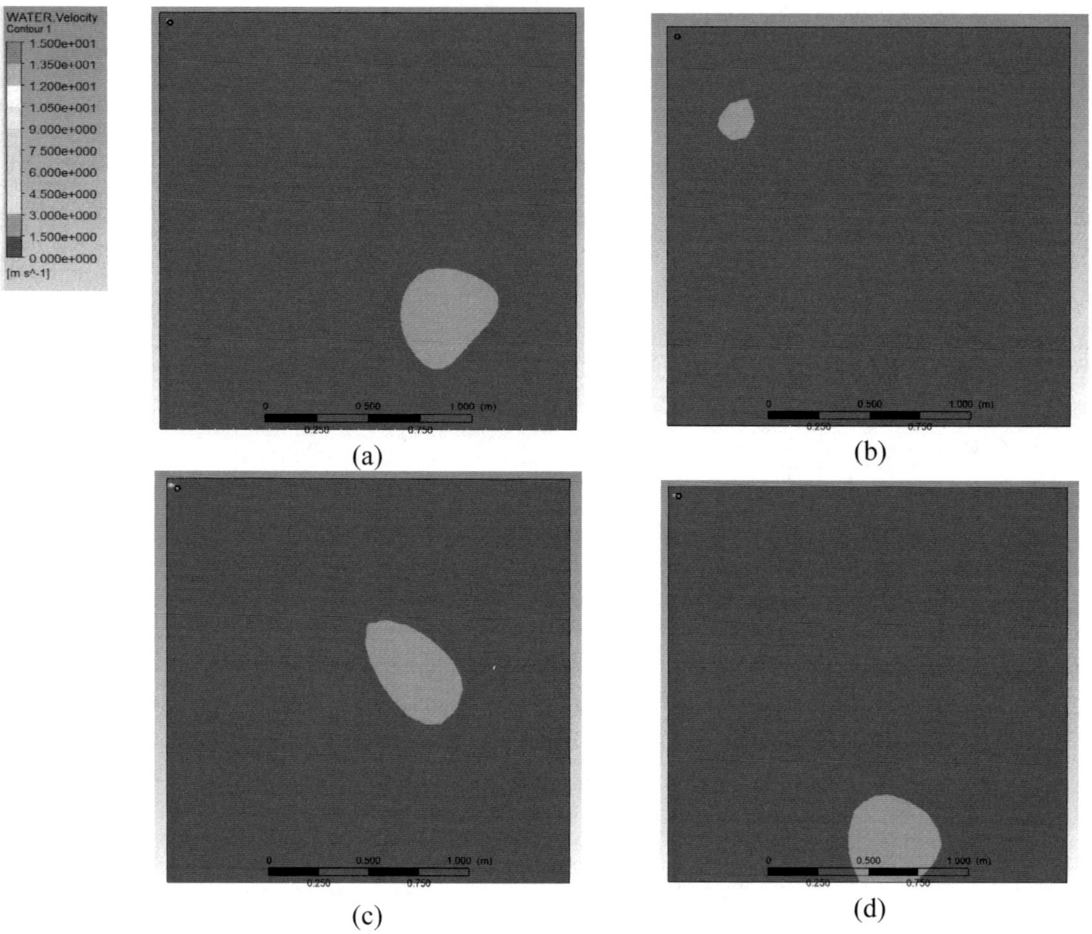

Figure 5. Contour of water velocity at the bottom of enclosure. The sprinkler head is located at top left corner. (a) Original model, (b) Enlarged nozzle, (c) 45-degree vane angle, (d) 5-degree vane curvature.

Table 4. Velocity of water jet in air

model	velocity of water jet (m/s)	percentage difference (%)	Rating
a	0.35913	0	0
b	0.29905	-16.73	2
c	0.41162	14.61	1
d	0.33658	-6.28	0

Model (a) and model (d) has similar water velocity while travelling in air. This shows that decreasing angle of curvature does not seem to have a large impact on water velocity in air. However, model (c) shows a significant decrease in velocity. Decreasing vane projectile angle does seem to make water lose more momentum when it collides with vane to be redirected. In short, model (b) has the best performance in terms of water velocity in air.

3.3. Watering distance

Distance of the water spray can be measured from the simulation result. Distance of water spray is the most important performance factor for water sprinkler as larger distance means lesser number of sprinkler need to be used for a same area. The result is shown in table 5.

Table 5. Distance of water jet

model	Distance (m)	difference (%)	Rating
a	1.983	0	0
b	0.5	-74.79	-2
c	1.573	-20.676	-1
d	2.117	6.757	1

Increasing diameter of nozzle decrease the distance of water spray enormously (-75%). Water velocity after the jet exit the nozzle in model (b) is too low to achieve a satisfactory range of water spray. Thus, the increase in nozzle diameter should be reduced to raise the distance of water spray to an appropriate value.

Decreasing the angle of projectile from 70 degrees to 45 degrees shows a decrease in the distance of water spray. This may be due to the angle of departure of water jet from vane is too low and it shorten the range of water jet distance. Although the angle of vane is 45 degrees, when water jet hit the vane, it is deflected to about 40 degrees. Thus, the angle of vane should be adjusted to 50 degrees to ensure that water jet have a projectile angle of 45 degrees.

Decreasing the curvature of vane increase the distance by 6.7%. When the vane has a lower curvature, a larger portion of initial kinetic energy is converted into linear kinetic energy instead of rotational kinetic energy at vane due to the lower curvature. Thus, water can travel further albeit the sprinkler spin slower. The curvature angle of vane can be reduced further to increase the watering distance more.

In conclusion, changing the curvature angle of vane from 10 to 5 degrees increase the distance of water spray slightly. However, increased diameter and 45 degrees' vane angle decrease the distance of water spray.

3.4. Watering area

The watering area is a measure of how dispersed is the water jet. In general, the larger the watering area, the smaller the water droplet. Besides that, a larger watering area can provide better coverage of crops.

Table 6. Angle of water stream.

model	Area (m^2)	percentage difference (%)	rating
a	9.348E-4	0	0
b	2.256E-3	141.34	2
c	1.442E-5	-98.46	-2
d	2.433E-4	-73.97	-2

Increased diameter of nozzle results in much larger cross sectional area of water jet. When the larger water jet is defected by vane, it splits into a large water spray covering a huge area. However, decrease vane projectile angle concentrates the water stream into a smaller diameter water jet that cover a small area only. On the other hand, decrease angle of curvature of vane did not diverge water jet as much as the original model.

In short, increased diameter can cover the largest area with a single water jet. Other variation does not do as well as the original model in diverging water jet.

3.5. Performance evaluation

The rating of original model as well as other variation is shown in the evaluation matrix below.

Table 7. Evaluation matrix for model (a) to (d).

		Model			
		a	b	c	d
Performance criteria	Weightage	Rating	Rating	Rating	Rating
Distance	0.6	0	-2	-1	1
Velocity in air	0.2	0	2	1	0
Area	0.2	0	2	-2	-2
Total rating	1	0	-0.4	-0.8	0.2

The table 7 above shows that model (b) and model (c) did not perform better than the original model, mainly because they perform worse in distance of water jet travelled.

Changing the angle of vane curvature from 70 degrees to 45 degrees and double the nozzle diameter had change the parameter too much such that it does not perform well. A smaller change in vane angle and nozzle diameter should

Modifying the angle of vane from 70 degrees to 45 degrees increase the velocity of water jet in air, making the spray less gentle and reduce area covered. Thus, grooves can be added to the vane to split up the water into a more diverged jet.

Increasing the nozzle diameter produce a beneficial result on the area covered by water spray and lower velocity of water jet in air, at the cost of greatly reducing the distance of watering. Hence, a slight increase in diameter of nozzle such as 4mm to 4.5mm is recommended as a compromise value between the three performance criteria.

Lowering the curvature of sprinkler can increase the watering distance. Hence, the curvature angle of vane can be reduced further to increase the watering distance more.

4. Conclusion and recommendation

In conclusion, the original model of water sprinkler and 3 variation of the original model is simulated. The result from simulation showed that:

a) Enlarge the nozzle diameter have a positive effect on the velocity of water jet in air and the area covered by water jet but it drastically decreases the watering distance of sprinkler.

b) Changing the angle of vane from 70 degrees to 45 degrees can decrease the watering distance slightly and it concentrates the water into a fine jet that cover a small area. To reduce the effect, grooves can be added to the vane to increase the divergence of water spray and the vane angle should be adjusted to 50 degrees.

c) Reducing the angle of curvature from 10 degrees to 5 degrees improves the watering distance. The angle of curvature can be reduced more to increase the watering distance further.

Acknowledgement
The author would like to express gratitude to Dr. Ishkrizat Bin Taib and Assoc. Prof. Dr. Norzelawati Binti Asmuin from UTHM, FKMP, for their guidance and support during the completion of this project. The support of the University Tun Hussein Onn Malaysia (UTHM), under the GPPS grant project, led by Dr. Ishkrizat Taib and under grant number **U723** is gratefully acknowledged. This paper was partly sponsored by Flow Analysis, Simulation and Turbulence (FAST) research group of UTHM.

References
[1] Cech T V 2010 *Principles of water resources: history, development, management, and policy* (United States of America: John Wiley & Sons)
[2] Maddaus L, Maddaus W and Maddaus M 2013 *Preparing Urban Water Use Efficiency Plans* (London, UK: Iwa Publishing)
[3] Panda H 2016 *The complete book on cultivation and manufacture of tea* (Kamla Nagar, Delhi: Asia Pacific Business Press Inc)
[4] Christiansen J E and Davis J R 1967 *Sprinkler irrigation systems*: Irrigation of agricultural lands vol 11 ed R M Hagan, H R Haise et al (Madison, Wisconsin: American Society of Agronomy) chapter 44 pp 885-904
[5] Narayanamoorthy A 2008 Drip irrigation and Rainfed Crop Cultivation Nexus: A Case of Cotton Crop. *Ind. Jn. of Agri. Econ.* **63**(3) 487
[6] Connellan G 2013 *Water Use Efficiency for irrigated turf and landscape* (Collingwood, Victoria, Australia: CSIRO PUBLISHING)
[7] Liu J, Zehnder A J and Yang H 2009 Global consumptive water use for crop production: The importance of green water and virtual water *Water Resour. Res.* **45**(5)
[8] Kah Jr C L, Kah III C L, Feng W, Hall F A and Carl LC Kah Jr III 2016 Arc and range of coverage adjustable stream rotor sprinkler. US patent 9,248,459
[9] Rosenberg P 1987 Water sprinkler. US patent 4,660,765
[10] Myers T, Marshall A and Baum H 2014 A Free-Surface Model of a Jet Impinging On a Sprinkler Head. *Fire Saf. Sci.* **11** 1184-95

IOP Conf. Series: Materials Science and Engineering **243** (2017) 012047 doi:10.1088/1757-899X/243/1/012047

Determine spray droplets on water sensitive paper (WSP) for low pressure deflector nozzle using image J

M F Sies[1], N F Madzlan[1], N Asmuin[1], A Sadikin[1] and H Zakaria[1]
[1] Faculty of Mechanical and Manufacturing Engineering, Universiti Tun Hussein Onn Malaysia (UTHM), 86400 Batu Pahat, Johor, Malaysia

E-mail: farids@uthm.edu.my

Abstract. In this study, determine of spray droplets size (SMD) using water sensitive paper (WSP) at low fluid pressure with deflector nozzle or tangential flow nozzle model Delavan AL75 and New Design Nozzle with two different type of swirl (ND2.5 A1.0 & ND2.5 B1.0). These three deflected flat sprays have used at different liquid mixing ratio. These liquid mixture ratios are pure water, 10% of lime juice + 90% of water (L10W90) and 30% of lime juice + 70% of water (L30W70). WSP is used to collect the spray droplets from nozzles. The operational liquid pressure of each nozzle is 3 bar, while air operational pressures are 3 bar and 6 bar. Then, the WSP were scanned using scanner then it was analyzed using ImageJ software. ImageJ can be used for determining the diameter of droplets size on the WSP. As the results from an experiment, the AL75 nozzle recorded the lowest Sauter mean diameter which is 193.69μm at 6 bar of pressurized air while ND2.5 A1.0 recorded the highest Sauter mean diameter which is 353.61μm at 3 bar of pressurized air. Summary from the experiment shows that the higher of droplet size is because of the lower air pressure (3 Bar). Then, increasing of liquid viscosity also increase the SMD. The orifice diameter for New Design nozzle (ND-2.5) is smaller than AL75, which are 2.5mm and 2.8mm respectively. The different nozzle design also gives effect the SMD. WSP is an alternative method to determine SMD for spray droplets with the low cost if compared to Phase Doppler Anemometry (PDA).

1. Introduction

Air pollution is a major environment-related health threat to humanity and a risk factor for both acute and chronic respiratory disease. Excessive burning of fuel which is a necessity of our daily lives for cooking, driving and other industrial activities. It will be releasing a huge amount of chemical substances in the air every day and these pollute the air. Under the Clean Air Act, Environmental protection Agency-US (EPA) establishes air quality standards to protect public health, including the health of "sensitive" populations such as people with asthma, children, and older adults. EPA also sets limits to protect public welfare. Department of Environment Ministry of Natural Resources and Environment (Malaysia) has recommended air qualities guidance following air emission standards for the control of air pollution and gaseous emissions [1][2]. The commercial cooking process is one of the major sources for air pollutions. EPA has reported that 21% gas emissions from industries include the cooking process. One of the solutions is installed atomization nozzles into kitchen hood ventilation system for cooling, cleaning or filtering the grease and smoke during the cooking process. The size of droplets of spray will give effect in the ventilation system, spray performance and result of emissions.

In this research, the main objective is to determine spray droplets using water sensitive paper. (WSP) is useful spray droplet sampling devices provided that their limitations are taken into account. This method has been used more than 40 years by Turner and Huntington at 1970. Currently, the big issue in Universiti Tun Hussein Onn Malaysia (UTHM) for spray research group is to determine spray droplet sizes which are UTHM don't have the apparatus to measure droplets size of spray such as Phase Doppler Anemometry (PDA). It is the economical method if compared with PDA and Particle Image Velocimetry (PIV). The color of WSP is yellow and change the blue color when the spray particles on this paper as shown as in Figure 1. This color transition is when to the reaction of Bromophenol Blue to a pH change from 2.8 to 4.6 and it is for spray deposition indicator. Most of the researcher using this method such as P. Jaeken, P. Lootens (2000), B. Panneton (2002), R. D. Fox, M. Salyani, J. A. Cooper (2001), M. Salyani, H. Zhu (2013), and R. D. Fox, R. C. Derksen (2003) [3]–[7]. The specification nozzle plays an important role to get a good of spray behavior such as spray angle, droplets size, etc.

This study is a focus to find the droplet size of water spray by using deflected nozzle. The mixture of water and organic lime water will be used for this experiment. It can be stated that the liquid ratio mixing, type of deflected nozzle and pressurized air can give an important role to reduce the hood from clogging. Through this experiment, it will give more understanding about the spray behavior after applying different parameters especially droplets sizes. Organic lime juice will be used in this experiment because it can reduce the cohesion of oil particles by breaking their bond strength. The scope of study, is the used nozzle are AL75 (atomizer), New Design of atomizer (ND2.5 A1.0) and New Design of atomizer (ND2.5 B1.0), the liquid used 100% of water (water), 10% of organic lime juice mixed with 90% of water (L10W90) and 30% of organic lime juice mixed with 70% of water (L30W70), the size of test rig is in range 60cm x 70cm x 90cm, analysis the WSP using ImageJ software, the pressurized liquid is 3 bar, while pressurized of air is 3 bar and 6 bar, and the experiment at atmospheric pressure (Patm= 1.01325 bar) and ambient temperature (T=28°C).

Semião [6]essentially use Sauter mean Diameter (SMD) as one of the methods to calculate the droplet sizes with available correlations for pressure-jet and pre-filming airblast atomizers. Rosin-Rammler and Nukiyama-Tanasawa use this function plus have modified some functions such as the upper-limit, log-normal and chi-squared functions [6]. Introducing the definition of SMD:

$$SMD = \frac{\int_0^\infty D^3\, dN}{\int_0^\infty D^2\, dN} \tag{1}$$

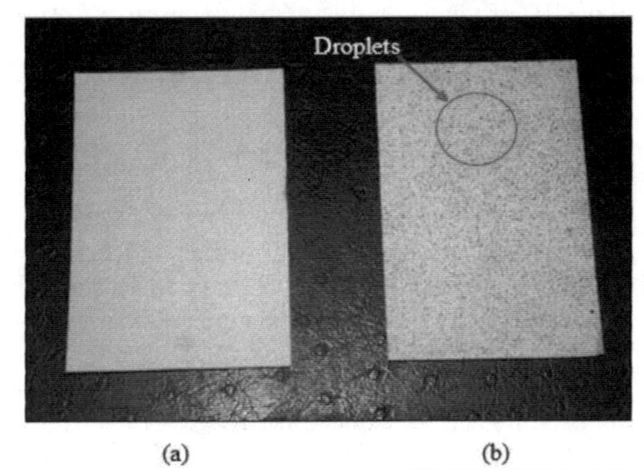

Figure 1. Water sensitive paper (a) before exposure and (b) after exposure to the spray

The sprays tested in this study were used by WSP. SMD correlation is an important parameter to identify the type of atomizers. According to N. Ashgriz (2011), AL75 and ND-2.5 are deflector nozzles with the tangential flow in category flat fan nozzle for cooling and cleaning process [9]. Base on the both designs, the type of nozzles is air assist atomizer because they need air to produces atomization with the internal mixing chamber. According to Kurt A. Estes and I. Mudawar (1995) and J. Kim (2007) [10]-[11], SMD correlation for cooling and cleaning process define as (Correlation A):

$$SMD = d_{32} = d_o \left[3.67 \left(We_{do}^{1/2} Re_{do}\right)^{-0.259}\right] \tag{2}$$

AL75 and ND-2.5 nozzles also have swirl in the nozzle assembly. According to Kennedy (1985), SMD correlation for pressure swirl define as below(Correlation B)[9]:

$$SMD = d_{32} = 10^{-3}\sigma\left(6.11 + 32000\frac{\dot{m}_L}{\Delta P_L^{0.5}} - 6.973 \times 10^{-3}\Delta P_L^{0.5} + 1.89 \times 10^{-6}\Delta P_L\right) \tag{3}$$

Correlation A and B will be comparable to determine the type of atomizer between nozzle AL75 and ND-2.5.

2. Methodology

The main objective of this study is to measure a droplets size by using three type of deflected nozzle with different ratio of liquid. The type of liquids are water, L10W90) and (L30W70). M. F. Sies, N. Asmuin (2017) had studied liquid properties such as density, viscosity, surface tension .etc [12]. The design of the deflector nozzles as shown as in Figure 2. Figure 2(a) is a front view of deflector nozzle AL75 with area mixing chamber 13.3 mm^2 and Figure 2 (b) is shown a front view new design deflector nozzle (ND-2.5) with swirl A1.0 (80 ^{O}) and B1.0 (80 ^{O}), while the area mixing chamber is 64.95 mm^2.

The pressurized air and liquid are fed through at (i) and (ii). Then the mixed of two phase flow (liquid and air) at mixing chamber. The experiment was conducted with water, L10W90, and L30W70 at 28oC as a test liquid. Figure 3is a side view of both deflector nozzle and the impingement surface. The function of surface impingement nozzle is a stream of liquid to impinge on a surface caused by a sheet of liquid that breaks up into droplets. Figure below shows a complete schematic diagram of the experimental setup and location (10 points) of WSP at the test rig for this research in Figure . The fluid flow control system is consists of the water pump and compressor to supply liquid and air at a certain pressure. Besides that, two flow meter and pressure gauge were mounted with 6mm air tube to monitor both liquid and air pressure. Air and liquid pressure will combine in the AL75, ND2.5 A1.0, and ND2.5 B1.0 deflector nozzles to produce the spray.

shows that the initial condition for experiment setup. From the previous study, the suitable liquid pressure for both deflector nozzle is 3 bar for produce full spray development with air pressure 3 bar and 6 bar.

After the WSP has been sprayed, the droplet size on the WSP needs to measure by using the image J software. It cannot be simply measured because of the tiny droplet size. Then, the WSP analyzed using ImageJ to measure size distribution, a number of spots per unit area, area mean diameter, an area covered by spots and covered area by enlarging the images of spots at different radial–distances from the individual spots [7].The detailed procedure using ImageJ software can refer at website http://www.smalldropsprays.info/stepbystepmethods/ImageJ-software.Unfortunately, the overlap of droplets from spray occur because of the high air pressure. The dense of droplets and unsuitable spots on the WSP image must be eliminated for easy to analyze and get the good result. The spot for shape features must be eliminated on WSP image. Therefore, researchers can eliminate non-circular spots manually [13].

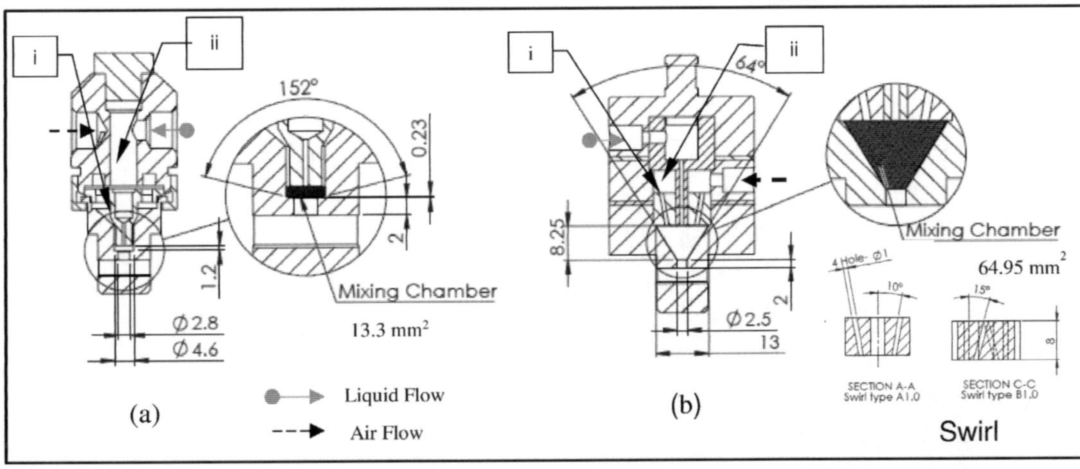

Figure 2. Deflector nozzles with different sizes (a) AL75 (b) ND-2.5 with swirl A1.0 & B1.0

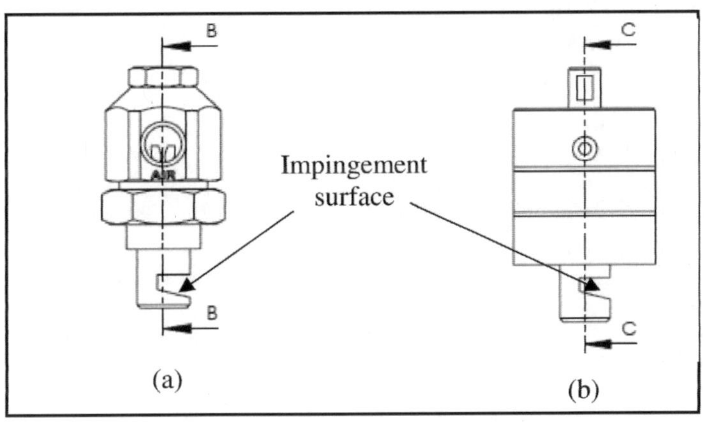

Figure 3. Side view of Deflector Nozzle (a) AL75 (b) ND-2.5

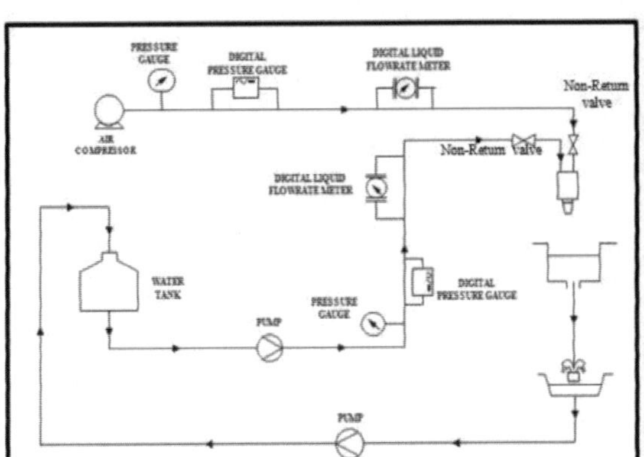

Figure 4. A schematic diagram of spray test rig

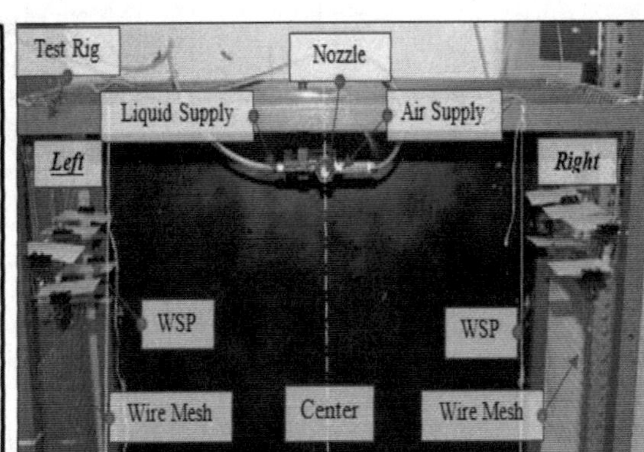

Figure 5. Water sensitive paper experiment setup

Table 1. Initial condition for experiments setup.

Exp. ID	Nozzle	Liquid	P Liquid (bar)	$Q_L\ (L/min)$	P Air (bar)	$Q_A\ (L/min)$
A	AL75	Water	3	0.140	3	40
B	AL75	L10W90	3	0.130	3	40
C	AL75	L30W70	3	0.123	3	40
D	AL75	Water	3	0.128	6	60
E	AL75	L10W90	3	0.110	6	60
F	AL75	L30W70	3	0.097	6	60
G	ND2.5-A1.0	Water	3	0.460	3	25
H	ND2.5-A1.0	L10W90	3	0.420	3	25
I	ND2.5-A1.0	L30W70	3	0.400	3	25
J	ND2.5-A1.0	Water	3	0.330	6	40
K	ND2.5-A1.0	L10W90	3	0.320	6	40
L	ND2.5-A1.0	L30W70	3	0.300	6	40
M	ND2.5-B1.0	Water	3	0.440	3	25
N	ND2.5-B1.0	L10W90	3	0.400	3	25
O	ND2.5-B1.0	L30W70	3	0.330	3	25
P	ND2.5-B1.0	Water	3	0.300	6	40
Q	ND2.5-B1.0	L10W90	3	0.230	6	40
R	ND2.5-B1.0	L30W70	3	0.200	6	40

Recently, Scanner printer can scan the WSP image to the Jpeg format (Joint Photographic Experts Group). The WSP had scanned to collect droplet sizes data include the spray distributions, the total of droplet number, droplet area, spray deposit amount and spray coverage percentage from ImageJ software [14]. After that, all data can transfer to MS excel to determine the droplets size using equation (1), equation (4), equation (5) and equation (6).

Next, there is an equation used to turn the spot to the actual droplet diameter. The equation is:

$$d = 0.95{d_s}^{0.910} \tag{4}$$

Where,

$$d_s = \sqrt{\frac{4A}{\pi}} \tag{5}$$

A = the spots area

The final equation to calculate actual drop diameter is:

$$d = 1.06\ A^{0.455} \tag{6}$$

By selecting the area of interest, ImageJ will process the image and calculate the droplet size as shown as in Figure 6 (a). Generate the result by analyzing the area of the interest as shown as in Figure 6 (b). Compare the experiment data with Correlation A using equation (2) and correlation B using equation (3).

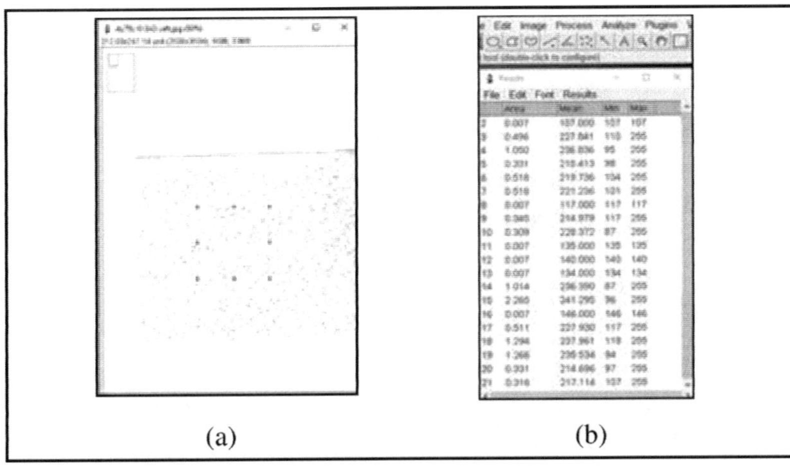

Figure 6. (a) An Example area of interest was drawn on the WSP (b) Result generated from ImageJ

3. Results and discussions

The SMD result for liquid pressure 3 bar and air pressure 3 bar and 6 bar using deflector nozzle type AL75, ND2.5 A1.0, and ND2.5 B1.0 as shown as in Figure 7. Figure 7 also shows that the SMD correlation for experiment data value. SMD Correlation A is higher than SMD Correlation B because of the different design and nozzle applied. Correlation A is for air assist nozzle, while Correlation B is for pressure swirl nozzle. The result can be seen that SMD is influenced by the different ratio of liquid, liquid properties and different type of spray nozzle. L30W70 shows that the highest value of SMD. The SMD value for ND2.5 A1.0 at 3 bar of pressurized air is 353.61μm and 6 bar is 342.20μm. The ND2.5 B1.0 spray nozzle was recorded 318.82μm for 3 bar and 306.63μm for 6 bar. At the AL75 spray nozzle, 3 bar air pressurized is 222μm and 6 bar is 220μm.

The second highest of SMD value is L10W90. For nozzle ND2.5 A1.0, 3 bar of air pressurized recorded 332.44μm and 6 bar is 314.09μm. At spray nozzle ND2.5 B1.0, 299.89μm is for 3 bar followed by 6 bar which is 293.48μm. For AL75, the value of SMD at 3 bar and 6 bar is 209.17 μm and 203.68μm respectively. The lowest value of SMD is water where for ND2.5 A1.0 are 312.87μm at 3 bar and 309.29μm at 6 bar. The SMD for spray nozzle ND2.5 B1.0 is 290.78μm and 288.81μm at 3 bar and 6 bar respectively. For spray nozzle AL75, the value of SMD at 3 bar is 193.69μm and 6 bar is 191.41μm. SMD value for AL75 is lower than ND-2.5 because the area of mixing chamber for AL75 smallest than ND-2.5.

The different of fluid flow rate is generally the different results of SMD. L30W70 the highest SMD value because of higher viscosity if compared with other liquid. From the observation during the experiment, L3070 was produced more droplets. The viscosity of liquid give effects of spray pattern formation and droplets size. The liquid viscosity higher than water viscosity will give higher SMD value for same liquid and air pressure. These results support the idea by Anisa et al. (2010), that different ratio of liquid will give a different viscosity. SMD value for L30W70 is the highest values for both pressurized air, which is 3 bar and 6 bar followed by L10W90. The lowest value of SMD is pure water. By increasing the air operating pressure, the SMD also increases [15].

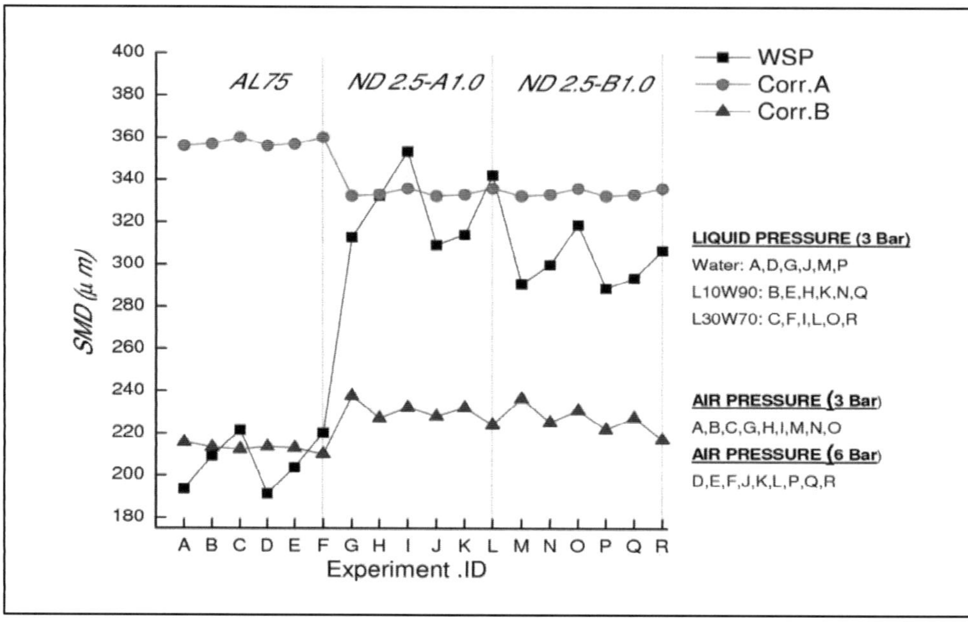

Figure 7. Relative of SMD experiment (WSP) and SMD correlation.

Table 2. SMD and percentage error between experiment (WSP), Correlation A and Correlation B

Exp. ID	Nozzle	Liquid	WSP	Correlation A		Correlation B	
			SMD(μm)	SMD(μm)	% error	SMD(μm)	% error
A	AL75	Water	193.69	356.26	84%	215.68	11%
B	AL75	L10W90	209.17	357.04	71%	213.38	2%
C	AL75	L30W70	221.56	360.03	62%	212.35	4%
D	AL75	Water	191.41	356.26	86%	213.63	12%
E	AL75	L10W90	203.68	357.04	75%	213.07	5%
F	AL75	L30W70	220.3	360.03	63%	210.10	5%
				Average	**74%**	**Average**	**6%**
G	ND2.5-A1.0	Water	312.87	332.40	6%	237.79	24%
H	ND2.5-A1.0	L10W90	332.44	333.13	0%	227.27	32%
I	ND2.5-A1.0	L30W70	353.61	335.93	5%	232.28	34%
J	ND2.5-A1.0	Water	309.29	332.40	7%	228.18	26%
K	ND2.5-A1.0	L10W90	314.09	333.13	6%	232.21	26%
L	ND2.5-A1.0	L30W70	342.2	335.93	2%	224.08	35%
				Average	**4%**	**Average**	**29%**
M	ND2.5-B1.0	Water	290.78	332.40	14%	236.41	19%
N	ND2.5-B1.0	L10W90	299.89	333.13	11%	225.21	25%
O	ND2.5-B1.0	L30W70	318.82	335.93	5%	230.90	28%
P	ND2.5-B1.0	Water	288.81	332.40	15%	221.94	23%
Q	ND2.5-B1.0	L10W90	293.48	333.13	14%	227.37	23%
R	ND2.5-B1.0	L30W70	306.63	335.93	10%	217.19	29%
				Average	**11%**	**Average**	**24%**

On this study, it shows that the higher of droplet size is because of the lower of liquid pressure and low pressurized air in 3 bar. It was explained that the air pressure 3 bar is the higher value of SMD than air pressure 6 bar. The air operational pressure increase will increase spray velocity and air can distribute to mixing chamber area [8]. The different between ND2.5 A1.0 and ND2.5 B1.0 is the swirl angle. For ND2.5 A1.0, the angle is 80°, while for ND2.5 B1.0 is 75°. Therefore, it can be said that bigger angle of swirl gives a high SMD value. Next, the difference between New Design nozzle and the AL75 nozzle is their orifice diameter. The orifice diameter for New Design nozzle is smaller than AL75, which is 2.5 mm and 2.8 mm respectively. These results agree with Santangelo's (2010) and L. Broniarz-press (2008) findings that the SMD is decreasing with increasing of orifice diameter [16][17].

Table 2 shows that the SMD value and percentage error between experiment, correlation A and correlation B. Correlation A are acceptable for ND-2.5 because the average of percentage error is smaller than AL75. Percentage error is 4% for ND-2.5-A1.0 and 11% for ND-2.5-B1.0. Nozzle AL75 has a 74% error after applied correlation A, but the smallest error is 6% after validated with correlation B. That means AL75 is the trend and predicted SMD with correlation B.

4. Conclusion

The present work has been focused on determining droplet size diameter using low cost operational using WSP technique. The different nozzle design, operational pressure, and liquid properties give the effect of droplets sizes (SMD). The increase of viscosity will increase the droplet size and decreasing droplet size is by increasing the operating pressure. Decrease area of mixing chamber and increase of orifice diameter contribute decreasing of SMD value. From the results, most of the droplets size can be categorized as medium sprays and coarse sprays because the range of droplets diameter for medium sprays is in between 90μm to 300μm. These deflector nozzles suitable for cleaning and cooling industrial process.

Nomenclature

SMD (d_{32})	Sauter mean diameter	σ	Liquid surface tension (N/m)
D	Diameter of droplet (m)	m_L	Liquid mass flow rate(kg/s)
d_o	Orifice diameter (m)	d_s	Spot diameter (mm)
We_{do}	Weber number at orifice diameter	ΔP_L	Differential of liquid pressure (Pa)
Re_{do}	Reynold number at orifice diameter	A	The spots area (mm^2)

Acknowledgements

The authors are indebted to Universiti Tun Hussein Malaysia (UTHM), Halton Manufacturing Sdn. Bhd, and Delavan Spray Technologies under the auspices of who's this work was carried out (Grant KTP No. 1440).

References

[1] Ministry of Natural Resources and Environment 2010 Environmental Requirements
[2] (DOSH) D 2014 On occupational safety and health for design, inspection, testing and examination of local exhaust ventilation system
[3] Panneton B 2002 Image Analysis of Water-Sensitive Cards for Spray Coverage Experiments *Appl. Eng. Agric.* **18**(2)
[4] Fox R D, Salyani M, Cooper J A and Brazee R D 2001 Spot Size Comparisons on OIL– and Water–Sensitive paper **17**(2) 131–36
[5] Salyani M, Zhu H, Sweeb R D and Pai N 2013 Assessment of spray distribution with water-sensitive paper *Agric. Eng. Int. CIGR J* **15**(2) 101–11
[6] Jaeken P, Lootens P, Vandecasteele P and de Moor M 2000 Image analysis of water sensitive paper as a tool for the evaluation of spray distribution of orchard sprayers *Asp. Appl. Biol.* **57** 1–9
[7] Fox R D, Derksen R C, Cooper JA, Krause C R and Ozkan H E 2003 Visual and image system

measurement of spray deposit using water sensitive paper *Appl Eng Agric* **19**(5) 549–552
[8] Semião V, Andrade P and Carvalho M D G 1996 Spray characterization: Numerical prediction of Sauter mean diameter and droplet size distribution *Fuel.* **75**(15) 1707–14
[9] Ashgriz N 2011 *Handbook of atomization and sprays* (New York Dordrecht Heidelberg London: Springer)
[10] Kurt A. Estes and Mudawar I 1995 Correlation of Sauter mean diameter and critical heat flux for spray cooling of small surfaces *Int. J. Heat Mass Transf.* **38** 2965–96
[11] Kim J 2007 Spray cooling heat transfer : The state of the art *Int. J. Heat Fluid Flow* **28** 753–67
[12] Sies M F, Asmuin N, Norsalim N A, Zakaria H, Pairan R and Sadikin A 2017 Determine physical properties of an organic citric acid (processed lime juice) dissolve with water using experimental *International Journal of Engineering and Technology (IJET)* **8(**6) 2530–36
[13] Sayinci B, Bastaban S and Sánchez-Hermosilla J 2012 Determination of optimal spot roundness variation interval for droplet size analysis on water sensitive paper *J. Agric. Sci. Technol.* **14(**2) 285–98
[14] Zhu H, Salyani M and Fox R D 2011 A portable scanning system for evaluation of spray deposit distribution *Comput. Electron. Agric.* **76(**1) 38–43
[15] Anisa A N I, Nour A H and Material A 2010 Affect of Viscosity and Droplet Diameter on water-in-oil (w / o) Emulsions : An experimental study *J World Academy of Science Engineering and Technology* **4**(2) 213–16
[16] Paolo E. Santangelo A W, Ning Ren and Paolo Tartarini 2008 Spray characterization of high pressure water mist injectors *ILASS* 8–10
[17] Broniarz-press L, Ochowiak M, Rozanski J and Woziwodzki S 2008 Effect of nozzle construction on the characteristics of twin-fluid atomization *ILASS* (3) 8–10

IOP Conf. Series: Materials Science and Engineering **243** (2017) 012048 doi:10.1088/1757-899X/243/1/012048

Application of design for six sigma methodology on portable water filter that uses membrane filtration system: A preliminary study

Mohd Fahrul Hassan*, Suhada Jusoh, Muhamad Zaini Yunos, A M T Arifin, A E Ismail, M Rasidi Ibrahim, M Zulafif Rahim

Faculty of Mechanical and Manufacturing Engineering, Universiti Tun Hussein Onn Malaysia, 86400 Parit Raja, Batu Pahat, Johor, Malaysia

* Email: fahrul@uthm.edu.my

Abstract. Portable water filter has grown significantly in recent years. The use of water bottles as a water drink stuff using hand pump water filtration unit has been suggested to replace water bottled during outdoor recreational activities and for emergency supplies. However, quality of water still the issue related to contaminated water due to the residual waste plants, bacteria, and so on. Based on these issues, the study was carried out to design a portable water filter that uses membrane filtration system by applying Design for Six Sigma. Design for Six Sigma methodology consists of five stages which is Define, Measure, Analyze, Design and Verify. There were several tools have been used in each stage in order to come out with a specific objective. In the Define stage, questionnaire approach was used to identify the needs of portable water filter in the future from potential users. Next, Quality Function Deployment (QFD) tool was used in the Measure stage to measure the users' needs into engineering characteristics. Based on the information in the Measure stage, morphological chart and weighted decision matrix tools were used in the Analyze stage. This stage performed several activities including concept generation and selection. Once the selection of the final concept completed, detail drawing was made in the Design stage. Then, prototype was developed in the Verify stage to conduct proof-of-concept testing. The results that obtained from each stage have been reported in this paper. From this study, it can be concluded that the application of Design for Six Sigma in designing a future portable water filter that uses membrane filtration system is a good start in looking for a new alternative concept with a completed supporting document.

1. Introduction

The use of the membrane as a water filter to treat water for aesthetic purposes by removing chlorine, bad taste, odor, and heavy metals such as lead and mercury [1]. Portable water filter can be used by recreational enthusiasts, military personnel, survivalists, and others who need to get drinking water from sources that are untreated (rivers, lakes, etc.) [2]. However, the use of the membrane as a water filter system is still at the beginning because it requires the design of appropriate filtration system to filter. Therefore, a comprehensive study is needed to design portable water filters using membrane systems.

This study aims to design a portable water filter that uses membrane filtration system using Design for Six Sigma methodology (DFSS). DFSS is a method that can be used to systematically designing a new product, service, or process. This method focuses on designing products, services or processes correctly the first time so as to reduce the time needed to be spent on improving the product, service, or process. Six Sigma can only take an organization so far, and organizations need to focus on product design and good product, less so where the need for them should they fix for avoidance of mistakes [3]. Quality and profitability of Deming cycle, increasing design quality can lead to higher value customers, which can contribute to increased market share, margins, earnings, and profitability [4, 5]. Most organizations can achieve a 4.5 sigma level through the implementation of Six Sigma projects to improve processes and enhance their products and services through the redesign of products or services [6]. DFSS is a powerful and complex mechanisms that can help to maximize market success of a new product or technology [7]. Nevertheless it must midweek that this concept analysis approach to product design. Design methods Six Sigma uses statistics to analyze different critical parameters specified.

2. Methodology

In this study, the research is about to design a portable water filter that uses membrane filtration system using Design for Six Sigma methodology (DFSS). DFSS that based on DMADV is used. DMADV is one of the most popular design and is defined as Define, Measure, Analyze, Design and Verify [8, 9]. Detail about DMADV as follows:

2.1. Define

Define is the first step in DFSS methods in which this step is to define the project goals and customer needs (internal and external). Thus the user's voice is a priority in proposing and tells the problems that has been faced by customer when using the existing product in the market. This information will be recorded for review.

2.2. Measure

The second step in DFSS is measure which this step is to measure and determine customer needs and specifications. The specification defines the product or service in a way that can be measured, which allows data to be collected and compared with the requirements set.

2.3. Analyze

At this stage, the proposed process or product is analyzed and reviewed to determine whether there is a better way to achieve the desired results. Areas that require adjustments or improvements, both in products and manufacturing processes, are identified. Prototype at this stage and analyzed. The purpose of this step is to come up with alternatives, analyze alternatives, and incorporating alternatives to improve the process or product.

2.4. Design

Based on what is learned in the analysis, it can design the new products and processes. To review the work that has been done, the analysis step is repeated to compare the new design with the features specified. Customers can test the product and the product can be used to test new designs to determine the extent to which they meet the expectations or needs of customers.

2.5. Verify

The fifth step involves confirming that the end result meets or exceeds customer requirements. This includes verifying the product was made as it should be done, and whether it meets the requirements of the specification. This is an ongoing process. Even after the product is released, customer feedback should be encouraged and incorporated into future designs. The desired result is to have products that perfectly meet the needs of the customer.

3. Application of DFSS on Portable Water Filter that uses Membrane Filtration System

In this section, a process of making a portable water filter using DFSS is conducted to illustrate the proposed approach. The DFSS that based on DMADV is presented as follows:

3.1. Define using Questionnaire approach

In continuation to know the needs of customers who use portable water filters, a survey has been conducted on 30 respondents. Survey forms was made to determine the effectiveness of portable water filters that available in the market and also to know the specifications and needs that the user want for a portable water filter in future. The results from this survey is shown in Figure 1.

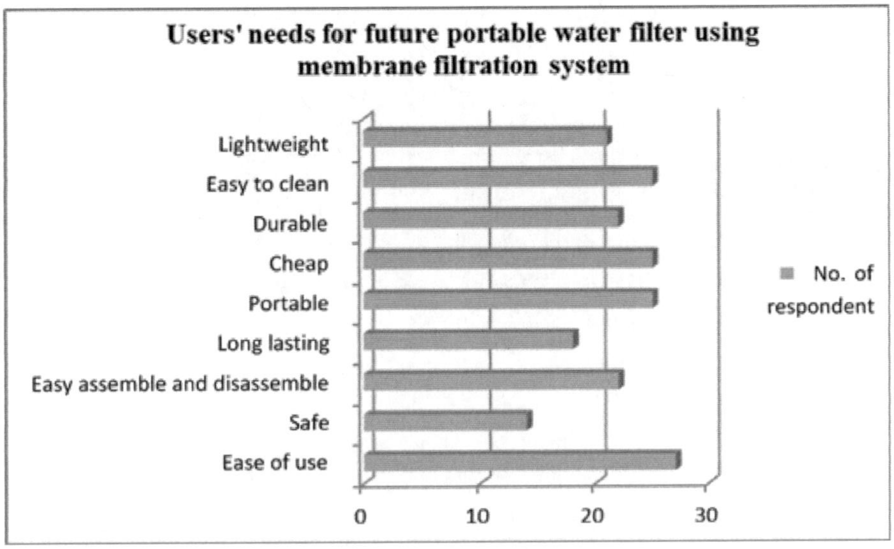

Figure 1. Portable water filter using membrane filtration system based on users' needs.

3.2. Measure using Quality Function Deployment (QFD)

Quality Function Deployment (QFD) is a systematic method in which the structure and development of information design problems. Structure House of Quality (HoQ) is started where the customer needs was define at first and then the weight of the interests of customers were determined. Next, engineering characteristics were identified. Once completed, correlation matrix between customer requirements and engineering characteristics were performed. Assessment was conducted to calculate the score given based on the correlation matrix. Table 1 shows the detail results of portable water filter using membrane filtration system using HoQ technique.

3.3. Analyze using Morphological chart and Weighted Decision Matrix

Morphological analysis is a method to represent and explore all the relationships in multidimensional problems. Morphological methods to help structure the problem to synthesize different components to meet the desired function. This process is made easier to find parts in one catalog. Morphological analysis assists in compiling the findings of research into the structure to enable the processing of information. Table 2 shows morphological chart of portable water filter for membrane and Figure 2 shows several initial sketches that have been generated based on the morphological analysis method.

The weighted decision matrix is a method in the evaluation of the concept. It is used to determine which of the components more suitable for the manufacture of the product. The selection is dependent on the suitability of components and demand based on market research conducted. The highest number of votes will become the weighting of the best and suitable. In this case, several criteria have

IOP Conf. Series: Materials Science and Engineering **243** (2017) 012048 doi:10.1088/1757-899X/243/1/012048

been assigned to select which one is the most suitable among the generated alternative solutions from the previous stage. The evaluation and selection using this tool was not presented in this paper due to limited of pages. Based on the results, the third concept of alternative solution has the highest score among the others and certainly was selected for the next stage.

Table 1. House of Quality for Portable Water Filter using membrane filtration system.

Assessment: 1 = Low; 3 = Moderate; 9 = High			Engineering Characteristics									
		Weightage (%)	Weight (kg)	Number of segments	Height (cm)	Safety (yes/no)	Warranty (year)	Pore size (µm)	Water filter type	Price (RM)	Number of parts	Material types
No	**Customer Requirements**	**%**	**1**	**2**	**3**	**4**	**5**	**6**	**7**	**8**	**9**	**10**
1	Easy to use	0.2	3	3				3				
2	Durable	0.075							3	1	9	
3	Long lasting	0.025				9	9	1	3			9
4	Cheap	0.2								9		3
5	Lightweight	0.15	9									
6	Easy to clean	0.1	3	3				3	3		1	
7	Portable	0.1	3		3							1
8	Safe	0.05				9		9				
9	Easy assemble and disassemble	0.1		9				1			1	
Total		**1.00**	**2.55**	**2.1**	**0.3**	**0.68**	**0.23**	**1.5**	**0.9**	**1.8**	**0.58**	**1.3**
Selected design			64	15	20	Y	2	15	UF	463	8	PP
Competitor A			22	6	15	Y	1	0.1	MF	142	2	PP
Competitor B			40	6	25	Y	2	0.2	MF	131	2	PP
Competitor C			57	5	18	Y	1	0.1	MF	170	3	PP
New product target			30	4	15	Y	1	0.1	MF	70	2	PP

IOP Conf. Series: Materials Science and Engineering **243** (2017) 012048 doi:10.1088/1757-899X/243/1/012048

Table 2. Morphological chart for selection of suitable components.

No.	Component	Alternative solutions			
		Alternative 1	Alternative 2	Alternative 3	Alternative 4
1.	Shutter				
2.	Filter	Microfiltration	Ultrafiltration	Nanofiltration	Reverse osmosis (RO)
3.	Bottle				
4.	Filter placement				
5.	Pump type			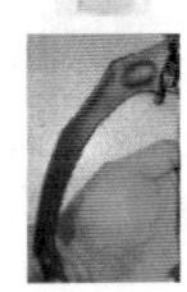	

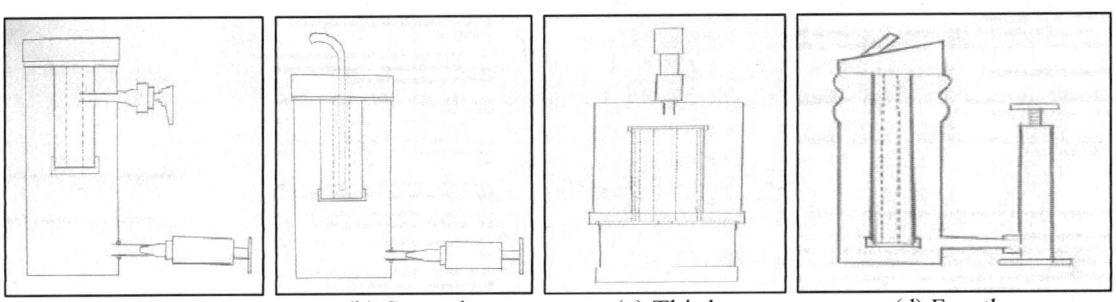

(a) First (b) Second (c) Third (d) Fourth

Figure 2. Initial sketches from the alternative combinations.

3.4. Design

SolidWorks software was used in this stage to assist in designing a 3D modeling of a portable water filter that based on the proposed concept in the previous stage. Figure 3 show the 3D modeling of the third concept. This stage is important where the 3D modeling was tested in terms of the suitability of the available components and for proof-of-concept test in the next stage. The drawing includes the actual size of each part and component and the detail assembly as well.

IOP Conf. Series: Materials Science and Engineering **243** (2017) 012048 doi:10.1088/1757-899X/243/1/012048

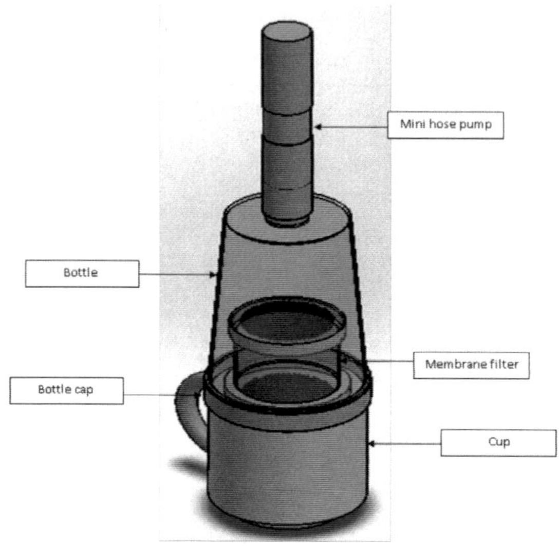

Figure 3. 3D modeling of the third concept for portable water filter.

3.5. Verify

In the last stage, verification on the selected design concept for portable water filter was made. This proof-of-concept test is critical since no design or concept has been made currently. For this task, the prototype has been developed. Figure 4 shows several components (mini pump, activated carbon filter, plastic bottle, membrane sheet unit, plastic cup and white garment) and was assembled into a complete prototype of the proposed design concept for the portable water filter that uses membrane as the filter.

After the complete prototype has been developed, several processes were conducted as preparation for proof-of-concept test. A bottle of dirty water from Sembrong river has been prepared. The water has been filled in the prototype for testing process. The water that has been filtered through this process was compared with the original water. Besides the color, turbidity testing also has been used using Genesys 10S UV-Vis Spectrophotometer machine as well. Figure 5 shows the comparison of turbidity readings between original water and filtered water, where the original water is 1.569A and the filtered water is 1.499A. The comparison was about 5% of the original water has been improved. From this result, it can be preliminary concluded that this concept can be further investigated in order to increase the percentage of improvement.

Figure 4. Prototype of portable water filter that uses membrane filtration system

4. Conclusion

In this paper, a portable water filter that uses membrane as a filter based on Design for Six Sigma approach has been successfully presented. The strength of this approach is the capability to guide a process of development systematically using DMADV: Define, Measure, Analyse, Design and Verify. Several tools were used to comply with these five stages for their specific objective. Define stage used questionnaire survey analysis to collect information from the user. For the Measure stage, QFD tool has been applied to convert the information from the users into engineering characteristics, so that those vague information from the users are capable to be understood in engineering terms. Morphological chart and weighted decision matrix have been used for concept generation and selection in the Analysis stage. For achieving the Design stage, 3D CAD modelling using SolidWorks software has been conducted as a pre-development for prototyping. In the last stage, verification of the proposed design concept has been made for proof-of-concept testing.

In summary, a new design of portable water filter using membrane filtration system has been developed and reported to overcome the stated problem. This study is a part of ongoing research which is about the use of membrane as a filtering system instead of the existing filter. Further works are needed to improve the selected design concept for better quality improvement of dirty water.

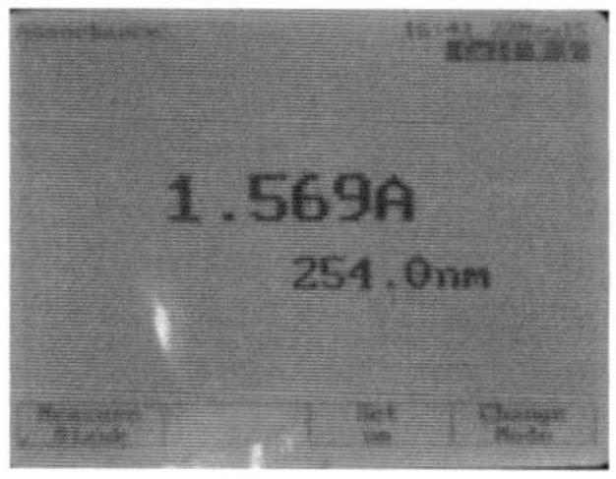

(a) Original water

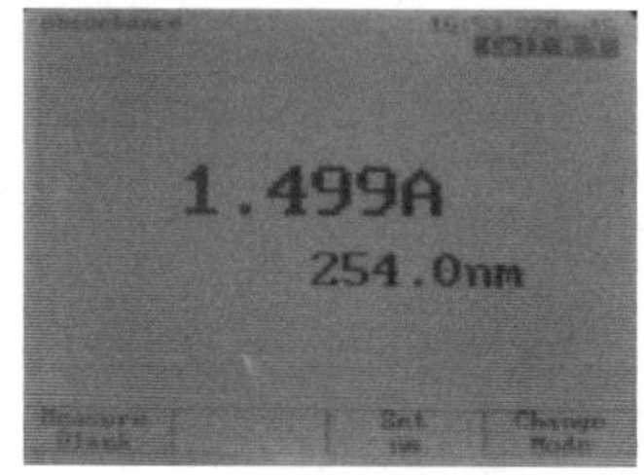

(b) Filtered water

Figure 5. Turbidity reading of water before and after filtering process

Acknowledgement
This research is funded by Universiti Tun Hussein Onn Malaysia (UTHM) under Incentive Grant Scheme for Publication (IGSP) with Vot no. U411.

References

[1] Mokhtar M F, Omar B, Nor N H M, Pauzi N F M, Hassan S and Mohamed W A W 2016 Social and Economic Concern of Supply Chain Sustainability (SCS) *IOP Conference Series: Materials Science and Engineering* **160**
[2] Sapuan S Z, Mohd Jenu M Z and Kazemipour A 2014 Issue on calibration of direct feed biconical antenna in a semi- anechoic chamber using standard antenna method *Advanced Material Research* **903** 273-278
[3] Elizabeth A C and Sandra L F 2012 *Design for Six Sigma in Product and Service Development.* (CRS Press, Boca Raton)
[4] Deming W E 2002 *Out of the Crisis* (MIT Press, Cambridge, MA)
[5] Mast J D, Diepstraten G and Does R J M M 2011 Quality quandaries: Design for six sigma: Method and application *Quality Engineering* **23** 204-211
[6] Chowdhury S 2002 *Design for Six sigma: the Revolution Process for Achieving Extraordinary Profits* (Chicago, IL: Dearbon Trade)
[7] Sapuan S Z and Jenu M Z M 2007 Shielding effectiveness of a rectangular enclosure with aperture and wire penetration *International Conference on Microwave and Millimeter Wave Technology, ICMMT '07* (Builin: China) 1-4
[8] Hassan M F, Keat M K, Yunos M Z, Adzila S, Arifin A M T, Rahman M N A and Haq R H A 2016 Investigation of Existing Laptop Computer Towards Sustainable Development Using Design for Six Sigma Approach *ARPN Journal of Engineering and Applied Sciences* **11** 7683
[9] Zainudin A, Sia C K, Ong P, Narong O L C and Nor N H M 2017 Taguchi design and flower pollination algorithm application to optimize the shrinkage of triaxial porcelain containing palm oil fuel ash *IOP Conference Series: Materials Science and Engineering* **166**

IOP Conf. Series: Materials Science and Engineering **243** (2017) 012049 doi:10.1088/1757-899X/243/1/012049

Analysis of high injection pressure and ambient temperature on biodiesel spray characteristics using computational fluid dynamics

Akasha Hashim[1], Amir Khalid [1,2], Norrizam Jaat [2], Azwan Sapit[1], Azahari Razali[1], Akmal Nizam[1]

[1]Combustion Research Group, Centre for Energy and Industrial Environment Studies (CEIES), Universiti Tun Hussein Onn Malaysia, Parit Raja, Batu Pahat, 86400 Johor, Malaysia.

[2]Automotive and Combustion Synergies Technology Group, Advanced Technology Centre (ATC), Faculty of Engineering Technology, Universiti Tun Hussein Onn Malaysia, Johor, Malaysia

E-mail : [1]kasha1213@gmail.com, [2] amirk@uthm.edu.my

Abstract. Efficiency of combustion engines are highly affected by the formation of air-fuel mixture prior to ignition and combustion process. This research investigate the mixture formation and spray characteristics of biodiesel blends under variant in high ambient and injection conditions using Computational Fluid Dynamics (CFD). The spray characteristics such as spray penetration length, spray angle and fluid flow were observe under various operating conditions. Results show that increase in injection pressure increases the spray penetration length for both biodiesel and diesel. Results also indicate that higher spray angle of biodiesel can be seen as the injection pressure increases. This study concludes that spray characteristics of biodiesel blend is greatly affected by the injection and ambient conditions.

1. Introduction

Fossil fuel shortage and environmental issues from continuously increasing global energy demand has gain concern by automotive manufacturers which resulted the industry to turn to the development of alternative fuels. Biodiesel is one of the promising candidates as alternative fuel for passenger car as it can be produced from many sources of biomass. Different spray and combustion characteristics may affect the mixing formation since many properties of biodiesel are different from the conventional diesel and thus engine performance and efficiency. The lower heating value of biodiesel, compared to that of convention diesel, and higher viscosity and density from long chain chemical formula might affect the injection process [1]. Moreover, biodiesel has higher oxygen concentration, it might yield more complete combustion [2]. Effect of spray behavior, such as the main penetration and spray angle, are influenced by air mixture formation, which related to engine combustion efficiency [3].

The key issue in using biodiesel fuel is oxidation stability, stoichiometric point, antioxidants on the degradation and much oxygen compared to conventional diesel. Although the application of the biodiesel fuel in the diesel engines is more attractive and more economical but it also creates problems of higher emissions compared to conventional diesel. Biodiesel fuel diesel engines still have problem of producing NOx and Particulate Matter (PM) into the atmosphere because of the oxidation stability, cetane number, stoichiometric point, bio-fuel composition and antioxidants on the degradation extremely viscous [4,5,7]. Thus, the improvement of emissions exhausted from biodiesel fuel engines is needed to meet the future stringent emission regulations. Biodiesel has higher cetane number than diesel fuel, no aromatics, almost no sulfur, and contains 10 to 11% oxygen by weight. These characteristics of the fuel decrease the emissions of carbon monoxide (CO), hydrocarbon (HC), and particulate matter (PM) emitted through combustion. In addition, higher viscosity and specific gravity are found to increase the fuel quantity, injection timing, and spray pattern will strongly influence the degree of initial premixing and combustion process.

Computational fluid dynamics (CFD) is a important research tool in order to study and predict the behavior of the injection of fuel spray. CFD is the science of predicting fluid flow, heat transfer, mass transfer, chemical reactions, and related phenomena by solving the mathematical equations which govern these processes using a numerical process. The result of CFD analysis is relevant engineering data such as conceptual studies of new designs, troubleshooting, detailed product development and redesign. CFD analysis complements testing and experimentation is reduces the total effort required in the laboratory [8].

With CFD, it clearly shown the influences of the injection pressure and fuel properties on the variant of spray characteristics of biodiesels and also conventional diesel. Different properties of fuel spray because of their significant influence on the spray atomization and air–fuel mixture affects the engine performance, combustion, and emission. Therefore, the research should investigate the effects of biodiesel on the spray characteristics is apparent and essential. To study the effects of different types of biodiesels on spray characteristics, numerical and experimental studies are carried out [9, 10].

In this study, the effects of high ambient and injection conditions of biodiesel blends on spray characteristics are investigated using Computational Fluid Dynamics (CFD). In this study, only one type of fuel injector is used with orifice 0.12 mm and cone angle of 60°. The ambient temperatures and injection pressures that were used are 850 K, 950 K and 1050 K and 220 MPa, 250 MPa and 280 MPa respectively. Ambient pressure was kept constant at 8 MPa throughout the simulation.

2. Simulation

Simulation in ANSYS FLUENT involves three main stages which are pre-processing, solver and post processing. The design of injector used in this study is only focused on the injector head and combustion chamber. A complete injector has six orifice hole with an angle of 60° between each other. Figure 1 shows the simulation flow chart for this study including the simulation of CFD. Figure 2 shows the 3D model of cross section of rapid compression machine (RCM) which can simulated the real behavior of internal combustion engine process. In addition, Figure 3 shows the geometry of injector with six orifice holes drawn using Solidworks. Figure 4 shows the section geometry of the combustion chamber. It is 1/6 section from the overall geometry because there are 6 nozzle orifices in the actual RCM and 1/6 section is adequate and considered sufficient for the simulation analysis.

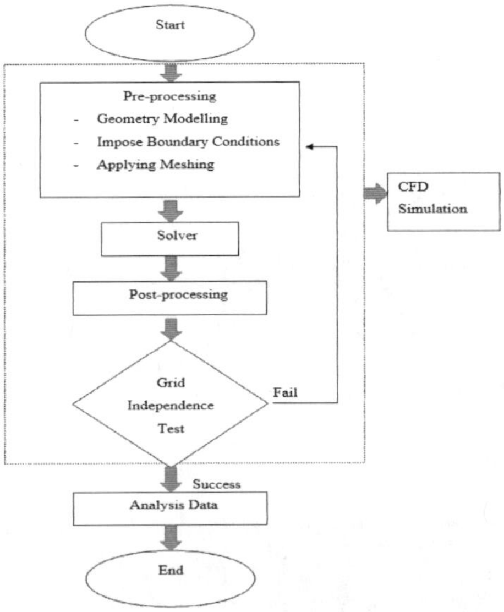

Figure 1. Modelling flowchart in ANSYS Fluent.

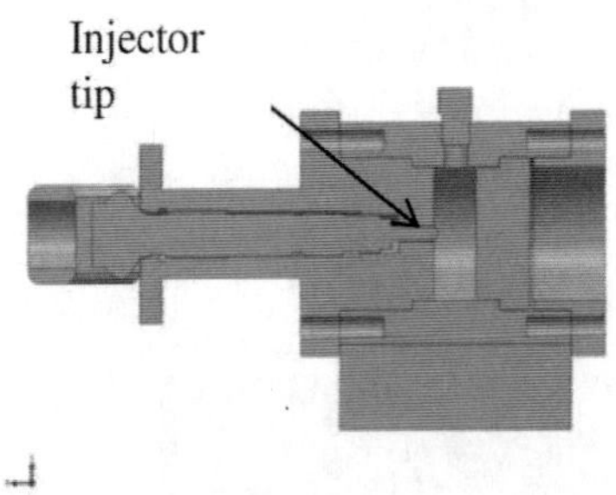

Figure 2. 3D model of cross section in internal combustion chamber of rapid compression machine (RCM).

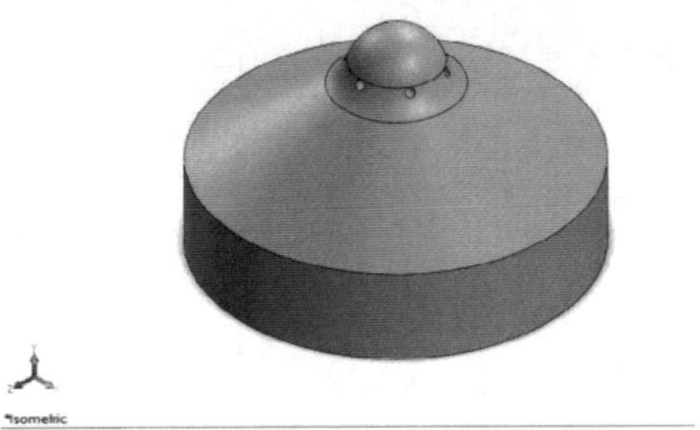

Figure 3. The geometry of injector with six orifice holes.

Figure 4. Geometry of 1/6 part of injector and its combustion chamber.

The meshing used in this can be seen in Figure 5. The boundary conditions used in this study were inlet, outlet and the wall. These boundary conditions are shown in Figure 6 , Figure 7 and Figure 8. Meanwhile, Table 1 shows the boundary conditions used in ANSYS Fluent to conduct the simulations.

IOP Conf. Series: Materials Science and Engineering **243** (2017) 012049 doi:10.1088/1757-899X/243/1/012049

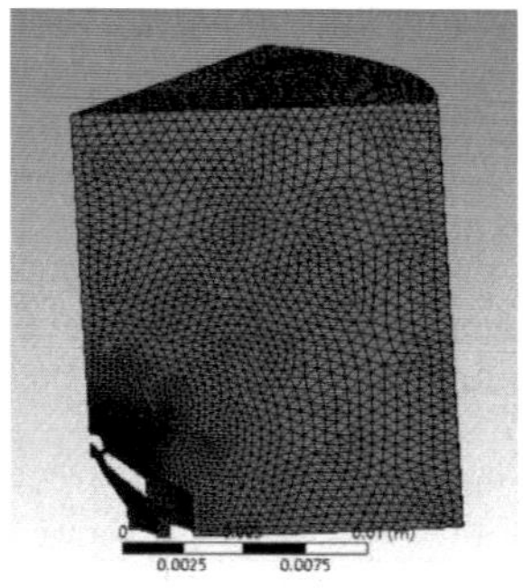

Figure 5. Meshing used on the injector.

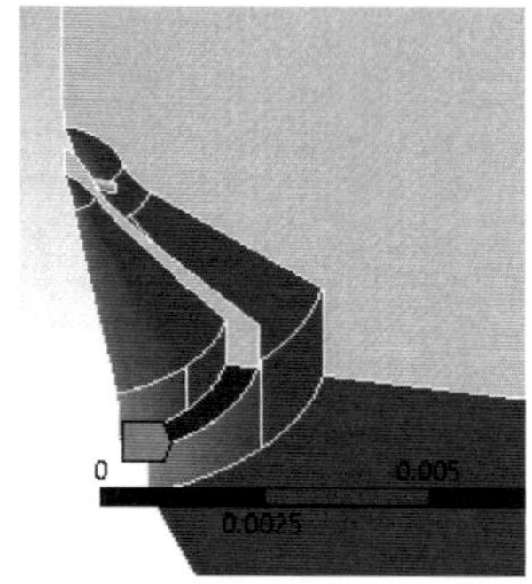

Figure 6. Inlet position at spray injector.

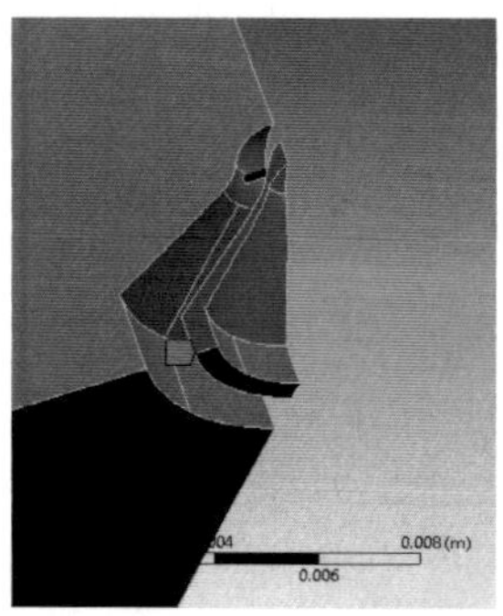

Figure 7. Outlet position at the spray chamber.

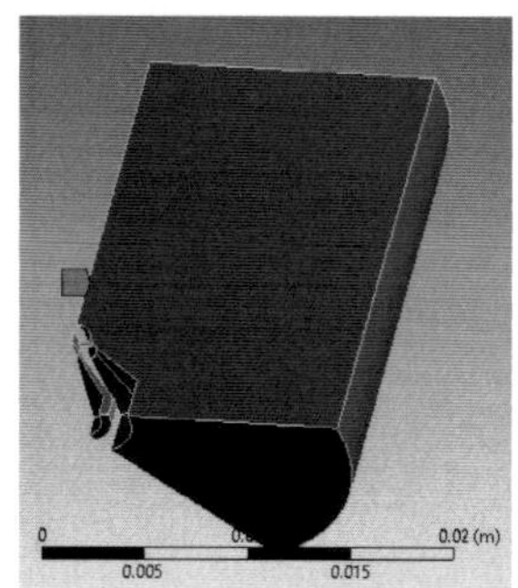

Figure 8. Wall position at injector and spray chamber.

Table 1. Boundary conditions used in ANSYS Fluent.

General	Pressure based with absolute velocity formulation	Transient time
Model	Species transport	Biodiesel
Viscous	k-epsilon (2 equation)	Realizable
Material	Air	Diesel-vapor
Boundary condition	Inlet	Outlet Wall
Parameters	Injection pressure	220MPa 250MPa 280MPa
	Ambient pressure	8MPa
	Ambient temperature	1050K

There are mainly three equations we solve in computational fluid dynamics problem. They are Continuity equation, Momentum equation (Navier Stokes equation) and Energy equation. Applying the mass, momentum and energy conservation, we can derive the continuity equation, momentum equation and energy equation as follows.

a) Continuity equation

$$\frac{D\rho}{Dt}+\rho\frac{\partial U_i}{\partial x_i}=0 \tag{1}$$

b) Momentum equation

$$\underbrace{\rho\frac{\partial U_j}{\partial t}}_{I}+\underbrace{\rho U_i\frac{\partial U_j}{\partial x_i}}_{II}=-\underbrace{\frac{\partial P}{\partial x_j}}_{III}-\underbrace{\frac{\partial \tau_{ij}}{\partial x_i}}_{IV}+\underbrace{\rho g_j}_{V} \tag{2}$$

Where

$$\tau_{ij}=-\mu\left(\frac{\partial U_j}{\partial x_i}+\frac{\partial U_i}{\partial x_j}\right)+\frac{2}{3}\delta_{ij}\mu\frac{\partial U_k}{\partial x_k} \tag{3}$$

I: Local change with time
II: Momentum convection
III: Surface force
IV: Molecular-dependent momentum exchange (diffusion)
V: Mass force

c) Energy equation

$$\underbrace{\rho c_\mu\frac{\partial T}{\partial t}}_{I}+\underbrace{\rho c_\mu U_i\frac{\partial T}{\partial x_i}}_{II}=\underbrace{-P\frac{\partial U_i}{\partial x_i}}_{III}+\underbrace{\lambda\frac{\partial^2 T}{\partial x_i^2}}_{IV}-\underbrace{\tau_{ij}\frac{\partial U_j}{\partial x_i}}_{V} \tag{4}$$

I: Local energy change with time
II: Convective term
III: Pressure work
IV: Heat flux (diffusion)
V: Irreversible transfer of mechanical energy into heat

3. Results and Discussion

This segment portrays the changes in spray characteristics with both injection pressure and ambient temperature increasing. The close relationship between injection pressure and ambient temperature on the formation of spray is explained in this section. Each injection pressures- 220MPa, 250MPa and 280MPa were simulated with each ambient temperatures- 850K, 950K and 1050K. From the simulation, the result develops from three different injection pressure and ambient temperature.

Figure 9 portrays the image results of different fuels at different ambient temperatures .The size of droplet diameter is usually dependent on the parameters that have been used. Based on the figure it is shown that ambient play a significant role on the changes of the droplet diameter. Based on the figure, it can be seen that the particle diameter of all fuels decreases as the ambient temperature. This is a result of fuel particles breaks to smaller particles as the difference of temperature between injector and combustion chamber is large.

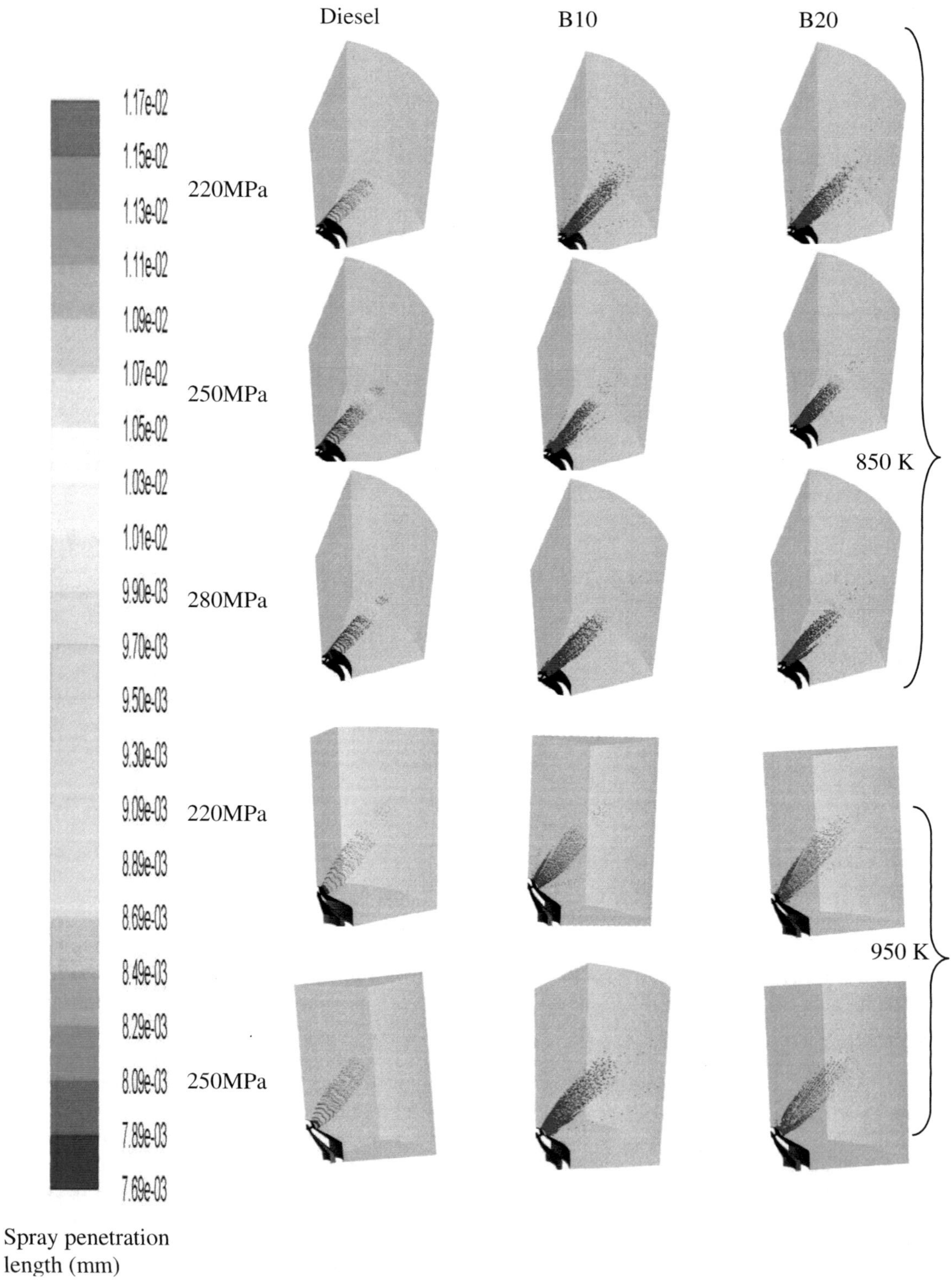

Figure 9. Spray characteristics under different ambient temperature and injection pressure (850 K and 950 K).

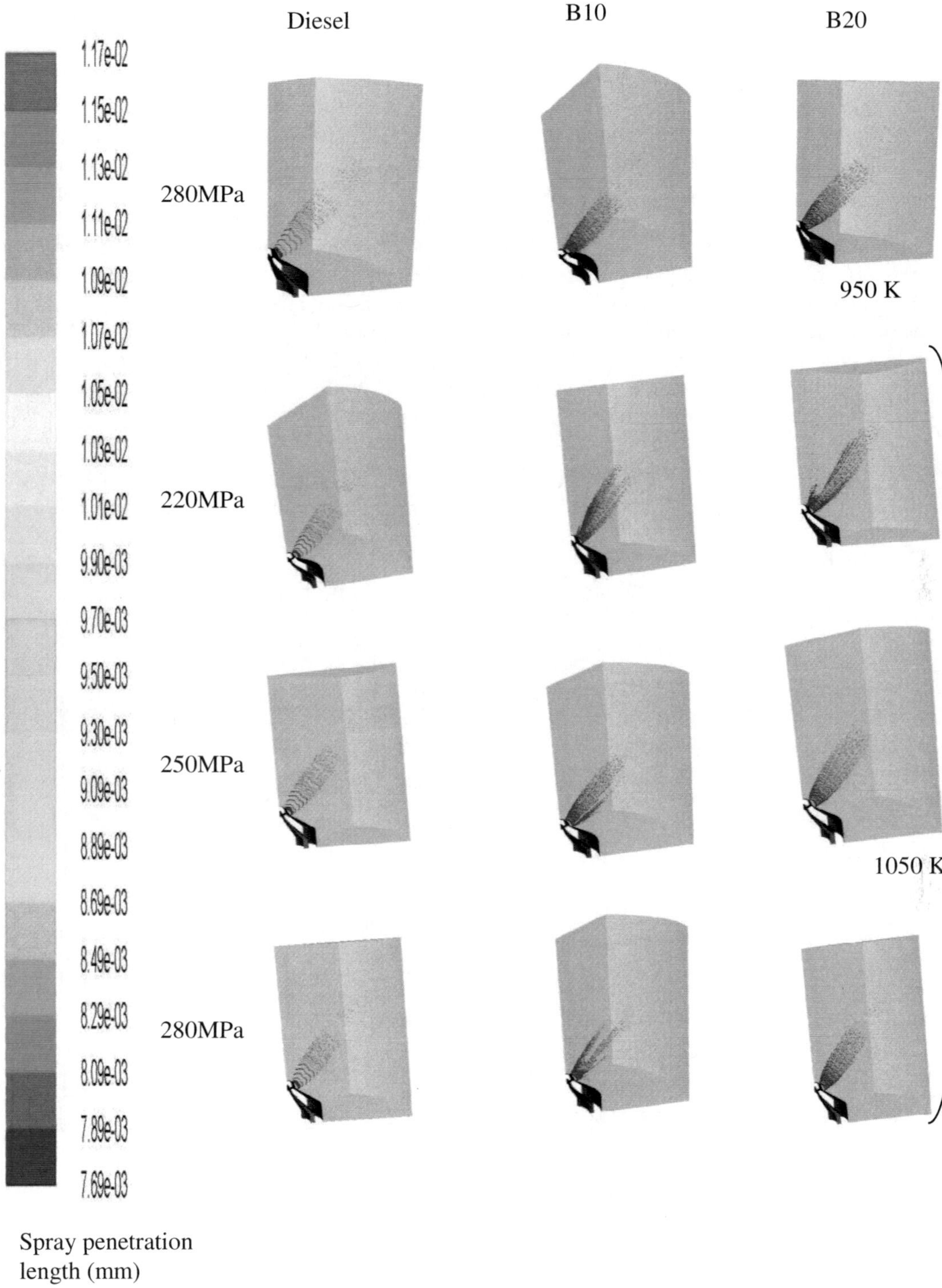

Figure 10. Spray characteristics under different ambient temperature and injection pressure (950 K and 1050 K).

Figure 11 shows the relationship of both ambient temperature and injection pressure and spray penetration length. Steady increase in spray penetration length as the injection pressure and ambient temperature increase. The longest penetration is achieved by B20 at injection pressure 280MPa and ambient temperature 1050K. On the other hand, shortest penetration length can be seen in pure diesel at injection pressure 220MPa and ambient temperature 850K. As for spray angle, the changes in spray characteristics can be seen in Figure 12. Unlike spray penetration, pure diesel has the highest spray angle compared to both B10 and B20 at injection pressure 280 MPa and ambient temperature 1050 K. Lowest spray angle can be seen in B20 at injection pressure 220 MPa and ambient temperature 850K. As the temperature and pressure is increased, a great variation in the spray structure is observed. It promotes mixture formation and distributes larger amount of fuel between sprays thus creates good spray atomization and exhibits a greater amount of fuel-air premixing prepared for combustion. Increment of the ambient temperature promotes the evaporation process or on the other hand, evaporation process is elevated by higher ambient temperature. Increases in the ambient temperature causes the rate of fuel evaporation increased. Fuel droplets with small diameter tend to evaporate easily while those with greater diameter take a much longer time to evaporate.

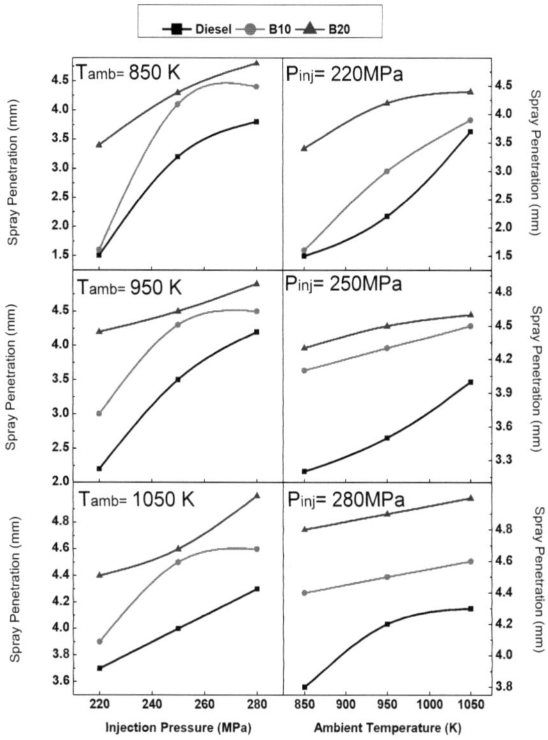

Figure 11. Effect of Injection Pressure on Spray Penetration Length.

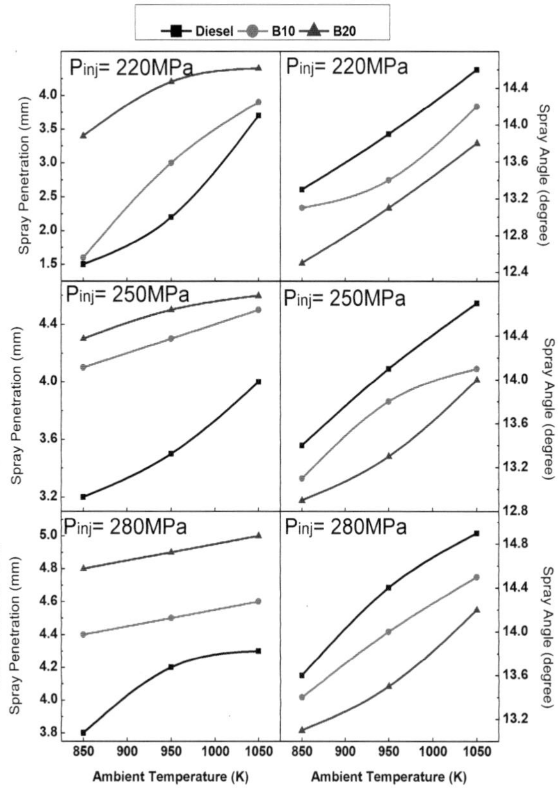

Figure 12. Effect of Injection Pressure on Spray Angle.

4. Conclusion

This study is to investigate the effects of injection pressure and ambient temperature on spray characteristics using constant orifice diameter in the constant volume chamber by using CFD. The results indicate that increase in injection pressure and ambient temperature increases the spray penetration length and spray angle of biodiesel fuel.This study has shown a simulation flow of the fuel flowing in the nozzle spray before combustion process. This simulation was performed on single nozzle orifice diameter which is 0.12mm. The ambient temperatures inside the chamber is 850, 950, 1050 K, the injection pressures are 220, 250, 280 MPa and the constant ambient pressure at 8 MPa.

Change in temperature affects the droplet diameter of fuel. As the ambient temperature increases, the spray tends to easily evaporate and disperse. This results in wider spray angle of the fuel. Thus, increase in ambient temperature increases the spray angle of the fuel used. In this study, it was found that diesel has a higher spray angle as the ambient temperature increases compared to B10 and B20 blends. This is due to high viscosity of B10 and B20 which causes the droplet hard to break into smaller size and disperse.

Different injection pressure influences the spray penetration length of the fuel. The higher the injection pressure, the longer the spray penetration length of fuel. This is the deduction made from this study. In all conditions, the penetration length of pure diesel, B10 and B20 blends increase as the injection pressure increases. The longest penetration length is achieved by B20 as it has the highest viscosity compared to B10 blend and diesel.

Acknowledgments

The authors also would like to thank the Ministry of Higher Education, Malaysia for supporting this research under Fundamental Research Grant Scheme (FRGS) vot.1054.

References

[1] Srichai P, Chollacoop N, Metal N, Ce M T, Mongkut K and T Ladkr T 2015 Spray Visualization of Biodiesel and Diesel in a High Pressure Chamber

[2] Khalid A, Anuar M D, Ishak Y, Manshoor B, Sapit A, Leman M, Zaman I 2014 Emissions characteristics of small diesel engine fuelled by waste cooking oil *MATEC Web of Conferences* **13** 06006 (DOI: 10.1051/matecconf/20141306006)

[3] Khalid A, Andsaler A R, Manshoor B and Jaat N 2016 Effect of High Pressure on the Flow Characteristics of Injector Using Computational Fluid Dynamics (CFD) **11**(12) 7503–7506.

[4] Bittle J A, Knight B M and Jacobs T J 2010 Interesting behavior of biodiesel ignition delay and combustion duration *Energy and Fuels* **24**(8) 4166–4177

[5] Miwa K, Ohmija T and Nishitani T 1988 A Study of the Ignition Delay of Diesel Fuel Spray Using a Rapid Compression Machine

[6] Khalid A, Yatsufusa T, Miyamoto T, Kawakami J and Kidoguchi Y 2009 Analysis of relation between mixture formation during ignition delay period and burning process in diesel combustion *SAE Technical Papers*

[7] Bakar R A, Semin and Ismail 2008 ARI Fuel Injection Pressure Effect on Performance of Direct Injection Diesel Engines Based on Experiment *Am. J. Appl. Sci.* **5**(3) 197–202

[8] Khalid A, Manshoor B 2013 Analysis of Mixture formation and Flame Development of Diesel Combustion using a Rapid Compression Machine and Optical Visualization Technique *Applied Mechanics and Materials* **315** 293-298

[9] Anand K, Avishek R, Sharma R P and Pramod S M *Experimental and Simulation Studies of Biodiesel Combustion in a C I Engine* (Chennai:Indian Institute of Technology Madras) pp 1–9

[10] Xie H, Song L, Xie Y, Pi D, Shao C and Lin Q 2015 An Experimental Study on the Macroscopic Spray Characteristics of Biodiesel and Diesel in a Constant Volume Chamber *Energies* **8**(6) 5952–5972

Experimental and CFD modelling for thermal comfort and CO_2 concentration in office building

H Kabrein[1,a], A Hariri[1,b], A M Leman[2], M Z M Yusof[1] and A Afandi[1]
[1]Industrial and Indoor Environment Research Group (IIERG), Centre for Energy and Industrial Environment Studies (CEIES), Faculty of Mechanical and Manufacturing Engineering Universiti Tun Hussein Onn Malaysia 86400 Parit Raja, Batu Pahat, Johor, Malaysia
[2] Faculty of Engineering Technology, Universiti Tun Hussein Onn Malaysia

E-Mail: hashimkabrein@gmail.com[a], azian@uthm.edu.my[b]

Abstract. Computational fluid dynamic CFD was used for simulating air flow, indoor air distribution and contamination concentration. Gases pollution and thermal discomfort affected occupational health and productivity of work place. The main objectives of this study are to investigate the impact of air change rate in CO_2 concentration and to estimate the profile of CO_2 concentration in the offices building. The thermal comfort and gases contamination are investigated by numerical analysis CFD which was validated by experiment. Thus the air temperature, air velocity and CO_2 concentration were measured at several points in the chamber with four occupants. Comparing between experimental and numerical results showed good agreement. In addition, the CO_2 concentration around human recorded high, compared to the other area. Moreover, the thermal comfort in this study is within the ASHRAE standard 55-2004.

1. Introduction

Computational Fluid Dynamics (CFD) is a numerical technique to solve fluid flow issues. Today's CFD able to perform use very sizable amount of calculations to simulate the behaviour of fluids in advanced environments and geometries [1]. The ventilation rates influence the concentration of carbon dioxide and also affected of uniformly of CO_2 distribution. Numerical analyses can be helpful in engineering to predict of air distribution and the average CO_2 concentration in selected samplings location in a place [2]. The numerical analysis can be positively used to estimate distribution of carbon dioxide exhaled by occupants in a room [3]. However, the CFD modelling should first identify the ventilation positions which air inlet, outlet, the room layout, number of occupants and types of equipment used. According to Bulińska et al. (2014) [3], numerical analysis can be positively used to estimate the distribution of carbon dioxide exhaled by occupants in a room.

Under a similar air supply volume and particle possession conditions, the displacement ventilated area has a lower deposition rate and bigger number of infiltrating particles than the mixing ventilation. Also, the average particle concentration within the displacement case is larger than the mixing one [4]. In the same condition, between stratum ventilation and displacement ventilation, the particle concentration in breathing zone were less in stratum ventilation compared to displacement ventilation [5]. Mixing ventilation had better results compared to displacement ventilation in terms of thermal comfort and condensation risk. This emphasises the sturdy influence of the air movement on heat and mass (humidity) transfer phenomena that turn up in airy rooms equipped with cooling ceilings [6]. Comparing between types of ventilations for gaseous contaminants, the results showed that the stratum

ventilation was better than the displacement ventilation in terms of the inhaled air quality by the occupant. However, when the source located around the office walls, and the supply air temperature is 19 °C the ventilation is better than to stratum ventilation for the inhaled air quality [7].
Study by Zhang & Chen, (2006) (8) showed the numerical analysis could be used to estimate the particle distribution during different ventilation system such as ceiling and side wall air supply system and under floor air distribution (UFAD) has more influence, that due to the profile of air flow inside the room. Also, he found that (UFAD) system has high efficiency for particle removal and efficiency reduction the air contamination.

Catalina, & Kuznik, (2009) [9] conducted the study about chilled ceiling panel, his CFD results showed that the air velocity discomfort was identified at feet/ankle zone but good values of air velocity were found of the rest of the test place. In addition, the results of air temperature as a vertical profile showed asymmetry is less than 1 °C and the maximum difference between glob temperature and the air temperature was 0.8 °C. Non uniform temperature distribution in office building appears with high record temperature, especially around the ceiling and above the ventilation zone. Also the air velocity in the centre of the office was 2 m/s that due the higher air flow rate compared to other place. The CO_2 concentration in the occupied zone was controlled within the target concentration level, but with approximately 20% defence with the CO_2 concentration at any point [10]. (Predicted mean vote) PMV is the greatest widely used thermal comfort index today. Six factors influence PMV index value are: air temperature, air velocity, metabolism of the human body from activity, clothing insulation, air humidity and mean radiant temperature [11]. Acceptable thermal comfort is measured by PMV and (Predicted Percentage Dissatisfied) PPD. The fluctuation of mean air temperature must be not too much and around 24 °C [12].

Increasing the air velocity, will not improve thermal comfort if the relative humidity and air temperature are too high. To get a comfortable sensation, the mean air velocity should be less than 1.5 m/s in the occupied zone. Whereas, the air velocity more than 0.2 m/s in the feet and ankles' level will effect comfort of the occupants [9]. According to Mahyuddin & Awbi, (2010) [2], the accurate estimation of CO_2 concentration in the room must be using several sensors' in varying level vertical or horizontal. The re-normalization k-ε (RNG k-ε) model can be used in a numerical simulation of striatum ventilation, the current model showed good results were compared between numerical and experimental data [13,14]. Q. Chen, (1995) [15], provides eight different kinds of turbulence model and he concluded that for all these models one is more accurate that is re- normalization group (RNG) k-ε model.

The main objectives of this study are to investigate the impact of air change rate in CO_2 concentration and to estimate the profile of CO_2 concentration in the offices building.

2. Method

This study used the results measured from experimental to model CFD simulation. The experimental method in this study was carried out using the environmental control chamber to simulate a typical office environment. The chamber dimension is 4.8 m x 4.8 m x 2.65 m, A schematic diagram of the chamber setup is shown in figure 1, the walls and the ceiling were of polyurethane insulation board of 10 cm thickness, the design of the joint, the groove and the tongue that made from stainless steel metal (5 mm) and located in the chamber on both sides. The chamber was equipped with HVAC system, compressor (10 HP), outdoor condenser and fan coil unit in the mixing room, which is located on the top of the chamber roof. The main supply duct was distributed in two main branches to provide air through ceiling diffusers (0.4 m x 0.4 m); moreover, there are two holes (0.55 m x 0.45 m) provided on the return air duct suspended on the top of the ceiling connecting the chamber and mixing box. The outdoor fresh air intake was in the mixing room through the flexible duct, exhaust fan, and damper. The damper was connected by a flexible duct to warrant that sufficient supply of outdoor fresh air compatible to ASHRAE standard 62.1 2007 [16]. Another damper was also connected to exhaust fan at the top of the ceiling to control the exhaust air.

The air flow through the system was adjusted to 560 CFM (951 m^3/h) to achieve a face velocity 0.9 m/s through any diffusers, with air changes rate about 3 h^{-1} (typical air change rate for an office building) as suggested by ASHRAE standard 55-2004, in this study the chamber occupancy by four occupants the fresh air volume was controlled of 84 CFM (142.7 m^3/h). The chamber environmental condition was adjusted between 23 °C to 26 °C and 78% for RH, which is the indoor environmental condition that suggested in Malaysia by the studies of [17,18]. The control of the air temperature and relative humidity by using AHU system, through temperature and relative humidity monitoring system to adjusting the temperature.

The chamber study was conducted in two cases. The first case involved in recording the environmental condition; daily in two months, simulating typical work hours from 8:00 am to 5:00 pm with four occupants in the chamber. Thermal environmental parameters were recorded. The second case involved the measuring CO_2 gas concentrations in the chamber. The temperature and relative humidity were recorded using four sensors (KIMO 100) that connected to the PC device using temperature and relative humidity monitoring system. Whereas, air velocity was measured using VelociCalc plus model 8324/8384. The instrument located at five levels of height. The first one at 0.5 m above the floor that represented the occupant seated at the knee high, while 1.1 m high presented the occupant level at sitting condition, and point is 1.5 m at standing level [11]. Additional points, 2 m, 2.5 m. The measurement of CO_2 concentration was recorded using an IAQCalc device model 8760/8762 at five levels of heights also in the chamber, the measurement at any location was repetitive three times to confirm the reliability of the test. The air temperature and relative humidity were recorded every 5 minutes interval using the chamber monitors / sensors.

2.1. Numerical method

2.1.1. Computational geometry and modeling

In this this study Ansys fluent 16.1 software was chosen to run CFD simulation for case study. The model geometry in this study (length x, high y, width z) is 4.8 *2.65 * 4.8 m as show in figure 1. The fresh air was supplied into the mixing room through an inlet located at the top of the mixing room ceiling, with dimension 0.3 m diameter. The exhaust air from the chamber through an outlet located on the top of ceiling with a dimension 0.3 m diameter, the centre of the circle (x= 0.3 m, z= 2.4 m), two holes inlet and 2 holes return air located in the top of the ceiling, rectangular (0.4 * 0.4 m2) and (0.55 *0.45 m2) inlet and return air respectively, the centre of inlet holes 1 and 2 positions are (x= 2.4 m, z= 0.28 m) and (x= 2.4 m, z= 2.4 m) respectively. Thus the centre of return air holes 1 and 2 positions are (x= 1.7 m, z= 4.52 m) and (x= 3.10 m, z= 4.52 m) respectively. The people in the test chamber were simulated by the human seated shape diminution in figure 2 and heated by 60 W /m^2/ person; depending on the human in offices building by ASHRAE standard 55-2004. The position of people in the test chamber (x= 1.5 m and 3.3 m) and (z= 1.03 m and 3.77 m). Human generated CO_2 through nostril and simulated by hole 12 mm diameter and height of 1.1 m above the floor [19], the calculation exhalation exit velocity can be calculated using the following equation 1,

$$V\ exit = Q/S = Q/\left((\pi D^{\wedge}2)/4\right) \tag{1}$$

Where: Q is flow rate around 8.5 l/m, S = cross section area, D = nostril diameter in this study 12 mm [19]. The exhalation through the nose was recognized to be directed downwards from the horizontal level with an angle of approximately 45°. However, the flow rate (Q) at a sitting level is around 8.5 l/m then the exit exhalation velocity is 1.25 m/s.

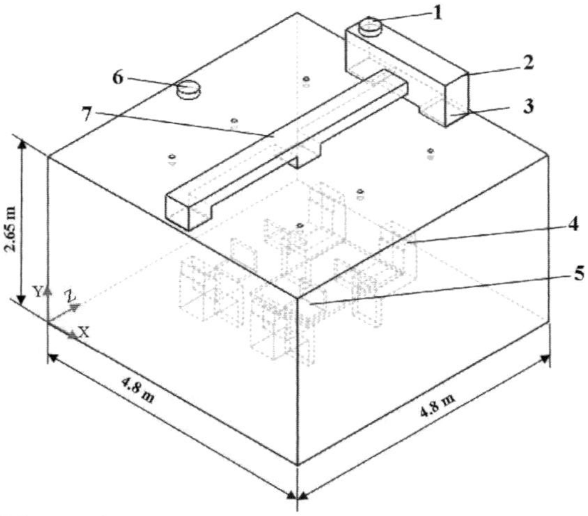

Figure 1 . schematic diagram of chamber
1- Fresh air, 2- Mixing room, 3- Return air, 4- Human body, 5- laptop,
6- Exhaust air, 7- air supply duct

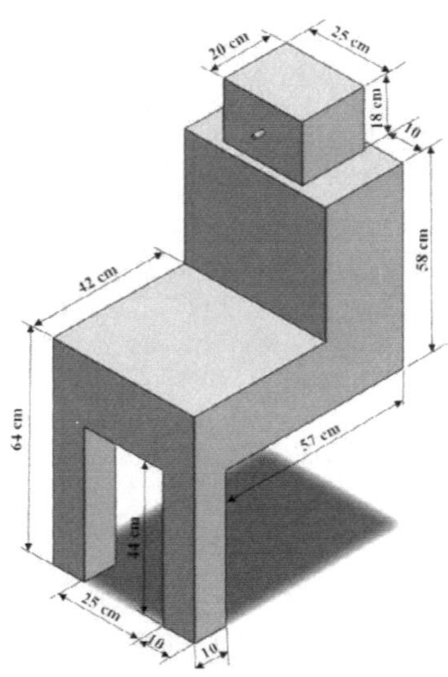

Figure 2. CFD model of a human seated

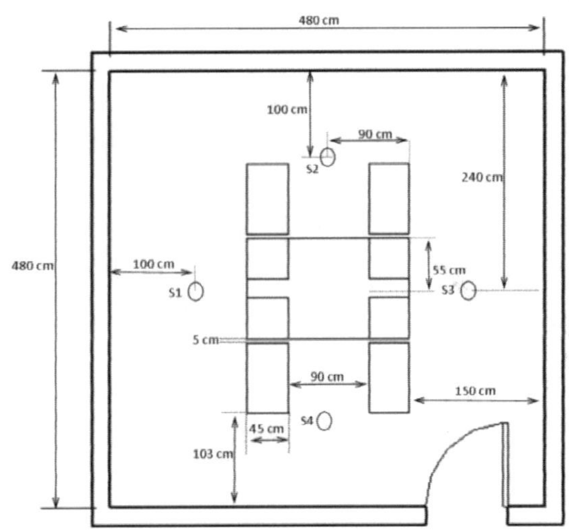

Figure 3. Layout of sensors and occupant position in chamber

A laptop shape 0.45 m X 0.45 m X 0.05 m, located in the table meeting in the middle of the room was simulated with three 60 W light bulbs. The laptops position 0.9 m above the floor and 0.05 m front of human, however, because the geometrical is quite complicated the legs of the table and chairs are not included that to decrease the computational grid [19]. 6 Fluorescent light with 0.07 m diameter and 0.12 m height installed in the ceiling and positioned at (x, z) (1.2 m, 1. 2 m), (1.2 m, 2.4 m), (1.2 m, 3.75 m), (3.6 m, 1.25 m), (3.6 m, 2.40 m), (3.6 m, 3.75 m). The light simulated with 12 W light bulbs and the total 6*12 W.

The record of temperature, velocity and carbon dioxide concentration along five plumb lines in the chamber are measured point by point and line to line. Figure 3 shows the sensors line distribution in the plan view in room.

2.1.2. Meshing
The overall outer mesh was dominated by a tetrahedrons mesh. Furthermore, the sizing of the mesh was adjusted in certain locations in order to get smooth and high quality of mesh. The grid sizing was applied to the human, inlets and laptops. The number of grids was about 2.4 million. In addition, grid independence solution was checked by using a fine mesh compared with coarse mesh. It was achieved by doubling the number of grid i.e. a fine mesh (2.4 million) and coarse mesh (1.2 million). The residual converge criteria of energy is set to 10^{-6}, while the turbulence kinetic energy, k turbulent dissipation rate, ε, and carbon dioxide, c were set to 10^{-3}.

2.1.3. Modelling and setup
Many resent research have used (RNG) k-ε because it can provide more accurate results for the turbulent airflow computation between eight different turbulence models that investigated by Chen [15]. In this study (RNG) k-ε was chosen. In addition, another modelling setup energy equation that the air movement and detached mass was directed to momentum and energy equations using a finite volume method.

General turbulence equations are as the flowing:
Momentum equation:

$$\frac{\partial(\rho u)}{\partial t} + \nabla.(\rho u V) = -\frac{\partial p}{\partial x} + \frac{\partial \tau_{xx}}{\partial x} + \frac{\partial \tau_{yx}}{\partial y} + \frac{\partial \tau_{zx}}{\partial z} + \rho f_x \quad (2)$$

$$\frac{\partial(\rho v)}{\partial t} + \nabla.(\rho v V) = -\frac{\partial p}{\partial y} + \frac{\partial \tau_{xy}}{\partial x} + \frac{\partial \tau_{yy}}{\partial y} + \frac{\partial \tau_{zy}}{\partial z} + \rho f_y \quad (3)$$

$$\frac{\partial(\rho w)}{\partial t} + \nabla.(\rho w V) = -\frac{\partial p}{\partial z} + \frac{\partial \tau_{xz}}{\partial x} + \frac{\partial \tau_{yz}}{\partial y} + \frac{\partial \tau_{zz}}{\partial z} + \rho f_z \quad (4)$$

Energy equation:

$$\frac{\partial}{\partial t}\left[\rho\left(e+\frac{V^2}{2}\right)\right] + \nabla.\left[\rho\left(e+\frac{V^2}{2}\right)V\right] = \rho\dot{q} + \frac{\partial}{\partial x}\left(k\frac{\partial T}{\partial x}\right) + \frac{\partial}{\partial y}\left(k\frac{\partial T}{\partial y}\right) + \frac{\partial}{\partial z}\left(k\frac{\partial T}{\partial z}\right) - \frac{\partial(up)}{\partial x} - \frac{\partial(vp)}{\partial y} - \frac{\partial(wp)}{\partial z} + \frac{\partial(u\tau_{xx})}{\partial x} + \frac{\partial(u\tau_{yx})}{\partial y} + \frac{\partial(u\tau_{zx})}{\partial z} + \frac{\partial(v\tau_{xy})}{\partial x} + \frac{\partial(v\tau_{yy})}{\partial y} + \frac{\partial(v\tau_{zy})}{\partial z} + \frac{\partial(w\tau_{xz})}{\partial x} + \frac{\partial(w\tau_{yz})}{\partial y} + \frac{\partial(w\tau_{zz})}{\partial z} + \rho f.V \quad (5)$$

Where: ρ = air density, V = velocity, f = force, P = pressure, τ = shear stress, $\dot{q}$ = heat flux, e = internal energy.

3. Validation
The article published by Tian et al (2010) [7], has been chosen as the validation model of this study. Figure 4 showed the schematic diagram of Tian model, which has resulted from his experimental and simulation study. The reason for selecting this model from this article is because the model was about an office room which has a several heat sources and gasses contamination. In addition, this model has focused on CO_2 concentration and distribution in the room.

In this study re-built the published model in ANSYS 16.1, and run with a setting close to the published model. The details of geometry of this model show in the Figure 4, the room (3.9 x 2.9 x 2.6 m), the air inlet (0.21 x 0.17 m) rectangular located in the middle of the right wall, (0.6 x 0.6 m) perforated ceiling exhaust with 15.3% as the effective area ratio. The human body was simulated by box (0.25 x 0.4 x 1.2 m), a personal computer simulated by box (0.4 x 0.4 x 0.4 m), located on the table. The air inlet is 1.19 m/s and total heat load is 399 W; in the validation model, the results were illustrated in Figure 5, for nine positions. Therefore, the results from validation model were compared with Tian's experimental and numerical results which show good agreement. The error ratio between validation line and measured line and between measured line and simulation line are 6.6% to 8.7% and 8.0% to 9.0% respectively. The simulation results are in acceptable agreement with experimental data.

Therefore, this re-built model from the distributed article can be applied to case study once modification in geometry and boundary condition that is described within the numerical method.

Figure 5 shows the numerical results compared with experimental and numerical simulation for [7] regarding of CO_2 concentration. It is found that the simulation results are in acceptable agreement with experimental data and simulation by Tian (2010).

4. Results and discussion

4.1. Grid independence test

The significant role in the simulation study is to check the mesh quality by grid independence test. This test is to confirm that the results are not sensitive with the mesh quantity. In another word, the solution is independent of the mesh. In this study, the process has been done by adjusting the mesh in which by doubling the number of grid mesh. However, the number of the grid was about 1.2 million with a coarse mesh and has been doubled to 2.4 million by applying a more fine mesh. Table 1 shows the results from a different number of elements.

Table 1. The different elements size with average of air parameters and CO_2.

Type of mesh	Number of elements	Air velocity m/s	Air temperature °C	CO_2 concentration ppm
Mesh 1	1288922	0.196698	297.390	694.75
Mesh 2	2450288	0.210158	298.258	711.50
Percentage difference		6.40%	0.29%	2.30%

From Table 1, the solver can simulate a relatively faster using coarse mesh. However, due to small variation obtained from both types of mesh, the Mesh 1 was chosen for the rest of the analysis in order to save time.

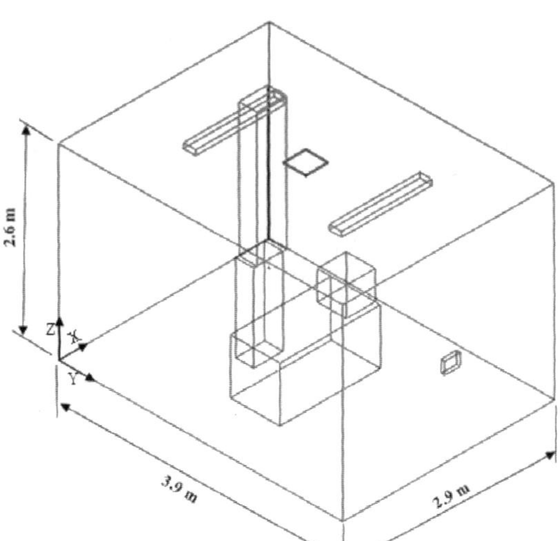

Figure 4. Geometric model designs of chamber (validation).

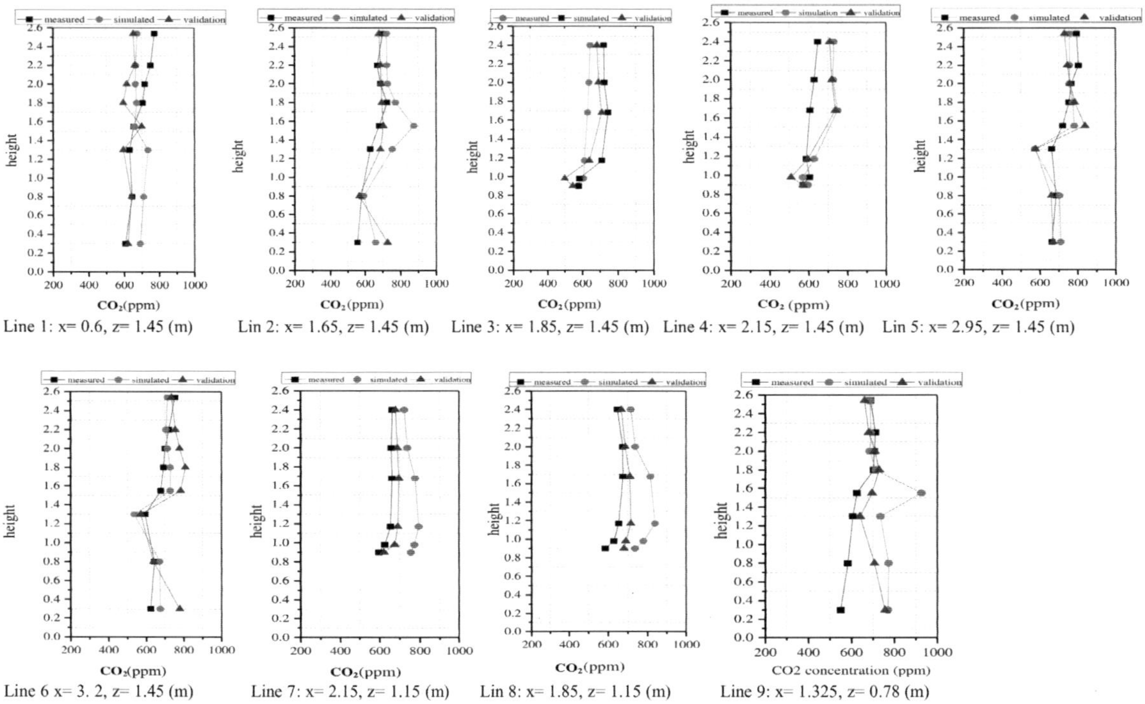

Figure 5. Measured, simulated (Tian et al. 2010) and simulation validation CO_2 concentration profile for various positions (ppm)

4.2. Air temperature distribution

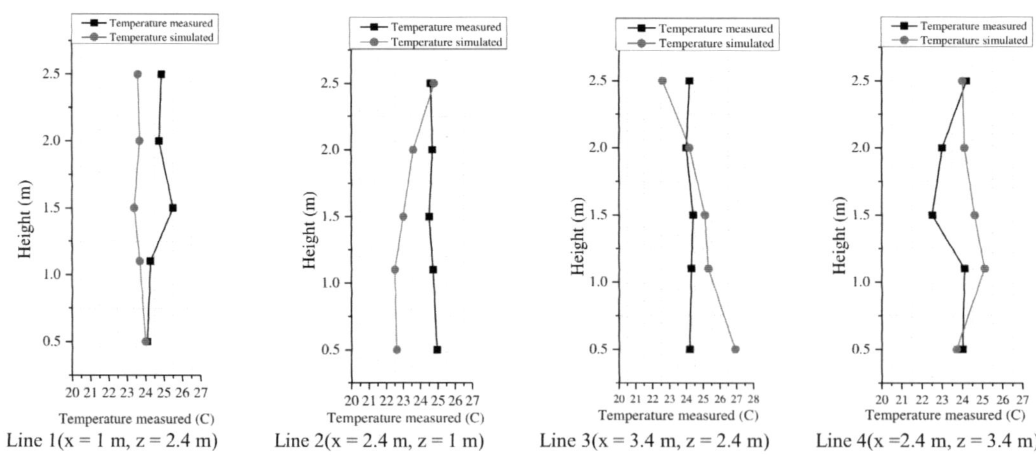

Figure 6. Experimental and simulation temperature profile in the chamber

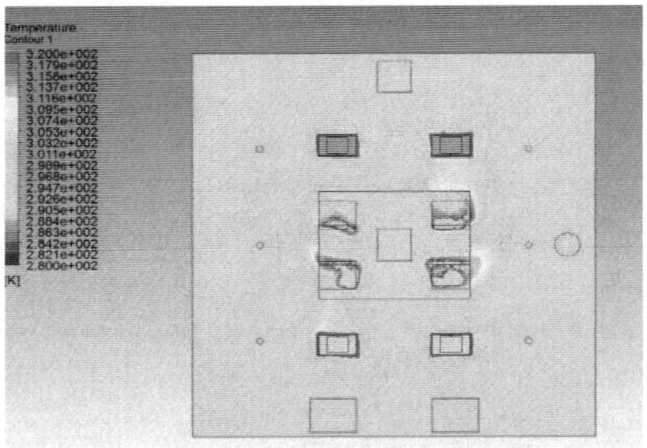

Figure 7. Air temperature distribution in horizontal plan

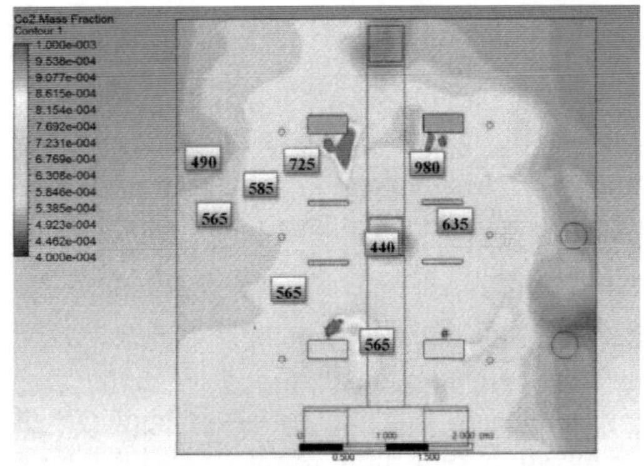

Figure 8. CO_2 concentration top plane (X Z); Y=1.1 m

Figure 6 compares the simulation findings with the experiment. It is observed that the air temperature decreases with height points. However, in line 2 the temperature had shown a little increase near to the ceiling. Some points in line 1 and 2 showed an agreement in the temperature degree between simulations and measurement. The cooler air entered to the room from two diffusers located in centre of the ceiling and the air then distributed to various types of heating sources such as humans, computers. Moreover, the lamps on the ceiling also contribute to the heating of the incoming air. The thermal plumb in line 1 of the simulation illustrated some similar degrees compared with the measurement.

The temperature from simulation was recorded in the range of 23.5 °C to 24.0 °C. Whereas, in Line 2 the results were indicated some agreement in the temperature degrees between measurement and simulation. Meanwhile, some temperature degrees show overestimation at level less than 1.1 m. In line 3, the higher degree was recorded at 0.5 m height and it was 26.9 °C, compared with 24.5 °C from measurement at the same level. This difference in temperature may be due into the error of measuring instrument as [20] Siamak R. Ardkapan *et al.* (2014) also found. In line 4, we can notice that the minimum degrees in level 1.1 m was 22.5 °C from measurement graph, whereas the result of the simulation at the same level was 24.0 °C. However, from the results of the simulation, air temperature profile indicated that the office area has a varying distribution of temperature. Generally the temperature near the floor was slightly higher. However, in this study, the effect of air diffusers in the centre line of the ceiling chamber also significantly influences the variation of the air temperature distribution.

The error of air temperature percentage between experimental and simulation is from 4.0% to 5.9%.

Figure 7 shows the air temperature distribution in horizontal plan at 0.8 m above the floor. The results indicate that the temperature at level 0.8 m was higher than other levels, especially around the heat sources.

4.3. Air velocity distribution

The air velocity profile from the simulation was obtained from the inlet velocity in experimental data. In addition, the air velocity inside the chamber was investigated at many locations and several heights. In this simulation, four positions and five heights were recorded line by line through shifting the sensors' point.

Figure 9 demonstrates the results of measurement and simulation for vertical air velocity at four plumes. In line 1, the air velocity in this line was lower than the other line (approximately 0.07 m/s), except in the level near to the ceiling, it was recorded 0.17 m/s at 2.5 m height. That is due to the air obtained from ceiling air flow inlet. Generally, the velocity in the lower level was low than the highest points.

IOP Conf. Series: Materials Science and Engineering **243** (2017) 012050 doi:10.1088/1757-899X/243/1/012050

The maximum average of air velocity recorded in line 4 that is located between two diffusers is 0.120 m/s, while the maximum velocity from the experimental results is 0.136 m/s. However, the distribution of air velocity depended on the type of ventilation and location of air diffuser. Ceiling ventilation affected the air distribution at low level of heights; the expected results of air velocity profile had already been discussed in the air temperature profile section which caused in reduction of the temperature in these levels.

4.4. CO_2 concentration distribution

The CO_2 concentration resulted from the simulation and experimental is shown in Figure 10. The figure indicates that the variation between these findings from the different background of CO_2 concentration, which is applied during the experiment and simulation. The background of CO_2 concentration in the chamber was measured during the working day, with the course of 16 hours for all the tests, which makes the time duration of each test is 4 hours.

Figure 10 shows that the CO_2 concentration in line 1 recorded the highest number at 0.5 m and decreased in the higher levels. In this line, the simulation graph recorded less value than the experimental results except at level 0.5 m above the floor that may be to the different background of CO_2 concentration. This is because in the real scenario the background of CO_2 varies with time, however, this variation was constant during the simulation process. From the simulation results, a little change of values occurred on line 2 and 4 at each level. Figure 8 shows the CO_2 concentration top plan at 1.1 m above the floor. The concentration of CO_2 recorded higher values in the centre of the chamber where the breathing zones at the level of 1.1 m was close to the nose of occupants. Whereas, the air velocity near to the walls was quite low, this means less concentration comparing with the CO_2 concentration at breathing zone.

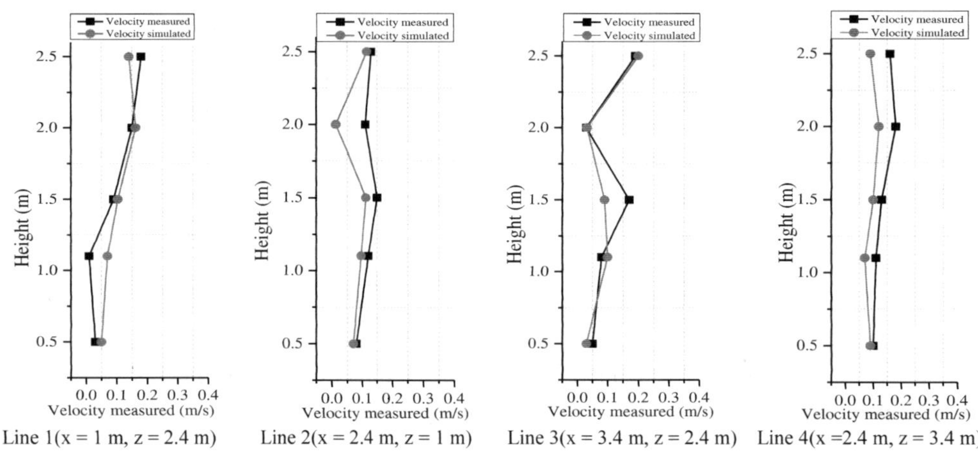

Figure 9. Experimental and simulation velocity profile in the chamber

This caused by the location of sensor holder of line 2 was near to return air hole, while for line 4 was located between two inlet diffusers. According to [10] Fan & Ito (2014), in the ceiling air supply system the concentration of CO_2, in the return air zone is relatively lower than the occupied zone. From the current result of the simulation of CO_2, the error ratio was 3.87 to 9%, however, this result agrees well with the findings of the experiment.

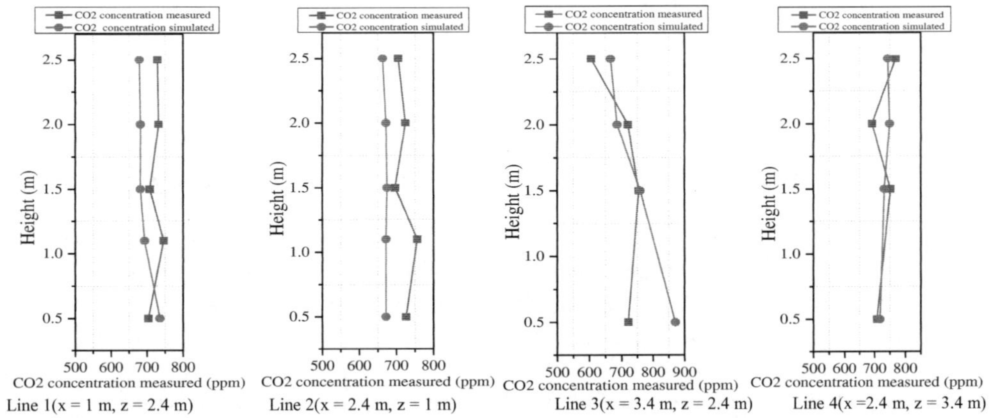

Figure 10. The CO_2 concentration at four level and deferent positions

5. Conclusion

The research was based on experimental study which had the support of a numerical analysis on the air flow type by using the CFD technique. The air velocity obtained from the experimental and CFD showed that the air diffuser in the centre line of the ceiling chamber significantly infuses the lowering values of air velocity at 0.5 m level. Thermal comfort in this study is within the ASHRAE stander 55-2004, local code and also accepted by occupants in the workplace. Controlling the CO_2 concentration depends on the ratio of fresh air divided by the total volume. This ratio can when increased will decrease the carbon dioxide concentration in work place, which causes the increase of energy consumption. The area around occupants is more polluted comparing with the upper ventilated zone. That is due the fresh air supplied through the ceiling openings was diffused in the upper zone but was not effectively transported to the occupied zone. Moreover, the assessment of CO_2 concentration demonstrates the maximum concentration at low level and around the human body, whereas, the concentration of the ceiling ventilation is very well, on one condition, when the air distributed is accurate.

The numerical analysis by CFD simulation in this study also showed the improvement IAQ of the current approach. The results from the numerical analysis showed good agreed with the experimental results. However, the advantage of the numerical simulation is the ability to depict the air velocity and air temperature distribution at any layers or locations in the place.

Acknowledgement

The authors would like to thank the staff of the Thermal Environmental Control laboratory in the Faculty of Mechanical & Manufacturing Engineering for their technical support and the Office for Research, Innovation, Commercialization and Consultancy Management (ORICC), Universiti Tun Hussein Onn Malaysia for financial support for this research (Grant U176).

References

[1] Brown K, Kalata W and Schick R 2014 Optimization of SO2 scrubber using CFD modeling *Procedia Eng.* **83** 170–80

[2] Mahyuddin N and Awbi H 2010 The spatial distribution of carbon dioxide in an environmental test chamber *Build Environ.* **45**(9) 1993–2001

[3] Bulińska A, Popiołek Z and Buliński Z 2014 Experimentally validated CFD analysis on sampling region determination of average indoor carbon dioxide concentration in occupied space *Build Environ.* **72** 319–31

[4] Zhao B, Zhang Z, Li X and Huang D 2004 Comparison of diffusion characteristics of aerosol particles in different ventilated rooms by numerical method *ASHRAE Trans.* **110**(1) 88–95
[5] Inthavong K, Tian Z F and Tu J Y 2009 Effect of ventilation design on removal of particles in woodturning workstations *Build Environ.* **44**(1) 125–36
[6] Teodosiu C, Ilie V and Teodosiu R 2016 Numerical prediction of thermal comfort and condensation risk in a ventilated office, equipped with a cooling ceiling *Energy Procedia* **85** 550–8
[7] Tian L, Lin Z and Wang Q 2010 Comparison of gaseous contaminant diffusion under stratum ventilation and under displacement ventilation *Build Environ.* **45**(9) 2035–46
[8] Zhang Z and Chen Q 2006 Experimental measurements and numerical simulations of particle transport and distribution in ventilated rooms *Atmos Environ.* **40**(18) 3396–408.
[9] Catalina T, Virgone J and Kuznik F 2009 Evaluation of thermal comfort using combined CFD and experimentation study in a test room equipped with a cooling ceiling *Build Environ.* **44**(8) 1740–50
[10] Fan Y and Ito K 2014 Integrated building energy computational fluid dynamics simulation for estimating the energy-saving effect of energy recovery ventilator with CO_2 demand-controlled ventilation system in office space *Indoor Built Environ.* **23**(6) 785–803
[11] ASHRAE 2005 *Standard 55 Thermal Environmental Conditions for Human Occupancy*
[12] Tian L, Lin Z, Liu J and Wang Q 2008 Numerical study of Indoor Air Quality and thermal comfort under stratum ventilation *Prog Comput Fluid Dyn An Int J* **8**(7/8) 541
[13] Lin Z, Tian L, Yao T, Wang Q and Chow T T 2011 Experimental and numerical study of room airflow under stratum ventilation *Build Environ.* **46**(1) 235–44
[14] Lin Z, Lee C K, Fong K F and Chow T T 2011 Comparison of annual energy performances with different ventilation methods for temperature and humidity control *Energy Build.* **43**(12) 3599–608
[15] Chen D Q, Muralidhar K and Munshi P 1995 Comparison of different k-ε models for indoor air flow computations *Numer Heat Transf Part B* **35** 353–69
[16] ASHRAE 2007 Standard. Ventilation for Acceptable IAQ *ANSI/ASHRAE Stand 621-2007* **62**(1) 41
[17] DOSH 2010 Industry Code of Practice on Indoor Air Quality *Minist Hum Resour Dep Occup Saf Heal* 1–50
[18] Jamaludin N, Mohammed N I, Khamidi M F and Wahab S N A 2015 Thermal Comfort of Residential Building in Malaysia at Different Micro-climates *Procedia - Soc Behav Sci* **170** 613–23
[19] Mahyuddin N, Awbi H B and Essah E 2014 A computational fluid dynamics modelling of the air movement in an environmental test chamber with a respiring manikin *J Build Perform Simul* 1493 1–16
[20] Ardkapan S R, Johnson M S, Yazdi S, Afshari A and Bergsøe N C 2014 Filtration efficiency of an electrostatic fibrous filter: Studying filtration dependency on ultrafine particle exposure and composition *J Aerosol Sci* **72** 14–20

Impact of the air filtration on indoor particle concentration by using combination filters in offices building

H Kabrein[1,a], A Hariri[1,b], A M Leman[2], N M R Noraini[3], M Z M Yusof[1] and A Afandi[1]
[1] [1]Industrial and Indoor Environment Research Group (IIERG), Centre for Energy and Industrial Environment Studies (CEIES), Faculty of Mechanical and Manufacturing Engineering Universiti Tun Hussein Onn Malaysia 86400 Parit Raja, Batu Pahat, Johor, Malaysia
[2] Faculty of Engineering Technology, Universiti Tun Hussein Onn Malaysia
[3]Industrial Hygiene Division, National Institute of Occupational Safety and Health (NIOSH), Bandar Baru, Bangi, Selangor, Malaysia

*E-mail: hashimkabrein@gmail.com[a], azian@uthm.edu.my[b]

Abstract. Heating ventilation and air conditioning system (HVAC) is very important for offices building and human health. The combining filter method was used to reduce the air pollution indoor such as that particulate matter and gases pollution that affected in health and productivity. Using particle filters in industrial HVAC systems (factories and manufacturing process) does not enough to remove all the indoor pollution. The main objective of this study is to investigate the impact of combination filters for particle and gases removal efficiency. The combining method is by using two filters (particulate filter pre-filter and carbon filter) to reduce particle matter and gases respectively. The purpose of this study is to use minimum efficiency reporting value (MERV filter) rating 13 and activated carbon filter (ACF) to remove indoor air pollution and controlling the air change rate to enhance the air quality and energy saving. It was concluded that the combination filter showed good removal efficiency of particle up to 90.76% and 89.25% for PM_{10} and $PM_{2.5}$ respectively. The pressure drop across the filters was small compared with the high- efficiency filters. The filtration efficiency of combination filters after three months' was better than efficiency by the new MERV filter alone.

1. Introduction

Air-conditioned office buildings objectives is to provide a thermally satisfactory environment for human comfort and work that would enable better work productivity and less thermal dissatisfaction [1]. Location and design of the fresh air inlet for the HVAC system can be significant to decrease the outdoor particle influence on indoor. Thus, the HVAC system design plays an important role in reducing the concentration of particles indoors [2]. In commercial buildings, the particle matter can affect the health of occupants. HVAC systems are used to decrease indoor particles and to provide extra ventilation and lower particle matter concentration. It is also effective even when outdoor particulate matter concentration is high [3]. Ventilation and filtration system is used to reduce the risk of pollutants infection and diseases transmission, that depends on the nature of contaminated the aerosols and building [4]. There are many studies investigate on how to reduce the particle matter in

IOP Conf. Series: Materials Science and Engineering **243** (2017) 012051 doi:10.1088/1757-899X/243/1/012051

the indoor environment. Those studies found that the application of large scales and real places involving the high airflow rate need to be considered [5].

Mixing air filters with outdoor air may reduce the influence particles source of indoor particulate concentration indoor [2]. However, the ratio of fresh outdoor air to recirculation air is very important. More outdoor air may lead to more outdoor particles. In addition, increase outdoor air lead to increasing the energy consumption for air treatment.

The study by Azimi [4], found that filtration of recirculation air may be used to reduce airborne infectious diseases especially in the indoor, However, there is no direct relationship proved between the usage of mechanical filters in buildings and the reduced asthma symptoms even when using specific common devices [6]. High-efficiency particulate air (HEPA) can be used in the combustion particles pollution area and it can provide health benefits by reduction of these particles [7]. However, the high efficiency filter such as HEPA filter not only has high efficiency for particle removal compared to the other filters but also have high air resistant that provided high-pressure drop.

The main objective of this study, to assess the combination filters method for high removal efficiency and low pressure drop.

The initial cost of HVAC system is the first criterion in its system selection. This is a limited approach because it can be costly in the building economic life cycle of the HVAC system operation and maintenance cost and the influence on office workers' health and productivity are not duly considered [1]. In addition, the utilisation of both outdoor and mixing air filters can significantly influence and keep indoor particle concentration levels lower when compared to mixing air filters alone [2]. The study by Chatoutsidou [8], found that mechanical ventilation and filtration may remove outdoor particle more than 50%, in addition, the mass concentration of outdoor PM_{10} was more than the indoor and the efficient removal was significant outdoor for the particulate matter. During the absence of occupants' activities, the outdoor air is a major source of particle matter and contributed total particles to indoor [9].

Many of researchers have discusses for measured particles in offices building [6], [8], [10]–[13]. Park [14], has found that mechanical filters can be removed up to 50% of particle outdoor exposes.

2. Method

This study was carried out using the environmental control chamber that can simulate a typical office environment. The chamber dimension is 4.8 m length, 4.8 m width and 2.65 m in height. The walls and the ceiling were made of polyurethane insulation board of 10 cm thickness, the design of the joint; the groove and the tongue were made from stainless steel metal (5 cm) and located in the chamber on both sides. The chamber was equipped with HVAC system; a compressor (10 HP), an outdoor condenser and a fan coil unit in the mixing room, which is located on the top of the chamber roof as the shown in Fig 1. The supply air duct located on the roof of the chamber. The main air supply duct links the fan coil unit to the chamber. The main supply duct was distributed in two main branches to provide air through ceiling diffusers (0.4 m x 0.4 m); moreover, there are two holes (0.55 m x 0.45 m) provided on the return air duct suspended on the top of the ceiling connecting the chamber and mixing box. A filter box (0.6 m x 0.3 m) was designed to house the new filters and to ensure good mixing of the air passage and cleaning through the filter before entering the fan coil unit. The outdoor fresh air intake was in the mixing room through the flexible duct, exhaust fan, and damper. The damper was connected by the flexible duct to warrant that sufficient supply of outdoor fresh air compatible to ASHRAE standard 62 [15]. Another damper was also connected to the exhaust fan at the top of the ceiling to control the exhaust air. The air flow through the system was adjusted to 560 CFM (951 m^3/h) to achieve a face velocity 0.9 m/s and 0.8 m/s through diffusers 1 and 2 respectively, with air changes rate about 3 h^{-1} (typical air change rate for an office building) as suggested by ASHRAE standard 62. The air flow rate was measured by using TSI 8371 AccuBalance 802082. The chamber environmental conditions were adjusted between 23 °C to 26 °C and 78% for RH, which is the indoor environmental condition that suggested in Malaysia by the previous studies of [16], [17]. The control of the air temperature and relative humidity is by using AHU system, through temperature and relative humidity monitoring system.

The chamber case study was conducted by recording the environmental condition for daily in six months and simulating typical work hours from 8:00 am to 5:00 pm with four occupants in the chamber. Combination filter of MERV 13 pre-filter and ACF filter (0. 6 x 0.3 x 0.025) m^3 were placed in a filter box before the fan coil unit. The filter box was made from wood and metal, and then installed in mixing room before fan coil unit.

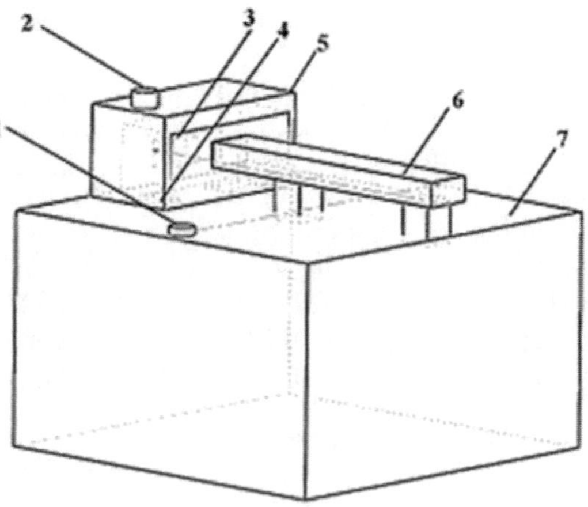

Figure 1 . Chamber schematic
1-Fresh air 2-Mixing room 3-filter box 4-Return air 5-supply duct 6- Chamber 7- Exhaust air

The filter cases (0.6 m x 0.3 m) were located in the middle of the box face and include two holes of filters thickness of 2.54 cm. The measuring device for particle concentration in the chamber, which measured PM_{10} and $PM_{2.5}$, is the TSI® DUSTTRAK Aerosol Monitor 8520 and TSI®TSI DUSTTRAK DRX-8534.

The parameters considered in this study were the contaminant concentration upstream and downstream, with filter and without the filter, air flow rate, pressure drop, humidity, temperature and air velocity. The investigator in the chamber was done to determine the efficiency of the filters and the ventilation system after modification on air handling unit (AHU), by installing box filter and flexible duct to added fresh air, and installing air damper to controlling of exhaust air from the chamber, which the exit of air using exhaust fan.

Airflow rate was controlled with duct cross section and air face velocity through filters. Temperature and relative humidity were measured during the period of the experiment. The pressure difference before and after the filter was measured using the pressure gauge. The scenario of the tests in this assessment was done by new filters, and after three months working and last after six months working. Thus, the analysis of results assessment including used combination filters with occupied and unoccupied, comparing with unused filters. In addition, the assessment included MERV 13 filter alone, compared with used combination filters. However, in this study, the results analysis was divided into two groups.

Particle concentration during occupied and unoccupied zone; was measured by using DustTrak Aerosol device inside the chamber. In this study, mass concentration mg/m^3 it has been employed in the study to check the efficiency of the filters. The pollutants removal efficiency of the filters in the duct was calculated using the following equation (1) [18].

$$Eff = \frac{(C_b - C_a)}{C_b} \quad (1)$$

Where: C_b and C_a concentration of pollutants before and after the filter respectively.

3. Results and discussion

3.1. Efficiency of the new filters

The average indoor particle concentration for PM $_{10}$ and PM $_{2.5}$ are listed in table 1 to 4.

Table 1. Particle concentration occupied/unoccupied test new combination filters

Test description	Occupied test		Unoccupied test	
	Average $PM_{2.5}$ concentration mg/m^3	Average PM_{10} concentration mg/m^3	Average $PM_{2.5}$ concentration mg/m^3	Average PM_{10} concentration mg/m^3
without filter	0.025	0.087	0.027	0.065
With new filter	0.0035	0.012	0.003	0.006
Particle different	0.0215	0.075	0.024	0.059
Efficiency	86%	86.2%	89.28%	90.76%

Table 1, shows the average indoor particle concentration inside the chamber. The average indoor particle concentration during occupied period was 0.025 mg/m^3 and 0.087 mg/m^3 for PM $_{2.5}$ and PM $_{10}$ respectively, the concentration of the installing new filter reduce also the concentration after installing new filters reduce slightly. Thus the filtration efficiency during occupied period for PM $_{10}$ and PM $_{2.5}$ was 86% and 86.2% respectively. The filter specification noted that filter removal efficiency for particle from (1-10) μm > 90%. In addition, during unoccupied period, the particle concentration for PM $_{10}$ and PM $_{2.5}$ was 0.065 and 0.027 respectively and filtration efficiency was 90.76% and 89.28% respectively. This result show that the filtration efficiency of PM $_{2.5}$ and PM $_{10}$ for occupied and unoccupied test were not much of different. These results are in agreement with the previous study by Azimi [19], which the median value of MERV 12 and MERV 14 are 66% - 71% respectively.

3.2. Efficiency of filters after three months

The second evaluation of combination filters was done after three months. Table 2 shows the downstream air cleaned during occupied and unoccupied period. Therefore, the air cleaning inside chamber according to the test after three months' work was recorded less than the air cleaning at the new filters; hence the particle removal efficiency by the three-month-old filter was less than the removal efficiency by new filters. This indicates that the efficiency of the filter was decreased because the filter has become dirty. However, the filtration efficiency of PM $_{10}$ and PM $_{2.5}$ was 85.38% and 85.18% with occupied period. Whereas, during unoccupied test the removal efficiency particulate matter was recorded 88.73% and 88% for PM $_{10}$ and PM $_{2.5}$ respectively. Noted that the filtration efficiency for both diameters particles was still not much of difference. This indicates that the filter after three months was working well. However, the difference values between filtration efficiency at new filter and three months old filter was 0.7% to 0.82% for PM $_{10}$ and PM $_{2.5}$ and 2% to 1.28% for PM $_{10}$ and PM $_{2.5}$ during occupied and unoccupied period respectively. Also, the results indicate that human contributed more particle matter during working hours compare with non-working hours conditions. The results show that filtration efficiency of combination filter from zero working days to three months-old, for PM $_{10}$ and PM $_{2.5}$ is noted much of difference.

Table 2. Particle concentration occupied/unoccupied test three months old combination filters

Test description	Occupied test		Unoccupied test	
	Average $PM_{2.5}$ concentration mg/m^3	Average PM_{10} concentration mg/m^3	Average $PM_{2.5}$ concentration mg/m^3	Average PM_{10} concentration mg/m^3
Without filter	0.027	0.130	0.015	0.071
With 3 months, old filter	0.004	0.019	0.0018	0.008
Particle different	0.023	0.111	0.0132	0.063
Efficiency	85.18%	85.38%	88%	88.73%

3.3. Efficiency of filter after six months

Table 3, shown the particle concentration inside the chamber measured after six months. The average concentration of PM $_{10}$ and PM $_{2.5}$ during working days was 0.016 mg/m^3 and 0.006 mg/m^3 respectively. Whereas, the concentration without the filter was recorded high, also during non-working period the average value of PM $_{10}$ and PM $_{2.5}$ was slightly low 0.007 mg/m^3 and 0.006 mg/m^3 compared with no filters the concentration were 0.05 mg/m^3 and 0.036 mg/m^3. Therefore, the particle removal efficiency recorded was 84.60% and 82.35% for removal of PM $_{10}$ and PM $_{2.5}$ respectively, lower than the filter efficiency with three months-old; the efficiency during unoccupied period was 86% and 83.33% for PM $_{10}$ and PM $_{2.5}$ respectively

Table 3. Particle concentration occupied/unoccupied test six months old combination filters

Test description	Occupied test		Unoccupied test	
	Average $PM_{2.5}$ concentration mg/m^3	Average PM_{10} concentration mg/m^3	Average $PM_{2.5}$ concentration mg/m^3	Average PM_{10} concentration mg/m^3
Without filter	0.034	0.104	0.036	0.05
With 6 months, old filter	0.006	0.016	0.006	0.007
Particle different	0.028	0.088	0.030	0.043
Efficiency	82.35%	84.60%	83.33%	86%

Table 4 shows the particle concentration for occupied and unoccupied period during the test with only MERV 13 filter. The efficiency of the filter during occupied period was recorded 85.20% and 83.33% for PM $_{10}$ and PM $_{2.5}$ respectively. However, the efficiency during unoccupied period was recorded 86.17% and 84% for PM $_{10}$ and PM $_{2.5}$; the results indicated that the filtration efficiency of new single MERV 13 filter was less than the efficiency of combination filter. Further, the filtration efficiency of combination filter after three months worked better than the new single MERV 13 filter. These results were in agreement with the study finding by Azimi [19]. In addition, the study by Michal P. Spilak et al., (2014) found that; when the HEPA filter was used the particle removal efficiency resulted more than the removal efficiency by combination filter. Although, HEPA filter has high efficiency for particle removal efficiency than the filters, but also have high air resistant that provided high-pressure drop.

Table 4. Particle concentration occupied/unoccupied test used MERV 13 filter

Test description	Occupied test		Unoccupied test	
	Average $PM_{2.5}$ concentration mg/m^3	Average PM_{10} concentration mg/m^3	Average $PM_{2.5}$ concentration mg/m^3	Average PM_{10} concentratio n mg/m^3
Without filter	0.09	0.115	0.05	0.094
With new filter	0.015	0.017	0.008	0.013
Particle different	0.075	0.098	0.042	0.081
Efficiency	83.33%	85.20%	84.00%	86.17%

Figure 2 shows the different filtration efficiency between combination filters and new MERV 13 filter. The diagram shows the efficiency after three months working better than new MERV 13 filter occupied and unoccupied period. The advantage of combination filter method, resulting from compiling two filters (MERV filter and ACF filter). ACF has a high capacity to absorb gases pollutants and small particle because of its advanced and high porosity and large surface area.

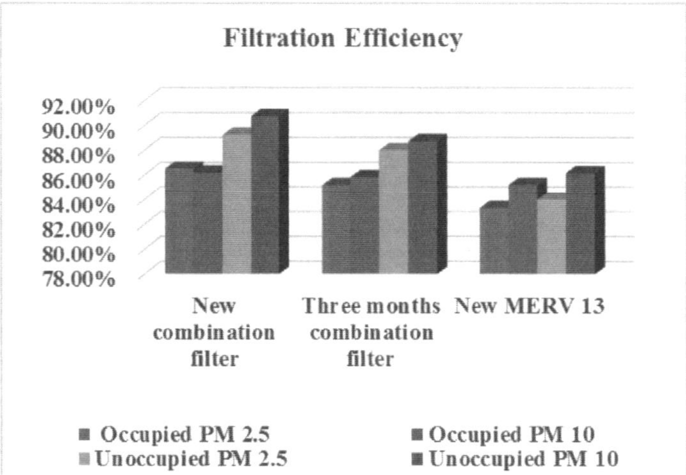

Figure 2. Different filtration efficiency combination filter and single MERV 13 filter

4. Conclusion

The indoor air quality can be improved by using a simplified method with the combination filter of MERV 13 filter and ACF. MERV filter cannot be reliable for removal of ultra-fine particle and $PM_{2.5}$ when each of them is applying to HVAC system as a single filter. Thus, to improve indoor air quality and efficient particle removal the combined filter was installed. The filtration efficiency for combined filter was 86.20% to 84.60% for PM_{10} and for $PM_{2.5}$ was 86% to 82.35%, during occupied period. While the filtration efficiency for unoccupied periods recorded 90.76% to 86% for PM_{10}, and 89.28% to 83.30% for $PM_{2.5}$. However, the filtration efficiency of the three-month-old filter was recorded less than the new filters by 0.82%, 0.82%, 1.28% and 2.03% for PM_{10} and $PM_{2.5}$ occupied and unoccupied periods respectively. However, the big value different was recorded after six months due the filter has become dirty, but the efficiency as in this range is acceptable because the decreased of efficiency was a little. Furthermore, the result showed that the efficiency after three months installation is better than new single MERV 13 filter during occupied and unoccupied period.

Acknowledgement
The authors would like to thank the staff of the Thermal Environmental Control laboratory in the Faculty of Mechanical & Manufacturing Engineering for their technical support and the Office for Research, Innovation, Commercialization and Consultancy Management (ORICC), Universiti Tun Hussein Onn Malaysia for financial support for this research (Grant U176).

References

[1] Cao G 2014 A review of the performance of different ventilation and airflow distribution systems in buildings *Build Environ* **73** 171–186
[2] Quang T N, He C, Morawska L, and Knibbs L D 2013 Influence of ventilation and filtration on indoor particle concentrations in urban office buildings *Atmos Environ* **79** 41–52
[3] Zaatari M, Novoselac A, and Siegel J 2014 The relationship between filter pressure drop, indoor air quality, and energy consumption in rooftop HVAC units *Build Environ* **73** 151–161
[4] Azimi P and Stephens B 2013 HVAC filtration for controlling infectious airborne disease transmission in indoor environments: Predicting risk reductions and operational costs *Build Environ* **70** 150–160
[5] Kabrein H, Yusof M Z M and Leman A M 2016 Progresses of filtration for removing particles and gases pollutants of indoor; limitations and future direction *ARPN J. Eng. Appl. Sci* **11** 6
[6] Xu Y *2010* Effectiveness of heating, ventilation and air conditioning system with HEPA filter unit on indoor air quality and asthmatic children's health *Build. Environ* **45** 2 330–337
[7] Peck R L, Grinshpun S A, Yermakov M, Rao M B, Kim J and Reponen T 2016 Efficiency of portable HEPA air purifiers against traffic related combustion particles *Build Environ* **98** 21–29
[8] Chatoutsidou S E, Ondráček J, Tesar O, Tørseth K, Ždímal V and Lazaridis M 2015 Indoor/outdoor particulate matter number and mass concentration in modern offices *Build Environ* **92** 462–474
[9] Spilak M P, Karottki G D, Kolarik B, Frederiksen M, Loft S and Gunnarsen L 2014 Evaluation of building characteristics in 27 dwellings in Denmark and the effect of using particle filtration units on PM2.5 concentrations *Build Environ* **73** 55–63
[10] Ardkapan S R, Afshari A, Bergsøe N C, Gunner A and Afshari J 2013 Combining active chilled beams and air-cleaning technologies to improve the indoor climate in offices: Testing of a low pressure mechanical filter in a laboratory environment *HVAC&R Res* **19** 81090–1094
[11] Rackes A and Waring M S 2013 Modeling impacts of dynamic ventilation strategies on indoor air quality of offices in six US cities *Build Environ* **60** 243–253
[12] Wolkoff P 2013 Indoor air pollutants in office environments: Assessment of comfort, health, and performance *Int J. Hyg. Environ. Health* **216** (4) 371–394
[13] Zuraimi M S and Tham K W 2009 Reducing particle exposures in a tropical office building using electrostatic precipitators *Build. Environ.* **44(**12) 2475–2485
[14] Park J S, Jee N Y and Jeong J W 2014 Effects of types of ventilation system on indoor particle concentrations in residential buildings *Indoor Air* **24** (6) 629–638
[15] ASHRAE 2010 ASHRAE *Standard 62-2001 Ventilation for acceptable indoor air quality*
[16] DOSH 2010 Industry Code of Practice on Indoor Air Quality *Minist. Hum. Resour. Dep. Occup. Saf. Heal.* 1–50
[17] Jamaludin N, Mohammed N I, Khamidi M F and Wahab S N A 2015 Thermal Comfort of Residential Building in Malaysia at Different Micro-climates *Procedia - Soc. Behav. Sci.* **170** 613–623
[18] Chen W, Gao Z, Jianshun Z S, Kosar D and Walker C E 2006 Reduced Energy Use Through Reduced Indoor Contamination in Residential Buildings *Energy*
[19] Azimi P, Zhao D, and Stephens B 2014 Estimates of HVAC filtration efficiency for fine and ultrafine particles of outdoor origin *Atmos. Environ.* **98** 337–346

IOP Conf. Series: Materials Science and Engineering **243** (2017) 012052 doi:10.1088/1757-899X/243/1/012052

Improving indoor air quality and thermal comfort in office building by using combination filters

H Kabrein[1,*], M Z M Yusof[1], A Hariri[1], A M Leman[2] and A Afandi[1]
[1] Industrial and Indoor Environment Research Group (IIERG), Centre of Energy and Industrial Environment Studies (CEIES),
Faculty of Mechanical and Manufacturing Engineering Universiti Tun Hussein Onn Malaysia 86400 Parit Raja, Batu Pahat, Johor, Malaysia
[2]Faculty of Engineering Technology, Universiti Tun Hussein Onn Malaysia

[*]E-Mail: hashimkabrein@gmail.com

Abstract. Poor indoor air quality and thermal comfort condition in the workspace affected the occupants' health and work productivity, especially when adapting the recirculation of air in heating ventilation and air-conditioning (HVAC) system. The recirculation of air was implemented in this study by mixing the circulated returned indoor air with the outdoor fresh air. The aims of this study are to assess the indoor thermal comfort and indoor air quality (IAQ) in the office buildings, equipped with combination filters. The air filtration technique consisting minimum efficiency reporting value (MERV) filter and activated carbon fiber (ACF) filter, located before the fan coil units. The findings of the study show that the technique of mixing recirculation air with the fresh air through the combination filters met the recommended thermal comfort condition in the workspace. Furthermore, the result of the post-occupancy evaluation (POE) and the environmental measurements comply with the ASHRAE 55 standard. In addition, the level of CO_2 concentration continued to decrease during the period of the measurement.

1. Introduction

One of the main missions of heating, ventilation and air-conditioning (HVAC) system is ensuring comfort inside buildings. The absence of comfort condition disturbs people and may cause some anxiety and health effects. HVAC system should be provided for controlling thermal comfort through controlling the humidity/dehumidify using humidifier system and controlling of temperature (heating or cooling) using fan coil [1]. According to American Society of Heating, Refrigerating, and Air-Conditioning Engineers ASHRAE-55 standard ASHRAE-55 standard , Thermal comfort is that state of mind that define the fulfillment of the thermal condition [2]. There are many distinctions such as psychologically and physiologically, which make it difficult to please all the occupants; due to the fact that there is a different impact of these variables on people and genders.

There have been many researches that carried out by researchers' (in various sites and climate characteristics) around the world in order to understand the comfort condition. ASHRAE conducted many studies to investigate the effect of users' thermal comfort in the tropical and cold climate [2]. There are two essential conditions that must be considered when we talk about the indoor environment such as the indoor air quality (IAQ) and thermal comfort. There are many attributes that have a significant impact on the sensation of thermal comfort that generally consist of air humidity, air temperature, mean radiant temperature, air velocity, activity levels and human clothing [3]. The main reason of ventilating a building is to ensure that the air circulation is covering all the occupied areas; in order, to eliminate contaminants. The effectiveness of ventilation systems is assessed by many attributes such as the thermal condition, contaminant removal effectiveness; air exchange effectiveness and local air mean age. These features guaranteed the quality of air in each building. In addition, indoor air quality and thermal comfort have a significant impact on the occupants output productivity [4]. According to Niu, thermal comfort could be achieved by varying the conditions of the supplied air [5]. It was also he stated that poor condition of IAQ could cause health problem such as sick building syndrome (SBS) to occupants.

Varied statistical data was gathered from a wide range of field studies and laboratory tests in order to gain an acceptable definition of thermal comfort. The preference of comfort condition varies in individuals. Section 5 of the ASHRAE-55 standard is designed to define the thermal environmental conditions in space, which are required to satisfy a specific percentage of the tenants of space [2]. The poor condition of IAQ was caused by the pollutants indoor. There are many components in the indoor air such as gaseous pollutants and particle pollutants [3]. Previous studies on the indoor thermal condition illustrated that the thermal dissatisfaction was related to higher air movements which did not meet the ASHRAE 55 standard [2]. The climate of Malaysia which is hot and humid has an impact on the indoor thermal comfort due to the effects of the outdoor condition.

From field measurement and thermal sensation votes the findings illustrated a significant reduction in CO_2 concentration while the parameters of thermal comfort slightly improved after the installation of a heat exchanger, although the energy consumption increased by small percent due to the hot and humid fresh air. Furthermore, good IAQ was achieved by the employment of heat exchanger located in the space. This procedure will not increase the energy consumption [6].

Relative humidity in the humid climates may not have a distinct impact on the thermal condition of building occupants. Subsequently, it is accepted that there is a high percent of relative humidity in air-conditioning zones. However, in the non-air conditioned area, there is no need to isolate humidity parameter and air temperature parameter as the two parameters are interrelated when dealing with indoor thermal comfort [7]. According to Lee & Brand [8], there exist a relation between indoor environment, comfort condition, health and the performance of office building users'. Whereas, the complaints and symptoms that reported by the users, physical indoor environment measurements and chemical are hard to achieve. The IAQ and the comfort of users' in traditional office buildings is achieved by a variable procedure such as an air handling unit (AHU) that have a control system for thermostatic and the air box, which collects and mixed air from an AHU and deliveries [2]. Wong [4] study the thermal condition in an office area, he mentioned that the thermal comfort should carefully apply due to reason that the office area was occupied by many users' who need to be satisfied with the environmental condition, which help to enhance the productivity of occupants. However, in tropical climates, the characteristic of the weather (hot and humid) may have a reverse influence on comfort of the occupants.

The adoption of natural ventilation in tropical climate through the openings in the building (door and window) may lead to increasing the percent of the indoor humidity and has a negative impact on the thermal condition. The status of poor IAQ is connected to the amount of relative humidity inside the building and one of many sources of discomfort condition; however, to solve this problem mechanical ventilation system is employed to enhance comfort, which leads to increase in the energy consumption. To control the air temperature and relative humidity, air-conditioning is used to provide more control and improve the IAQ for the users' [9]. Dear [10], conducted a field study in two high-rise office building that used natural ventilation and mechanical cooling system (air- conditioning); their findings that the employment of natural ventilation makes the indoor condition not complying with ISO standard. It mentioned that the office buildings that used air-conditioning system, did not require much cooling; on the other hand, in the natural ventilated office building the operating temperature was 31.5 °C, which according to the conventional comfort standard, this temperature was higher than the recommended temperature [10], [11]. In residential buildings in Malaysia, thermal

IOP Conf. Series: Materials Science and Engineering **243** (2017) 012052 doi:10.1088/1757-899X/243/1/012052

comfort was assessed and the neutrality of occupants' thermal condition had been conducted by Zain [12]. The result showed that the neutral temperature was 26.19 °C. In this range, the interior space does not need much cooling to maintain thermal comfort for building occupants.

Wan et, al [13], conducted thermal comfort study in an office building in southern China, using the principle of effective temperature. The findings showed an increase in the cooling load of air-conditioning and the energy consumption when the indoor air temperature was high. They suggested a solution to this problem by fixing a high relative humidity and lower indoor air temperature, due to the fact that the human body could adapt the higher humidity more than the temperature. By studying the comfort air temperature in an office building for staff and visitors, the result showed that temperature range of 24.5 - 28.2 °C was comfortable as perceived by the office staff; whereas the visitors were more comfortable with the temperature ranging between 24 - 27.8 °C. Comparison between the maximum temperature that was acceptable by both staff and visitors was higher than the recommended temperature based on ASHRAE 55 standard by 2 °C. By highlighting the range of accepted, comfortable temperature in hot and humid climate it could be noticed that the temperature is generally not complying with the comfort zone as that recommended in ASHRAE standard 55-2004 [14]. According to Kwong, Adam, & Sahari [15], it could meet the thermal comfort in natural ventilation residential building if these areas are not exposed to be under the requirement of building safety; but to achieve thermal comfort in these areas the occupants must be able to adjust their comfort condition. However, in an office building in the tropics the operative temperature was listed as 24.2 - 28.3 °C, and between 23.3 – 29.2 °C in other building types; however, the temperature varies in the natural ventilation building; where by the temperature is higher than other building. ASHRAE standard 55 investigated on the impact of the air velocity on thermal condition, the influence of air velocity on thermal comfort is connected with other environmental variables such as radiant temperature, air temperature, and relative humidity.

In ASHRAE standard the use of PMV model is governed by the air velocity not exceeding 0.20 m/s (40 fpm). The air velocity that is more than the suggested range could help to improve the boundaries of the comfort zone in specific situations. In the last decade, the demand for energy has increased significantly in Southern Asian countries. The urgent and decisive demand on air-conditioner use in these countries was a cause for the increase of energy consumption for cooling the building [10]. In hot and humid climate, the non-residential building, especially the commercial buildings, consume the most of the energy as the fact that these buildings are normally provided with a mechanical ventilation and air-conditioning system to improve indoor thermal comfort and indoor air quality. However, this system consumes 30 to 60% of the total energy use among all building services, as in many studies which attempt to identify the best method to save energy [16].

Many experts in the field of building construction such as operators, air-conditioner manufacturers and building engineers in hot and humid climate adopted the idea that the low operating air temperature of the cooling system (air-conditioning) is essential in providing a thermally comfort conditioned space for building users'. Consequently, this leads to waste of valuable energy and raising the cost of maintenance and operating in buildings. Previous studies that focused on thermal comfort in commercial buildings recommended that the setup of upper air-conditioning temperature, could achieve thermal comfort of users'; whereas, the high air speed may adjust the cooling condition to gain less energy consumption without the increase in the expenses. On the other hand, the studies on thermal comfort focus on the impact on several aspects on the thermal condition such as physical condition, educational level, age and gender of occupants; in addition, more investigation are needed on the non-thermal parameters that influence the sensation of thermal comfort [16].

The objectives of this study are to assess the effectiveness of indoor thermal comfort in the office buildings, and to control the concentration of CO_2 by using suitable ventilation and air filtration system. Xu et, al [17], studied the concentration of CO_2 and TVOC in a residential building, using the HEPAiRx. The results showed a reduction of about 210 mg/m^3 and 710 ppm of both CO_2 and TVOC concentration levels respectively; thus, lower CO_2 concentration can be achieved by controlling the fresh air and recirculation air. On the other hand, in a tropical climate the fresh air intake will increase temperature and relative humidity of the supply air, which led to increasing the energy consumption due to the reason that the fresh air needs to be cool to the design supply condition.

2. Methods

The study is carried out using the environmental control chamber that is able to simulate a typical office environment. The chamber dimension is (5 m x 5 m x 2.8 m), A schematic diagram of the chamber setup is shown in Figure. 1, Figure 2 show the layout of sensors measurement and occupants position inside the chamber, the walls and the ceiling were of polyurethane insulation board of 10 cm thickness, the design of the joint, the groove and the tongue that made from stainless steel metal (5 mm) and located in the chamber on both sides. The chamber was equipped with HVAC system, a compressor (10 HP), an outdoor condenser and fan coil unit in the mixing room, which is located on the top of the chamber roof. The supply air duct located at the roof of the chamber. The main air supply duct links the fan coil unit to the chamber. The main supply duct was distributed in two main branches to provide air through ceiling diffusers (0.4 m x 0.4 m); moreover, there are two holes (0.55 m x 0.45 m) provided on the return air duct suspended on the top of the ceiling connecting the chamber and mixing box. A filter box (0.6 m x 0.3 m) was designed to house the new filters and to ensure good mixing of the air passage and cleaning through the filter before entering the fan coil unit. The outdoor fresh air intake was in the mixing room through the flexible duct, exhaust fan, and damper. The damper was connected by a flexible duct to warrant that sufficient supply of outdoor fresh air compatible to ASHRAE standard 62 [18]. Another damper was also connected to the exhaust fan at the top of the ceiling to control the exhaust air.

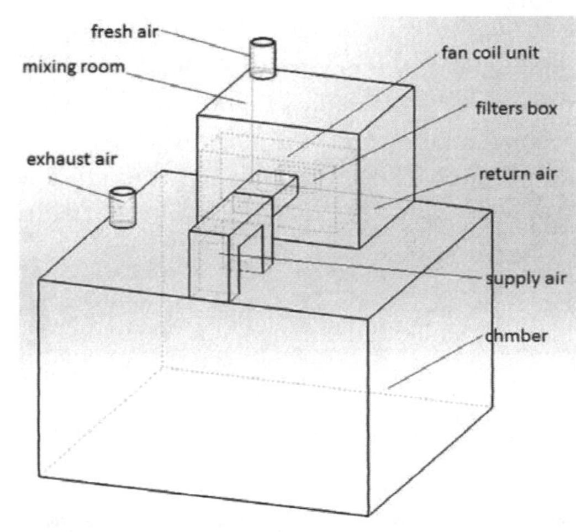

Figure 1. A schematic diagram of test chamber

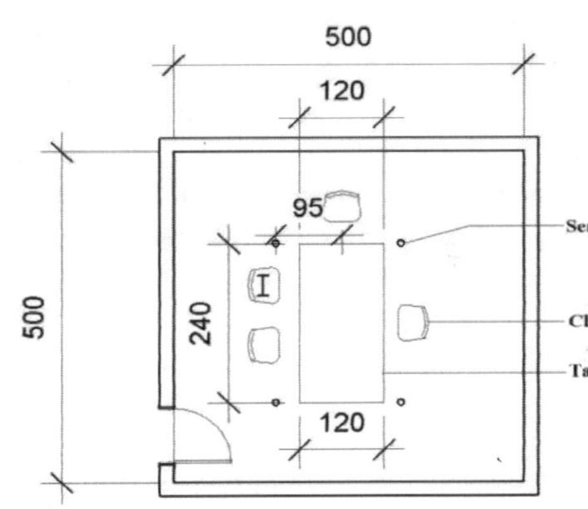

Figure 2. Plan view layout of contaminant sources and sensors

The air flow through the system was adjusted to 560 CFM (951 m^3/h) to achieve a face velocity 1 m/s through any diffusers, with air changes rate about 3 h^{-1} (typical air change rate for an office building) as suggested by ASHRAE standard 62. The chamber environmental condition was adjusted between 23 ºC to 26 ºC and 78% for RH, which is the indoor environmental condition that suggested in Malaysia by the studies of [19, 20]. The control of the air temperature and relative humidity by using AHU system, through temperature and relative humidity monitoring system to adjusting the temperature.

The chamber study was conducted in two cases. The first case involved in recording the environmental condition; daily in two months, simulating typical work hours from 8:00 am to 4:00 pm with four occupants in the chamber. A MERV 13 pre-filter and ACF filter (0. 6 x 0.3 x 0.025) m^3

IOP Conf. Series: Materials Science and Engineering **243** (2017) 012052 doi:10.1088/1757-899X/243/1/012052

placed in a filter box before the fan coil unit. Thermal environmental parameters were measured and recorded. The second case involved the measuring of (CO_2) gas concentrations in the chamber. The temperature and relative humidity were recorded using four sensors (KIMO 100) that connected to the computer device using temperature and relative humidity monitoring system. Whereas, air velocity was measured using VelociCalc plus model 8324/8384 the instrument sat at three heights points, the first one at 0.5 m above the floor that represented the occupant seated at the knee height, while 1.1 m high represented the occupant level at sitting condition, and the last point is 1.6 m at standing level [2]. The measurement of CO_2 concentration was recorded using an IAQCalc device model 8760/8762 at four locations in the chamber, the measurement at any location was repeated three times to ensure. The air temperature and relative humidity were recorded every 5 minutes interval using the chamber monitors / sensors.

Another assessment tool used to determine the occupant's comfort condition in the experimental space is the post- occupancy evaluation (POE) survey. The occupants were asked in the evaluation to assess their perception of the indoor condition. The survey questionnaire designed based on (ASHRAE standard 55, MM 040 NA offices and Industry Code of Practice on Indoor Air Quality 2010) [2, 20] and divided into two sections. The first section consists of general information such as gender, age, nationality, activity, clothing value (CLO value) and the thermal sensation; while, the second section consists of the perception on environment satisfaction rating of air temperature, the time of dissatisfaction and the source of dissatisfaction, and occupants' additional comments related to their satisfaction of the chamber environment condition.

3. Results and discussion

The measurement was carried out from January 25 to March 25, 2016. The outdoor air temperature was between 24 °C in the morning and 32 °C (at the peak during the experimental period); whereas, the relative humidity was in the range of 87 ± 7%. The thermal comfort condition in the chamber was assessed by environmental measurement and POE questionnaire; the results of the experiment were compared with ASHRAE 55 standard. The CO_2 concentration in the chamber log within recommended level that would not affected the users' health. The environmental variables (air temperature, mean radiant temperature, relative humidity and air velocity) are amongst the variables / parameters that affect thermal comfort in the space.

3.1. Results from Measurement

3.1.1. Environmental Assessment

Figure 3 shows the fluctuating of temperature and relative humidity on a working day (8 hours). From Figure 3 it is noticed that the temperature varies between 23.2 °C to 25.1 °C. In this study the temperature was controlled by adjusting the thermostat range for controlling the temperature to 25 °C. According to [21], the indoor and outdoor climates both have influences on human adaptability; furthermore, the occupants' activity had affected the internal heat load and hence increasing the temperature.

The study also cited that the outdoor relative humidity was fluctuating between 78 to 95% average humidity in Malaysia during the northeast season. In the chamber, the relative humidity fluctuated between 69 to 89% and the average was 80%. Study on the perception of thermal comfort in air-conditioning building in Thailand by [22], focused on two non- measurable factors such as adaptation and educational level. The study showed that the air temperature settings of 26 °C and 50 – 60% relative humidity result in a thermally comfortable condition to the building users / occupants.

IOP Conf. Series: Materials Science and Engineering **243** (2017) 012052 doi:10.1088/1757-899X/243/1/012052

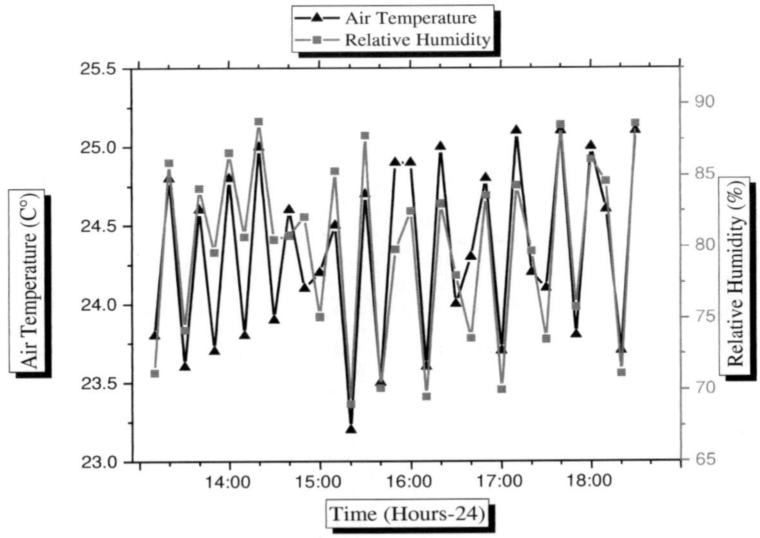

Figure 3. Air temperature and relative humidity variation in office (with four occupants)

The air movement in the chamber ranged between 0.13 to 0.2 m/s, (measurement conducted at three levels and six points in the chamber. According to Malaysian standard 2007 had recommended the indoor air velocity be within the range of 0.15-0.5 m/s [23]. Another assessment tool used to evaluate air velocity through using predicted percentage dissatisfied (PPD) and predicted mean vote (PMV) indices. Figure 4 presents the PMV and PPD plot in the chamber on a comfort chart. A comfort chart is a tool that used to define building's comfort condition according to ASHRAE. It is noticed that the PMV value was (-0.08), which indicate that overall thermal sensation inside the chamber was slightly cool. However, questionnaire study by participants voted that the general condition inside the chamber was neutral.

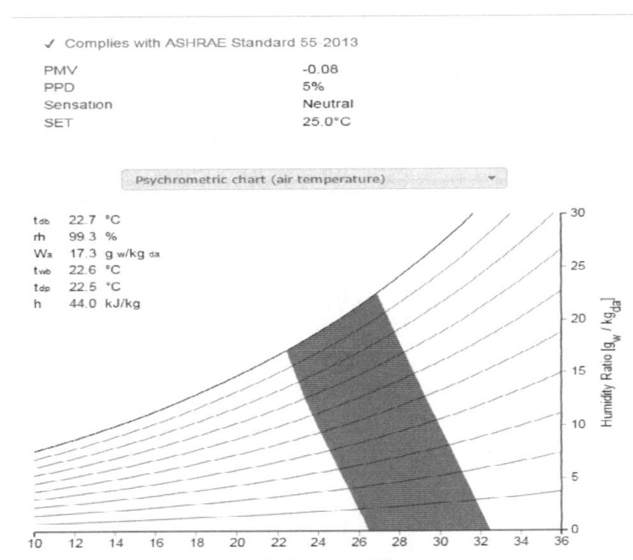

Figure 4. CBE Thermal Comfort Tool

3.1.2. Air Quality Assessment

The concentration of CO_2 as in Figure 5, recorded high level of 650 ppm when the chamber was occupied by four persons, and observed that the decrease of concentration as passage of the time. From Figure 5 it is noticed that the concentration of CO_2 dropped to 620 ppm after 20 minutes, and to 595 after 60 minutes. Furthermore, the CO_2 concentration depend on the ratio of fresh air to recirculation air; according to study of Lee [24], explained that the effect of the makeup air ratio to recirculation air ratio and evaluated the optimum ratio of fresh air and recirculated air for acceptable indoor air quality and energy consumption. In this study it was controlled by air damper to ensure the ratio of 0.2; this ratio complied with ASHRAE standard 55 for air change rate in office building.

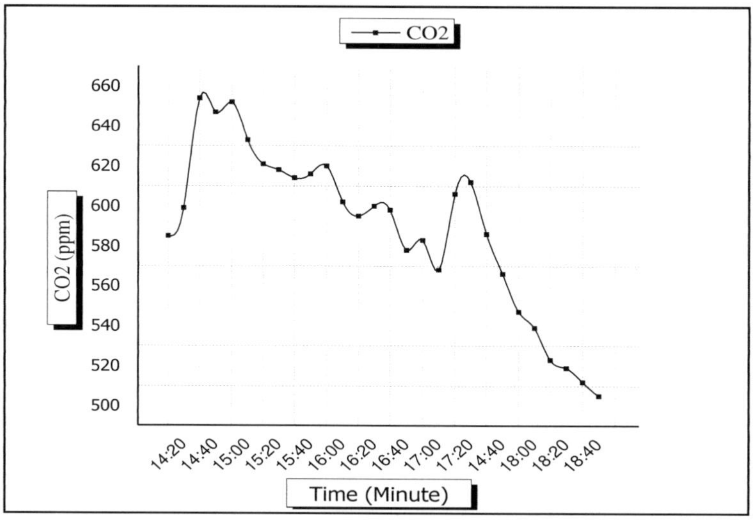

Figure 5. CO_2 concentrations in chamber with four occupants

3.2. Results from Questionnaire

The thermal comfort, including air velocity, was evaluated by questionnaires and the measurement of physical parameters compared with ASHRAE 55 standard. The questionnaires were conducted on occupants inside the chamber. The results of questionnaire analysis are shown in Figures 6 (A, B and c), on the basis of temperature and thermal comfort, the participants rated/ voted as cool (30%), as slightly cool (35%), Neutral (19%), slightly warm and cold (4%). The overall thermal sensation is shown in Figure 6B; the percentage of occupants comfortable (38%), comfortably worm, (4%) much too cool (2%), too cool (10%) and comfortably cool (40%). Whereas, the indoor thermal comfort for satisfied occupants with temperature in the test chamber as shows in Figure 6C; consisting of satisfied (48%) followed by normal (19%), quite dissatisfied (15%), very satisfied (10%) and quite satisfied (4%). However, the POE asked the participants to vote their satisfaction with the indoor air temperature, the composition of participants were local Malaysian students and international students as in Figure 7B. The final findings on the perception of indoor air temperature by nationality noticed that more than 45% of local Malaysian participants feel comfortably cool, whereas almost 50 percent of international occupants voted as comfortable as shown in Figure 7A. Daghigh [10] stated that Malaysian subject feel comfortable with the indoor air temperature not greater than 28 °C. Lastly, the participants were asked to describe the source cause of discomfort in the chamber, consisting of 6 parameters such as humidity, air movement, incoming solar and incoming noise; the most distinct cause of discomfort was due to the air movement. Whereby, more than 58% of participants voted that the air movement was normal, while 27% complaining that the air movement was high and 16% of occupants considered that the air movement was too low. The results were compared with ASHRAE 55 standard. The environmental parameters and the level of CO_2 were in agreement with the results of the study of F.-J. Wang [7].

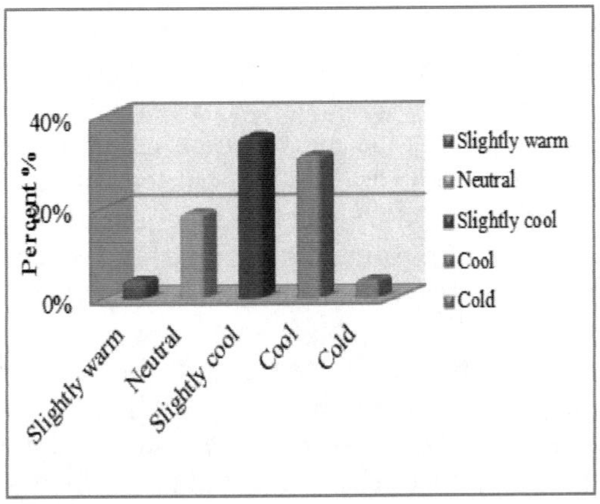

Figure 6A. Occupants votes on air temperature

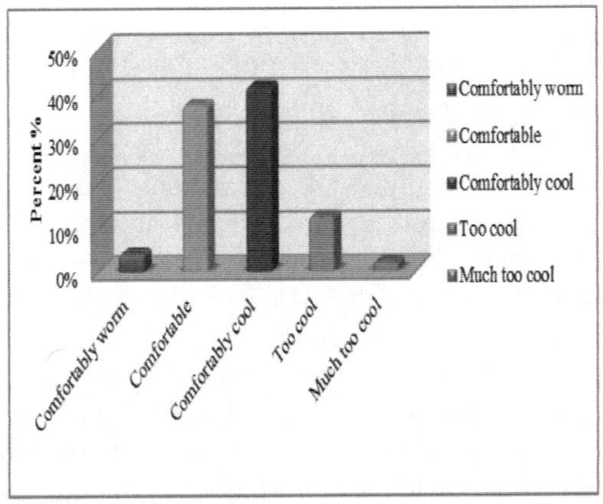

Figure 6B. Occupants votes on comfort sensation

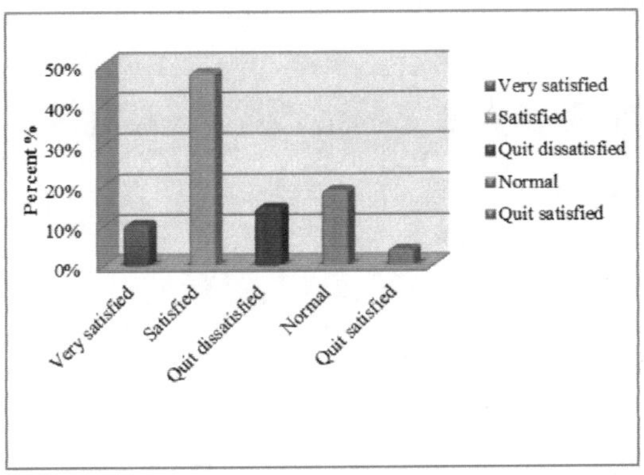

Figure 6C. Occupant's sensation acceptability

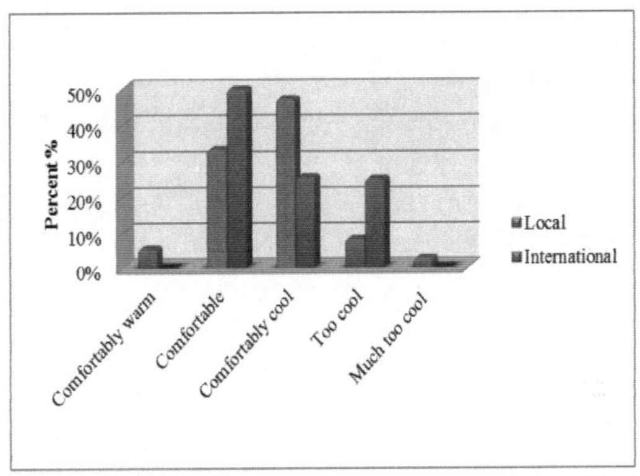

Figure 7A. Occupants votes on comfort sensation

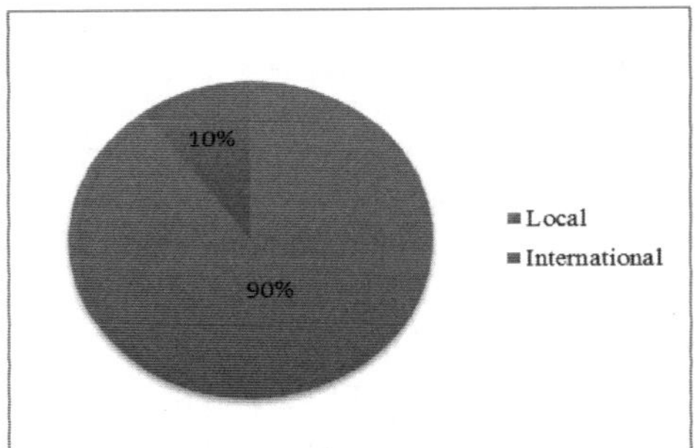

Figure 7B. Occupant test subject by nationality

4. Conclusions

This study investigated the indoor thermal comfort in an office building in tropical climates, besides improving the indoor air quality by controlling the concentration of CO_2 by using ventilation and filtration system (combination filters). The environmental parameters (air temperature, relative humidity, air velocity, PMV, PPD, and CO_2) were measured and compared with ASHRAE standard 55-2004. Occupants' perception on the thermal environment and air quality were carried out.

The PMV value was recorded as (-0.08), which indicating that the indoor environment inside the chamber was slightly cool. This result was experienced when the outdoor air temperature was recorded low. According to ASHRAE standard 55, the range of PMV value that recommended for the office building is between (-0.5 to 0.5) and percentage of the PPD value less than 10%. Based on this suggestion the results for PMV and PPD values in the chamber were within the recommended values of ASHRAE standard 55-2004.

Results from measurement showed good agreement for thermal comfort even when using the combination filter. The acceptability of the thermal comfort as recommended in ASHRAE standard. However, the air temperature achieved in the chamber was 24 °C, whereas the relative humidity ranged between 69 to 89%. Furthermore, the air movement was between 0.13 to 0.2 m/s, which was in the range of ASHRAE standard recommendation.

The CO_2 concentration decreased during the measurement period, the results showed that that average of CO_2 concentration varied during the working hours; which were high, when the participants were doing some activities. A recommendation for future work is to find the balance among improving indoor air quality, thermal comfort and energy consumption in offices building in tropical climates.

Acknowledgment

The authors would like to thanks the staff of the Thermal Environmental Control laboratory in the Faculty of Mechanical & Manufacturing Engineering; the Office for Research, Innovation, Commercialization and Consultancy Management (ORICC), University Tun Hussein Onn Malaysia for supporting this research project through the U176 research grant.

References

[1] Md Yusof M Z 2011 *Getting to know our buildings better: Ventilation, thermal comfort, air freshness, indoor air quality (IAQ), sick building syndrome (SBS)*

[2] Ansi/Ashrae 2004 ANSI/ASHRAE 55:2004 Thermal Environmental Conditions for Human Occupancy *Ashrae* 2004 30

[3] Yu B F, Hu Z B, Liu M, Yang H L, Kong Q X, and Liu Y H 2009 Review of research on air-conditioning systems and indoor air quality control for human health *Int. J. Refrig.* **32**(1) 3–20

[4] Wong N H and Khoo S S 2003 Thermal comfort in classrooms in the tropics *Energy Build.* **35**(4) 337–351

[5] Niu J 2004 Some significant environmental issues in high-rise residential building design in urban areas *Energy Build.* **36**(12) 1259–1263

[6] Wang F J, Lee M C, Chang T B, Chen Y S and Jung R C 2014 Improving indoor air quality and thermal comfort by total heat exchanger for an office building in hot and humid climate *HVAC&R Res.* **20**(7) 731–737

[7] Djamila H, Chu C and Kumaresan S 2014 Effect of Humidity on Thermal Comfort in the Humid Tropics *J. Build. Constr.* June 109–117

[8] Lee S Y and Brand J L 2005 Effects of control over office workspace on perceptions of the work environment and work outcomes *J. Environ. Psychol.* **25**(3) 323–333

[9] Daghigh R 2015 Assessing the thermal comfort and ventilation in Malaysia and the surrounding regions *Renew. Sustain. Energy Rev.* **48** 682–691

[10] Dedear R J, Leow K G and Foo S C 1991 Thermal Comfort in the Humid Tropics - Field

Experiments in Air-Conditioned and Naturally Ventilated Buildings in Singapore *Int. J. Biometeorol.* **34**(4) 259–265
[11] Busch J F 1992 A tale of two populations: thermal comfort in air-conditioned and naturally ventilated offices in Thailand *Energy Build.* **18**(3–4) 235–249
[12] Zain Z M, Taib M N and Baki S M S 2007 Hot and humid climate: prospect for thermal comfort in residential building *Desalination* **209**(1–3) 261–268
[13] Wan J W, Yang K, Zhang W J and Zhang J L 2009 A new method of determination of indoor temperature and relative humidity with consideration of human thermal comfort *Build. Environ.*, **44**(2) 411–417
[14] Jitkhajornwanich K and Pitts A C 2002 Interpretation of thermal responses of four subject groups in transitional spaces of buildings in Bangkok *Build. Environ.* **37**(11) 1193–1204
[15] Kwong Q J, Adam N M, and Sahari B B 2014 Thermal comfort assessment and potential for energy efficiency enhancement in modern tropical buildings: A review *Energy Build.* **68** 547–557
[16] Habeebullah B A 2007 Economic feasibility of thermal energy storage systems *Energy Build.* **39**(3) 355–363
[17] Xu Y 2010 Effectiveness of heating, ventilation and air conditioning system with HEPA filter unit on indoor air quality and asthmatic children's health *Build. Environ.* **45**(2) 330–337
[18] ASHRAE 2010 ASHRAE Standard 62 . 1 *Ventilation for acceptable indoor air quality*
[19] DOSH 2010 Industry Code of Practice on Indoor Air Quality *Minist. Hum. Resour. Dep. Occup. Saf. Heal.* 1–50
[20] Jamaludin N, Mohammed N I, Khamidi M F and Wahab S N A 2015 Thermal Comfort of Residential Building in Malaysia at Different Micro-climates *Procedia - Soc. Behav. Sci.* **170** 613–623
[21] Zomorodian Z S, Tahsildoost M and Hafezi M 2016 Thermal comfort in educational buildings: A review article *Renew. Sustain. Energy Rev.* **59** 895–906
[22] Yamtraipat N, Khedari J and Hirunlabh J 2005 Thermal comfort standards for air conditioned buildings in hot and humid Thailand considering additional factors of acclimatization and education level *Sol. Energy* **78**(4) 504–517
[23] Depatment of Standard Malaysia 2007 Malaysian Code of Practice on Energy *Code of Practise On Energy Efficiency and Use of Renewable Energy for Non-Residential Buildings* 1–46
[24] Joo J, Zheng Q, Lee G, Kim J T and Kim S 2012 Optimum energy use to satisfy indoor air quality needs *Energy Build.* **46** 62–67

Institute of Physics
Dirac House, Temple Back
Bristol BS1 6BE UK

ISSN: 1757-8981
ISBN 978-1-5108-4894-8

CURRAN ASSOCIATES INC.
proceedings.com